Smart Innovation, Systems and Technologies

Volume 76

Series editors

Robert James Howlett, Bournemouth University and KES International,
Shoreham-by-sea, UK
e-mail: rjhowlett@kesinternational.org

Lakhmi C. Jain, University of Canberra, Canberra, Australia;
Bournemouth University, UK;
KES International, UK
e-mails: jainlc2002@yahoo.co.uk; Lakhmi.Jain@canberra.edu.au

About this Series

The Smart Innovation, Systems and Technologies book series encompasses the topics of knowledge, intelligence, innovation and sustainability. The aim of the series is to make available a platform for the publication of books on all aspects of single and multi-disciplinary research on these themes in order to make the latest results available in a readily-accessible form. Volumes on interdisciplinary research combining two or more of these areas is particularly sought.

The series covers systems and paradigms that employ knowledge and intelligence in a broad sense. Its scope is systems having embedded knowledge and intelligence, which may be applied to the solution of world problems in industry, the environment and the community. It also focusses on the knowledge-transfer methodologies and innovation strategies employed to make this happen effectively. The combination of intelligent systems tools and a broad range of applications introduces a need for a synergy of disciplines from science, technology, business and the humanities. The series will include conference proceedings, edited collections, monographs, handbooks, reference books, and other relevant types of book in areas of science and technology where smart systems and technologies can offer innovative solutions.

High quality content is an essential feature for all book proposals accepted for the series. It is expected that editors of all accepted volumes will ensure that contributions are subjected to an appropriate level of reviewing process and adhere to KES quality principles.

More information about this series at http://www.springer.com/series/8767

Giuseppe De Pietro · Luigi Gallo
Robert J. Howlett · Lakhmi C. Jain
Editors

Intelligent Interactive Multimedia Systems and Services 2017

 Springer

Editors
Giuseppe De Pietro
National Research Council of Italy
  (CNR-ICAR)
Institute for High-Performance Computing
  and Networking
Naples
Italy

Luigi Gallo
National Research Council of Italy
  (CNR-ICAR)
Institute for High-Performance Computing
  and Networking
Naples
Italy

Robert J. Howlett
Bournemouth University
Poole
UK

and

KES International
Shoreham-by-Sea
UK

Lakhmi C. Jain
University of Canberra
Canberra, ACT
Australia

and

Bournemouth University
Poole
UK

and

KES International
Shoreham-by-Sea
UK

ISSN 2190-3018   ISSN 2190-3026 (electronic)
Smart Innovation, Systems and Technologies
ISBN 978-3-319-86633-8   ISBN 978-3-319-59480-4 (eBook)
DOI 10.1007/978-3-319-59480-4

Preface

Dear Readers,

We introduce to you a series of carefully selected papers presented during the 10th KES International Conference on Intelligent Interactive Multimedia Systems and Services (IIMSS-17).

At a time when computers are more widespread than ever, and computer users range from highly qualified scientists to non-computer expert professionals, intelligent interactive systems are becoming a necessity in modern computer systems. The solution of "one-fits-all" is no longer applicable to wide ranges of users of various backgrounds and needs. Therefore, one important goal of many intelligent interactive systems is dynamic personalization and adaptivity to users. Multimedia systems refer to the coordinated storage, processing, transmission, and retrieval of multiple forms of information, such as audio, image, video, animation, graphics, and text. The growth rate of multimedia services has become explosive, as technological progress matches consumer needs for content.

The conference took place as part of the Smart Digital Futures 2017 multi-theme conference, which groups AMSTA, IDT, InHorizons, InMed, SEEL with IIMSS in one venue. It was a forum for researchers and scientists to share work and experiences on intelligent interactive systems and multimedia systems and services. It included a general track and eight invited sessions.

The invited session "Processing visual data in intelligent systems: methods and applications" (Chaps. 1–8) specifically focuses on processing and understanding visual data in intelligent systems. The invited session "Cognitive Systems and Robotics" (Chaps. 9–20) focused on two main research areas, strictly related among them: adaptive and human-like cognitive systems, and artificial intelligence systems and cognitive robotics. The invited session "Big Data Management & Metadata" (Chaps. 21–24) focuses on models, techniques, and algorithms capable of dealing with the volume, velocity, variety, veracity, and value of big data. Differently, the invited session "Intelligent Big Data Analytics: Models, Techniques, Algorithms" (Chapter 25) discusses models, techniques, and algorithms for supporting intelligent analytics over big data in critical application contexts. The invited session

"Autonomous System" (Chaps. 26–29) considers technical and non-technical issues for what concerns intelligent, autonomous systems. The invited session "Mobile Data Analytics" (Chaps. 30–43) focuses on modeling, processing, and analyzing data generated by mobile devices, positioning technologies, and mobile users' activities. The invited session "Smart Environments and Information Systems" (Chaps. 44–49) provides insight into the most recent efforts in the field of information systems operating in dynamic environments. The invited session "Innovative Information Services for Advanced Knowledge Activity" (Chaps. 50–53) focuses on novel functionalities for information services. Finally, the general track (Chaps. 54–57) focuses on topics related to image processing algorithms and image processing-based rehabilitation and recommender systems.

Our gratitude goes to many people who have greatly contributed to putting together a fine scientific program and exciting social events for IIMSS 2017. We acknowledge the commitment and hard work of the program chairs and the invited session organizers. They have kept the scientific program in focus and made the discussions interesting and valuable. We recognize the excellent job done by the program committee members and the extra reviewers. They evaluated all the papers on a very tight schedule. We are grateful for their dedication and contributions. We could not have done it without them. More importantly, we thank the authors for submitting and trusting their work to the IIMSS conference.

We hope that readers will find in this book an interesting source of knowledge in fundamental and applied facets of intelligent interactive multimedia and, maybe, even some motivation for further research.

The editors

Organization

Honorary Chairs

Toyohide Watanabe   Nagoya University, Japan
Lakhmi C. Jain   University of Canberra, Australia and Bournemouth
                      University, UK

Co-General Chairs

Giuseppe De Pietro   National Research Council of Italy, Italy
Luigi Gallo   National Research Council of Italy, Italy

Executive Chair

Robert J. Howlett   University of Bournemouth, UK

Programme Chair

Antonino Mazzeo   University of Naples Federico II, Italy

Publicity Chair

Giuseppe Caggianese   National Research Council of Italy, Italy

Invited Session Chairs

Processing Visual Data in Intelligent Systems: Methods and Applications

Francesco Bianconi	Università degli Studi di Perugia, Italy
Elena González	Universidade de Vigo, Spain
Manuel Ángel Aguilar	Universidad de Almería, Spain

Cognitive Systems and Robotics

Ignazio Infantino	National Research Council of Italy, Italy
Massimo Esposito	National Research Council of Italy, Italy

Big Data Management and Metadata

Flora Amato	University of Naples Federico II, Italy
Vincenzo Moscato	University of Naples Federico II, Italy

Intelligent Big Data Analytics: Models, Techniques, Algorithms

Alfredo Cuzzocrea	University of Trieste, and ICAR-CNR, Italy

Autonomous System

Milan Simic	RMIT University, Australia

Mobile Data Analytics

Jalel Akaichi	University of Tunis, Tunisia, and King Khalid University, Saudi Arabia

Smart Environments and Information Systems

Rafael H. Bordini	FACIN-PUCRS, Brazil
Massimo Cossentino	National Research Council of Italy, Italy
Marie-Pierre Gleizes	University Paul Sabatier of Toulouse, France
Luca Sabatucci	National Research Council of Italy, Italy

Innovative Information Services for Advanced Knowledge Activity

Koichi Asakura	Daido University, Japan
Toyohide Watanabe	Nagoya Industrial Research Institute, Japan

International Programme Committee

Manuel Ángel Aguilar	Universidad de Almería, Spain
Flora Amato	Università degli Studi di Napoli Federico II, Italy
Marco Anisetti	Università degli Studi di Milano, Italy
Koichi Asakura	Daido University, Japan
Jalel Akaichi	University of Tunis, Tunisia, and King Khalid University, Saudi Arabia
Vivek Bannore	KES UniSA, Australia
V. Bellandi	Università degli Studi di Milano, Italy
Monica Bianchini	University of Perugia, Italy
Rafael H. Bordini	FACIN-PUCRS, Brazil
Helder Coelho	Mind-Brain College, BioISI, University of Lisbon, Portugal
Luigi Coppolino	Università degli Studi di Napoli "Parthenope", Italy
Massimo Cossentino	National Research Council of Italy, Italy
Giovanni Cozzolino	Università degli Studi di Napoli Federico II, Italy
Alfredo Cuzzocrea	University of Trieste and ICAR-CNR, Italy
Salvatore D'Antonio	Università degli Studi di Napoli "Parthenope", Italy
Ernesto Damiani	Università degli Studi di Milano, Italy
Mario Doeller	University of Applied Science Kufstein Tirol, Austria
Dinu Dragan	University of Novi Sad, Faculty of Technical Sciences, Novi Sad, Serbia
Massimo Esposito	National Research Council of Italy, Italy
Margarita Favorskaya	Siberian State Aerospace University, Russia
Marie-Pierre Gleizes	University Paul Sabatier of Toulouse, France
Christos Grecos	Central Washington University, USA
Elena González	Universidade de Vigo, Spain
Vincent Hilaire	Université de Belfort-Montbeliard, France
Katsuhiro Honda	Osaka Prefecture University, Japan
Hsiang-Cheh Huang	National University of Kaohsiung, Taiwan
Ignazio Infantino	National Research Council of Italy, Italy
Gwanggil Jeon	Xidian University, China
Dimitris Kanellopoulos	Department of Mathematics, University of Patras, Greece
Chengjun Liu	New Jersey Institute of Technology, USA
Marian Cristian Mihaescu	University of Craiova, Romania

Contents

Hand-Designed Local Image Descriptors vs. Off-the-Shelf CNN-Based Features for Texture Classification: An Experimental Comparison

Raquel Bello-Cerezo[1(✉)], Francesco Bianconi[1], Silvia Cascianelli[1],
Mario Luca Fravolini[1], Francesco di Maria[1], and Fabrizio Smeraldi[2]

[1] Department of Engineering, Università degli Studi di Perugia,
Via G. Duranti 93, 06135 Perugia, PG, Italy
{raquel.bellocerezo,silvia.cascianelli}@studenti.unipg.it,
bianco@ieee.org, {mario.fravolini,francesco.dimaria}@unipg.it
[2] School of Electronic Engineering and Computer Science,
Queen Mary University of London, Mile End Road, London E1 4NS, UK
f.smeraldi@qmul.ac.uk

Abstract. Convolutional Neural Networks have proved extremely successful in object classification applications; however, their suitability for texture analysis largely remains to be established. We investigate the use of pre-trained CNNs as texture descriptors by tapping the output of the last fully connected layer, an approach that has proved its effectiveness in other domains. Comparison with classical descriptors based on signal processing or statistics over a range of standard databases suggests that CNNs may be more effective where the intra-class variability is large. Conversely, classical approaches may be preferable where classes are well defined and homogeneous.

Keywords: Convolutional Neural Networks · Image classification · Texture · Local Binary Patterns

1 Introduction

Texture, along with colour, shape and gloss, is a fundamental visual feature of objects, materials and scenes. As a consequence, texture analysis plays an important role in several computer vision applications, such as image classification, content-based image retrieval, medical image analysis, surface inspection and remote sensing. Research on texture has been intense for more than forty years now: ideally, we could trace its origin as far back as 1973, when Haralick's seminal work on co-occurrence matrices [12] was first published. Since then a lot of different textures descriptors have been proposed in the literature: so many that Xie and Mirmehdi referred to them as 'a galaxy' [29]. Among them, methods based on signal processing like Gabor filters and wavelets dominated the scene

© Springer International Publishing AG 2018
G. De Pietro et al. (eds.), *Intelligent Interactive Multimedia Systems and Services 2017*,
Smart Innovation, Systems and Technologies 76, DOI 10.1007/978-3-319-59480-4_1

for a while, whereas in the last two decades statistical and rank-based features have become more popular. The bag-of-features paradigm [30] has also become the prominent aggregation strategy.

In recent years the appearance of Convolutional Neural Networks (CNNs) [14] represented a major breakthrough that changed the outlook for the pattern recognition field. This new paradigm for image analysis proved to consistently outperform pre-existing methods in a number of applications including object image recognition and scene classification [14,26]. Central to this scheme is the ability to learn complex image-to-object or image-to-feature mappings starting from very large datasets of labelled images. More importantly, pre-trained CNNs have also showed to be able to generalise quite well to datasets different from those they are trained on [8,26,31], a feature that makes them amenable to being used 'out of the box' in a potentially large number of applications. Yet the real effectiveness of CNN-based methods with *fine-grained* images – such as texture – is still subject of debate. Most of the related literature, that we briefly review in Sect. 2, is in fact rather new, and the results are far from being consolidated.

In this work we investigate the effectiveness of CNNs compared with classic local image descriptors such as Local Binary Patterns and variants, Gabor filters and grey-level co-occurrence matrices for texture classification. Specifically, we are interested in determining the potential of pre-trained CNNs when used as feature extractors in an off-the-shelf manner, relying directly on the pooling effect of the fully connected layers of the network. This avoids the added complexity of the separate pooling stages appearing in some related studies, that we review in Sect. 2). In the remainder of the paper we describe the materials (Sect. 3) and methods (Sect. 4) used in this study. We discuss the experimental set-up and the results in Sect. 5 and conclude the paper with some final considerations (Sect. 6) and directions for future studies (Sect. 7).

2   Related Research

Convolutional Neural Networks have been attracting increasing research interest in the computer vision community: suffice it to say that Krizhevsky *et al.*'s milestone work [14] has been so far cited more than 2700 times[1] since its publication in 2012.

In the field of texture analysis CNN-based methods have been receiving increasing attention. Cimpoi *et al.* [8] is the first in-depth investigation of the transferability of CNN models to the texture domain. The proposed solution (FV-CNN), however, entails complex and time-consuming pre- and post-processing procedures (respectively repeated image rescaling and Fisher vector pooling) that actually make it a new texture descriptor on its own rather than a direct application of CNNs to textures. A potential drawback of this solution is also the huge number of features produced (65K) which may represent a limit in many practical applications. Andrearczy and Whelan [1] recently improved on this idea and proposed a pooling scheme which relies on a lower number of features.

---

[1] Source: Scopus®; visited on Januray 18, 2017.

Table 1. Round-up table of the datasets used in the experiments.

ID	Name	No. of classes	No. of samples per class	Sample images
1	KTH-TIPS	10	81	
2	KTH-TIPS2b	11	432	
3	Kylberg	28	160	
4	Kylberg-Sintorn	25	6	
5	MondialMarmi	25	16	
6	Outex-00013	68	20	
7	Outex-00014	68	60	
8	PerTex	334	16	
9	RawFooT	68	184	
10	UIUC	25	40	

An interesting comparison between LBP variants and CNN-based features – though once again obtained by vector pooling – was recently presented by Liu *et al.* [19]. Here the authors find that the best performance is obtained by an LBP variant known as Median Robust Extended Local Binary Patterns (MRELBP). Of late, an experimental evaluation of colour texture descriptors under variable lighting conditions – including CNN-based features – was proposed by Cusano *et al.* [11]. Their approach consists of generating a texture descriptor by using, as image features, the output of the last fully-connected layer of a CNN. The main advantage of this strategy is that it generates significantly fewer features than the pooling method, and can be considered the model that best fits the idea of off-the-shelf use of CNNs for texture analysis. Finally, it is worth noting that in later experiments the same Cusano *et al.* [10] found that Fisher vector pooling produced worse results than were obtained by directly using CNNs features, probably due to the high number of features generated by FV-CNN.

3 Materials

We considered 10 datasets of texture images: (1) KTH-TIPS; (2) KTH-TIPS2b; (3) Kylberg Texture Dataset; (4) Kylberg-Sintorn Rotation Dataset; (5) MondialMarmi; (6) Outex-00013; (7) Outex-00014; (8) Pertex; (9) RawFooT and (10) UIUC. The main features of each dataset are detailed in Sect. 3.1 and summarised in Table 1.

3.1 Datasets

KTH-TIPS [13,15] features 10 classes of materials: aluminum foil, bread, corduroy, cotton, cracker, linen, orange peel, sandpaper, sponge and styrofoam. Images of each material were taken under different viewpoints and illumination conditions, giving 81 images for each class.

KTH-TIPS2b [6,15] is an extension of KTH-TIPS and contains 11 types of materials: aluminum foil, brown bread, corduroy, cork, cotton, cracker, lettuce, linen, white bread, wood and wool. Four samples for each class were acquired under varying scale, illumination and pose resulting in 432 images for each class.

Kylberg Texture Dataset (v. 1.0) [16] contains 28 texture classes such as fabric, natural stone, grains and seeds. There are 160 images for each class; the samples contain no variation in scale, rotation or illumination.

Kylberg-Sintorn Rotation Dataset [17,18] is a collection of 25 classes of heterogeneous materials including food (seeds and sugar), textiles (wool and knitwear) and tiles, with one image per class. The image samples used in our experiments contain no variation in scale, rotation or illumination.

MondialMarmi (v 2.0) [2,21] is a visual catalogue of polished natural stone products (marble and granites) featuring 25 classes of commercial denominations (e.g., *Azul Platino*, *Bianco Sardo*, *Rosa Porriño*, etc.) with four samples per class – each sample representing one tile. The images were acquired at fixed scale, in controlled illumination conditions and under different rotation angles. In our experiments we only used non-rotated images and subdivided each of them into four non-overlapping sub-images, thus obtaining 16 samples per class.

Outex-00013 contains the same 68 texture classes as Outex's test suite TC-00013, i.e. a collection of heterogeneous materials such as grains, fabric, natural stone and wood (see also Ref. [22] for details). There are 20 image samples for each class which were acquired at fixed scale, rotation angle and under invariable illumination conditions.

Outex-00014 is composed of the same classes as Outex-00013; in this case, however, each sample was acquired under three different lighting sources. As a consequence there are 60 samples for each class instead of 20. Note that in order to maintain the same evaluation protocol for all the datasets (see Sect. 5) the splits used in our experiments are different from those provided respectively with the TC-00013 and TC-00014 test suites.

PerTex [9,24] includes 334 texture classes representing heterogeneous materials such as embossed vinyl, woven wall coverings, carpet, rugs, fabric, building materials, product packaging, etc. The images were obtained by calculating the height-maps of the samples first, then by relighting them in order to remove variations due to reflectance. The results is a dataset of highly homogeneous textures – some of which are very similar to each other, a feature that makes this a very challenging dataset.

Table 2. Summary table of the image descriptors considered in the experiments

Method	Acronym	No. of features
Hand-designed local image descriptors		
Completed Local Binary Patterns	CLPB	324
Gradient-based Local Binary Patterns	GLBP	108
Improved Local Binary Patterns	ILBP	213
Local Binary Patterns	LBP	108
Local Ternary Patterns	LTP	216
Texture Spectrum	TS	2502
Gabor Filters	$\text{Gabor}_{4,6}^{rw}$	48
	$\text{Gabor}_{4,6}^{cn}$	48
	$\text{Gabor}_{5,7}^{rw}$	70
	$\text{Gabor}_{5,7}^{cn}$	70
Grey-level co-occurrence matrices	GLCM	60
CNN-based features		
CNN-imagenet-caffe-alex	Caffe-AlexNet	4096
CNN-imagenet-vgg-fast	VGG-F	4096
CNN-imagenet-vgg-medium	VGG-M	4096
CNN-imagenet-vgg-slow	VGG-S	4096
CNN-imagenet-vgg-medium-128	VGG-M-128	128
CNN-imagenet-vgg-medium-1024	VGG-M-1024	1024
CNN-imagenet-vgg-medium-2048	VGG-M-2028	2048
CNN-imagenet-vgg-verydeep-16	VGG-VD-16	4096
CNN-vgg-face	VGG-Face	4096

RawFooT [11,25] contains 68 classes of different types of food such as grain, fish, fruit and meat. There are 46 image samples for each class, each sample having been acquired under 46 different lighting conditions, whereas scale and rotation angle are invariable. In our experiments we subdivided each sample into four non-overlapping images of smaller size, therefore obtaining $46 \times 4 = 184$ samples for each class.

UIUC features 25 classes of heterogeneous materials and objects such as bark, wood, water, granite, marble, floor, pebbles, wall, brick, glass, carpet, upholstery, wallpaper, fur, knit, corduroy and plaid. There are 40 samples for each class, and within each class there is a lot of variability due to significant changes in the imaging conditions (i.e. rotation, scale and viewpoint) and warped surfaces.

4   Methods

We included in the experiments 11 hand-designed local image descriptors – specifically: six variants of Local Binary Patterns, four sets of features from

Gabor filters and one from grey-level co-occurrence matrices. On the network side we had nine sets of CNN-based features from as many pre-trained CNNs. The comparison was carried out on grey-scale images, therefore discarding colour information altogether. Details about settings and implementation are provided in the following subsections. Table 2 summarises the whole set of image descriptors and lists the number of features generated by each method.

4.1   Hand-Designed Local Image Descriptors

We took into account the following LBP variants: Completed Local Binary Patterns, Gradient-based Local Binary Patterns, Improved Local Binary Patterns, Local Binary Patterns, Local Ternary Patterns and Texture Spectrum (please refer to Ref. [5] for details). For each descriptor we concatenated the rotation-invariant features (e.g. LBP^{ri}) computed over three concentric, non-interpolated, eight-pixel circles respectively of radius 1px, 2px and 3px.

Gabor features [3] were computed using two filter banks: one with four frequencies and six orientations, and the other with five frequencies and seven orientations, which in the remainder we respectively indicate as $Gabor_{4,6}$ and $Gabor_{5,7}$. In both cases we set the maximum frequency to $0.5px^{-1}$, the frequency spacing to half-octave, the spatial frequency bandwidth and the aspect ratio to 0.5. We considered both raw and contrast-normalised filter output (in the latter case the filter responses for one point in all frequencies and rotations were normalized to sum one). In the reminder we indicate the two versions respectively with superscripts 'rw' and 'cn'. Image features were in all cases the mean and standard deviation of the magnitude of the Gabor-transformed images.

For the co-occurrence features we used 12 displacement vectors resulting from combining three distances (i.e. 1px, 2px and 3x – just as for LBP variants) and four standard orientations (i.e. $0°$, $45°$, $90°$ and $135°$). From each matrix we extracted the following global statistics as image features: *contrast, correlation, energy, entropy* and *homogeneity* (see also Ref. [4] for details).

4.2   CNN-Based Features

CNN-based features were computed using nine pre-trained Convolutional Neural Networks. The image processing pipeline included a pre-processing step whereby the input images were converted to grey-scale first, then resized through bicubic interpolation to fit the input dimension of each network – which for all the networks considered here was 224px × 224px. The nets were fed by dealing the resized, grey-scale images to the three colour input channels. Following the approach proposed by Cusano *et al.* [11] we used as texture features the L_2-normalised output of the last fully-connected layer. The implementation was based on the MatConvNet platform [20, 28]. The main features of each network are summarised here below.

  – **Caffe-AlexNet**: a MatConvNet porting of AlexNet, the architecture originally proposed by Krizhevsky *et al.* [14]. It is composed of eight layers, of which the first five are convolutional and the remaining three fully-connected.

Table 3. Overall accuracy by descriptor and dataset. Boldface figures indicate the best result for each dataset. Dataset IDs are listed in Table 1.

Descriptor	Dataset ID									
	1	2	3	4	5	6	7	8	9	10
Hand-designed local image descriptors										
CLBP	90.5	93.0	99.2	92.5	97.5	77.7	79.9	96.5	90.0	76.5
GBLBP	86.9	89.3	98.2	95.4	97.2	81.9	82.9	95.7	88.1	60.0
ILBP	89.9	91.7	99.1	95.8	**97.7**	**83.6**	**85.7**	96.5	93.1	72.0
LBP	87.7	89.7	98.0	90.1	96.7	78.2	80.5	95.5	90.1	60.5
LTP	87.8	89.8	98.1	90.1	96.7	79.0	81.7	95.6	90.5	60.6
TS	85.7	91.4	98.6	91.8	96.4	77.8	80.5	**97.3**	91.6	67.9
$Gabor_{4,6}^{rw}$	75.9	82.7	94.0	85.4	87.9	64.3	67.5	90.9	72.3	51.1
$Gabor_{5,7}^{rw}$	77.8	84.9	96.3	88.7	89.7	66.8	70.0	92.4	74.0	53.5
$Gabor_{4,6}^{cn}$	75.1	78.6	93.9	83.3	82.8	70.3	77.1	91.8	88.6	40.4
$Gabor_{5,7}^{rw}$	75.6	79.9	96.2	86.0	88.0	71.7	78.8	92.7	92.2	42.2
GLCM	75.4	80.5	97.2	92.9	89.9	65.3	68.2	92.9	74.4	52.8
CNN-based features										
Caffe-AlexNet	94.4	96.5	97.8	99.6	91.7	79.9	84.2	89.0	96.7	82.9
VGG-M	95.1	97.0	99.5	98.6	92.1	80.6	84.9	94.2	97.9	89.8
VGG-F	93.3	96.0	98.8	**99.7**	91.4	80.1	84.2	91.1	97.3	86.5
VGG-S	94.5	97.3	**99.7**	98.7	92.8	79.4	84.7	93.3	97.8	91.0
VGG-M-128	90.5	93.2	98.1	96.9	85.2	76.6	81.7	86.6	97.0	81.3
VGG-M-1024	94.4	96.6	99.4	99.2	91.0	79.2	84.3	93.3	97.7	88.3
VGG-M-2048	94.4	96.8	99.4	98.7	92.4	79.6	84.4	94.0	97.8	89.1
VGG-VD-16	**96.8**	**97.8**	99.5	99.5	93.8	80.6	85.6	93.4	**98.3**	**93.3**
VGG-face	86.5	87.1	92.1	97.5	85.0	71.4	82.0	68.7	95.5	57.7

- **VGG-F**, **VGG-M** and **VGG-S**: three networks all consisting of five convolutional and three fully-connected layers. The main differences are the size of the filters, the stride and the dimension of the pooling windows ('F', 'M' and 'S' respectively stand for *fast*, *medium* and *slow* – see also Ref. [7] for details).
- **VGG-M-128**, **VGG-M-1024** and **VGG-M-2048**: three variations of VGG-M with a lower-dimensional last fully-connected layer [7].
- **VGG-VD-16**: a deep network featuring 13 convolutional and three fully-connected layers [27].
- **VGG-Face**: a network designed for face recognition composed of eight convolutional and three fully-connected layers [23].

Apart from VGG-Face, which understandably was trained on faces [23], all the other networks were trained for object recognition.

5   Experiments and Results

To comparatively assess the effectiveness of the image descriptors presented in Sect. 4 we ran a supervised image classification experiment using the 1-NN classifier with L_1 distance. Accuracy estimation was based on split-sample validation with stratified sampling where $1/4$ of the samples of each class was used to train the classifier and the remaining $3/4$ to test it. The estimated accuracy was the ratio between the number of samples correctly classified and the total number of samples of the test set. For a stable estimation of the classification error we averaged the results (see Table 3) over 100 random splits.

The results are interesting and show a trend strongly dependent on the dataset used. In six datasets out of 10, CNN-based features outperformed the hand-designed methods (though in dataset #3 the margin is narrow); whereas the reverse occurred in four datasets out of 10 (though again by a narrow margin in dataset #7). CNN-based features seemed to be more effective when there was high intra-class variability due to changes in viewpoint/scale/appearance: paradigmatic and impressive is the 93.3% attained by VGG-VD-16 on dataset UIUC – a notoriously difficult one. By contrast, hand-designed image descriptors appeared to be more comfortable with homogeneous, fine-grained textures and little intra-class variability – as for instance in datasets #5 and #8. Within this group of methods, LBP variants clearly outperformed Gabor filters and GLCM.

6   Conclusions

Convolutional Neural Networks represented a major breakthrough in computer vision, having significantly improved the state of the art in many applications. Originally developed for object and scene classification, the approach proved effective in other domains as well, for example face recognition. It is however still a subject of debate whether this paradigm is amenable to being successfully applied to fine-grained images – i.e. texture. In this work we have carried out a comparison between some classic local image descriptors and off-the-shelf CNN-based features from an array of pre-trained nets. Our results were split, showing that though CNN-based features performed generally well, they were in some cases outperformed by state-of-the-art hand-designed descriptors. More specifically, our findings seem to suggest that CNNs performed better when there was high intra-class variability, whereas LBP variants provided better results with homogeneous, fine-grained textures with low intra-class variability.

7   Limitations and Future Work

The results presented here are promising and should be validated in a broader cohort of experiments. Importantly, our investigation was limited to grey-scale images, therefore the contribution of colour to image classification wasn't considered. Likewise, disturbing effects such as rotation and noise were not investigated. Assessing the effectiveness of more complex pooling schemes for CNN-based features (e.g. Fisher vectors) is also another important question for future studies.

Acknowledgements. This work was partially supported by the Department of Engineering at the Università degli Studi di Perugia, Italy, under project *BioMeTron* – Fundamental research grant D.D. 20/2015 and by the Spanish Government under project AGL2014-56017-R.

References

1. Andrearczyk, V., Whelan, P.F.: Using filter banks in convolutional neural networks for texture classification. Pattern Recogn. Lett. **84**, 63–69 (2016)
2. Bianconi, F., Bello, R., Fernández, A., González, E.: On comparing colour spaces from a performance perspective: application to automated classification of polished natural stones. In: Murino, V., Puppo, E., Sona, D., Cristani, M., Sansone, C. (eds.) New Trends in Image Analysis and Processing, ICIAP 2015 Workshops, Genoa, Italy. LNCS, vol. 9281, pp. 71–78. Springer (2015)
3. Bianconi, F., Fernández, A.: Evaluation of the effects of Gabor filter parameters on texture classification. Pattern Recogn. **40**(12), 3325–3335 (2007)
4. Bianconi, F., Fernández, A.: Rotation invariant co-occurrence features based on digital circles and discrete Fourier transform. Pattern Recogn. Lett. **48**, 34–41 (2014)
5. Bianconi, F., Fernández, A.: A unifying framework for LBP and related methods. In: Brahnam, S., Jain, L.C., Nanni, L., Lumini, A. (eds.) Local Binary Patterns: New Variants and Applications. Studies in computational intelligence, vol. 506, pp. 17–46. Springer (2014)
6. Caputo, B., Hayman, E., Mallikarjuna, P.: Class-specific material categorisation. In: Proceedings of the Tenth IEEE International Conference on Computer Vision (ICCV 2005), vol. II, pp. 1597–1604 (2005)
7. Chatfield, K., Simonyan, K., Vedaldi, A., Zisserman, A.: Return of the devil in the details: delving deep into convolutional nets. In: Proceedings of the British Machine Vision Conference 2014, Nottingham, United Kingdom, September 2014
8. Cimpoi, M., Maji, S., Vedaldi, A.: Deep filter banks for texture recognition and segmentation. In: Proceedings of the IEEE Computer Society Conference on Computer Vision and Pattern Recognition, Boston, USA, pp. 3828–3836, June 2015
9. Clarke, A.D.F., Halley, F., Newell, A.J., Griffin, L.D., Chantler, M.J.: Perceptual similarity: a texture challenge. In: Proceedings of the British Machine Vision Conference 2011, Dundee, UK, August–September 2011
10. Cusano, C., Napoletano, P., Schettini, R.: Combining multiple features for color texture classification. J. Electron. Imaging **25**(6) (2016)
11. Cusano, C., Napoletano, P., Schettini, R.: Evaluating color texture descriptors under large variations of controlled lighting conditions. J. Opt. Soc. Am. A **33**(1), 17–30 (2016)
12. Haralick, R.M., Shanmugam, K., Dinstein, I.: Textural features for image classification. IEEE Trans. Syst. Man Cybern. **3**(6), 610–621 (1973)
13. Hayman, E., Caputo, B., Fritz, M., Eklundh, J.-O.: On the significance of real-world conditions for material classification. In: Proceedings of the 8th European Conference on Computer Vision (ECCV 2004), Prague, Czech Republic. LNCS, vol. 3024, pp. 253–266. Springer, May 2004
14. Krizhevsky, A., Sutskever, I., Hinton, G.E.: ImageNet classification with deep convolutional neural networks. In: Proceedings of Advances in Neural Information Processing Systems, Lake Tahoe, USA, vol. 2, pp. 1097–1105 (2012)

15. The KTH-TIPS and KTH-TIPS2 image databases (2004). http://www.nada.kth.se/cvap/databases/kth-tips/. Last Accessed 17 Oct 2016
16. Kylberg, G.: The Kylberg texture dataset v. 1.0. External report (Blue series) 35, Centre for Image Analysis, Swedish University of Agricultural Sciences and Uppsala University, Uppsala, Sweden, September 2011
17. Kylberg, G., Sintorn, I.-M.: On the influence of interpolation method on rotation invariance in texture recognition. EURASIP J. Image Video Process. **2016**(1) (2016)
18. Kylberg Sintorn Rotation dataset (2013). http://www.cb.uu.se/~gustaf/KylbergSintornRotation/. Last Accessed 17 Oct 2016
19. Liu, L., Fieguth, P., Wang, X., Pietikäinen, M., Hu, D.: Evaluation of LBP and Deep Texture Descriptors with a New Robustness Benchmark. In: Proceedings of the 14th European Conference on Computer Vision (ECCV 2016). LNCS, Amsterdam, The Netherlands, vol. 9907, pp. 69–86. Springer (2016)
20. MatConvNet: CNNs for MATLAB (2016). http://www.vlfeat.org/matconvnet/. Last Accessed 25 Oct 2016
21. MondialMarmi: A collection of images of polished natural stones for colour and texture analysis. version 2.0 (2015). http://dismac.dii.unipg.it/mm. Last Accessed 17 Oct 2016
22. Ojala, T., Pietikäinen, M., Mäenpää, T., Viertola, J., Kyllönen, J., Huovinen, S.: Outex - new framework for empirical evaluation of texture analysis algorithms. In: Proceedings of the 16th International Conference on Pattern Recognition (ICPR 2002), Quebec, Canada, vol. 1, pp. 701–706. IEEE Computer Society (2002)
23. Parkhi, O.M., Vedaldi, A., Zissermann, A.: Deep face recognition. In: Proceedings of the British Machine Vision Conference 2015, Swansea, United Kingdom, September 2015
24. Pertex database (2011). http://www.macs.hw.ac.uk/texturelab/resources/databases/pertex/. Last Accessed 17 Oct 2016
25. RawFooT, D.B.: Raw food texture database (2015). http://projects.ivl.disco.unimib.it/rawfoot/. Last Accessed 17 Oct 2016
26. Razavian, A.S., Azizpour, H., Sullivan, J., Carlsson, S.: CNN features off-the-shelf: an astounding baseline for recognition. In: Proceedings of IEEE Computer Society Conference on Computer Vision and Pattern Recognition Workshops, Columbus, USA, pp. 512–519, June 2014
27. Simonyan, K., Zisserman, A.: Very deep convolutional networks for large-scale image recognition. CoRR, abs/1409.1556 (2014)
28. Vedaldi, A., Lenc, K.: MatConvNet: convolutional neural networks for MATLAB. In: MM 2015 - Proceedings of the 2015 ACM Multimedia Conference, Brisbane, Australia, pp. 689–692, October 2015
29. Xie, X., Mirmehdi, M.: A galaxy of texture features. In: Mirmehdi, M., Xie, X., Suri, J. (eds.) Handbook of Texture Analysis, pp. 375–406. Imperial College Press (2008)
30. Zhang, J., Marszałek, M., Lazebnik, S., Schmid, C.: Local features and kernels for classification of texture and object categories: a comprehensive study. Int. J. Comput. Vision **73**(2), 213–238 (2007)
31. Zhong, Y., Sullivan, J., Li, H.: Face attribute prediction using off-the-shelf CNN features. In Proocedings of the 2016 International Conference on Biometrics, ICB 2016, Halmstad, Sweden, 6 2016

Images Selection and Best Descriptor Combination for Multi-shot Person Re-identification

Yousra Hadj Hassen[(✉)], Kais Loukil, Tarek Ouni, and Mohamed Jallouli

National School of Engineers of Sfax, Computer and Embedded Systems Laboratory,
University of Sfax, 4.7 km Street of Soukra, 3038 Sfax, Tunisia
hadjhassen.yousra@gmail.com
http://www.ceslab.org
http://www.enis.rnu.tn

Abstract. To re-identify a person is to check if he/she has been already seen over a cameras network. Recently, re-identifying people over large public cameras networks has become a crucial task of great importance to ensure public security. The vision community has deeply studied this area of research. Most existing researches rely only on the spatial appearance information extracted from either one (single-shot) or multiple images (multi-shot) for each person. Actually, the real person re-identification framework is a multi-shot scenario. However, to efficiently model a person's appearance and to select the most informative samples remain a challenging problem. In this work, an extensive comparison of descriptors of state of art associated to the proposed frame selection method is considered. Specifically, we evaluate the samples selection approach using different known descriptors. For fair comparisons, two standard datasets PRID 2011 and iLIDS-VID are used showing the effectiveness and advantages of the proposed method.

Keywords: Camera network · Descriptor · Model · Multi-shot · Person re-identification · Selection

1 Introduction

Recently, person re-identification (re-id) in non-overlapped cameras network presents a crucial task for many real applications like video surveillance, multimedia applications, behavior recognition,... [1]. To re-identify a person is to match his identity across camera views despite the changes that may occur. Actually, many environmental constraints can alter a persons appearance over different cameras views such as luminance variations, different point of view, scale zooming as shown in Fig. 1.

The proposed approaches can be classified either as single-shot or multiple-shot methods depending on the number of images used to construct the person identity. Contrary to the use of a single image to re-identify a person, the multi-shot based methods have been largely studied and significant results are achieved

© Springer International Publishing AG 2018
G. De Pietro et al. (eds.), *Intelligent Interactive Multimedia Systems and Services 2017*,
Smart Innovation, Systems and Technologies 76, DOI 10.1007/978-3-319-59480-4_2

[2]. They mainly focus on designing discriminative feature descriptors collected over many images of the same person. Whereas, real objectives of person re-id steel far from being reached because both relatively reduced execution time and robust feature descriptor are required.

In this paper, the trade-off between the use of robust descriptor and the reduction of execution time is considered. Multiple-shot proposed methods give potential results for person re-id while ignoring the selection of shots used for the re-id. Actually, most of multiple-shot methods select randomly the images forming the identity of the person but try to generate sufficiently robust descriptors that handle appearance changes caused by scale, lighting variations, view angles conditions and occlusions (Fig. 1). Since, for the multi-shot case, the results of images selection methods have a strong impact not only on the descriptor used for re-id, but also on the overall processing time of the system, the selection of both robust descriptor and discriminative frames to construct representative identity for each person is studied.

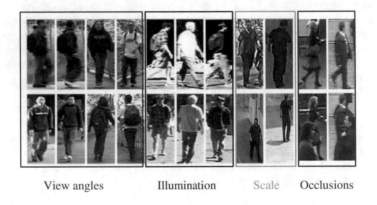

View angles   Illumination   Scale   Occlusions

Fig. 1. Re-identification system constraints.

A key frame selection method is proposed and a rich comparison of recent robust proposed descriptors is associated to find the most performing combination for a real person re-id system.

The paper is organized in 5 sections. The first section is the introduction, followed by Sect. 2 of the related work. Section 3 describes the overall framework. The evaluation and the results comparison are detailed in Sect. 4. The final section is the conclusion and the future works.

2   Related Work

The contribution of this work may be considered on two fields; person re-identification descriptors as well as samples selection methods.

2.1 Descriptors

In person description, the most commonly used features are color and texture. Symmetry Driven Accumulation of Local Features (SDALF) [3] exploits the symmetric property of a person through obtaining head, torso, and leg positions to handle view variations. Gheissari et al. [4] propose a spatial-temporal segmentation method to detect stable foreground regions. For a local region, an HS histogram and an edgel histogram are computed. The latter encodes the dominant local boundary orientation and the RGB ratios on either sides of the edgel.

Hirzer et al. propose a generic descriptive statistical model in [5] and the appearance is modeled by a set of region covariance descriptors. Gray and Tao [6] use 8 color channels (RGB, HS, and YCbCr) and 21 texture filters on the luminance channel, and the pedestrian is partitioned into horizontal stripes. Similarly, Mignon et al. [7] build the feature vector from RGB, YUV and HSV channels and the LBP texture histograms in horizontal stripes. In [8–12], the 32-dim LAB color histogram and the 128-dim SIFT descriptor are extracted from each 10 × 10 patch densely sampled with a step size of 5 pixels. Das et al. [13] apply HSV histograms on the head, torso and legs from the silhouette. Pedagadi et al. [14] extract color histograms and moments from HSV and YUV spaces before dimension reduction using PCA. Liu et al. [15] extract the HSV histogram, gradient histogram and the LBP histogram for each local patch. In [16], Liao et al. propose the local maximal occurrence (LOMO) descriptor, which includes the color and SILTP histograms. Bins in the same horizontal stripe undergo max pooling and a three-scale pyramid model is built before a log transformation.

Ayedi et al. propose a multi-scale covariance descriptor in [17] using a quad-tree feature to tackle scale zooming appearance and occlusions in person re-id. In [18], Zheng et al. propose extracting the 11-dim color names descriptor for each local patch, and aggregating them into a global vector through a Bag-of-Words (BoW) model. In [19], a hierarchical Gaussian feature is proposed to describe color and texture cues, which models each region by multiple Gaussian distributions. Each distribution represents a patch inside the region. Liu et al. [20] improve the latent Dirichlet allocation (LDA) model using annotated attributes to filter out noisy LDA topics. In [21], Su et al. embed the binary semantic attributes of the same person but different cameras into a continuous low-rank attribute space, so that the attribute vector is more discriminative for matching. Shi et al. [22] propose learning a number of attributes including color, texture, and category labels from existing fashion photography. In [23], where the image is divided into a number of subblocks, each with its associated color histogram, multi-precision similarity matching is granted despite scale and lighting variations.

2.2 Sample Selection

Video summarization is the most known field in representative selection applications. The problem has been treated using clustering, vector quantization [24–26] or sparse selection [27–29]. In [29], the Sparse Modeling Representative Selection (SMRS), based on the summary of videos by considering the proximity of the selected frames in the timeline, removes redundant frames. Intra and inter iteration redundancy splitting is proposed in [30] and significant re-id results are achieved.

Most of proposed representative samples selection methods rely on appearance variation in time, however, time information is almost unavailable in person re-id datasets. The proposed framework, based on key clusters and key frames selection, takes care of this issue. This enables the selection of as many informative samples as possible to improve the identification performance but at the same time avoids tedious training and useless gallery images. Multi-shot are outperforming single-shot re-id approaches. Incorporating supervision using training data leads to superior performance, which is the goal of metric learning. However, for a multi-shot metric learning person re-id approach with superior real scenario performance, the amount of data presents a hard issue for training time. Thats why, we propose a multi-shot metric learning person re-id framework based on continuous representative samples selection.

3   Proposed Approach

In Fig. 2, the input of the proposed framework is a large gallery of images mostly formed by large sequence of frames of different persons captured in one camera view.

The proposed overall scheme of person re-id is composed of three main parts; first, features are extracted from the images in order to project the visual appearance into concrete parameters (color, texture, position, pose view). After that, a key frame selection algorithm is introduced [31]. The goal is to keep only informative images for each person so that all the informative appearance variations over time and space are summarized and useless noisy and redundant frames are removed. The major contribution of this work is to get the best combination of the descriptor (set of features) and the key frame selection algorithm so that a relatively speed and robust (accurate) re-id is achieved. To that end, the inter and intra descriptors distances are computed. The feed-back enables the evaluation of the used features and the algorithm converges to the best (descriptor, key frame selection) combination. Thus, a new gallery is formed containing key frames descriptors. Finally, the matching block allows the mapping of a newly unknown observed individual in another camera view (a probe) to one of the identity stored in the gallery yet constructed.

3.1   Features Extraction

As detailed in section two, different descriptors are proposed and excellent performances are reached. It is pretty promising to construct robust identities for each person. However, execution time and memory consumption present crucial constraints that must be treated for real video surveillance applications. Therefore,

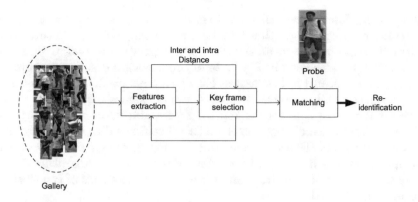

Fig. 2. Overall proposed approach.

person re-id systems have to take into account the trade-off between memory consumption (and so time consumption) and accuracy. Generally, robust descriptors can not be applied for a practical scenario due to their complex mathematical principals and large memory enquiries. So, basic descriptors such as covariance [5] treating lighting variations constraint, HOG [23] dealing with color and pose variations and multi-scale covariance [17] studying the scale zooming and occlusions, are tested. These descriptors are complementary in treating the different re-id system challenging issues. The impact of the key frame selection algorithm is evaluated on both re-id rates and memory gain.

3.2 Key Frame Selection

To concisely summarize long sequences of moving persons in one camera view (tracks), a representative frames selection method is proposed in Fig. 3.

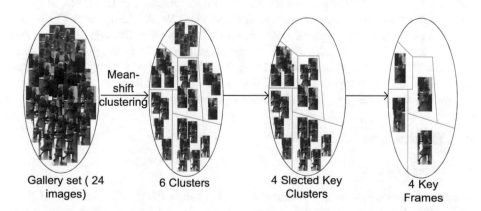

Fig. 3. Detailed example of key frame selection for person 1 camera 1 of PRID_2011.

Since redundancy presents a hard issue for most of person re-id systems, the use of a clustering algorithm may be a performing solution to surpass such problem. To that end, the mean shift clustering algorithm is used to form groups of similar images; as shown in Fig. 3, 24 different images describe 6 appropriate visual appearances. In Fig. 3, 2 useless clusters are removed. Then, non-frequent frames forming small clusters, in term of size or frames number, are removed and only informative clusters are kept. A representative frame is selected as a head, for each yet selected key cluster, so that it replaces the whole cluster in the person identity. As in the example presented in Fig. 3, instead of extract features of 24 images, only 4 images will form the person identity. Algorithm 1 details the key frame selection algorithm and conditions of the choice of key clusters or key frames are defined.

```
Algorithm1: Key frame selection based clustering
 Input: set of N target captures
 Output: set of discriminative target captures
 Initialization: set radius for mean shift clustering


   1: Obtain a number of clusters by performing
   the mean-shift clustering process in the
   feature space  among the N samples.
   2: Find the key cluster such;
       Size (key cluster)> Mean (Size (clusters))/2
   3: Find the key frame such;
       Distance (key frame, center cluster) is minimal.
```

3.3   Matching

To re-identify an unknown newly captured person is to match his image to an identity stored in the gallery. To that end, a multi-class SVM classifier, yet trained by the gallery set, is tested for the query image. So, the SVM map the unknown person to the most similar trained identity. As detailed in Fig. 4, the tracked persons in one camera view are modeled and their identity based on key frames are stored and fed, in a first step, into the SVM through a file '.train' representing the gallery. In a second step, the descriptor of a query image, saved in a '.test' file, is mapped to one of the classes of the gallery by the SVM. The output of the matching phase is a label that re-identify the tested image.

4   Evaluation

4.1   Datasets

For experiments, we use multi-shot standard datasets PRID_2011 [32] and iLIDS-VID [33]. These two datasets are very challenging due to clothing similarities among people, lighting and view point variations across camera views, cluttered background and occlusions.

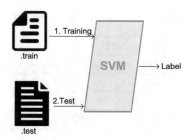

Fig. 4. Matching based SVM classifier.

PRID_2011 dataset: PRID_2011 dataset is formed of images of 200 and 749 people captured by two cameras A and B respectively. Each person has 5 to 675 images available. This dataset is hard because it presents real images with noisy background and illumination variations.

iLIDS-VID dataset: it presents different frames of 300 people captured by two non-overlapped cameras in an airport arrival hall. It is a challenging dataset due to the huge amount of images per person with clothing similarities and both partial and total occlusions.

4.2   Test Design

The results of re-id rates and memory consumption are computed in Tables 1 and 2 for respectively PRID_2011 and iLIDS-VID datasets for the three basic descriptors covariance [5], histogram of gradient (HOG) [23] and multi-scale covariance [17]. Of course, the person re-id rates are not very promising using the proposed discriminative selection but they still similar to the competing rates given by the robust descriptors in the full training case i.e. using the whole dataset for training. Moreover, the memory gain is significant thanks to the notable reduction of the trained data for the SVM classifier. Thus, real world re-id scenarios could be efficiently treated.

Table 1. Memory gain and re-id rate for 3 descriptors for PRID_2011

Descriptors	Re-id rates		Memory gain
	Full training	Key frame training	
Covariance	84.2%	74.3%	80%
HOG	56.7%	43.1%	95%
Multi-scale covariance	97%	72.8%	96%

Table 2. Memory gain and re-id rate for 3 descriptors for iLIDS-VID

Descriptors	Re-id rates		Memory gain
	Full training	Key frame training	
Covariance	84.9%	78.1%	88%
HOG	50.2%	48.5%	97%
Multi-scale covariance	81.9%	72.9%	94%

4.3   Results Discussion

The three basic descriptors compared are frequently used in person re-id. The results shown in Tables 1 and 2 demonstrate that the covariance descriptor outperforms HOG and Multi-scale covariance descriptors for both tested datasets. Thanks to the robust extracted features for the covariance descriptor, mainly colors and texture, the reached re-id rate is about 84%. Actually, it is an efficient result suitable for a practical real video surveillance application. The evaluation prooves that to model some representative frames by a selected robust descriptor seems to be a promising solution to guaranty both re-id efficiency and memory consumption gain. However, a real world person re-id scenario steels far from being solved. In fact, the use of the robust classifier SVM shows that training key frames significantly outperforms full training (i.e. train all the images available for each person) in terms of execution time and memory consumption. However, the re-id rate steel greater for the latter case. Thats why, an extensive comparison of different classifier (as Adaboost) and the impact in re-id results may be proposed in future works.

5   Conclusion

Person re-id, has become an inherently task, of extensive interest, for real video surveillance applications. Having proposed a novel algorithm of representative images selection, the goal, in this paper, is to associate the most robust descriptor giving best re-id results. A comparative analysis of frequently used descriptors for re-id frameworks is conducted by computing the re-id rate. The impact of the key frames selection algorithm is highlighted thanks to the significant reduction of memory consumption and of course execution time. Finally, the combination of a selected descriptor with selected frames leads to an efficient multi-shot person re-id system. The proposed approach will be deeply studied to reach more performing real-world results and further evaluations will be delivered in future works.

References

1. Karanam, S., Gou, M., Wu, Z., Rates-Borras, A., Camps, O., Radke, R.J.: A comprehensive evaluation and benchmark for person re-identification: features, metrics, and datasets. arXiv preprint arXiv:1605.09653 (2016)

2. Zheng, L., Yang, Y., Hauptmann, A.G.: Person re-identification: past, present and future. arXiv preprint arXiv:1610.02984 (2016)
3. Bazzani, L., Cristani, M., Murino, V.: Symmetry-driven accumulation of local features for human characterization and re-identification. Comput. Vis. Image Underst. **117**, 130–144 (2013)
4. Gheissari, N., Sebastian, T.B., Hartley, R.: Person re-identification using spatiotemporal appearance. In: 2006 IEEE Computer Society Conference on Computer Vision and Pattern Recognition, pp. 1528–1535. IEEE Press, New York (2006)
5. Hirzer, M., Beleznai, C., Roth, P.M., Bischof, H.: Person re-identification by descriptive and discriminative classification. In: Scandinavian Conference on Image Analysis, pp. 91–102. Springer, Heidelberg (2011)
6. Gray, D., Tao, H.: Viewpoint invariant pedestrian recognition with an ensemble of localized features. In: European Conference on Computer Vision, pp. 262–275. Springer, Marseille (2008)
7. Mignon, A., Jurie, F.: PCCA: a new approach for distance learning from sparse pairwise constraints. In: IEEE Conference on Computer Vision and Pattern Recognition, pp. 2666–2672. IEEE Press, Providence (2012)
8. Zhao, R., Ouyang, W., Wang, X.: Unsupervised salience learning for person re-identification. In: Proceedings of the IEEE Conference on Computer Vision and Pattern Recognition, pp. 3586–3593. IEEE Press, Portland (2013)
9. Li, Z., Chang, S., Liang, F., Huang, T.S., Cao, L., Smith, J.R.: Learning locally-adaptive decision functions for person verification. In: Proceedings of the IEEE Conference on Computer Vision and Pattern Recognition, pp. 3610–3617. IEEE Press, Portland (2013)
10. Chen, D., Yuan, Z., Chen, B., Zheng, N.: Similarity learning with spatial constraints for person re-identification. In: Proceedings of the IEEE Conference on Computer Vision and Pattern Recognition, pp. 1268–1277. IEEE Press, Las Vegas (2016)
11. Zhao, R., Ouyang, W., Wang, X.: Person re-identification by salience matching. In: Proceedings of the IEEE International Conference on Computer Vision, pp. 2528–2535. IEEE Press, Sydney (2013)
12. Zhao, R., Ouyang, W., Wang, X.: Learning mid-level filters for person re-identification. In: Proceedings of the IEEE Conference on Computer Vision and Pattern Recognition, pp. 144–151. IEEE Press, Columbus (2014)
13. Das, A., Chakraborty, A., Roy-Chowdhury, A.K.: Consistent re-identification in a camera network. In: European Conference on Computer Vision, pp. 330–345. Springer, Zurich (2014)
14. Pedagadi, S., Orwell, J., Velastin, S., Boghossian, B.: Local fisher discriminant analysis for pedestrian re-identification. In: Proceedings of the IEEE Conference on Computer Vision and Pattern Recognition, pp. 3318–3325. IEEE Press, Portland (2013)
15. Liu, X., Song, M., Tao, D., Zhou, X., Chen, C., Bu, J.: Semi-supervised coupled dictionary learning for person re-identification. In: Proceedings of the IEEE Conference on Computer Vision and Pattern Recognition, pp. 3550–3557. IEEE Press, Columbus (2014)
16. Liao, S., Hu, Y., Zhu, X., Li, S.Z.: Person re-identification by local maximal occurrence representation and metric learning. In: Proceedings of the IEEE Conference on Computer Vision and Pattern Recognition, pp. 2197–2206. IEEE Press, Boston (2015)
17. Ayedi, W., Snoussi, H., Abid, M.: A fast multi-scale covariance descriptor for object re-identification. Pattern Recogn. Lett. **33**, 1902–1907 (2012)

18. Zheng, L., Shen, L., Tian, L., Wang, S., Wang, J., Tian, Q.: Scalable person re-identification: a benchmark. In: Proceedings of the IEEE International Conference on Computer Vision, pp. 1116–1124. IEEE Press, Chile (2015)
19. Matsukawa, T., Okabe, T., Suzuki, E., Sato, Y.: Hierarchical gaussian descriptor for person re-identification. In: Proceedings of the IEEE Conference on Computer Vision and Pattern Recognition, pp. 1363–1372. IEEE Press, Las Vegas (2016)
20. Liu, X., Song, M., Zhao, Q., Tao, D., Chen, C., Bu, J.: Attribute restricted latent topic model for person re-identification. Pattern Recogn. **45**, 4204–4213 (2012)
21. Su, C., Yang, F., Zhang, S., Tian, Q., Davis, L.S., Gao, W.: Multi-task learning with low rank attribute embedding for person re-identification. In: Proceedings of the IEEE International Conference on Computer Vision, pp. 3739–3747. IEEE Press, Chile (2015)
22. Shi, Z., Hospedales, T.M., Xiang, T.: Transferring a semantic representation for person re-identification and search. In: Proceedings of the IEEE Conference on Computer Vision and Pattern Recognition, pp. 4184–4193. IEEE Press, Boston (2015)
23. Lin, S., Ozsu, M.T., Oria, V., Ng, R.: An extensible hash for multi-precision similarity querying of image databases. In: Proceedings of the 27th International Conference on Very Large Databases, Italy, pp. 221–230 (2001)
24. De Avila, S.E.F., Lopes, A.P.B., da Luz, A., de Albuquerque Arajo, A.: VSUMM: a mechanism designed to produce static video summaries and a novel evaluation method. Pattern Recogn. Lett. **32**, 56–68 (2011)
25. Frey, B.J., Dueck, D.: Clustering by passing messages between data points. Science **315**, 972–976 (2007)
26. Garcia, S., Derrac, J., Cano, J., Herrera, F.: Prototype selection for nearest neighbor classification: taxonomy and empirical study. IEEE Trans. Pattern Anal. Mach. Intell. **34**, 417–435 (2012)
27. Cong, Y., Yuan, J., Luo, J.: Towards scalable summarization of consumer videos via sparse dictionary selection. IEEE Trans. Multimedia **14**, 66–75 (2012)
28. Elhamifar, E., Sapiro, G., Vidal, R.: Finding exemplars from pairwise dissimilarities via simultaneous sparse recovery. In: Advances in Neural Information Processing Systems, pp. 19–27 (2012)
29. Elhamifar, E., Sapiro, G., Vidal, R.: See all by looking at a few: sparse modeling for finding representative objects. In: IEEE Conference on Computer Vision and Pattern Recognition, pp. 1600–1607. IEEE Press, Providence (2012)
30. Das, A., Panda, R., Roy-Chowdhury, A.K.: Continuous adaptation of multi-camera person identification models through sparse non-redundant representative selection. Comput. Vis. Image Underst. **156**, 66–78 (2016)
31. Hadj Hassen, Y., Ayedi, W., Ouni, T., Jallouli, M.: Multi-shot person re-identification approach based key frame selection. In: Proceedings of the Eighth International Conference on Machine Vision, International Society for Optics and Photonics, Barcelona, p. 98751H (2015)
32. Corvee, E., Bremond, F., Thonnat, M.: Person re-identification using spatial covariance regions of human body parts. In: Proceedings of the Seventh IEEE International Conference on Advanced Video and Signal Based Surveillance, pp. 435–440. IEEE Press, Boston (2010)
33. Wang, T., Gong, S., Zhu, X., Wang, S.: Person re-identification by video ranking. In: European Conference on Computer Vision, pp. 688–703. Springer International Publishing, Zurich (2014)

Dimensionality Reduction Strategies
for CNN-Based Classification
of Histopathological Images

Silvia Cascianelli[1(✉)], Raquel Bello-Cerezo[1], Francesco Bianconi[1],
Mario L. Fravolini[1], Mehdi Belal[1], Barbara Palumbo[2], and Jakob N. Kather[3]

[1] Department of Engineering, Università degli Studi di Perugia,
Via G. Duranti 93, 06135 Perugia, PG, Italy
{silvia.cascianelli,raquel.bellocerezo,mehdi.belal}@studenti.unipg.it,
mario.fravolini@unipg.it, bianco@ieee.org
[2] Department of Surgery and Biomedical Sciences,
Università degli Studi di Perugia, Piazza L. Severi 1, 06132 Perugia, PG, Italy
barbara.palumbo@unipg.it
[3] Computer Assisted Clinical Medicine, Medical Faculty Mannheim,
Heidelberg University, Theodor-Kutzer-Ufer 1-3, 68167 Mannheim, Germany
jakob.kather@nct-heidelberg.de

Abstract. Features from pre-trained Convolutional Neural Newtorks
(CNN) have proved to be effective for many tasks such as object, scene
and face recognition. Compared with traditional, hand-designed image
descriptors, CNN-based features produce higher-dimensional feature vectors. In specific applications where the number of samples may be limited – as in the case of histopatological images – high dimensionality
could potentially cause overfitting and redundancy in the information to
be processed and stored. To overcome these potential problems feature
reduction methods can be applied, at the cost of a moderate reduction in
the discrimination accuracy. In this paper we investigate dimensionality
reduction schemes for CNN-based features applied to computer-assisted
classification of histopathological images. The purpose of this study is to
find the best trade-off between accuracy and dimensionality. Specifically,
we test two well-known techniques (i.e.: Principal Component Analysis
and Gaussian Random Projection) and propose a novel reduction strategy based on the cross-correlation between the components of the feature
vector. The results show that it is possible to reduce CNN-based features
by a high ratio with a moderate decrease in accuracy with respect to the
original values.

Keywords: Convolutional Neural Networks · Feature reduction ·
Histopathological images · Image classification

1   Introduction

Pathologists routinely examine tissue samples and classify them on the basis
of their appearance for diagnosis, patient stratification and follow-up. In recent

© Springer International Publishing AG 2018
G. De Pietro et al. (eds.), *Intelligent Interactive Multimedia Systems and Services 2017*,
Smart Innovation, Systems and Technologies 76, DOI 10.1007/978-3-319-59480-4_3

years the advent, improvement and proliferation of digital slide scanners represented a major breakthrough in the field. On the one hand the availability of high-resolution images in digital format not only enables data sharing, remote consultation and cooperation among experts, but also makes it possible to process the data through image processing and machine learning techniques. On the other hand, the ever increasing amount of data inevitably gives rise to a bottleneck in manual, microscopy-based evaluation. Besides, manual procedures can be affected by marked intra-and/or inter-observer variability, as for instance demonstrated in [23]. Therefore, it is not surprising that computer-assisted analysis of histopathological images has been attracting increasing research interest in recent years. Notably, the tremendous advancements in the image acquisition methods have been accompanied by the parallel improvements in image processing techniques. In particular, the recent advent of Convolutional Neural Networks has been recognized as a major breakthrough in the area. The effectiveness of CNN has been demonstrated in a number of tasks such as object, scene and face recognition [14,19], and their use has been recently proposed also for the classification of histological images [17,24]. Particularly important is the remarkable generalization capacity that allows CNN to operate in contexts different from those used in the training phase. This last aspect makes CNN usable as generic feature extractors for a wide range of applications [10,20]. A potential drawback is that the number of features generated by a CNN can be rather high. This last aspect is typically the responsible of undesired effects such as increased computational load, overfitting and redundancy in the information to be processed and stored.

To tackle the above mentioned issues in this work we investigate three feature reduction strategies. In particular, we evaluate two state-of-the-art techniques - i.e.: Gaussian Random Projection (GRP) and Principal Component Analysis (PCA) - and propose a novel approach based on the cross-correlation between the feature vector components. The three approaches have been validated on six datasets of histopathological images, showing that CNN-based features can be reduced by a large amount with only a slight decrease in the classification accuracy. In the remainder of the paper, after giving an overview of the datasets used (Sect. 2), we describe the feature reduction schemes in Sect. 3. Then, we provide details on the experimental set-up and summarise the results in Sect. 4. Finally, the conclusions and general considerations are drawn in Sect. 5.

2   Materials

We considered six different databases of histopathological images that will be described below. The datasets are presented in chronological order and samples from each dataset are shown in Figs. 1, 2 and 3.

- **Epistroma.** This two-class dataset (Fig. 1) contains images of variable size representing well-defined regions of tumour *epithelium* (825 samples) or *stroma* (551 samples). The samples come from a series of 643 patients with colorectal cancer who underwent surgery at Helsinki University Central Hospital (Helsinki, Finland) from 1989 to 1998 [7,15]. Further clinico-pathological characteristics of the patients series are available in [16].

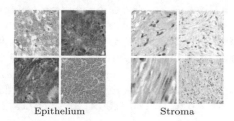

Epithelium   Stroma

Fig. 1. Samples from the Epistroma dataset.

- **Multi-class Kather's.** This is a multi-class dataset containing images of fixed size (150px × 150px – approx. 74 μm × 74 μm) representing the following seven classes of sub-tissue structures (Fig. 2): *epithelium, simple stroma, complex stroma, lymphocites, debris, mucosa* and *adipose tissue* plus one class for *background* [12,13]. The samples come from 10 patients affected by low- and high-grade colorectal cancer primary tumours, and were obtained from the pathology archive at the University Medical Center Mannheim (Heidelberg University, Mannheim, Germany). There is a total of 625 images for each class, therefore 625 × 8 = 5000 images in the whole dataset.

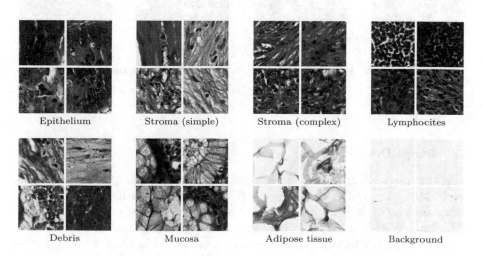

Epithelium   Stroma (simple)   Stroma (complex)   Lymphocites

Debris   Mucosa   Adipose tissue   Background

Fig. 2. Samples from multi-class Kather's dataset.

24   S. Cascianelli et al.

- **Breast cancer histopathology database (BreaKHis).** The BreakHis
database is composed of 7909 microscopic biopsy images of benign and malignant breast tumors collected from 82 patients from January to December 2014
at the P&D Laboratory – Pathological Anatomy and Cytopathology (Parana,
Brazil) [3,22]. There are 2480 samples of four different types of benign breast
tumors, i.e.: *adenosis, fibroadenoma, phyllodes tumor* and *tubular adenoma*;
and 5429 samples of malignant breast tumors, i.e.: *ductal carcinoma, lobu-
lar carcinoma, mucinous carcinoma* and *papillary carcinoma*. All the images
have a dimension of 700px × 460px and were acquired using a series of four
different magnifying factors, i.e.: 40×, 100×, 200× and 400×.

  Herein we considered each magnification factor as making up a different database, and therefore derived four datasets from the original one. The
distribution of benign/malignant samples in the four datasets is: 625/1370
(40×), 644/1437 (100×), 623/1390 (200×) and 588/1232 (400×).

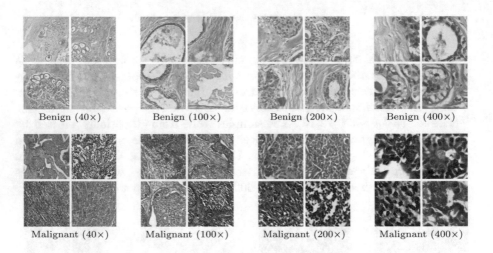

Benign (40×)   Benign (100×)   Benign (200×)   Benign (400×)

Malignant (40×)   Malignant (100×)   Malignant (200×)   Malignant (400×)

Fig. 3. Breast cancer histopathology database (BreaKHis)

3   Methods

3.1   Image Descriptors

Different image descriptors were computed from three pre-trained CNN nets,
specifically: a fast (VGG-F) and a slow (VGG-S) network as described in [4], and
a very deep network (VGG-veryDeep) as proposed by Simonyan and Zissermann
[21]. Both VGG-F and VGG-S consist of five convolutional and three fully-
connected layers, although the former uses a larger stride and a smaller pooling
window to speedup the processing. The VGG-VeryDeep-16 has a more complex structure featuring 13 convolutional and three fully-connected layers [21].

In this study the image descriptors (features) for the three CNNs were the L_2-normalised outputs of the last fully-connected layer. This approach, which is usually referred to as the 'FC' configuration [5], generates 4096 features. Pre-processing was requested to adapt the format and size of the input images to the CNNs input layer. Specifically, the images were converted to grey-scale and resized to a fixed size of 224px × 224px.

For comparison purposes we also considered three hand-designed image descriptors, namely: Local Binary Patterns (LBP) [18], Improved Local Binary Patterns (ILBP) [11] and Texture Spectrum (TS) [9]. For each of them the feature vector was the concatenation of the rotation-invariant features computed over digital circles of radius 1px, 2px and 3px respectively. Thus we obtained feature vectors of length 108 for LBP, 213 for ILBP and 2502 for TS (see also Ref. [8] for details).

It is worth noticing that the CNN-based (FC version) and the hand-designed features differ significantly in that the former are orderless, whereas the latter are order-sensitive – i.e.: they also capture the spatial distribution of the local features over the whole image (for a discussion on this point see also Refs. [5,6]).

3.2 Strategies for Dimensionality Reduction

We investigated three dimensionality reduction schemes, namely: Principal Component Analysis (PCA), Gaussian Random Projection (GRP) and Correlation-based Feature Selection (CBFS). As detailed below, there are two important differences between the three methods. The first is that PCA and CBFS require a training step (therefore some training data, which in our experiments were represented by the images of the train set – see Sect. 4), whereas GRP does not. The second is that both PCA and GRP generate new features through linear combination of the original ones, whereas CBFS only selects a subset of the original features (no new features are generated in this case).

Principal Component Analysis. The first technique is the standard Principal Component Analysis (PCA). Dimensionality reduction via PCA consists of projecting the data onto a lower-dimensional space such that the variance of the projected data is maximized [2, Sect. 12.1]. The projecting sub-spaces are determined via singular value decomposition over the train data $\mathbf{X} \in \mathbb{R}^{M \times N}$, where M is the number of samples and N is the number of components of the feature vector ($N = 4096$ in this case).

Gaussian Random Projection. Gaussian Random Projection (GPR) also consists of projecting the original data onto a lower-dimensional space; in this case, however, the projection relies on a random matrix whose columns have unit length [1]. In the reduced dimensionality space the data maintain their original relative distance.

Correlation-Based Feature Selection. This method relies on the assumption that similar (i.e.: highly correlated) components of the feature vector can be represented by just one of them. We therefore consider each column of the feature matrix of the training set $\mathbf{X}$ as an individual signal and compute the pairwise cross-correlation among each pair of signals as a first step; then discard all but one of those components whose cross-correlation was higher then a certain threshold τ. The higher the threshold the fewer the components discarded and vice-versa.

4 Experiments and Results

To investigate the effects of the dimensionality reduction schemes discussed in Sect. 3 we performed a set of supervised image classification experiments using the nearest-neighbour classifier with L_1 distance. Accuracy estimation was based on split sample validation with stratified sampling and a training ratio of 1/4; this means that 25% of the samples of each class were used to train the classifier and the remaining 75% to test the accuracy. The estimation was averaged over 100 different splits into train and test set.

The overall results are summarised in Tables 1, 2, 3 and 4. In all the tables we use the following abbreviations to refer to the datasets: '1' = Epistroma; '2a' = BreaKHis (40×); '2b' = BreaKHis (100×); '2c' = BreaKHis (200×); '2d' = BreaKHis (400×) and '3' = Multi-class Kather's.

Table 1 reports the accuracy obtained with the full-length feature vectors, and this represents the baseline of our experiments. The other tables report the performance obtained with feature reduction by PCA, GRP and CBFS in terms of difference with respect to the baseline. In each table the best result by dataset is highlight in boldface, whereas the values between parentheses indicate the number of features retained.

Table 1. Overall accuracy by datasets (original descriptors).

	Dataset ID					
	1	2a	2b	2c	2d	3
Hand-crafted features						
LBP	91.1 (108)	75.4 (108)	71.9 (108)	70.6 (108)	67.5 (108)	72.8 (108)
ILBP	91.6 (213)	77.5 (213)	77.5 (213)	74.0 (213)	70.6 (213)	79.6 (213)
TS	90.9 (2502)	77.6 (2502)	75.2 (2502)	73.5 (2502)	70.2 (2502)	77.1 (2502)
CNN-based features						
VGG-F	94.4 (4096)	84.5 (4096)	82.5 (4096)	84.0 (4096)	**82.8** (4096)	83.8 (4096)
VGG-S	93.8 (4096)	84.8 (4096)	82.6 (4096)	83.6 (4096)	81.0 (4096)	84.0 (4096)
VGG-VD-16	**94.7** (4096)	**87.0** (4096)	**85.2** (4096)	**85.0** (4096)	81.3 (4096)	**84.0** (4096)

Table 2. Accuracy gain/loss with respect to the baseline (Table 1) in case of dimensionality reduction via PCA.

Descriptor	Dataset ID					
	1	2a	2b	2c	2d	3
VGG-F	+0.3 (10)	−4.9 (10)	−5.3 (10)	−3.4 (10)	−3.9 (10)	−5.6 (10)
	+0.3 (25)	−0.6 (25)	−0.6 (25)	−0.4 (25)	−0.3 (25)	+0.2 (25)
	−1.0 (50)	**+0.4 (50)**	+0.6 (50)	+0.7 (50)	**+0.7** (50)	**+1.0** (50)
	−1.0 (100)	0.0 (100)	0.3 (100)	+0.2 (100)	+0.2 (100)	+0.7 (100)
	−2.5 (200)	−2.0 (200)	−0.7 (200)	−1.1 (200)	−1.3 (200)	−0.6 (200)
	−3.1 (300)	−4.1 (300)	−2.0 (300)	−2.6 (300)	−3.1 (300)	−3.0 (300)
VGG-S	−0.4 (10)	−5.7 (10)	−5.2 (10)	−4.1 (10)	−3.3 (10)	−3.8 (10)
	−2.4 (25)	−1.9 (25)	−0.7 (25)	−0.8 (25)	−0.3 (25)	+0.1 (25)
	−1.2 (50)	−0.2 (50)	+0.1 (50)	−0.1 (50)	+0.4 (50)	+0.6 (50)
	−0.5 (100)	−0.6 (100)	−0.1 (100)	−0.4 (100)	−0.2 (100)	+0.1 (100)
	−0.1 (200)	−2.6 (200)	−1.6 (200)	−1.7 (200)	−2.0 (200)	−1.8 (200)
	−0.4 (300)	−4.7 (300)	−3.2 (300)	−3.4 (300)	−3.9 (300)	−5.3 (300)
VGG-VD-16	−0.2 (10)	−5.4 (10)	−4.9 (10)	−4.3 (10)	−6.039 (10)	−5.4 (10)
	+0.3 (25)	−1.9 (25)	−1.2 (25)	−1.4 (25)	−1.9 (25)	−0.8 (25)
	0.0 (50)	−0.7 (50)	−0.3 (50)	−0.3 (50)	−0.3 (50)	+0.1 (50)
	−0.4 (100)	−1.2 (100)	−0.6 (100)	−0.3 (100)	−0.6 (100)	−0.1 (100)
	−1.0 (200)	−3.1 (200)	−1.8 (200)	−1.3 (200)	−1.7 (200)	−1.0 (200)
	−1.5 (300)	−4.9 (300)	−2.7 (300)	−2.4 (300)	−2.6 (300)	−2.9 (300)

In the case of GRP (Table 3) a different random projection matrix was computed for each subdivision into train and test set. The standard deviation of the accuracy (not reported in the table) varied from 0.3% with 2500 projection planes to 3.0% with 10 projection planes. In Correlation-based Feature Reduction (Table 4) we used four different values for τ, i.e. 0.1, 0.3, 0.4 and 0.7, which respectively produced the number of features shown in Table 4.

In general the results show that significant compression ratios can be achieved at the cost of only a slight decrease in the discrimination accuracy, and that in some cases there can even be an improvement. Among the three methods considered, the one proposed here (CBFS) provided on average the best accuracy with a compression ratio between 1/3 and 1/2. Higher compression ratios could be achieved through PCA and GRP, but at the cost of a more marked decrease in the accuracy.

The trend observed with GRP and CBFS is that accuracy correlates positively with the number of features retained, as one would expect. By contrast, reduction by PCA showed to reach a local maximum approximately centred between 50 and 100 features.

Table 3. Accuracy gain/loss with respect to the baseline (Table 1): dimensionality reduction via GRP.

Descriptor	Dataset ID					
	1	2a	2b	2c	2d	3
VGG-F	−11.6 (10)	−15.7 (10)	−15.1 (10)	−14.5 (10)	−16.0 (10)	−27.9 (10)
	−4.0 (25)	−8.7 (25)	−8.6 (25)	−7.8 (25)	−9.3 (25)	−13.4 (25)
	−1.7 (50)	−4.7 (50)	−4.6 (50)	−4.0 (50)	−4.9 (50)	−6.5 (50)
	−0.7 (100)	−2.2 (100)	−2.1 (100)	−1.8 (100)	−2.1 (100)	−2.7 (100)
	−0.3 (200)	−0.8 (200)	−0.7 (200)	−0.5 (200)	−0.5 (200)	−0.8 (200)
	+0.1 (2500)	+0.6 (2500)	+0.7 (2500)	+0.8 (2500)	**+1.1** (2500)	**+1.0** (2500)
VGG-S	−14.1 (10)	−17.27 (10)	−16.24 (10)	−16.02 (10)	−16.3 (10)	−29.2 (10)
	−5.3 (25)	−10.3 (25)	−9.8 (25)	−9.2 (25)	−9.9 (25)	−14.7 (25)
	−2.3 (50)	−5.9 (50)	−5.6 (50)	−5.0 (50)	−5.5 (50)	−7.5 (50)
	−1.0 (100)	−3.1 (100)	−2.7 (100)	−2.4 (100)	−2.4 (100)	−3.4 (100)
	−0.4 (200)	−1.4 (200)	−1.0 (200)	−0.9 (200)	−0.7 (200)	−1.2 (200)
	+ 0.2 (2500)	+0.2 (2500)	+0.6 (2500)	+0.7 (2500)	+1.0 (2500)	**+0.8** (2500)
VGG-VD-16	−12.0 (10)	−17.5 (10)	−15.9 (10)	−15.6 (10)	−15.5 (10)	−28.7 (10)
	−4.0 (25)	−9.9 (25)	−8.7 (25)	−8.8 (25)	−9.4 (25)	−14.0 (25)
	−1.5 (50)	−5.7 (50)	--4.7 (50)	−4.9 (50)	−5.4 (50)	−7.0 (50)
	−0.4 (100)	−3.1 (100)	−2.3 (100)	−2.4 (100)	−2.9 (100)	−3.4 (100)
	+0.1 (200)	−1.7 (200)	−1.0 (200)	−1.1 (200)	−1.4 (200)	−1.5 (200)
	+0.5 (2500)	−0.3 (2500)	**+0.3** (2500)	**+0.2** (2500)	−0.1 (2500)	+0.1 (2500)

Table 4. Accuracy gain/loss with respect to the baseline (Table 1): dimensionality reduction via CBFS.

Descriptor	Dataset ID						
	τ	1	2a	2b	2c	2d	3
VGG-F	0.1	+0.1 (1442)	−0.3 (1385)	−0.6 (1416)	−0.4 (1436)	−1.2 (1447)	−2.4 (1370)
	0.3	−0.4 (1529)	−0.5 (1570)	−0.2 (1614)	−1.0 (1596)	−0.2 (1693)	−1.3 (1467)
	0.5	0.0 (1950)	0.0 (2720)	−0.3 (2882)	−0.2 (2776)	−0.3 (3208)	−1.0 (1558)
	0.7	−0.1 (3363)	+0.1 (3698)	+0.1 (3850)	−0.1 (3812)	0.0 (3971)	−0.4 (2961)
VGG-S	0.1	+0.1 (1344)	−0.1 (1408)	−1.0 (1324)	−0.9 (1346)	−1.2 (1365)	−2.2 (1158)
	0.3	+0.2 (1483)	−0.1 (1503)	−0.4 (1520)	−0.5 (1547)	−0.5 (1586)	−0.8 (1466)
	0.5	−0.3 (1900)	−0.2 (2706)	−0.2 (2697)	−0.3 (2953)	+0.1 (3156)	−0.2 (1557)
	0.7	+0.1 (3698)	+0.1 (3915)	−0.2 (3921)	+0.1 (3990)	−0.1 (4013)	−0.1 (3339)
VGG-VD-16	0.1	−0.1 (1360)	−1.8 (1407)	−1.6 (1379)	−1.2 (1393)	−1.5 (1423)	−0.5 (1248)
	0.3	**+0.1** (1472)	−0.3 (1559)	−0.5 (1562)	−0.2 (1562)	−0.3 (1556)	−1.0 (1462)
	0.5	0.0 (1935)	−0.1 (2852)	−0.1 (2870)	**+0.1** (2750)	0.0 (2869)	−0.4 (1608)
	0.7	−0.2 (3581)	**0.0** (3990)	−0.2 (3928)	0.0 (3873)	**+0.1** (3981)	−0.4 (2296)

5   Conclusions

The use of pre-trained Convolutional Neural Networks as generic features extractors for image classification tasks has generated considerable research interest lately. This approach seems particularly promising in digital pathology, outperforming in some cases classical, hand-crafted descriptors. Feature vectors from

CNNs, however, tend to be bigger than those generated by traditional descriptors. Potential side-effects of this are the higher computational demand and 'curse of dimensionality'.

In this work we have investigated three dimensionality reduction strategies applied to CNN-based features for the classification of histological images. The results show that significant compression ratios can be achieved at the cost of very slight decrease in the overall accuracy. Among the three methods considered in the study GRP and CBFS proved to be more stable than PCA, therefore more appropriate for the task.

Acknowledgements. This work was partially supported by the Department of Engineering at the Università degli Studi di Perugia, Italy, under project *BioMeTron* – Fundamental research grant D.D. 20/2015.

References

1. Bingham, E., Mannila, H.: Random projection in dimensionality reduction: applications to image and text data. In: Proceedings of the Seventh ACM SIGKDD International Conference on Knowledge Discovery and Data Mining, San Francisco, United States, pp. 245–250, August 2001
2. Bishop, C.M.: Pattern Recognition and Machine Learning. Springer, Secaucus (2006)
3. Breast Cancer Histopathological Database (BreakHis) (2016). http://web.inf.ufpr.br/vri/breast-cancer-database. Last Accessed 16 Dec 2016
4. Chatfield, K., Simonyan, K., Vedaldi, A., Zisserman, A.: Return of the devil in the details: delving deep into convolutional nets. In: Proceedings of the British Machine Vision Conference 2014, Nottingham, United Kingdom, September 2014
5. Cimpoi, M., Maji, S., Kokkinos, I., Vedaldi, A.: Deep filter banks for texture recognition, description, and segmentation. Int. J. Comput. Vision **118**(1), 65–94 (2016)
6. Cusano, C., Napoletano, P., Schettini, R.: Combining multiple features for color texture classification. J. Electron. Imaging **25**(6), 061402 (2016)
7. egfr colon stroma classification - WebMicroscope (2012). http://fimm.webmicroscope.net/Research/Supplements/epistroma. Last Accessed 6 Sep 2016
8. González, E., Bianconi, F., Fernández, A.: An investigation on the use of local multi-resolution patterns for image classification. Inf. Sci. **361–362**, 1–13 (2016)
9. He, D.-C., Wang, L.: Texture unit, texture spectrum, and texture analysis. IEEE Trans. Geosci. Remote Sens. **28**(4), 509–512 (1990)
10. Hertel, L., Barth, E., Kaster, T., Martinetz, T.: Deep convolutional neural networks as generic feature extractors. In: Proceedings of the International Joint Conference on Neural Networks, Killarney, Ireland, July 2015
11. Jin, H., Liu, Q., Lu, H., Tong, X.: Face detection using improved LBP under Bayesian framework. In: Proceedings of the 3rd International Conference on Image and Graphics, Hong Kong, China, pp. 306–309, December 2004
12. Kather, J.N., Weis, C., Bianconi, F., Melchers, S.S., Schad, L.R., Gaiser, T., Marx, A., Zöllner, F.G.: Multi-class texture analysis in colorectal cancer histology. Sci. Rep. **6**, 27988 (2016)
13. Kather, J.N., Zöllner, F.G., Bianconi, F., Melchers, S.M., Schad, L.R., Gaiser, T., Marx, A., Weis, C.-A.: Collection of textures in colorectal cancer histology, May 2016. doi:10.5281/zenodo.53169. Last Accessed 19 Sep 2016

14. Krizhevsky, A., Sutskever, I., Hinton, G.E.: ImageNet classification with deep convolutional neural networks. In: Proceedings of Advances in Neural Information Processing Systems, Lake Tahoe, USA, vol. 2, pp. 1097–1105, December 2012

15. Linder, N., Konsti, J., Turkki, R., Rahtu, E., Lundin, M., Nordling, S., Haglund, C., Ahonen, T., Pietikäinen, M., Lundin, J.: Identification of tumor epithelium and stroma in tissue microarrays using texture analysis. Diagn. Pathol. 7(22), 1–11 (2012)

16. Linder, N., Martelin, E., Lundin, M., Louhimo, J., Nordling, S., Haglund, C., Lundin, J.: Xanthine oxidoreductase - clinical significance in colorectal cancer and in vitro expression of the protein in human colon cancer cells. Eur. J. Cancer 45(4), 648–655 (2009)

17. Litjens, G., Sánchez, C.I., Timofeeva, N., Hermsen, M., Nagtegaal, I., Kovacs, I., Hulsbergen-Van De Kaa, C., Bult, P., Van Ginneken, B., Van Der Laak, J.: Deep learning as a tool for increased accuracy and efficiency of histopathological diagnosis. Sci. Rep. 6 (2016)

18. Ojala, T., Pietikäinen, M., Mäenpää, T.: Multiresolution gray-scale and rotation invariant texture classification with local binary patterns. IEEE Trans. Pattern Anal. Mach. Intell. 24(7), 971–987 (2002)

19. Parkhi, O.M., Vedaldi, A., Zissermann, A.: Deep face recognition. In: Proceedings of the British Machine Vision Conference 2015, Swansea, United Kingdom, September 2015

20. Razavian, A.S., Azizpour, H., Sullivan, J., Carlsson, S.: CNN features off-the-shelf: an astounding baseline for recognition. In: Proceedings of IEEE Computer Society Conference on Computer Vision and Pattern Recognition Workshops, Columbus, USA, pp. 512–519, June 2014

21. Simonyan, K., Zisserman, A.: Very deep convolutional networks for large-scale image recognition. CoRR, abs/1409.1556 (2014)

22. Spanhol, F.A., Oliveira, L.S., Petitjean, C., Heutte, L.: A dataset for breast cancer histopathological image classification. IEEE Trans. Biomed. Eng. 63(7), 1455–1462 (2016)

23. Courrech Staal, E.F.W., Smit, V.T.H.B.M., van Velthuysen, M.-L.F., Spitzer-Naaykens, J.M.J., Wouters, M.W.J.M., Mesker, W.E., Tollenaar, R.A.E.M., van Sandick, J.W.: Reproducibility and validation of tumour stroma ratio scoring on oesophageal adenocarcinoma biopsies. Eur. J. Cancer 47(3), 375–382 (2011)

24. Xu, J., Luo, X., Wang, G., Gilmore, H., Madabhushi, A.: A deep convolutional neural network for segmenting and classifying epithelial and stromal regions in histopathological images. Neurocomputing 191, 214–223 (2016)

Optimizing Multiresolution Segmentation for Extracting Plastic Greenhouses from WorldView-3 Imagery

Manuel A. Aguilar[1(✉)], Antonio Novelli[2], Abderrahim Nemamoui[1],
Fernando J. Aguilar[1], Andrés García Lorca[3], and Óscar González-Yebra[1]

[1] Department of Engineering, University of Almería, 04120 Almería, Spain
`{maguilar,an932,faguilar,oglezyebra}@ual.es`
[2] DICATECh, Politecnico di Bari, Via Orabona 4, 70125 Bari, Italy
`antonio.novelli@poliba.it`
[3] Department of Geography, University of Almería, 04120 Almería, Spain
`aglorca@ual.es`

Abstract. Multiresolution segmentation (MRS) has been pointed out as one of the most successful image segmentation algorithms within the object-based image analysis (OBIA) framework. The performance of this algorithm depends on the selection of three tuning parameters (scale, shape and compactness) and the bands combination and weighting considered. In this work, we tested MRS on a World-View-3 bundle imagery in order to extract plastic greenhouse polygons. A recently published command line tool created to assess the quality of segmented digital images (AssesSeg), which implements a modified version of the supervised discrepancy measure named Euclidean Distance 2 (ED2), was used to select both the best aforementioned MRS parameters and the optimum image data source derived from WorldView-3 (i.e., panchromatic, multispectral and atmospherically corrected multispectral orthoimages). The best segmentation results were always attained from the atmospherically corrected multispectral World-View-3 orthoimage.

Keywords: Segmentation · Multiresolution algorithm · Object based image analysis · WorldView-3 · AssesSeg

1   Introduction

The latest breed of very high resolution (VHR) commercial satellites successfully launched over the last decade (e.g., GeoEye-1, WorldView-2, WorldView-3 and World-View-4) has marked a turning point in the field of remote sensing. These satellites provide improved capability to acquire impressive high spatial resolution images with ground sample distances (GSD) of 0.5 m or even less, being able to capture four, eight or even more multispectral (MS) bands. VHR satellite images are being increasingly used in remote sensing. Moreover, most of the Land Use/Land Cover (LULC) remote sensing classification research works from this type of satellite images were conducted using object-based image analysis (OBIA) techniques [1–7].

© Springer International Publishing AG 2018
G. De Pietro et al. (eds.), *Intelligent Interactive Multimedia Systems and Services 2017*,
Smart Innovation, Systems and Technologies 76, DOI 10.1007/978-3-319-59480-4_4

OBIA techniques are based on aggregating similar pixels to obtain homogenous segments (often referred to as objects). Then the image classification is performed on objects (rather than pixels) by using meaningful features related to spectral (e.g. mean spectral values), shape, texture and context information associated with each object, so having a great potential to efficiently handle more difficult image analysis tasks [8–10], especially when working on VHR satellite images. The quality of the segmentation significantly influences the final results of OBIA approaches [10, 11] since it is this first stage that generates image objects and determines their corresponding attributes.

Image segmentation is influenced by many factors such as image quality, number of spectral bands, spatial resolution and scene complexity [12, 13]. There exist several types of image segmentation algorithms which largely depend on the specified parameters, so implying that segmentation is not an easy task. Among existing algorithms, multiresolution segmentation (MRS) available in the eCognition software (Trimble, Sunnyvale, California, United States) has been the most widely and successfully employed under the context of remote sensing OBIA applications [9, 14]. Scale, shape, compactness and bands combination and weighting are the main tuning parameters that affect the algorithm performance. More details about MRS can be found in the work published by Baatz and Schäpe [15].

The selection of the optimum MRS parameters is often a tedious trial-and-error process. Fortunately, a few tools have been recently addressed to help the user with this task. For instance, Estimation of Scale Parameters tool for a single band (ESP tool) [16] and for multiband images (ESP2 tool) [12] are being widely applied as an unsupervised method to estimate the optimum scale parameter of MRS algorithm. More recently, Novelli et al. [17] have published a new free of charge command line tool named AssesSeg, thought to assess the quality of segmented digital images. It implements a modified version of the Euclidean Distance 2 (ED2) supervised discrepancy measure proposed by Liu et al. [18]. AssesSeg tool has been already successfully tested to estimate the best segmentation from Sentinel-2, Landsat 8 and WorldView-2 imagery [17, 19]. Moreover, Aguilar et al. [20] tested both ESP2 tool and ED2 method for extracting plastic greenhouses by means of MRS segmentation from an atmospherically corrected MS WorldView-2 orthoimage. ED2 metric was also used by Aguilar et al. [21] to select the optimum MRS parameters from a couple of WorldView-2 MS images geometrically and atmospherically corrected.

In this way, many researches dealing with segmentation within OBIA framework have been conducted on different VHR image sources such as (i) panchromatic (PAN) images [22], (ii) VHR pansharpened images [12], (iii) VHR atmospherically corrected MS images [17, 19–21] and, finally, (iv) VHR MS images preserving the original digital numbers [23].

At this point, it is worth highlighting that this work takes part in a research project aimed at extracting plastic greenhouses from satellite imagery. The segmentation stage is faced by estimating the optimum tuning parameters (i.e., scale, shape, compactness and bands combination) of MRS algorithm in order to delineate plastic greenhouses from a WorldView-3 bundle image (PAN and MS images) under an OBIA framework. To the best knowledge of the authors, this is the first research work headed up to test several VHR image sources (WorlView-3 PAN, MS and atmospherically corrected MS orthoimages) to search for the best segmentation results in the case study of plastic greenhouses.

2   Study Site and Datasets

2.1   Study Site

The test area is located in the "Sea of Plastic", province of Almería (Southern Spain) characterized by the greatest concentration of greenhouses in the world. The study area comprised a rectangle area of about 8000 ha centered on the WGS84 geographic coordinates of 36.7824°N and 2.6867°W (Fig. 1).

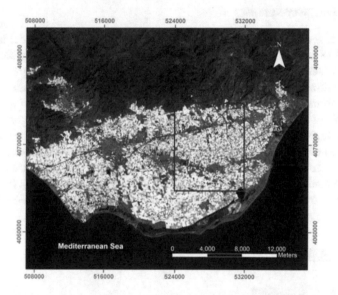

Fig. 1.   Location of the study area. Coordinate system: ETRS89 UTM Zone 30N.

2.2   WorldView-3 Data

WorldView-3 (WV3) is a VHR satellite successfully launched on August 13, 2014. This sensor provides optical images with 0.31 m and 1.24 m GSD at nadir in PAN and MSmode, respectively. The MS image is composed by 8 bands: coastal (C, 400–450 nm), blue (B, 450–510 nm), green (G, 510–580 nm), yellow (Y, 585–625 nm), red (R, 630–690 nm), red edge (RE, 705–745 nm), near infrared-1 (NIR1, 770–895 nm) and near infrared-2 (NIR2, 860–1040 nm).

A single WV3 image taken on July 5, 2016 on the study area was acquired. It was collected in Ortho Ready Standard Level-2A (ORS2A) format, containing both PAN and MS 8 bands imagery. This satellite image had a mean off-nadir view angle of 22.2°, mean collection azimuth of 273.6° and 0% of cloud cover. The final product GSD turned out to be 0.3 m and 1.2 m for PAN and MS images, respectively. All delivered products were ordered with a dynamic range of 11-bit and without the application of the dynamic range adjustment preprocessing.

Three different orthoimages were generated from this WV3 ORS2A bundle image by using seven very accurate ground control points and a medium resolution 10 m grid spacing photogrammetric-derived DEM with a vertical accuracy of 1.34 m (root mean square error; RMSE): (i) PAN othoimage with 0.3 m GSD and retaining the original digital numbers in its single band; (ii) MS orthoimage with 1.2 m GSD and retaining the original digital numbers in all the 8 bands; (iii) MS ATCOR orthoimage with 1.2 m GSD and atmospherically corrected (ground reflectance) by using the ATCOR (atmospheric correction) module included in Geomatica v. 2016.

2.3   Reference Greenhouses

This work is focused on optimizing automatic plastic greenhouses delineation from applying MRS algorithm on WV3 satellite imagery, so it has been only considered one land cover or class. Up to 400 polygons representing individual plastic greenhouses were manually digitized on the whole working area onto the WV3PAN orthoimage, but also using the information provided by the WV3 MS orthoimage so that to have a representative sample of all the greenhouses of the study area. These 400 polygons or reference geometries were grouping in four sets of 100, 200, 300 and 400 greenhouses, all of them presenting an even spatial distribution around the study area. These sets of reference geometries were applied to study the influence of the number of references on the supervised segmentation quality assessment undertaken by using AssesSeg. In this regards, it is important to know that only 30 polygons per class have been considered in previous segmentation quality studies [18, 23].

3   Methodology

3.1   Image Segmentation

The image segmentation method tested in this research was the MRS algorithm included into the OBIA software eCognition v. 8.8. The outcome of MRS algorithm is controlled by three main factors: (i) scale parameter (Scale), determining the maximum allowed heterogeneity for the resulting segments, (ii) weight of color and shape criteria in the segmentation process (Shape), and (iii) weight of the compactness and smoothness criteria (Compactness). The users also have to decide the bands combination and their corresponding weights to be applied in the segmentation process.

This segmentation approach is a bottom-up region-merging technique starting with one-pixel objects or seeds. In numerous iterative steps, two smaller objects are merged into larger one [15] if the corresponding fusion factor results to be less than the square of scale factor, given that local mutual best fitting is true. This heuristic based on mutual best fitting allows finding the best fitting pair of objects in the local vicinity of a seed object following the gradient of homogeneity.

Fusion factor (f) is computed from the weighted combination of shape and color heterogeneity (Eq. 1), while Δh_{color} expresses difference in spectral heterogeneity (Eq. 2).

$$f = w_{color}\Delta h_{color} + w_{shape}\Delta h_{shape} \tag{1}$$

$$\Delta h_{color} = \sum_c w_c \left(n_{merge}\sigma_{c,merge} - \left(n_{Obj1}\sigma_c^{Obj1} + n_{Obj2}\sigma_c^{Obj2} \right) \right) \tag{2}$$

Where n_{merge}, n_{Obj1} and n_{Obj2} being the number of pixels in the merged object, object 1 and object 2, respectively. The terms $\sigma_{c,merge}$, σ_c^{Obj1} and σ_c^{Obj2} would represent standard deviations of the merged object, object 1 and object 2, while w_c being the weight chosen for the c spectral band.

Thousands of segmentations from applying MRS algorithm were computed by means of a semi-automatic eCognition rule set characterized by a looping process varying the aforementioned MRS tuning parameters. From the results provided in Aguilar et al. [20], the tested Shape values ranged from 0.1 to 0.5 with a step of 0.1, whereas Compactness was fixed to 0.5 in all cases. Regarding Scale parameter, it ranged from 175 to 250, 30 to 95 and 1050 to 1200, always setting a step of 1, for the cases of MS orthoimage, MS ATCOR corrected orthoimage and PAN orthoimage, respectively. According to Novelli et al. [17], the bands combination for MS and MS ATCOR corrected orthoimages was fixed to Blue-Green-NIR2 bands equally weighted. In the case of PAN orthoimage, only the PAN single band was used.

3.2   Segmentation Assessment

Although there are several supervised methods and metrics to quantitatively asses segmentation quality, the ED2 measure proposed by Liu et al. [18] has provided very good results working on plastic greenhouses [20]. In a nutshell, ED2 aims to optimize, onto a two dimensional Euclidean space, both the geometrical discrepancy (potential segmentation error (PSE)) and the arithmetic discrepancy between image objects and reference polygons (number-of-segmentation ratio (NSR)).

In this work, the selection of the best three MRS parameters for each WV3 image data was carried out through a modified version of ED2 including in a command line tool named AssesSeg. Full details about the modified ED2 measure as well as the stand-alone command line tool (AssesSeg.exe) can be found in a recently published work by Novelli et al. [19]. As a supervised segmentation quality metric, the modified ED2 works with a set of reference objects (i.e., those reference greenhouses or geometries explained in Sect. 2.3) to evaluate segmentation goodness.

It is important to note that a modified ED2 value of zero would indicate an optimal combination of both geometric and arithmetic match. An optimum geometric match would be related to the absence of over-segments or under-segments. The best arithmetic match would occur when a reference polygon only matches a calculated object MRS. The ideal segmentation will be pointed out by the minimum value of modified ED2 measure.

4   Results and Discussion

The optimum segmentations for each image data (i.e., PAN, MS and MS ATCOR orthoimages) were based on the minimum value of the modified ED2 metric computed for each set of reference geometries (i.e., 100, 200, 300 and 400 polygons). The modified ED2 presented a very good agreement with the visual quality of the greenhouse segmentations for all the studied cases. For instance, Fig. 2 depicts the values of modified ED2 computed for each segmentation extracted from the WV3 MS ATCOR orthoimage against the 100 reference geometries set. The fixed parameters were band combination (Blue-Green-NIR2) and Compactness (0.5), while Shape and Scale were kept variable. In this case, the minimum value of modified ED2 was obtained for Shape and Scale values of 0.4 and 50, respectively (Fig. 2). This figure also allows to appreciate the importance of testing a wide range of parameters to find out the ideal segmentation. Notice that ESP tool or ESP2 tool [16, 17] only search for the optimum scale parameter of MRS algorithm.

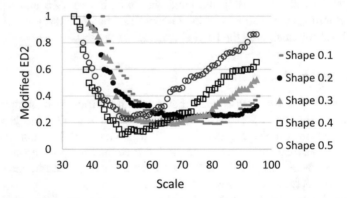

Fig. 2. Modified ED2 computed by using AssesSeg for all MRS outputs (MS ATCOR orthoimage and 100 reference geometries). Best estimated segmentation: Scale = 50 and Shape = 0.4

Table 1 depicts the best MRS parameters according to the final values of modified ED2 obtained for each image source and reference geometries set. We can make out that regarding the best plastic greenhouse segmentations, the WV3 MS ATCOR orthoimage turned out to be the best image source with modified ED2 values ranging from 0.112 to 0.146. The WV3 PAN orthoimage yielded modified ED2 values between 0.178 and 0.203, clearly worse than those provided by MS ATCOR corrected orthoimage in spite of presenting higher spatial resolution. Finally, the WV3 MS original orthoimage, with 1.2 m GSD, produced modified ED2 measures slightly worse than PAN image data (0.193–0.221). In addition, the optimum Shape parameter seems to be related to the image data source. Because of that, the recommendations by Aguilar et al. [20, 21] about Shape parameter selection for plastic greenhouses segmentation should be taken carefully.

Table 1. Optimum MRS outputs (i.e., minimum ED2 values) for the different image data tested and every set of reference geometries.

Image source	No. reference geometries	Segmentation parameters			Modified ED2
		Scale	Shape	Compactness	
MS	400	210	0.5	0.5	0.205
(bands: Blue-	300	221	0.5	0.5	0.207
Green-NIR2)	200	220	0.5	0.5	0.193
	100	195	0.5	0.5	0.221
MS ATCOR	400	60	0.4	0.5	0.141
(bands: Blue-	300	68	0.3	0.5	0.146
Green-NIR2)	200	68	0.3	0.5	0.129
	100	50	0.4	0.5	0.112
PAN	400	1152	0.4	0.5	0.183
(band: PAN)	300	1150	0.4	0.5	0.178
	200	1099	0.4	0.5	0.179
	100	1101	0.4	0.5	0.203

Regarding the number of polygons involved in computing the modified ED2 metric to be applied in plastic greenhouses, MRS parameters keeps stable as from 200. It points out to the necessity to count on a high number of reference geometries bearing in mind that only 30 references have been considered in previous segmentation quality studies [18, 23].

Figure 3 depicts a visual comparison restricted to a detailed area between the optimum segmentations attained by using the 100 reference geometries and modified ED2. Figure 3a shows the reference geometries (red polygons), each one representing a single plastic greenhouse. Figures 3b, c and d display the corresponding PAN, MS and MS ATCOR derived segmentations, respectively. We can see that the reference geometry marked with an orange ellipse represents a greenhouse which is showing strong strip shapes corresponding to ventilation roof windows. These windows resulted to be individually segmented when PAN orthoimage, with high spatial resolution (0.3 m) and pixel values given as digital numbers ranging from 225 to 2366, was used as image source (Fig. 3b). These strips were only partially segmented when using the MS ortho-image (Fig. 3c), having worse geometric resolution (1.2 m) but also presenting pixel values given as digital numbers (ranging from 201 to 2154 in the case of the Green band). Finally, the roof windows were completely ignored when employing the MS ATCOR corrected orthoimage (Fig. 3d) with 1.2 m GSD and pixel values expressing ground reflectance ranging from 0 to 100%. It is worth noting that the application of atmospheric correction in the ATCOR corrected orthoimage involved a substantial reduction in the quantitative range of values or pixel relative mapping positions available for assigning pixel content (from 1 to 100% in the case of ground reflectance). This numerical effect, together with the mathematical formulation of the fusion factor or threshold employed for grouping pixels in the MRS algorithm (see Eq. 2), would imply that the higher the range of the pixel mapping content the larger the number of objects would be segmented for a certain Scale parameter. In fact, increasing heterogeneity, measured through standard deviation of neighboring pixels/objects, can be expected when dealing with images presenting higher relative differences in pixel content values. This effect would also explain why the optimum Scale parameters result to be significantly higher in the case of MS orthoimages as compared to MS ATCOR corrected ones (Table 1). In this sense, the WV3 MS ATCOR product achieved a more realistic segmentation of the individual greenhouses, avoiding the over-segmentation due to the existence of roof windows.

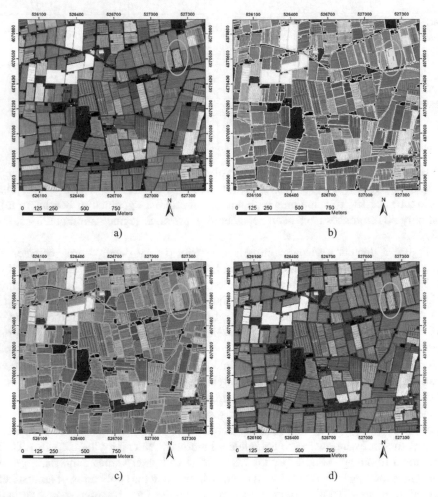

Fig. 3. Visual comparison over the RGB WV3 MS ATCOR orthoimage of the best segmentations (100 reference geometries): (a) Reference geometries (Red polygons); (b) Optimum segmentation from WV3 PAN orthoimage (Yellow polygons); (c) Optimum segmentation from WV3 MS orthoimage (Green polygons); (d) Optimum segmentation from WV3 MS ATCOR orthoimage (Blue polygons).

5   Conclusions

As far as the authors knowledge, this work is the first attempt to identify the optimum image data source derived from VHR bundle satellite imagery (e.g., GeoEye-1 and WorldView-2/3/4) to perform MRS algorithm on a plastic greenhouse area.

In this regards we found that the WV3 MS ATCOR corrected orthoimage was the best image data source to attain the best greenhouses segmentation according to the modified ED2 metric. This image product presented a geometric resolution of 1.2 m and digital values expressed as ground reflectance.

Modified ED2 metric presented a very good agreement with the visual quality of the greenhouse segmentations. Moreover, the command line tool AssesSeg allowed easily checking a high number of MRS parameters combinations.

Finally, the number of reference geometries to compute ED2 should be much higher than 30. In fact, when the class to be segmented is very heterogeneous, as in our case dealing with plastic greenhouses, sets of references between 200 and 300 should be considered in order to obtain reliable results.

Acknowledgments.   This work was supported by the Spanish Ministry of Economy and Competitiveness (Spain) and the European Union FEDER funds (Grant Reference AGL2014-56017-R). It takes part of the general research lines promoted by the Agrifood Campus of International Excellence ceiA3.

References

1. Carleer, A.P., Wolff, E.: Urban land cover multi-level region-based classification of VHR data by selecting relevant features. Int. J. Remote Sens. **27**(6), 1035–1051 (2006)
2. Stumpf, A., Kerle, N.: Object-oriented mapping of landslides using random forests. Remote Sens. Environ. **115**, 2564–2577 (2011)
3. Pu, R., Landry, S., Yu, Q.: Object-based urban detailed land cover classification with high spatial resolution IKONOS imagery. Int. J. Remote Sens. **32**(12), 3285–3308 (2011)
4. Pu, R., Landry, S.: A comparative analysis of high spatial resolution IKONOS and WorldView-2 imagery for mapping urban tree species. Remote Sens. Environ. **124**, 516–533 (2012)
5. Aguilar, M.A., Saldaña, M.M., Aguilar, F.J.: GeoEye-1 and WorldView-2 pan-sharpened imagery for object-based classification in urban environments. Int. J. Remote Sens. **34**(7), 2583–2606 (2013)
6. Fernández, I., Aguilar, F.J., Aguilar, M.A., Álvarez, M.F.: Influence of data source and training size on impervious surface areas classification using VHR satellite and aerial imagery through an object-based approach. IEEE J. Sel. Top. Appl. Earth Observ. Remote Sens. **7**(12), 4681–4691 (2014)
7. Heenkenda, M.K., Joyce, K.E., Maier, S.W.: Mangrove tree crown delineation from high-resolution imagery. Photogramm. Eng. Remote Sens. **81**(6), 471–479 (2015)
8. Marpu, P.R., Neubert, M., Herold, H., Niemeyer, I.: Enhanced evaluation of image segmentation results. J. Spat. Sci. **55**(1), 55–68 (2010)
9. Blaschke, T.: Object based image analysis for remote sensing. ISPRS-J. Photogramm. Remote Sens. **65**, 2–16 (2010)

10. Blaschke, T., Hay, G.J., Kelly, M., Lang, S., Hofmann, P., Addink, E., Feitosa, R.Q., van der Meer, F., van der Werff, H., van Coillie, F., Tiede, D.: Geographic object-based image analysis-towards a new paradigm. ISPRS-J. Photogramm. Remote Sens. **87**, 180–191 (2014)
11. Witharana, C., Civco, D.L.: Optimizing multi-resolution segmentation scale using empirical methods: exploring the sensitivity of the supervised discrepancy measure Euclidean Distance 2 (ED2). ISPRS-J. Photogramm. Remote Sens. **87**, 108–121 (2014)
12. Drăguţ, L., Csillik, O., Eisank, C., Tiede, D.: Automated parameterisation for multi-scale image segmentation on multiple layers. ISPRS-J. Photogramm. Remote Sens. **88**, 119–127 (2014)
13. Belgiu, M., Drăguţ, L.: Comparing supervised and unsupervised multiresolution segmentation approaches for extracting buildings from very high resolution imagery. ISPRS-J. Photogramm. Remote Sens. **96**, 67–75 (2014)
14. Neubert, M., Herold, H., Meinel, G.: Assessing image segmentation quality –concepts, methods and application. In: Blaschke, T., Hay, G., Lang, S. (eds.) Object-Based Image Analysis – Spatial Concepts for Knowledge-Driven Remote Sensing Applications. Lecture Notes in Geoinformation & Cartography, vol. 18, pp. 769–784. Springer, Berlin (2008)
15. Baatz, M., Schäpe, M.: Multiresolution segmentation - an optimization approach for high quality multi-scale image segmentation. In: Strobl, J., Blaschke, T., Griesebner, G. (eds.) Angewandte Geographische Informations-Verarbeitung XII, pp. 12–23. Wichmann Verlag, Karlsruhe (2000)
16. Drăguţ, L., Tiede, D., Levick, S.: ESP: a tool to estimate scale parameters for multiresolution image segmentation of remotely sensed data. Int. J. Geogr. Inf. Sci. **24**(6), 859–871 (2010)
17. Novelli, A., Aguilar, M.A., Aguilar, F.J., Nemmaoui, A., Tarantino, E.: AssesSeg—a command line tool to quantify image segmentation quality: a test carried out in Southern Spain from satellite imagery. Remote Sens. **9**, 40 (2017)
18. Liu, Y., Biana, L., Menga, Y., Wanga, H., Zhanga, S., Yanga, Y., Shaoa, X., Wang, B.: Discrepancy measures for selecting optimal combination of parameter values in object-based image analysis. ISPRS-J. Photogramm. Remote Sens. **68**, 144–156 (2012)
19. Novelli, A., Aguilar, M.A., Nemmaoui, A., Aguilar, F.J., Tarantino, E.: Performance evaluation of object based greenhouse detection from Sentinel-2 MSI and Landsat 8 OLI data: a case study from Almería (Spain). Int. J. Appl. Earth Obs. Geoinf. **52**, 403–411 (2016)
20. Aguilar, M., Aguilar, F., García Lorca, A., Guirado, E., Betlej, M., Cichon, P., Nemmaoui, A., Vallario, A., Parente, C.: Assessment of multiresolution segmentation for extracting greenhouses from WorldView-2 imagery. Int. Arch. Photogramm. Remote Sens. Spat. Inf. Sci. **XLI-B7**, 145–152 (2016)
21. Aguilar, M.A., Nemmaoui, A., Novelli, A., Aguilar, F.J., García Lorca, A.: Object-based greenhouse mapping using very high resolution satellite data and Landsat 8 time series. Remote Sens. **8**, 513 (2016)
22. Lefebvre, A., Corpetti, T., Moy, L.H.: Segmentation of very high spatial resolution panchromatic images based on wavelets and evidence theory. In: Bruzzone, L. (ed.) Image and Signal Processing for Remote Sensing XVI, 78300E. SPIE, vol. 7830 (2010)
23. Witharana, C., Civco, D.L.: Optimizing multi-resolution segmentation scale using empirical methods: Exploring the sensitivity of the supervised discrepancy measure Euclidean Distance 2 (ED2). ISPRS J. Photogramm. Remote Sens. **87**, 108–121 (2014)

A New Threshold Relative Radiometric Correction Algorithm (TRRCA) of Multiband Satellite Data

Antonio Novelli[1]([⊠]), Manuel A. Aguilar[2], and Eufemia Tarantino[1]

[1] Politecnico di Bari, via Orabona 7, 70125 Bari, Italy
{antonio.novelli,eufemia.tarantino}@poliba.it
[2] Department of Engineering, University of Almería, 04120 Almería, Spain
maguilar@ual.es

Abstract. It is well known that remote sensed scenes could be affected by many factors and, for optimum change detection, these unwanted effects must be removed. In this study a new algorithm is proposed for PIF (Pseudo Invariant Features) extraction and relative radiometric normalization. The new algorithm can be labeled as a supervised one and combines three methods for the detection of PIFs: Moment distance index (MDI), Normalized Difference Vegetation Index (NDVI) masks morphological erosion and dilate operators. In order to prove its effectiveness, the algorithm was tested by using Landsat 8 scenes of the "Mar de Plstico" landscape of the Andalusian Almería. Many tests were performed in order to provide a set of valid input parameters for the chosen environments. Lastly, the results were statistically assessed with parametric and non-parametric tests showing very good and stable results in the four different study areas.

Keywords: Relative radiometric normalization · PIF · Multispectral imagery · Landsat 8 · Change detection

1 Introduction

In the last decades, satellite image analysis has provided invaluable data for environment monitoring and change detection (CD) analysis [1–3]. However, remote sensing observations are instantaneous and affected by many factors (e.g. such as atmospheric conditions) [4]. These unwanted effects must be removed for radiometric consistency among temporal images. To detect measurable landscape changes, it is necessary to carry out a radiometric correction. Two approaches to radiometric correction are possible: absolute and relative radiometric normalization (RRN) [5]. In absolute radiometric correction, atmospheric radiative-transfer codes (e.g. 6Sv, MODTRAN) are used to obtain the reflectance at the Earth's surface from the measured spectral radiances. These techniques depend on in situ data and sensor. Consequently, for most historically remote scenes, absolute surface reflectance retrieval may not always be practical [4].

© Springer International Publishing AG 2018
G. De Pietro et al. (eds.), *Intelligent Interactive Multimedia Systems and Services 2017*,
Smart Innovation, Systems and Technologies 76, DOI 10.1007/978-3-319-59480-4_5

An alternative to absolute radiometric correction is relative correction, which is commonly used in one of two ways; adjusting bands of data within a single image and normalizing bands in images of multiple dates relative to a Reference (R) image [6]. Relative methods are applied at least on two scenes, the reference and one or more target (T) (e.g. the Dark-object subtraction (DOS) [7] and the histogram matching (HM) [8]). Many researchers opt for a linear radiometric normalization method for multi temporal analysis (e.g. [9]). The common form for linear radiometric image normalization is Eq. 1

$$Y_k^N = g_k \times X_k + o_k \tag{1}$$

Here X_k is the reflectance from the kth band of the target scene X, Y_k^N is the normalized reflectance of the kth band of the reference scene Y, g_k and o_k are respectively the evaluated gain and offset implemented for the kth band of the target scene. Several methods have been proposed for the radiometric normalization with linear regression of multitemporal images (e.g. [9,10]). The first attempts were based on simple regression considering all pixels of multitemporal images [6,11]. Subsequently, normalization was performed considering landscape elements with reflectance values that are nearly constant over time. These areas belong to the so called pseudo-invariant features (PIF) [12,13].

This paper shows a new PIF selecting algorithm combining Moment Distance Index (MDI) [14,15] thresholding, Normalized Difference Vegetation Index (NDVI) masks and morphological erosion and dilation operators. The proposed method was called Threshold Relative Radiometric Correction Algorithm (TRRCA) and was designed in a Python 2.7 environment. The algorithm was tested in the Mar de Plástico landscape of Almería (Spain) in which were defined a set of optimal input parameters to statistically assess the TRRCA results. To the author's knowledge this is the first method that takes into account both the MDI and morphological operators (typically implemented in image filtering problems, e.g. [16,17]).

2   Study Area and Data

The test area falls in the so-called "Sea of Plastic" (Mar de Plástico in Spanish), in the province of Almería (Southern Spain Fig. 1). The main economic activity is agriculture under plastic covered greenhouses [15] that implements different typologies of plastic materials to cover greenhouse structures. The climate is semi-arid and plastic covered greenhouses are coupled with the use of groundwater [18] as often happens in other semi-arid Mediterranean regions

Table 1. Reference (R) and Target (T) scenes used in this study.

Acquisition date	Scene ID	Subset
13 July 2014	LC82000342014194LGN00	R-S
30 June 2016	LC82000342015181LGN00	T-S

(e.g. [19,20]). The "Sea of Plastic" test area was chosen to test the TRRCA with homogenous artificial areas (high number of common artificial reflectors) with two Landsat 8 scenes (Table 1).

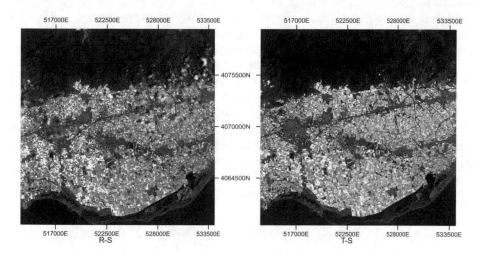

Fig. 1. Reference-Spanish (R-S) and Target-Spanish (T-S) scenes used in this study. Coordinate System UTM WGS 84 zone 30N.

Landsat 8 satellite takes images covering the entire Earth every 16 days and carries a two-sensor payload, the Operational Land Imager (OLI) and the Thermal Infrared Sensor (TIRS). The OLI and TIRS spectral bands remain broadly comparable to the Landsat 7 Enhanced Thematic Mapper plus (ETM+) bands. Further details on OLI and TIRS band specification (e.g. data format, level of processing, etc.) can be found in [21]. The OLI digital numbers of the eight Landsat 8 images were linearly converted to sensor Top of Atmosphere (TOA) reflectance and then corrected for the sun angle using gains, offsets and local sun elevation values stored in each scene metadata (as defined in the Landsat 8 (L8) Data User Handbook). The integrity of the subsets extracted from the two scenes was checked through the Landsat 8 Quality Assessment (QA) Band and the L-LDOPE Toolbelt, a no-cost tool available from the USGS Landsat-8 website [22].

3   Method

The PIF selection algorithm combines the extraction of dark (local minimum) and bright (local maximum) targets through morphological operators (M_{morph}), NDVI masks (M_{NDVI}) and MDI measures (M_{MDI}). Only the pixels positive to all the imposed conditions are selected as PIF (Eq. 2). The method was not designed to deal with the panchromatic, the cirrus and the two TIRS bands.

$$PIF = M_{morph} \cap M_{NDVI} \cap M_{MDI} \qquad (2)$$

According to [23], if the selected scenes own common local maximum and minimum values within specific bands, then the considered locations can be considered as candidate PIF. In this paper, the blue band was used to locate local minimum (often corresponding to water bodies) while bare soil or man-made objects were located through the red band. They were respectively found by using the morphological erosion and dilation. The erosion (dilation) of a digital greyscale image A by a flat structuring element S (at any location) is defined as the minimum (maximum) value of the image within the region coincident with S [8,24]. In the subsequent equations, S is defined by a square $n \times n$ matrix with all elements equal to 1. The influence of the structuring element size over the achieved PIF extraction was tested considering seven n in the closed interval $[3; 15]$ *with* $n \in 2\mathbb{N} + 1$. Bright and dark PIF were found by means of Eqs. 3 and 4, considering the morphological dilation ($\oplus$ for M_{max}) and erosion ($\ominus$ for M_{min}) common outputs for the considered bands.

$$M_{max} = [(Band_4^R \oplus S) \cap Band_4^R] \cap [(Band_4^T \oplus S) \cap Band_4^T] \tag{3}$$

$$M_{min} = [(Band_2^R \ominus S) \cap Band_2^R] \cap [(Band_2^T \ominus S) \cap Band_2^T] \tag{4}$$

$$M_{morph} = M_{max} \cup M_{min} \tag{5}$$

Pixels selected through Eqs. 3 and 4 were used to create the M_{morph} mask (Eq. 5). PIF selected through Eq. 5 could fall within vegetated areas. Moreover, they could not be invariant pixel for other bands. To increase the quality of the selected PIF, the TRRCA introduces a vegetation mask derived from the very well-known NDVI [25] computed from both R and T scenes. The selection of these areas was achieved fixing three NDVI thresholds ($NDVI_{max} > NDVI_{mid} > NDVI_{min}$) by creating a further NDVI mask with Eq. 6.

$$\begin{aligned} M_{NDVI} = \{&[(NDVI_R < NDVI_{max}) \cap (NDVI_T < NDVI_{max})] \cap \\ &[(NDVI_R > NDVI_{mid}) \cap (NDVI_T > NDVI_{mid})]\} \cup \\ &[(NDVI_R < NDVI_{min}) \cap (NDVI_T < NDVI_{min})] \end{aligned} \tag{6}$$

To test the influence of these thresholds the following interval of values were tested with a step of 0.05: $0.00 \leq NDVI_{max} < 0.26$; $-0.10 \leq NDVI_{mid} < 0.16$; $-0.60 \leq NDVI_{min} < -0.096$.

In addition to the aforementioned masks, the TRRCA takes advantage of the MDI. To the best knowledge of the authors, it is the first time that MDI is tested in RRN problems. The MDI is designed to describe the distribution of reflectance values associated with a pixel by calculating the moment distances among the bands (further details can be found in [14]): for this reason, its contribution was added in the proposed algorithm. Computationally, the MDI is calculated, for each pixel of the reference and target subset, from the difference (Eq. 7) of the moment distance (MD) between a right and a left pivot (λ_{RP} and λ_{LP} expressed in μm).

$$MDI = \sum_{i=\lambda_{LP}}^{\lambda_{RP}} \sqrt{[\rho^2 + (i - \lambda_{LP})^2]} - \sum_{i=\lambda_{RP}}^{\lambda_{LP}} \sqrt{[\rho^2 + (\lambda_{LP} - i)^2]} \qquad (7)$$

Where ρ is the reflectance for the band centred on the ith wavelength. Only the pixels in which the absolute value of the difference between the two MDI (Eq. 8) is less than a specific threshold (l) were considered as potentials PIF.

$$M_{MDI} = |MDI^R - MDI^T| < l \rightarrow MDI_{diff} < l \qquad (8)$$

Several values of l within the semi-opened interval $[0.01; 0.31[$, with a step of 0.03, were tested. Considering the implemented values, a $MDI_{diff} = 0.30$ corresponds to $9 - 12\%$ of difference between MDI_R and MDI^T.

Lastly, the radiometric normalization coefficients were evaluated for each band by means of the Orthogonal Distance Regression (ODR) algorithm implemented in the ODRPACK library (further details on the solution implemented can be found in [26]). The ODR algorithm outputs are the gains and the offsets for each band (as shown in Eq. 1), the root mean square error ($RMSE$), the correlation coefficient (r) and the coefficient of determination ($R2$) between PIF belonging to the R and the T scenes. Almost twenty thousand combinations were performed considering the selected parameter $S, l, NDVI_{max}, NDVI_{mmid}, NDVI_{min}$. Only tests characterized by an elevated number of retrieved PIF, with high R-T band by band ODR $R2$ and low ODR $RMSE$ were selected as potentials High Quality PIF (HQ-PIF) extractions. Particularly, the above PIF extractions were considered HQ-PIF only if the two-sample t test (for equal sample means), the two-sample F test (for equal sample variances) and two-sample Wilcoxon rank sum test (for equal sample medinas) between reference PIF and corrected target PIF were contemporary satisfied at 5% confidence level.

4   Results and Discussion

Table 2 summarizes the results for the tested parameters related to HQ-PIF extractions. The computations executed for the study area showed 1503 HQ-PIF extraction with a mean number of PIF equal to 984 pixels. This result shows the sensitivity of the proposed algorithm to the abundance of natural/artificial reflectors.

All the tested $NDVI_{mid}$ and $NDVI_{min}$ values are present in HQ-PIF extractions. $NDVI_{mid}$ and l are the parameters that exhibit the major variability. Probably, this occurs since the test area is less sensitive to these parameters. Indeed, the effect on NDVI is masked by the presence of plastic coverings above the vegetation. For the $NDVI_{max}$ the peak of frequency is equivalent to the maximum value tested. This shows that the M_{NDVI} performs as a coarse PIF filter for a subsequent and improved PIF selection through M_{MDI} and M_{morph}. The same occurs for the l parameter which great variability depicts the spectral difference form the R and the T scenes. Lastly, the tests demonstrated that lower Kernels sizes (S) are coupled with a greater number of extracted PIF (Fig. 2).

Table 2. Statistical parameters of the of the implemented TRRCA parameters and of the number of HQ-PIF extractions.

	S	l	$NDVI_{max}$	$NDVI_{mid}$	$NDVI_{min}$	N_{pif}
Min	3	0.01	0.10	−0.10	−0.60	491
Max	13	0.28	0.25	0.15	−0.10	4203
Mode	5	0.04	0.25	0.10	−0.60	599
Median	7	0.10	0.20	0.10	−0.35	774
Dev. stand	3	0.08	0.04	0.07	0.16	532
Mean	7	0.11	0.21	0.08	−0.35	984

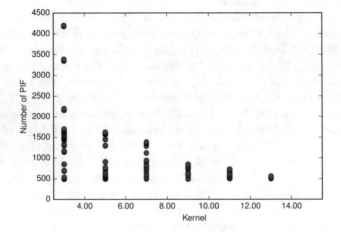

Fig. 2. Scatter plot Kernel - number of PIF.

This was also an expected result since morphological operators consider the local minimum and local maximum over the Kernel area. Because of this, the probability to find a corresponding singular value over the same Kernel areas is higher in smaller areas than in greater ones.

Since the proposed algorithm is dependent from user driven parameters the results showed in Table 2 have been used to select one single test combination. The combination was randomically extracted from the test parameters characterized by a high frequency of occurrence and removing the restriction adopted during the test phase to save computation time (i.e. the step). Table 3 shows the selected parameters and the related number of PIF.

Table 3. Implemented thresholds and kernel size.

S	l	$NDVI_{max}$	$NDVI_{mid}$	$NDVI_{min}$	N_{pif}
7	0.03	0.221	0.100	−0.503	893

Figure 3 compares the RGB visualizations of a magnified area of R and corrected T with overlapped TRRCA HQ-PIF. In the test area, the major part of the PIF falls within artificial pools, bare soil, built-up areas and highways. Although the large amount of plastic covered greenhouses, PIF do not fall within them. Indeed, greenhouses are generally covered by plastic sheets characterized by different spectral signatures over time. This is mainly due to their different spectral properties, thickness and local agricultural practices [27,28].

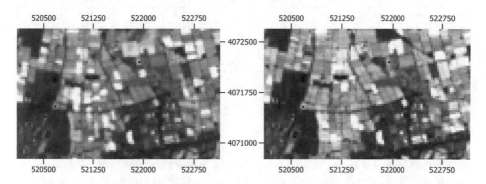

Fig. 3. Comparison of RGB visualizations of Reference (R) image (on the left) and corrected Target (T) image (on the right) with overlapped PIF: Coordinate System UTM WGS 84 zone 30N.

Table 4 compares the achieved gains, offsets, correlation coefficient (r), $RMSE$, two-sample t test, two-sample F test and two-sample Wilcoxon rank sum test results achieved for the selected parameters before (Pre) and after (Post) the correction. The table shows that the TRRCA was able to find PIF with a strong linear agreement in each test area. For each computed gains the significance of the linear relationship was tested against the null hypothesis of absence of slope. All tests strongly rejected the null hypothesis. It is thus possible to conclude that the method produces a feasible linear regression model.

Considering the evaluated r and $RMSE$, Table 4 always shows high quality and stable results (small variance). After the application of the correction all the performed tests failed to reject the null hypothesis. Moreover, p-values indicate that all the null hypothesis would be accepted by using a level of confidence far beyond the default one. Considering each typology of test, this respectively indicates a strong statistical similarity between the variance, the means and medians of the reflectance values of test PIF extracted from the reference and corresponding normalized reflectance values of PIF extracted from the target. This was an expected result since the definition of HQ-PIF imposes the respect of all statistical tests. The effects of the corrections were bigger on the median of the distributions of the selected PIF. This is demonstrated by the pre/post Wilcoxon rank sum test.

Table 4. Evaluated band-by-band gain, offset, correlation r and RMSE; F p-value, t p-value, W p-value are p-values for the two sample F test, two sample t test and the Wilcoxon rank sum test. The result h = 1 indicates a rejection of the null hypothesis, and h = 0 indicates a failure to reject the null hypothe-sis at the 5% significance level.

Band	gain	offset	r	RMSE	F p		F h		t p		t h		W p		W h	
					Pre	Post	Pre	Post	Pre	Post	Pre	Post	Pre	Post	Pre	Post
1	1.076	−0.006	0.96	0.01	0.0	0.9	1	0	0.0	1.0	1	0	0.0	0.8	1	0
2	1.063	−0.006	0.97	0.01	0.1	1.0	0	0	0.0	1.0	1	0	0.0	0.7	1	0
3	1.081	−0.013	0.96	0.01	0.0	0.9	1	0	0.7	1.0	0	0	0.0	0.3	1	0
4	1.116	−0.023	0.96	0.01	0.0	0.9	1	0	0.8	1.0	0	0	0.0	0.4	1	0
5	1.190	−0.051	0.95	0.02	0.0	0.8	1	0	0.7	1.0	0	0	0.7	0.5	0	0
6	1.272	−0.066	0.96	0.02	0.0	0.8	1	0	0.0	1.0	1	0	0.0	0.7	1	0
7	1.255	−0.045	0.96	0.02	0.0	0.8	1	0	0.0	1.0	1	0	0.0	0.5	1	0

5   Conclusions

This paper shows a new PIF selecting algorithm combining spectral momentum measures, NDVI masks and extraction of local maximum and minimum through morphological operators. The method was tested with Landsat-8 images but its design is suitable for other passive sensors with a similar spectral resolution (e.g. Sentinel-2). Due to its dependence by user driven thresholds, many combinations were tested in the selected test area characterized by the presence of an extreme anthropic impact. All the tests performed were driven to obtain the distribution of thresholds able to perform a good relative radiometric normalization. To show the capabilities of the method was randomically selected a set of parameters. In this case, the proposed algorithm recognized a great number of PIF and performed a correction on the selected PIF able to eliminate statistical differences between reference PIF and corrected target PIF. The results confirm the effectiveness of the method as new relative radiometric normalization technique and as a valid alternative to established methods from scientific literature. Future development will be focused on the improvement of the quality of selected PIF, on the analysis of a greater spectral range and in an improved reduction of user driven parameter.

Acknowledgement. This work was supported by the Spanish Ministry of Economy and Competitiveness (Spain) and the European Union FEDER funds (Grant Reference AGL2014-56017-R). It takes part of the general research lines promoted by the Agrifood Campus of International Excellence ceiA3.

References

1. Janzen, D.T., Fredeen, A.L., Wheate, R.D.: Radiometric correction techniques and accuracy assessment for landsat TM data in remote forested regions. Can. J. Remote Sens. **32**(5), 330–340 (2006)
2. Novelli, A., Caradonna, G., Tarantino, E.: Evaluation of relative radiometric correction techniques on landsat 8 OLI sensor data. In: Fourth International Conference on Remote Sensing and Geoinformation of the Environment, International Society for Optics and Photonics 968808 (2016)
3. Tarantino, E., Novelli, A., Aquilino, M., Figorito, B., Fratino, U.: Comparing the MLC and JavaNNS approaches in classifying multi-temporal LANDSAT satellite imagery over an ephemeral river area. Int. J. Agric. Environ. Inf. Syst. (IJAEIS) **6**(4), 83–102 (2015)
4. Du, Y., Teillet, P.M., Cihlar, J.: Radiometric normalization of multitemporal high-resolution satellite images with quality control for land cover change detection. Remote Sens. Environ. **82**(1), 123–134 (2002)
5. Yang, X., Lo, C., et al.: Relative radiometric normalization performance for change detection from multi-date satellite images. Photogram. Eng. Remote Sens. **66**(8), 967–980 (2000)
6. Jensen, J.: Image preprocessing: Radiometric and geometric correction. In: Introductory Digital Image Processing, 30p (1996). Chapter 6
7. Mandanici, E., Franci, F., Bitelli, G., Agapiou, A., Alexakis, D., Hadjimitsis, D.: Comparison between empirical and physically based models of atmospheric correction. In: Third International Conference on Remote Sensing and Geoinformation of the Environment, International Society for Optics and Photonics 95350E (2015)
8. Gonzalez, R.C., Woods, R.E., Eddins, S.: Morphological image processing. Digit. Image Process. **3**, 627–688 (2008)
9. Canty, M.J., Nielsen, A.A.: Automatic radiometric normalization of multitemporal satellite imagery with the iteratively re-weighted mad transformation. Remote Sens. Environ. **112**(3), 1025–1036 (2008)
10. de Carvalho, O.A., Guimarães, R.F., Silva, N.C., Gillespie, A.R., Gomes, R.A.T., Silva, C.R., de Carvalho, A.P.F.: Radiometric normalization of temporal images combining automatic detection of pseudo-invariant features from the distance and similarity spectral measures, density scatterplot analysis, and robust regression. Remote Sens. **5**(6), 2763–2794 (2013)
11. Tokola, T., Löfman, S., Erkkilä, A.: Relative calibration of multitemporal LANDSAT data for forest cover change detection. Remote Sens. Environ. **68**(1), 1–11 (1999)
12. Caselles, V., Garcia, M.L.: An alternative simple approach to estimate atmospheric correction in multitemporal studies. Int. J. Remote Sens. **10**(6), 1127–1134 (1989)
13. Schott, J.R., Salvaggio, C., Volchok, W.J.: Radiometric scene normalization using pseudoinvariant features. Remote Sens. Environ. 26(1) 1–14, IN1, 15–16 (1988)
14. Salas, E.A.L., Boykin, K.G., Valdez, R.: Multispectral and texture feature application in image-object analysis of summer vegetation in eastern tajikistan pamirs. Remote Sens. **8**(1), 78 (2016)
15. Aguilar, M.A., Nemmaoui, A., Novelli, A., Aguilar, F.J., Lorca, A.G.: Object-based greenhouse mapping using very high resolution satellite data and landsat 8 time series. Remote Sens. **8**(6), 513 (2016)
16. Tarantino, E., Figorito, B.: Steerable filtering in interactive tracing of archaeological linear features using digital true colour aerial images. Int. J. Digit. Earth **7**(11), 870–880 (2014)

17. Tarantino, E., Figorito, B.: Extracting buildings from true color stereo aerial images using a decision making strategy. Remote Sens. **3**(8), 1553–1567 (2011)
18. Van Cauwenbergh, N., Pinte, D., Tilmant, A., Frances, I., Pulido-Bosch, A., Vanclooster, M.: Multi-objective, multiple participant decision support for water management in the andarax catchment, almeria. Environ. Geol. **54**(3), 479–489 (2008)
19. Giordano, R., Milella, P., Portoghese, I., Vurro, M., Apollonio, C., D'Agostino, D., Lamaddalena, N., Scardigno, A., Piccinni, A.: An innovative monitoring system for sustainable management of groundwater resources: Objectives, stakeholder acceptability and implementation strategy. In: 2010 IEEE Workshop on Environmental Energy and Structural Monitoring Systems (EESMS), pp. 32–37. IEEE (2010)
20. Giordano, R., D'Agostino, D., Apollonio, C., Scardigno, A., Pagano, A., Portoghese, I., Lamaddalena, N., Piccinni, A.F., Vurro, M.: Evaluating acceptability of groundwater protection measures under different agricultural policies. Agric. Water Manage. **147**, 54–66 (2015)
21. Roy, D.P., Wulder, M., Loveland, T., Woodcock, C., Allen, R., Anderson, M., Helder, D., Irons, J., Johnson, D., Kennedy, R., et al.: Landsat-8: Science and product vision for terrestrial global change research. Remote Sens. Environ. **145**, 154–172 (2014)
22. Roy, D.P., Borak, J.S., Devadiga, S., Wolfe, R.E., Zheng, M., Descloitres, J.: The modis land product quality assessment approach. Remote Sens. Environ. **83**(1), 62–76 (2002)
23. Hall, F.G., Strebel, D.E., Nickeson, J.E., Goetz, S.J.: Radiometric rectification: Toward a common radiometric response among multidate, multisensor images. Remote Sens. Environ. **35**(1), 11–27 (1991)
24. Soille, P.: Morphological Image Analysis: Principles and Applications. Springer, Heidelberg (2013)
25. Tucker, C.J.: Red and photographic infrared linear combinations for monitoring vegetation. Remote Sens. Environ. **8**(2), 127–150 (1979)
26. Boggs, P.T., Byrd, R.H., Schnabel, R.B.: A stable and efficient algorithm for nonlinear orthogonal distance regression. SIAM J. Sci. Stat. Comput. **8**(6), 1052–1078 (1987)
27. Novelli, A., Aguilar, M.A., Nemmaoui, A., Aguilar, F.J., Tarantino, E.: Performance evaluation of object based greenhouse detection from sentinel-2 MSI and landsat 8 OLI data: A case study from Almería (Spain). Int. J. Appl. Earth Obs. Geoinformation **52**, 403–411 (2016)
28. Novelli, A., Tarantino, E.: Combining ad hoc spectral indices based on LANDSAT-8 OLI/TIRS sensor data for the detection of plastic cover vineyard. Remote Sens. Lett. **6**(12), 933–941 (2015)

Greenhouse Detection Using Aerial Orthophoto and Digital Surface Model

Salih Celik[1] and Dilek Koc-San[2,3(✉)]

[1] Department of Space Sciences and Technologies, Akdeniz University, 07058 Antalya, Turkey
[2] Department of City and Regional Planning, Akdeniz University, 07058 Antalya, Turkey
dkocsan@akdeniz.edu.tr
[3] Remote Sensing Research and Applications Centre, Akdeniz University, 07058 Antalya, Turkey

Abstract. Detection of greenhouse areas from remote sensing imagery is important for rural planning, yield estimation and sustainable development. This study is focused on plastic and glass greenhouse detection and discrimination using true color orthophoto and Digital Surface Model (DSM). Initially, the greenhouse areas were detected from true color aerial photograph and the additional normalised Digital Surface Model (nDSM) band using Support Vector Machines (SVM) classification technique. Then, an approach was developed for discriminating plastic and glass greenhouses by utilising the nDSM. The developed approach was implemented in a selected study area from Kumluca district of Antalya, Turkey that includes intensive plastic and glass greenhouses. The obtained results show the effectiveness of the SVM classifier for greenhouse detection with overall accuracy value of 96.15%. In addition, the plastic and glass greenhouses were discriminated efficiently using the developed approach that use nDSM.

Keywords: Plastic and glass greenhouses · Orthophotos · nDSM · SVM

1 Introduction

Greenhouse cultivation is the cultivation of fruits, vegetables and ornaments by creating necessary environmental conditions and removing the effect of external climate factors. The greenhouse cultivation has become one of the most important agricultural activities and the numbers of greenhouses increase day by day. Therefore, monitoring and mapping greenhouse areas is a challenging task. The usage of remote sensing data sources for greenhouse detection instead of traditional techniques, decreases the necessary time, manpower and costs. When we look at the used remote sensing data sources in greenhouse detection studies, it can be stated that the satellite images are the most common data sources. In the literature, there are studies to detect the greenhouses from different satellite images that have various resolutions. The most widely used satellite images are Landsat [1–5], IKONOS and/or Quickbird [6–10], and WorldView-2 and/or GeoEye-1 [11–13]. However the usage of aerial photos and DSMs or nDSMs in the greenhouse detection studies is quite rare. To the best of our knowledge, the only study that uses aerial photos for greenhouse detection is [14]. On the other hand, the usage of

© Springer International Publishing AG 2018
G. De Pietro et al. (eds.), *Intelligent Interactive Multimedia Systems and Services 2017*,
Smart Innovation, Systems and Technologies 76, DOI 10.1007/978-3-319-59480-4_6

DSM in this study is beside the point. Actually, the solely study that use nDSM for greenhouse detection is the study performed by Aguilar et al. in 2014 [12].

From the greenhouse detection studies analyzed it can be seen that the Maximum Likelihood Classification is the mostly used pixel-based classification technique. However, up to now, few studies were conducted for greenhouse detection using machine learning algorithms [10, 11, 13, 15].

In Turkey, glass is a commonly used cover material for greenhouses just after plastic. The plastic and glass greenhouse cover materials have different light transmittance, cost, life cycle and yield. For this reason, in addition to the greenhouse detection from remote sensing data, the automatic determination of the covering material of these greenhouses is also important for greenhouse database generation and yield estimation. As stated by [11, 13] most of the greenhouse detection studies focused on the plastic greenhouses. The studies that considers both plastic and glass greenhouses and their discrimination are [7, 11, 13]. Among them, Sonmez and Sari [7] differentiated glass and plastic greenhouses by visual image analysis. On the other hand, in the studies [11, 13], the plastic and glass greenhouses were detected and differentiated using machine learning algorithms.

Different to the greenhouse detection studies performed so far, in this study orthophoto and DSM generated from stereo aerial images were used. The novelty of this approach is the integrated usage of orthophoto and nDSM for greenhouse detection purpose and discriminating glass and plastic greenhouses with the developed approach by utilizing nDSM.

The main goals of this study can be summarized as (i) to detect greenhouse areas from orthophoto and additional nDSM band by utilizing a machine learning algorithm and (ii) to discriminate plastic and glass greenhouses with the aid of nDSM data.

2   Study Area and Used Data

The greenhouse cultivation has an important place in Turkey's agriculture. Approximately 6.7 million tons vegetable-fruit production is obtained from land under protective cover in Turkey in 2015 [16]. The greenhouse cultivation intensively placed in Mediterranean coast of Turkey, because of the climatic properties and ecological conditions of the region. Antalya, which is located in the Mediterranean region, is the most prominent city in greenhouse cultivation. In this study, the Kumluca district of Antalya, Turkey was selected as study area, due to the fact that this district includes intensive plastic and glass greenhouse areas (Fig. 1). Kumluca is known as the capital of greenhouse in Turkey. The determined study area covers approximately 3200 ha area.

The data sets used in this study are digital aerial photos and DSM. The digital orthophotos were collected with a 0.30 m Ground Sampling Distance (GSD) and obtained in 2012. The image includes RGB bands and referenced to UTM-WGS84 system. The DSM used in this study had been generated from 2012 dated stereo aerial images with 0.30 m GSD by automatic matching having 5 m spatial resolution and ±3 m vertical accuracy in 90% confidence interval. This DSM includes 3D man-made objects and vegetation as well as the topography.

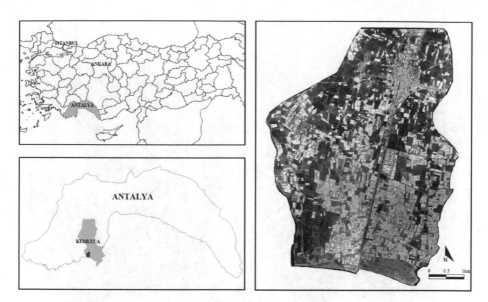

Fig. 1. The location of the study area in Kumluca district of Antalya, Turkey (a) and the natural color composite orthophoto of the selected study area (b)

3   Methodology

There are basically two stages in this study, which are (i) greenhouse detection and (ii) plastic and glass greenhouse discrimination. For greenhouse detection, the orthophoto was classified using Support Vector Machines (SVM) classifier. The SVM machine learning algorithm was preferred in this study, because the recent studies show that it is used efficiently and successfully in remote sensing researches [11, 17–20]. During the classification, the nDSM band was used as additional band. To generate nDSM initially a DTM was generated from DSM, since we do not have any DTM or data to generate a DTM. Therefore, in this case DTM was generated from the DSM image by applying DSM2DTM algorithm of PCI Geomatica® image processing software. This algorithm produces bare-Earth DTM by obtaining local area minimum/maximum values and then operating a moving polynomial function utilising the local values in the specified tile size [21]. After that, nDSM, which is a valuable data source to separate 3D objects from the ground, was generated by subtracting the DTM from the DSM [12]. The DSM and generated nDSM of the study area are given in Fig. 2. To decide the usage of nDSM data as additional band during the classification, separability analysis was performed for the data set that include only the RGB bands of orthophoto and the other data set that include additional nDSM band as well as the RGB bands of orthophoto.

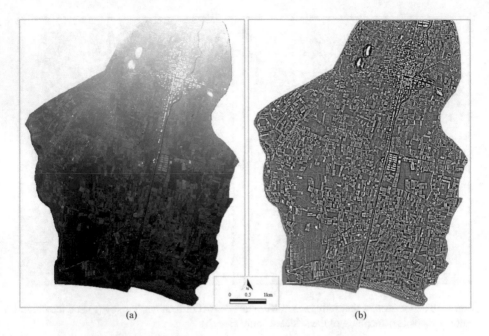

(a)   (b)

Fig. 2. The DSM (a) and nDSM (b) of the study area.

After deciding the band combinations, binary SVM algorithm was performed to classify the image as greenhouse and non-greenhouse areas. At the beginning, the classes in the selected study area were determined by visual analysis as plastic greenhouse, glass greenhouse, brick roof building, concrete roof building, bare land, vegetation, road, shadow, water to distribute the training and test pixels to all classes. The training and testing areas were roughly collected for these classes from independent areas. Then determined numbers of samples were selected from these areas using stratified random sampling and the sub-classes plastic greenhouse and glass greenhouse were merged as greenhouse class, while the other classes were merged as non-greenhouse class. Finally, for each class 500 training and 1000 testing samples were selected. During the SVM classification, Radial Basis Function was selected as kernel method. The optimum values for gamma and penalty parameters were found by utilizing 3-fold cross validation technique. The classified image may have misclassified pixels and/or too small areas. To remove them, opening and closing morphological operations were performed, respectively.

After detecting the greenhouse areas, in the second stage an approach was developed to discriminate plastic and glass greenhouses with the aid of nDSM data using Python Programming Language. The height of a greenhouse is about 2 m. However, during the field studies we realised that glass greenhouses are generally higher than plastic greenhouses and their heights are usually higher than 3.5 m (Fig. 3). Therefore, we thought that the nDSM can be used to differentiate the glass greenhouses from plastic greenhouses. In the developed approach, initially the non-greenhouse areas were masked out from the classification result and each greenhouse pixel were labeled as plastic or glass greenhouse by looking at the corresponding nDSM value. If the nDSM value is lower than 3.5 m then this pixel was assigned to plastic greenhouse. On the other hand, if the nDSM is higher than 3.5 m then this pixel was assigned to glass greenhouse. In the next step, Watershed Segmentation was applied to the red band in which the non-greenhouse areas were masked out for separating each greenhouse patch. Then, the plastic and glass greenhouse pixels were counted within each greenhouse patches for assigning it to class that have majority pixels (plastic or glass greenhouse). In the final step, morphological operations were applied to the image to remove too small areas.

Fig. 3. The height differences between plastic and glass greenhouses. The photographs are taken from the study area, Kumluca.

4   Experimental Results

To evaluate the signature separabilities of training areas and the effect of additional nDSM band the Transformed Divergence (TD) spectral separability values were computed. The range of divergence values is between 0.000 and 2.000. The computed TD values are 1.7026 and 1.8866 for the data set 1 that include RGB bands of orthophoto and data set 2 that include additional nDSM band as well as RGB bands of orthophoto. These values indicate that using nDSM as additional band increases the separabilities. Therefore, nDSM band as well as RGB bands were used during the classification. The greenhouse and non-greenhouse areas obtained by SVM classification are given in Fig. 4. By analyzing the classification result visually, it can be said that the greenhouses were classified accurately.

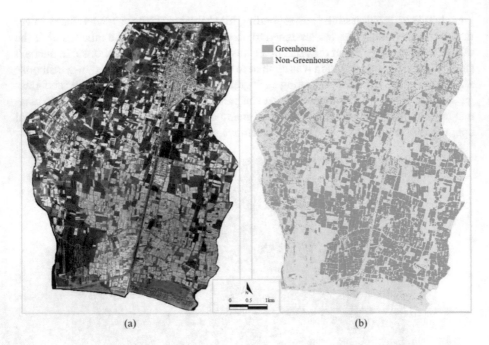

(a)   (b)

Fig. 4. The true color orthophoto (a) and the SVM classification result that use DSM as additional band.

To quantify the accuracies of the classification results error matrix, which compare the relationship between the test data and corresponding classification results, were utilized. The error matrix of the SVM classification result is given in Table 1. The misclassified areas and too small areas were removed by applying morphological operations to the SVM classification result. The error matrix of morphological operations applied SVM result is given in Table 2. Based on the error matrices, it can be stated that the accuracy of SVM classification is quite high with overall accuracy value of 95.20% and kappa coefficient value of 90.40%. The usage of morphological operations as a post classification analysis increase the overall accuracy value about 1%. After applying the morphological operations the overall accuracy and kappa coefficient values were computed as 96.15% and 92.30%, respectively.

Table 1. The classification accuracies of orthophoto images and nDSM

Class	Producer's accuracy (%)	User's accuracy (%)
Greenhouse	94.50	95.80
Non-greenhouse	95.90	94.58
Overall accuracy: 95.20%		
Kappa coefficient:90.40%		

Table 2. The classification accuracies after applying morphological operations

Class	Producer's accuracy (%)	User's accuracy (%)
Greenhouse	95.60	96.66
Non-greenhouse	96.70	95.65
Overall accuracy:96.15%		
Kappa coefficient:92.30%		

Both quantitative and visual results show the effectiveness of the proposed procedure for greenhouse detection from orthophotos and nDSM image.

The aerial photo used in this study have only RGB bands. Therefore, discriminating glass and plastic greenhouses in the classification stage is difficult. In addition, because of the transparent structures of the glass and plastic cover materials, the reflectance

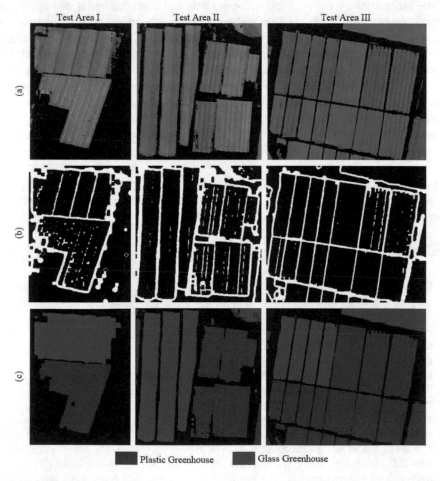

Fig. 5. The discrimination of glass and plastic greenhouses: The detected greenhouse patches (a), the watershed segmentation results (b), and the obtained results after applying morphological operations (c)

values were changed according to the crop type in the greenhouses. To overcome this problems, an approach that use nDSM was developed. The results after applying the developed approach to discriminate the plastic and glass greenhouses are given in Fig. 5. As can be seen in the obtained results, the glass and plastic greenhouses discrimination is performed efficiently using this approach. Using the segmentation algorithm, the detected greenhouses, even they are very close to each other, were separated and can be analyzed separately. After assigning the separated areas to class that have majority pixels, using morphological operations improved the obtained results and the small holes were removed.

5 Conclusions

In the present study, the greenhouses were detected from the aerial orthophoto and nDSM image using SVM classifier and an approach was proposed for discriminating plastic and glass greenhouses with the aid of nDSM image. The obtained results indicate that the proposed greenhouse detection procedure can be used effectively for plastic and glass greenhouse detection and discrimination. The integrated usage of true color orthophoto and nDSM for greenhouse detection provided quite accurate results with overall accuracy values of 96.15%. However, using orthophotos that have NIR band and using more accurate DSMs with improved geometric resolution and vertical accuracy would be more beneficial and would increase the accuracies especially for the glass and plastic greenhouse discrimination.

Acknowledgements. The authors would like to thank General Command of Mapping (GCM) for providing the data used in this study. The authors are grateful to Mustafa Kaynarca (Geomatics Engineer-ASAT) for his help during the DEM editing.

References

1. Picuno, P., Tortora, A., Capobianco, R.L.: Analysis of plasticulture landscapes in southern Italy through remote sensing and solid modelling techniques. Landscape Urban Plann. **100**, 45–56 (2011)
2. Novelli, A., Tarantino, E.: Combining Ad Hoc spectral indices based on landsat-8 OLI/TIRS sensor data for the detection of plastic cover vineyard. Remote Sens. Lett. **6**(12), 933–941 (2015)
3. Hasituya, Chen, Z., Wang, L., Wu, W., Jiang, Z., Li, H.: Monitoring plastic-mulched farmland by landsat-8 OLI imagery using spectral and textural features. Remote Sens. **8**, 1–16 (2016). 353, rs8040353
4. Novelli, A., Aguilar, M.A., Nemmaoui, A., Aguilar, F.J., Tarantino, E.: Performance evaluation of object based greenhouse detection from sentinel-2 MSI and landsat 8 OLI data: a case study from almeria (Spain). Int. J. Appl. Earth Obs. Geoinf. **52**, 403–411 (2016)
5. Wu, C.F., Deng, J.S., Wang, K., Ma, L.G., Tahmassebi, A.R.S.: Object-based classification approach for greenhouse mapping using landsat-8 imagery. Int J Agric. Biol. Eng. **9**(1), 79–88 (2016)

6. Agüera, F., Aguilar, M.A., Aguilar, F.J.: Detecting greenhouse changes from quickbird imagery on the mediterranean coast. Int. J. Remote Sens. **27**(21), 4751–4767 (2006)
7. Sonmez, N.K., Sari, M.: Use of remote sensing and geographic information system technologies for developing greenhouse databases. Turk. J. Agric. For. **30**, 413–420 (2006)
8. Agüera, F., Aguilar, F.J., Aguilar, M.A.: Using texture analysis to improve per-pixel classification of very high resolution images for mapping plastic greenhouses. ISPRS J. Photogrammetry Remote Sens. **63**, 635–646 (2008)
9. Agüera, F., Liu, J.G.: Automatic greenhouse delineation from quickbird and IKONOS satellite images. Comput. Electr. Agric. **66**, 191–200 (2009)
10. Carjaval, F., Agüera, F., Aguilar, F.J., Aguilar, M.A.: Relationship between atmospheric corrections and training-site strategy with respect to accuracy of greenhouse detection process from very high resolution imagery. Int. J. Rem. Sens. **31**(11), 2977–2994 (2010)
11. Koc-San, D.: Evaluation of different classification techniques for the detection of glass and plastic greenhouses from WorldView-2 satellite imagery. J. Appl. Remote Sens. 7 (2013). 073553-1-20
12. Aguilar, M.A., Bianconi, F., Aguilar, F.J., Fernandez, I.: Object-based greenhouse classification from geoeye-1 and worldview-2 stereo imagery. Remote Sens. **6**, 3554–3582 (2014)
13. Koc-San, D., Sonmez, N.K.: Plastic and glass greenhouses detection and delineation from worldview-2 satellite imagery. In: The International Archives of the Photogrammetry, Remote Sensing and Spatial Information Sciences, vol. XLI-B7, XXIII ISPRS Congress, 12-19 July, pp. 257–262, Prague, Czech Republic (2016)
14. Tarantino, E., Figorito, B.: Mapping rural areas with widespread plastic covered vineyards using true color aerial data. Remote Sens. **4**, 1913–1928 (2012)
15. Carvajal, F., Crizanto, E., Aguilar, F.J., Agüera, F. Aguilar, M.A.: Greenhouses detection using an artificial neural network with a very high resolution satellite image. In: International Archives of Photogrammetry, Remote Sensing and Spatial Information Sciences, vol. XXXVI, part 2, pp. 37–42, Vienna, Austria (2006)
16. Turkish Statistical Institute (TSI), Vegetable and fruit production for land under protective cover, 1995–2015 (2016). https://biruni.tuik.gov.tr/bitkiselapp/bitkisel.zul
17. Mathur, A., Foody, G.M.: Crop classification by support vector machine with intelligently selected training data for an operational application. Int. J. Rem. Sens. **29**(8), 2227–2240 (2008)
18. Watanachaturaporn, P., Arora, M.K., Varshney, P.K.: Multisource classification using support vector machines: an empirical comparison with decision tree and neural network classifiers. Photogramm. Eng. Rem. Sens. **74**(2), 239–246 (2008)
19. Koc-San, D., Turker, M.: A model-based approach for automatic building database updating from high resolution space imagery. Int. J. Rem. Sens. **33**(13), 4193–4218 (2012)
20. Koc-San, D., Turker, M.: Support vector machines classification for finding building patches from ikonos imagery: the effect of additional bands. J. Appl. Remote Sens. 8 (2014). 083694-1-17
21. Geomatica, P.C.I.: Software User's Manual. PCI Geomatics Enterprises Inc., Richmond Hill (2013)

Comparison of Mesh Simplification Tools in a 3D Watermarking Framework

Francesca Uccheddu[✉], Michaela Servi, Rocco Furferi, and Lapo Governi

University of Florence, Florence, Italy
francesca.uccheddu@unifi.it

Abstract. Given a to-be-watermarked 3D model, a transformed domain analysis is needed to guarantee a robust embedding without compromising the visual quality of the result. A multiresolution remeshing of the model allows to represent the 3D surface in a transformed domain suitable for embedding a robust and imperceptible watermark signal. Simplification of polygonal meshes is the basic step for a multiresolution remeshing of a 3D model; this step is needed to obtain the model approximation (coarse version) from which a refinement framework (i.e. 3D wavelet analysis, spectral analysis, ...) able to represent the model at multiple resolution levels, can be performed. The simplification algorithm should satisfy some requirements to be used in a watermarking system: the repeatability of the simplification, and the robustness of it to noise or, more generally, to slight modifications of the full resolution mesh. The performance of a number of software packages for mesh simplification, including both commercial and academic offerings, are compared in this survey. We defined a benchmark for testing the different software in the watermarking scenario and reported a comprehensive analysis of the software performances based on the geometric distortions measurement of the simplified versions.

Keywords: 3D watermarking · Wavelets 3D · Mesh simplification · Mesh comparison

1 Introduction

Digital watermarking has been considered a potential efficient solution for copyright protection of various multimedia contents. This technique carefully hides some secret information in the functional part of the cover content. A trade-off between imperceptibility and robustness to geometric attacks, is given by tuning up the watermark power in order to balance the two requirements. A multiresolution analysis of the 3D models allows the embedding of the watermarking noise in a lower level of resolution thus letting the higher resolution levels vertices to cover the embedded defects. This analysis permits realizing a process during which multiple representations with different complexities (i.e. resolutions) can be created. Such an analysis finally produces a very coarse mesh that represents the basic shape (low frequencies) and a set of details information at different resolution levels (median and high frequencies).

© Springer International Publishing AG 2018
G. De Pietro et al. (eds.), *Intelligent Interactive Multimedia Systems and Services 2017*,
Smart Innovation, Systems and Technologies 76, DOI 10.1007/978-3-319-59480-4_7

The coarse version of the full resolution model is usually the result of a simplification (or decimation) algorithm.

Unfortunately, most of the 3D watermark detectors are prone to fail when the protected asset changes its topology or the vertex/triangle are re-ordered, due to a signal desynchronization. A stable simplification algorithm guarantees to obtain the same multiresolution representation of the to-be-watermarked model. This study addresses a comparative evaluation of both commercial and academic available software for mesh simplification and their characterization in terms of stability in extracting the coarse mesh (e.g. the simplified version as a collection of feature points) and their robustness to noise.

2 Multiresolution 3D Watermarking

A digital watermarking system is able to protect digital works after the transmission phase and the legal access. Usually, a robust watermark is required for copyright protection purposes, being able to go through common malicious attacks. In authentication applications, the watermark is intentionally designed to be fragile, even to very slight modifications. There still exist few watermarking methods for 3D meshes, in contrast with the relative maturity of the theory and practices of image, audio and video watermarking. This situation is mainly caused by the difficulties encountered while handling the arbitrary topology and irregular sampling of 3D meshes, as well as the complexity of the possible attacks on watermarked meshes.

Most of the successful robust image watermarking algorithms are based on spectral analysis; a better imperceptibility can be gained because these approaches spread the watermarking information in all the spatial and spectral parts of the cover content. A better robustness can also be achieved if the watermark is inserted in the low and median frequency parts. Unfortunately, for 3D meshes, there does not yet exist an efficient and robust spectral analysis tool. However, a multiresolution analysis is possible [1], and allows a flexible usage of 3D models for watermarking. These methods permit realizing a synthesis process during which multiple representations with different complexities (i.e. resolutions) can be created. Such an analysis finally produces a very coarse mesh that represents the basic shape (low frequencies) and a set of details information at different resolution levels (median and high frequencies).

The main advantage of the multiresolution analysis for watermarking is its flexibility: there are different available locations to different application requirements. For example, insertion in the coarsest-level representation ensures a good robustness, while embedding in the details parts provides an excellent capacity. Under the same watermark signal intensity, insertion in the mesh low resolution component can be both more robust and more imperceptible because it makes the object expand or contract a little, while keeping its basic shape.

Wavelets are a common tool for such a multiresolution analysis. The mathematical formulation of the wavelet analysis and synthesis of 3D meshes was introduced by Lounsbery et al. [2]. A group of four triangles is merged into one, and three of the six initial vertices are conserved in the lower resolution. The wavelet coefficients are calculated as the prediction errors for all the deleted vertices, and they are 3D vectors associated with each edge of the coarser mesh.

Based on the regular wavelet analysis tool Kanai et al. [3] proposed a non-blind algorithm that modifies the ratio between a wavelet coefficient norm and the length of its support edge, which is invariant to similarity transformations. Uccheddu et al. [4] described a blind watermarking algorithm with the hypothesis of the statistical independence between the wavelet coefficients norms and the inserted watermark code.

To achieve robustness against hard connectivity attacks a remeshing step at both the insertion and extraction sides [5]. First of all, the cover mesh is remeshed to generate, for instance, a semi-regular mesh with a similar geometrical shape. This procedure is supposed to include two steps: simplification, then subdivision and rectification (i.e. vertices displacements). For the extraction, the input mesh is also remeshed, and the extraction is executed on the obtained semi-regular mesh. The resistance to connectivity attacks is achieved by devising a remeshing scheme that is independent of and insensitive to connectivity changes. Actually, Alface and Macq [6] have made some efforts in this area. They have devised a remeshing scheme based on mesh feature points, which are umbilical points obtained by curvature analysis.

3   Mesh Simplification Approaches

Polygonal meshes are the representation of choice for 3D visualizations and simulations. Very complex and highly detailed models are usually generated in practical applications by either solid modeling CAD systems [7] or in reverse engineering approach like body scanners [8] or as a pre-processing strategy for mesh segmentation [9]. To faithfully represent these complex and detailed models the resulting meshes may have an extremely large number of polygons or triangles. If a polygonal mesh conveys colour information though an image texture [10, 11], the simplification, by reducing the polygons number, allows to achieve a significant speedup of the rendering of large scenes in dynamic simulation without compromising the visual quality of the scene appearance. Current mesh reduction techniques primarily reduce mesh size by iteratively merging/collapsing edges into vertices with the goal of geometric and topological feature preservation; mesh decimation techniques are distinguished by the cost function/metric they attempt to minimize as they collapse edges in the mesh. Each approach uses different error criteria to measure the fitness of the approximated surfaces and the simplification process itself is guided by a given error metric. Most of these metrics use geometric measurements of distance or curvature.

Schroeder et al. [12] use a vertex to plane distance as decimation criterion. Luebke et al. [13] uses a function based on curvature to guide its simplification process. Klein et al. [14] use an error metric based on the Hausdorff distance. Boubekeur et al. [15] proposes a geometry-aware random sampling technique. Hoppe [16] uses energy functions, which preserve surface geometry, scalar attributes and discontinuity curves. Garland et al. [17] use a quadric error metric giving vertex-to-plane distances.

In this study, we analysed the simplification tools implemented in some popular 3D processing software both free and commercial. In particular, the capability of each of them to reproduce the same simplified model in front of perturbation of the corresponding full resolution model can be assessed and used to determine which software implementation is the best choice for a multiresolution representation of the 3D model in a watermarking scheme.

4   Test-Bed

4.1   The Software Packages

Over the last decades many commercial software packages for mesh processing have appeared. Some of them are dedicated software, whose only purpose is mesh simplification. On the other hand, since simplification is so commonly used in mesh processing, most of the large commercial systems for mesh manipulation have a builtin simplification tool.

Moreover, there exist also software packages developed in academia as implementations of simplification algorithm development. Many of them are just prototypes, but there exist robust and efficient implementations as well. The software packages differ in applicability, usability, flexibility, efficiency, but ultimately the main goal of all the packages is to produce a mesh with the best possible tradeoff between a good approximation of the original mesh and the resulting mesh triangle count.

This study treats the software as a black box without analysing the algorithms or technology behind the software packages [18, 19]. We refer the reader to the surveys on mesh simplification algorithms in [20–22]. Here we verify the robustness of such simplification software for being used as robust features points computation tools in watermarking applications.

In this work, we examined a representative subset of the available packages including six software packages for mesh simplification, both commercial and free solution. The packages are classified by their application domains and listed in Table 1.

Table 1.   List of the evaluated software packages

Name	Application domain	Website
Meshlab	Reverse engineering	www.meshlab.net/
Geomagic design X	Reverse engineering	www.geomagic.com
Polyworks	Reverse engineering	www.innovmetric.com/
Siemens NX	CAD/CAM design tools	www.plm.automation.sie mens.com/en_us/ products/nx/index.shtml
Blender	3D Modeling	www.blender.org/

Reverse engineering: In this domain, we selected three popular 3D processing software mainly oriented to the management and processing of large meshes; all of them provides a set of tools for editing, cleaning, healing, inspecting, rendering, and converting the meshes. Design X and Polyworks are commercial solution, while Meshlab is the most famous free and open source software for 3D data handling.

Modeling: In this domain, we selected Blender, a professional free and open-source computer graphics software 3D modeling, animation and rendering.

CAD/CAM design tools: Even some CAD environment integrates some basic mesh processing including the mesh decimation as the case of Siemens NX.

Most of the packages provide various options to control the simplification process; the nature of these options varies from one package to another, some are more intuitive than others. In our use of the software packages, when possible we chose the option that favoured the shape quality, while preserving for the remaining vertices, the same original coordinates.

4.2 Experimental Results

We tested the software on the following 3D models of different sizes, properties and acquisition sources (see Fig. 1).

Fig. 1. The three representative 3D models: a high-quality part of a statue, the Stanford Bunny and a CAD derived valve model.

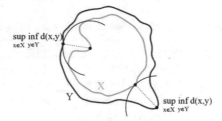

Fig. 2. The Hausdorff distance

Stanford Bunny Model, is a collection of 69,451 triangles that was assembled from range images of a clay bunny; it is widely used in the 3D processing community.

The Statue from the Baptistery Door in Florence; it is a huge model consisting in 2,028,487 triangles; it has been surveyed with a professional 3D laser scanner.

The valve model is designed in a CAD 3D environment, and then converted in a mesh. It consists in 30,734 triangles.

Each model was simplified down to 40%, 4% and 0.4% to cover a range of possible decimation ratios. In Fig. 3 is reported a visual evaluation of the effect of the several simplification tools on the Bunny model.

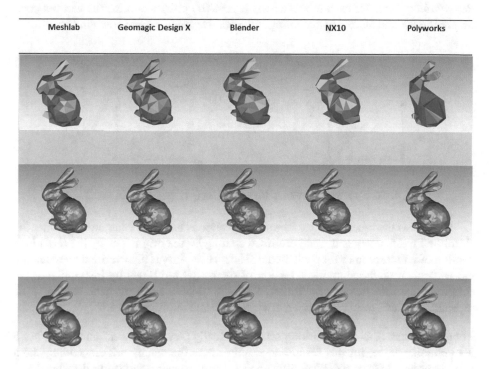

Meshlab	Geomagic Design X	Blender	NX10	Polyworks

Fig. 3. Visual evaluation of the considered mesh simplification schemes

For every simplified mesh, we measured its Hausdorff distance from the original [23] by using the free version of Metro tool [24].

The Hausdorff distance corresponds to the maximal Euclidean distance between a point on one surface and the closest point on the other surface.

Let X and Y be two non-empty subsets of 3D points of a metric space (M, d), the Hausdorff distance is defined as:

$$d_H(X, Y) = \max\left\{sup_{x \in X} inf_{y \in Y} d(x, y), sup_{y \in Y} inf_{x \in X} d(x, y)\right\} \tag{1}$$

Where sup represents the supremum and inf the infimum.

Metro tool implements the Hausdorff distance tool; the average distances are taken in both directions: when the first mesh is sampled the distances to the second mesh are computed, and vice versa. To symmetrize this measure, we considered the maximum of the two Hausdorff distances. Metro employs several types of sampling methods; we used the subdivision sampling with 10 samples per face.

The Hausdorff distance provides a good estimate of the quality of approximation and also of the visual appearance of the simplified models. Other measures of geometric differences exist, such as volumetric deviation from the original surface. However, they provide a less robust estimate of the visual and geometric fidelity (Fig. 2).

Repeatability test: for each model a copy is made by just scrambling the vertices position in the file and adjusting accordingly the relative triangles as shown in Fig. 4.

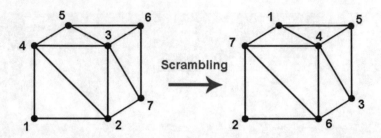

Fig. 4. The vertex scrambling does not change the geometry of the object

The test on the repeatability of the simplification is based on examining whether the simplified mesh changes if it is performed starting by the original or by the scrambled version of it for ten times for each model. The repeatability is also verified by repeating the evaluation of the simplified version of the model ten times by just applying the simplification tool in different times. All the evaluated software have shown to be repeatable in each test thus proving always identical simplified model.

Robustness test: the watermarking schemes usually perceptually act on a mesh has a 3D Gaussian noise; for this reason, to simulate both the watermark embedding and a generic noise attack, some 3D additive noise is applied using a uniform distribution.

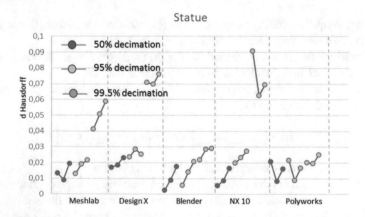

Fig. 5. Hausdorff distance of the Statue model (2,028,487 triangles) at various simplification stages and three increasing power perturbation.

By using the random vertex displacement filter in Meshlab [24] we generate a random displacement of the vertex positions and recompute the normal vectors (we use a random noise of a magnitude of 0.02%, 0.04% and 0.2% of the object size); then we calculate the Hausdorff distance between the noisy versions and the original. The robustness is tested by verifying how the simplified model changes by varying the noise power.

As it can be noticed in Figs. 5, 6 and 7, the trend is that the Hausdorff distance increases at increasing decimation and/or perturbation power.

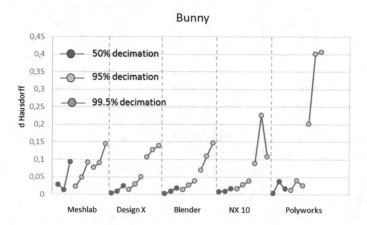

Fig. 6. Hausdorff distance of the Bunny model (69,451 triangles) at various simplification stages and three increasing power perturbation.

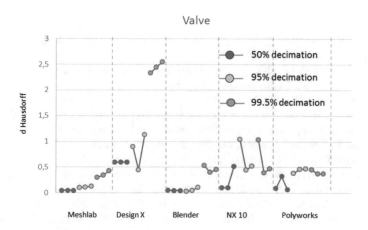

Fig. 7. Hausdorff distance of the CAD derived Valve model (30,734 triangles) at various simplification stages and three increasing power perturbation.

In particular, for the model derived from a CAD environment the decimation step outputs unexpected geometries, thus resulting in high Hausdorff distances.

5   Conclusion and Future Works

In this study a specific evaluation of both some commercial and academic available software for mesh simplification is performed. The considered simplification tools have been devised in terms of coarse mesh computation tools (e.g. the simplified mesh is devised as a collection of robust feature points) and their robustness to noise in terms of Hausdorff distance has been tested. The Siemens NX decimator tool appears to be very sensitive to the noise perturbation; usually such CAD design software implements very basic algorithms for Reverse Engineering process. On the other hand, Polyworks that is a RE dedicated software appears to be more robust to mesh perturbations when used to decimate the CAD derived model, but still show a non-robust behaviour for the decimation of the objects from the 3D scanner. Meshlab and Blender show the most regular trend of the Hausdorff distance, while remaining in a very robust area thus making them the best choice for building a robust multiresolution framework to be integrated both in the embedder and detector of a 3D watermarking system.

In order to evaluate the robustness in watermarking applications, a further analysis is needed; by considering some multiresolution based watermarking algorithms, an analysis of the detector is needed to evaluate if the simplification tool of some among the considered software allow a multiresolution analysis without losing the synchronism with the embedded watermark code.

References

1. Dodgson, N.A., Floater, M.S., Sabin, M.A.: Advances in multiresolution for geometric modelling. Springer-Verlag (2005)
2. Lounsbery, M., DeRose, T.D., Warren, J.: Multiresolution analysis for surfaces of arbitrary topological type. ACM Trans. Graph. **16**(1), 34–73 (1997)
3. Kanai, S., Date, H., Kishinami, T.: Digital watermarking for 3D polygons using multiresolution wavelet decomposition. In: Proceedings of the International Workshop on Geometric Modeling: Fundamentals and Applications 1998, pp. 296–307 (1998)
4. Uccheddu, F., Corsini, M., Barni, M.: Wavelet-based blind watermarking of 3D models. In: Proceedings of the ACM Multimedia and Security Workshop 2004, pp. 143–154 (2004)
5. Wang, K., Lavouè, G., Denis, F., Baskurt, A.: A comprehensive survey on three-dimensional mesh watermarking. IEEE Trans. Multimedia **10**(8), 1513–1527 (2008)
6. Alface, P.R., Macq, B.: Blind watermarking of 3D meshes using robust feature points detection. In: Proceedings of the IEEE International Conference on Image Processing 2005, vol. 1, pp. 693–696 (2005)
7. Puggelli, L., Volpe, Y., Giurgola, S.: Analysis of deformations induced by manufacturing processes of fine porcelain whiteware. In: Advances on Mechanics, Design Engineering and Manufacturing. Springer International Publishing, pp. 1063–1072 (2017)
8. Furferi, R., Governi, L., Uccheddu, F., Volpe, Y.: A RGB-D based instant body-scanning solution for compact box installation. In: Advances on Mechanics, Design Engineering and Manufacturing. Springer International Publishing, pp. 387–396 (2017)
9. Buonamici, F., Carfagni, M., Volpe, Y.: Recent strategies for 3D reconstruction using reverse engineering: a bird's eye view. In: Advances on Mechanics, Design Engineering and Manufacturing. Springer International Publishing, pp. 841–850 (2017)

10. Pelagotti, A., Ferrara, P., Uccheddu, F.: Improving on fast and automatic texture mapping of 3D dense models. In: 2012 18th International Conference on Virtual Systems and Multimedia (VSMM). IEEE (2012)
11. Uccheddu, F., Pelagotti, A., Picchioni, F.: A greedy multiresolution method for quasi automatic texture mapping. In: SPIE Optical Metrology (2011). International Society for Optics and Photonics
12. Schroeder, W.J., Zarge, J.A., Lorensen, W.E.: Decimation of triangle meshes. ACM Siggraph Comput. Graph. **26**(2), 65–70 (1992). ACM
13. Luebke, D., Hallen, B.: Perceptually driven simplification for interactive rendering. In: Rendering Techniques 2001. Springer Vienna, pp. 223–234 (2001)
14. Klein, R., Liebich, G., Straßer, W.: Mesh reduction with error control. In: Proceedings of the 7th Conference on Visualization 1996. IEEE Computer Society Press (1996)
15. Boubekeur, T., Alexa, M.: Mesh simplification by stochastic sampling and topological clustering. Comput. Graph. **33**(3), 241–249 (2009)
16. Hoppe, H.: New quadric metric for simplifying meshes with appearance attributes. In: Proceedings of the Conference on Visualization 1999: Celebrating Ten Years. IEEE Computer Society Press (1999)
17. Garland, M., Heckbert, P.S.: Surface simplification using quadric error metrics. In: Proceedings of the 24th Annual Conference on Computer Graphics and Interactive Techniques. ACM Press/Addison-Wesley Publishing Co. (1997)
18. Cignoni, P., Montani, C., Scopigno, R.: A comparison of mesh simplification algorithms. Comput. Graph. **22**(1), 37–54 (1998)
19. Heckbert, P.S., Garland, M.: Survey of polygonal surface simplification algorithms. Siggraph 1997, Course Notes (1997)
20. Luebke, D.P.: A developer's survey of polygonal simplification algorithms. IEEE Comput. Graph. Appl. **21**(3), 24–35 (2001)
21. Mocanu, B., Tapu, R., Petrescu, T., Tapu, E.: An experimental evaluation of 3D mesh decimation techniques. In: 2011 10th International Symposium on Signals, Circuits and Systems (ISSCS). IEEE (2011)
22. Taime, A., Saaidi, A., Satori, K.: Comparative study of mesh simplification algorithms. In: Proceedings of the Mediterranean Conference on Information & Communication Technologies 2015. Springer International Publishing (2016)
23. Aspert, N., Santa-Cruz, D., Ebrahimi, T.: Mesh: measuring errors between surfaces using the hausdorff distance. In: Proceedings 2002 IEEE International Conference on Multimedia and Expo, ICME 2002, vol. 1. IEEE (2002)
24. Cignoni, P., Rocchini, C., Scopigno, R.: Metro: measuring error on simplified surfaces. Comput. Graph. Forum **17**(2), 167–174 (1998). Blackwell Publishers

A Smart-CA Architecture for Opencast Matterhorn

Vicente Goyanes[1,2(✉)], Rubén González[1], Anxo Sánchez[3],
and Domingo Docampo[2]

[1] Teltek S.A., Praza Miralles, 36310 Vigo, Spain
vgoya@teltek.es
[2] Departamento de Teoría do Sinal e Comunicacións, ATLANTIC,
Universidade de Vigo, Vigo, Spain
[3] Departamento de Enxeñería Química, Universidade de Vigo,
Campus Universitario, 36310 Vigo, Spain

Abstract. Opencast is a flexible and customizable video capture and distribution system that facilitates mass videotaping of lectures and classes, an open source alternative to proprietary automated recording systems. The current Opencast architecture is highly centralized, integrating very simple recording devices within the classrooms with a powerful central processing system. As a natural evolution of Opencast we describe in this paper a new architecture in which much of the image processing tasks are transferred from the central servers to a new generation of powerful recording devices. Those devices, smart capture agents, contribute to enhance the capabilities of the Opencast systems along different directions: better scalability, lower complexity and lower purchasing and operating costs.

Keywords: Web Tv · Opencast · Smart capture agent · Massive recording · Automatic class recording

1 Introduction

Opencast is an open capture and distribution system, primarily used in university settings, designed to allow mass recording of lectures and classes [1]. The project was born as an initiative of UCBerkeley in 2005 to build a community of institutions interested in creating a system of massive (automatic) recording of classes. From this community, a consortium of 13 institutions worldwide, supported by the Hewlett and Melon foundations under the leadership of UC Berkeley and ETH Zurich, managed to develop the first operating version of the Opencast platform.

The platform has since evolved into a free software project under the guidance of the Apereo foundation; it arguably constitutes the open source alternative to well established, proprietary automated recording systems. As an example of its success within the realm of Higher Education it is worth mentioning that the largest class recording facility which encompasses more than 350 classrooms,

© Springer International Publishing AG 2018
G. De Pietro et al. (eds.), *Intelligent Interactive Multimedia Systems and Services 2017*,
Smart Innovation, Systems and Technologies 76, DOI 10.1007/978-3-319-59480-4_8

launched and managed by the University of Manchester [2], uses the Opencast capturing and distributing system.

The current Opencast architecture continues to be the one designed at the beginning of the project almost ten years ago, a highly centralized architecture that integrates very simple recording devices within the classrooms along with a powerful central processing unit, the Opencast core. A description of the Opencast project can be found in [3,4]; a fully operative account of a classic Opencast system implementation can be found in [5].

With prices of hardware going steadily down in the last few years, we think it is worth exploring the alternative of shifting a large chunk of the processing load to the recording devices, thus streamlining the Opencast core. Hence, as a natural evolution of the Opencast setting, we describe in this paper a new architecture, in which much of the image processing tasks are transferred from the central servers to the recording devices, a new generation of smart capture agents.

2   Background

The goal of the Opencast system is to record two synchronous videos in the classroom; the first video is focused on the professor, while the second one tapes, directly from the computer, the contents of the screen that provide the students with multimedia support. Both videos are then processed to create an enriched multimedia object, see Fig. 1, namely, an audiovisual educational object that

Fig. 1. Enriched multimedia object in Opencast

enables the viewer to watch the two original synchronous videos (professor and computer slides) along with a set of additional functions designed to improve the learning experience, such as:

- Displaying a temporal index through the use of small images captured from the computer, as well as through timeline tagging of the presentation content changes. The tagging is particularly useful when indexing presentation slides, in which the use of an OCR enables text tagging linked to the slide time-stream, thus allowing the viewer to look up the video in search of specific content.
- Both videos can be resized in real time, allowing the viewer to focus on specific details of the video from the computer.

The system performs an automatic indexing of the contents, by first detecting the change of slides in the second video and then applying an OCR on those slides. To create a fully integrated viewing experience, it is then necessary to post-process the original recording. First, a series of layers of isochronous metadata are produced, along with graphic elements, such as captures in the form of images of the different slides detected in the second video. All these elements are collected within a media package (MP). Once the processing of the content has taken place, the media package is then complete and ready to be reproduced (see Fig. 2).

Fig. 2. Media package in Opencast

Original Opencast Arquitecture. As it has already been pointed out, the current architecture of the Opencast platform is strongly centralized. It was heavily dependent on the hardware availability at the time of its inception. Hence, it was rooted in the use of affordable recording devices [6], Light Capture Agents (Light-CA). Those Light-CA, based on PC hardware, were only in charge of recording the videos to be delivered to the Opencast core, where all the processing of the multimedia content would be carried out at the central servers to produce the finished media packages, as Fig. 3 shows.

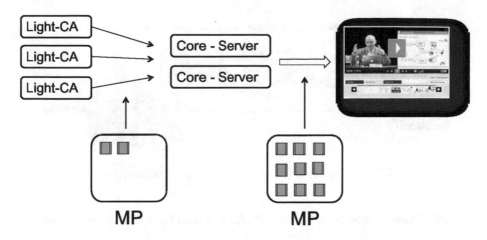

Fig. 3. Original Opencast architecture

To process the multimedia content coming from the Light-CAs the Opencast core configures a complex Workflow composed of 22 individual operations that can be grouped into the 9 functional blocks shown in Fig. 4

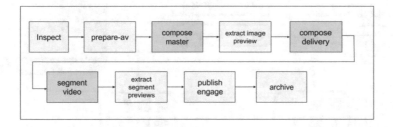

Fig. 4. Standard Opencast workflow

Inside a Light Capture Agent. The internal structure of a Light-CA or light recorder consists of a double pipeline video processing that captures the two synchronous video sources and then compresses the recordings at low computational cost. It comes with the logic needed to create a media package and to upload it to the Opencast core in which further processing will take place.

The mission of a Light-CA is to capture the two synchronous videos and encode the recordings by means of a low computational cost algorithm. It is therefore important to note that the final product delivered by Light-CAs to the central system will become the master copies to which any further processing shall be applied; hence, it will no longer be possible to produce a copy of the recording with a higher quality than the one captured and delivered by the Light-CAs. The architecture of a light capture agent is outlined in Fig. 5.

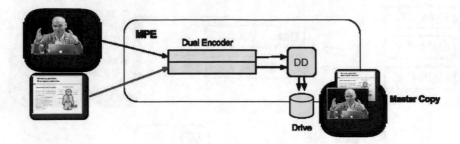

Fig. 5. Outline of the light capture agent architecture

3   A New Smart-CA Based Architecture for Opencast

In light of the shortcomings of the current architecture, as regards the limited capability of the capture agents, this paper analyzes a new architecture, see Fig. 6, in which most of the processing tasks take place in real time within the recorders.

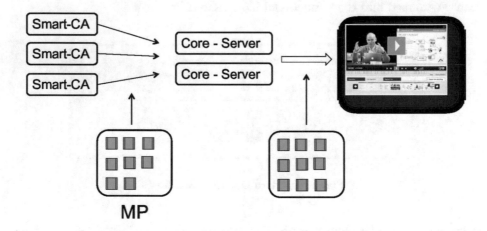

Fig. 6. New Smart-CA based Opencast architecture

This radical change of the Opencast setting has been made possible by the evolution of the general purpose hardware of consumer PCs, which has resulted in the availability of highly efficient, very affordable equipment. The new architecture proposed in this paper relies upon the use of smart capture agents that perform most of the necessary processing, delivering a virtually complete media package to the Opencast core ready for publication and archiving purposes.

Using Smart-CAs the Workflow to be run on central servers is radically reduced and the number of operations streamlined from 9, three of which computationally demanding, to just 3 simple operations, as seen in Fig. 7

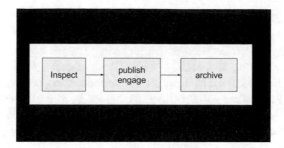

Fig. 7. Ultrafast Opencast workflow for Smart-CA

3.1 Inside a Smart Capture Agent

The internal structure of a Smart-CA consists of three video processing pipelines, the Media Processing Engines (MPE; two of them acting as well as additional modules, Media Enrichment Engines, MEE), which perform in real time metadata extraction tasks, slide tracking, and QR code lecturing among other activities. The new structure is equivalent to inserting three light CAs working in parallel within a single capture agent, as Fig. 8 shows.

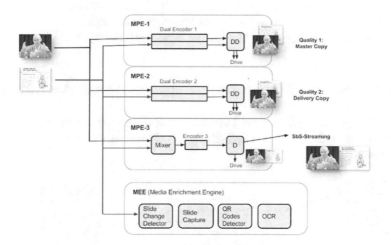

Fig. 8. Outline of the smart capture agent architecture

3.2 Smart Capture Agent's Functionalities

A Smart-CA such as the one described in this paper[1], running on an Intel i7 platform, can perform the following tasks, executing them concurrently in real time:

---

[1] A working version of the Smart CA is currently marketed as Galicaster-PRO by the Teltek company.

- Master copy: By using the first Media Processing Engine (MPE-1), the Smart-CA captures two High Definition synchronous videos and encodes them in a broadband format, to be used as high quality master copies.
- Emission copy: Using the MPE-2 two HD synchronous videos are captured and encoded in a bandwidth suitable for broadcasting on the Internet. Those two videos will be used as broadcasting copies by the Opencast system
- Mono-stream and Streaming Live Copying: Using the MPE-3, two synchronous videos are captured and combined into a single side-by-side video ready to be deployed through streaming services.
- Slide shift detection: Early access to raw video (as it is being recorded) enables very effective real-time slide show change detection.
- High quality slide shows: Slides are captured directly from raw video offering the highest possible quality. Since the Opencast core does not have to capture the slides from previously compressed video, Smart-CAs help to deliver superior quality.
- Text extraction from slides: The quality of the captured slides enables the OCR software to produce a more reliable extracted text.
- Detection of QR codes within the video that facilitate automatic actions such as:

    Stop recording when the Smart CA detects a code indicating that the material inserted in a slide may not be publicly reproduced.

    Segment content when a top-of-chapter QR is detected.

    Catalog a recording when the QR of a professor/subject ID is detected.

4 Advantages of the Smart Capture Agent Architecture

The architecture described, based upon the use of the new smart capture agents, contributes to enhance the capabilities of Opencast systems along different directions.

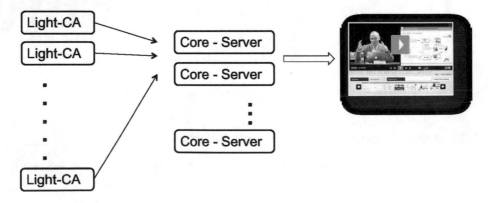

Fig. 9. Scaling of the original architecture

4.1   Better Scalability

The original Opencast architecture suffers from a noticeable drawback, an insufficient scalability of the solution: as the number of light capture agents increases, a number of central servers will have to be added to the Opencast core, see Fig. 9, resulting in a number of database loading problems; besides, the size of the workforce in charge of video processing operations will also need to be enlarged.

Using the architecture based on Smart-CAs described in this paper, Opencast system can grow in a much simpler way: each new capture agent contributes to both the generation of a larger number of videos to be processed, and to the addition of more computing power to the whole operation, thus alleviating the need to enlarge the Opencast core with additional central servers, as Fig. 10 shows.

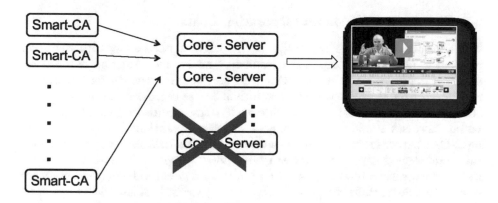

Fig. 10. Scaling of the new architecture

The lower burden on the central servers to process a media package created by a Smart-CA versus a Light-CA is remarkable. On a test system using the Materhorn Workflow Browser [7], recordings of one hour from both a Smart-CA and a Light-CA have been processed. The workflow of the joint processes is shown in Fig. 11. While the first one required only thirteen minutes of further processing time to be ready for publication, the second recording from the old Light-CA required two and a half hours.

4.2   Lower Complexity

The current central server system of a medium or large Opencast installation turns out to be very complex, difficult to deploy and maintain properly. The Smart-CA architecture contributes to streamline the Opencast core by shifting computing burden to the capture agents, thus turning the whole system easier to deploy and operate.

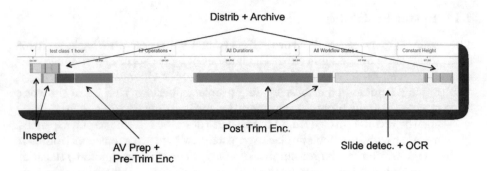

Fig. 11. Light-CA vs Smart-CA workflows at the Opencast core server

4.3 Lower Acquisition and Operating Costs

The simplification of the complexity of the whole service along with the lower costs of the hardware components result in a noticeable reduction of the operating costs of a medium or large Opencast service. Whereas within the original architecture framework the overall performance was sacrificed in order to ensure the operation of low cost light capture agents, the architecture proposed in this paper increases the global performance at lower operational costs. The use of light-CAs increases the burden on central servers and data center infrastructures (air conditioning capacity, uninterruptible power supply). Smart-CAs are a little more expensive; however, as all consumer hardware products, their costs are bound to substantially decrease in the coming years; besides, their large scale deployment does not require associated data center infrastructure.

The exploitation of the system is also more economical, due to the aforementioned reduction in complexity and associated energy savings. It should be noted that just the need for air conditioning in the data center doubles the amount of electrical energy needed to operate the system. In the proposed architecture the air conditioning is not necessary since the equipment in charge of most of the processing is not concentrated in a technical room but is instead distributed among the classrooms.

5 Conclusions

In this paper we have described a new architecture for Opencast, a video capture and distribution system that facilitates mass recording of lectures and classes. Due to reasons related to hardware availability at the time of its inception, the original Opencast architecture turned out to be highly centralized, integrating very simple recording units located inside the classrooms with powerful central processing systems at the Opencast core. The architecture described in this paper constitutes a natural evolution of the Opencast systems towards shifting much of the image processing tasks from the central servers to more computationally powered recording devices.

The proposed architecture relies upon the introduction of (now) affordable smart capture agents, consisting of three video processing pipelines, Media Processing Engine modules, with two of them doubling as Media Enrichment Engine modules. The new smart capture agents perform in real time operations that are currently reserved to the central servers at the Opencast systems core, such as meta data extraction tasks, slide tracking or QR code lecturing. Thanks to the introduction of the smart capture agents, we can set up streamlined capturing and distributing Opencast systems with enhanced capabilities: better scalability, lower complexity and lower acquisition and operating costs.

References

1. Opencast Consortium. http://www.Opencast.org
2. Reece, R.R.: Lecture capture at The University of Manchester. Opting (-out) to enhance the student experience, Opencast Conference Manchester (2015). Available from the Apereo Foundation server at. https://www.apereo.org/content/video-lecture-capture-university-manchester-opting-out-enhance-student-experience
3. Ketterl, M., Schulte, O.A., Hochman, A.: Opencast Matterhorn: A community-driven open source software project for producing, managing, and distributing academic video. Int. J. Interact. Technol. Smart Educ. **7**(3), 168–180 (2010). Emerald Group Publishing Limited
4. Ketterl, M., Brooks, C., Schulte, O.A., Rolf, R.: Opencast Matterhorn-Open source lecture capture and video management. ACM Records 5(3), September 2013
5. Jonach, R., Ebner, B., Grigoriadis, Y.: Automatic system for producing and distributing lecture recordings and livestreams using opencast matterhorn. J. Educ. Issues **1**(2), 149–158 (2015). ISSN: 2377-2263
6. Opencast Capture Agent Configuration. https://Opencast.jira.com/wiki/display/mh16/Capture+Agent+Configuration. Retrieved 3 Jan 2017
7. Harvard Division of Continuing Education: Materhorn Workflow Browser. https://github.com/harvard-dce/mh-workflowbrowser

An Effective Corpus-Based Question Answering Pipeline for Italian

Emanuele Damiano(✉), Raffaele Spinelli, Massimo Esposito,
and Giuseppe De Pietro

National Research Council of Italy, Institute for High Performance Computing
and Networking - ICAR, Via Pietro Castellino 111, 80131 Naples, Italy
emanuele.damiano@cnr.icar.it
http://www.icar.cnr.it/

Abstract. Question Answering is a longevous field in computer science, aimed at realizing systems able to answer questions expressed in natural language. However, building Question Answering systems for Italian and able to extract answers from a corpus pertaining a closed domain is still an open research problem. Indeed, extracting clues from a question to generate a query for the information retrieval engine as well as determining the likelihood that a candidate answer is correct are two very thorny tasks. To face these issues, the paper presents a Question Answering pipeline for Italian and based on a corpus of documents pertaining a closed domain. In particular, this pipeline exhibits functionalities for: (i) analyzing natural language questions in Italian by using lexical features; (ii) handling both factoid and description answer types and, depending on them, filtering contextual stop words from questions; (iii) scoring and selecting candidate answers with respect to their type in order to determine the best one. The proposed solution has been subject to an evaluation of its performance using standard metrics, showing promising results.

Keywords: Question Answering · Cognitive Systems · NLP

1 Introduction

Question Answering (QA) is a computer science discipline born from the Information Retrieval (IR) field and aimed at providing exact answers to natural language questions. Even if QA is a long-standing research problem, recently, it has gained increasing attention since it has been advocated as the basic function in digital assistants like Siri, Google Assistant, Alexa, and Cortana. Furthermore, it has been identified as fundamental to build the next generation of *Cognitive Systems* [10] able to offer complex analytic algorithms as composite services in the cloud [1–3].

Generally speaking, a QA system processes a question and extracts as many clues as possible to understand what it is being asked for (*Question Processing*). Then, a *knowledge base/corpus* is queried to retrieve relevant information about

© Springer International Publishing AG 2018
G. De Pietro et al. (eds.), *Intelligent Interactive Multimedia Systems and Services 2017*,
Smart Innovation, Systems and Technologies 76, DOI 10.1007/978-3-319-59480-4_9

the question (*Information Retrieval*). Finally, all the possible answers are evaluated in order to select the most confident one (*Answer Processing*).

Driven by the observation that knowledge bases are far from complete and, thus, they may not always contain the information required to answer questions, this work studies corpus-based QA systems, with the focus on directly mining answers from a collection of documents pertaining a closed domain. Building such systems, however, has difficulties mostly due to some challenging issues.

First, in the *Question Processing* phase, all the clues collected so far are used to generate a query aimed at retrieving relevant documents. Therefore, the policy employed to extract relevant terms from the question has to be carefully defined, according to the type of information the question is asking for. Second, in the *Answer Processing* phase, determining the likelihood that a candidate answer is correct is a very thorny task. Most of the corpus-based QA systems use filtering algorithms based on patterns to score the set of candidates and determine the most pertinent one. However, filtering needs more complex heuristics to be applied to a corpus made of unstructured documents and closed domain knowledge. Finally, while a huge literature exists for English QA systems, few studies are available for corpus-based QA on closed-domain for non-English languages.

For these reasons, this paper proposes a novel pipeline for corpus-based QA in Italian, developed within C³, a Cognitive Computing Center in Naples and involving the National Research Council of Italy and the Italian Labs of the IBM Research and Development Division.

The first distinctive element of this pipeline relies on its capacity of analyzing a question expressed in Italian and determining, by exploiting only lexical features, the best useful terms to be used for IR into a closed domain. Secondly, no restriction is made on the expected answer type that can be handled. In particular, two main types are considered, namely *Factoid* for precise words, and *Description* for short paragraphs talking about concepts or named entities. Factoid type is further organized into a two-level taxonomy in order to finely classify the expected answers. Finally, the third novel functionality is represented by the manner in which candidate answers are evaluated and selected. In particular, a set of filtering procedures is proposed to score evidence for each candidate with respect to its type, i.e. factoid or description.

The rest of the paper is organized as follows. Section 2 describes the common approaches in literature, highlighting differences with the proposed one. Section 3 describes the pipeline, the choices made and its implementation. Section 4 describes the experimental evaluation and presents the results achieved. Section 5 concludes the work with several considerations and future activities.

2   Related Work

In the last decade, literature about QA has been very flourishing. Plenty of works are available spanning across quite different approaches.

Most studies are limited to the English language only. The most relevant solution is represented by *Watson* [12], realized in the IBM facilities. Among the other proposals available in this field and working with the English language, three notable studies are: *YodaQA* [4], *Watsonsim* [8] and *OpenEphyra* [18].

The former is an end-to-end QA pipeline designed to answer factoid questions. It has been built by exploiting UIMA (Unstructured Information Management Architecture)[1] and it relies on different NLP tools, like Stanford CoreNLP, OpenNLP, and LanguageTool. *Watsonsim* is a class project realized at the University of North Carolina. It is inspired by IBM Watson and based on Lucene and OpenNLP. It is meant to answer questions from Jeopardy!. The latter is a QA pipeline that relies on *Patterns Learning* techniques to interpret questions and extract answers from structured and unstructured knowledge bases.

With reference to other languages, few examples exist, which address multilingual QA [19] or focus on a specific language [5,13,14]. For what concerns Italian, two studies stand out [14,16]. The former proposed a QA system named *Question Cube*, relying on NLP algorithms for both *Question Processing* and *Answer Processing*, machine learning for question classification and probabilistic IR models for *Information Retrieval*. The latter, named *Quasit*, is a QA system that leverages the power of a previously known linguistic knowledge, and it bases its reasoning process on both rules and ontologies. It uses the structured knowledge known as DBpedia[2].

More recently, novel end-to-end QA approaches have been proposed, essentially based on deep learning algorithms [7,23]. In particular, Yu et al. [25] carry a matching between the semantic representations of questions and candidate answers, by means of *Convolutional Neural Networks*. In another case, the feedback of the users is taken into account [22] to improve the quality of the answers.

Furthermore, most of the existing QA systems are meant to be *open-domain* [4,8,9,12,21] while very few are meant to be *closed-domain* [20]. Of the studies on closed domains, only a fraction uses unstructured knowledge base. The majority of works are on structured knowledge bases [7,16,23], whereas few others exploit unstructured knowledge bases [11,24] or use hybrid knowledge bases [15].

Summarizing, most of the studies are meant for the English language, and regard open-domain, while few systems work on non-English languages and closed-domain. Moreover, to the best of our knowledge, no system exists that answers factoid and description questions expressed in Italian by operating with unstructured documents in a closed domain. Comparing to these previous works, our contribution consists in: (i) supporting both *Factoid* and *Description* questions in Italian; (ii) modeling the expected answer type by means of a two-level taxonomy; (iii) employing contextual *stop words* to generate the query from the question; (iv) filtering candidate answers by means of novel scoring procedures.

3 The Proposed Question Answering Pipeline

The proposed QA pipeline, depicted in Fig. 1 extends the one defined in [6] and operates on a corpus of documents in Italian related to a closed domain. It consists of four main phases: (i) *Indexing*; (ii) *Question Processing*; (iii) *Information Retrieval*; (iv) *Answer Processing*.

---

[1] https://uima.apache.org/.

[2] http://wiki.dbpedia.org/.

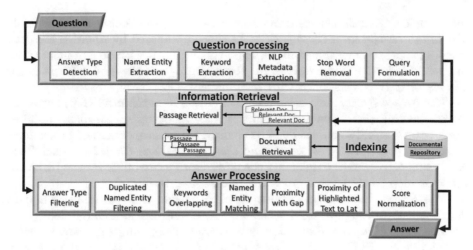

Fig. 1. The proposed QA pipeline.

3.1 Indexing

This phase consists into an off-line procedure that elaborates the documents of the corpus and produces *annotations* representing the semantic types of pieces of text. In particular, the technique of *Predictive Annotations* [17] has been here employed, which essentially indexes not only the textual contents of the documents for the IR phase, but also the semantic classes of entities or pieces of text in order to focus the search on the candidate answers of the right type.

According to this idea, annotations are designed to represent classes of information that people ask questions about, overlapping with the expected answer types. Two different approaches have been here used for generating annotations with reference to factoid and description answer types. In the first case, the annotations are anchored to entities and are used to univocally refer to them, whereas in the second one, they are anchored to entities belonging to sentences or short paragraphs and are used to univocally refer to these pieces of text.

A two-level hierarchy of semantic classes has been proposed for annotating precise entities in the text (i.e. factoid answer types), where *Person, Entity, Location, Date* are examples of top level categories, and *Address, City, Region, State, Artifact* are examples of bottom level categories. On the other hand, a particular semantic class, named *Description*, has been defined to annotate sentences or short paragraphs in the text (i.e. description answer types).

3.2 Question Processing

This phase employs the following set of procedures to elaborate the question text and extract a set of features to ultimately generate a query for the IR phase.

The *Answer Type Detection* identifies the expected answer type of a question, generally referred as Lexical Answer Type (LAT), in accordance with the hierarchy of semantic classes defined for annotating the documents. This task, also known as *Question Classification*, has been performed by adopting the deep learning approach mentioned in [6], opportunely tuned with an extended training set in order to face novel LATs. For instance, the question *"Where is the Sistine Chapel?"*, is classified with *Location* as LAT.

The *Named Entity Extraction* identifies all the Named Entity (NE) occurring in the question text, intended as entities, such as persons, locations, organizations, artifacts, and so on, that can be denoted with a proper name. For instance, in the question *"Where is the Sistine Chapel?"*, *Sistine Chapel* is a NE. Such NE is successively used to generate the most appropriate query for the IR phase. For more details about this procedure, please refer to [6].

The *Keyword Extraction* extracts significant terms from the question and, in detail, nouns and high-content verbs. Further details are available in [6].

The *NLP Metadata Extraction* is in charge of applying the standard NLP techniques like POS Tagging, Lemmatisation and Stemming to the question terms in order to determine their role. More details are reported in [6].

The *Stop Word Removal - SWR* determines and removes stop words, intended as terms that add no or little information to the text and, therefore, are useless in the query for the IR phase. Two types of stop words have been here considered. The first one is represented by the most common words in the Italian language, such as prepositions, determiners, conjunctions, question words and so on, in accordance with a classical perspective. The second one includes words that are correlated to the LAT of the questions, and, for this reason, are superfluous. For instance, in the question *"In what year did Michelangelo Buonarroti paint the Sistine Chapel?"*, the word *year* is considered as a stop word and, thus, removed, since tied to its LAT, i.e. *Date*. It is worth noting that, for this second type, the list of stop words changes depending on the LAT considered.

The *Query Formulation* generates the query to pass to the IR phase, by using a subset of the features collected by the aforementioned procedures, depending on the expected answer type. In detail, for factoid type, all the features are used, whereas, for description type, only two features are employed: *LAT* and *NE*.

3.3   Information Retrieval

This phase retrieves a set of documents (default to 5) satisfying the query. Each returned document has a score and a list of annotations added in the *Indexing* phase. Each document is then split into sentences for further processing.

3.4   Answer Processing

This phase first filters and scores sentences containing candidate answers, intended as pieces of text annotated with the expected LAT and, next, selects the best candidate answer. Each sentence can have zero or more candidate answers.

Two typologies of *Answer filtering procedures* have been developed, namely *Cutting filtering* and *Scoring filtering*.

The *Cutting filtering* removes sentences without eligible candidate answers. Two procedures of this typology have been implemented and are described below.

The *Answer Type Filtering* removes all the sentences that do not contain the expected LAT. For instance, given the question *"When was the Sistine Chapel built?"*, with *Date* as LAT, the sentence *"The Sistine Chapel is a chapel in the Apostolic Palace, the official residence of the Pope, in Vatican City"* is discarded, since it does not include a piece of text annotated with the expected LAT.

The *Duplicated Named Entity Filtering - DNEF* removes sentences containing a candidate answer annotated with the expected LAT but represented by a named entity also present in the question. For instance, given the question *"Who was the father of Leonardo da Vinci?"*, with *Person* as LAT and the sentence *"Leonardo was the firstborn son of the notary Piero from Vinci"*, it ensures that *Leonardo* is discarded as a possible answer, even if its LAT is equal to *Person*.

With reference to the second typology of *Answer filtering procedures*, the *Scoring filtering* gives a score to each candidate answer. A collection of filtering procedures of this typology has been proposed, where each of them is independent from all the others and may be applied depending on the type of answer. All the different scores calculated by them are normalized and exhibit a monotonic behavior, i.e. they are increased or decreased with the likelihood of the answer being correct. This second type of procedures is described as follows.

The *Keywords Overlapping* scores each candidate answer in relation to the number of keywords (not including stop words and NEs) in both the sentence containing it and the question. For instance, given the question *"Which famous artwork did Michelangelo realize based on a commission from the Pope?"* and the sentences *"The Sistine Chapel is the most famous artwork realized by Michelangelo, commissioned by the Pope Julius II..."* and *"Pope Julius II requested Michelangelo to paint the ceiling of the Sistine Chapel, that now is considered one of the most notorious artworks of Michelangelo"*, the procedure assigns a higher score to the candidate answer *"Pope Julius II"* in the first sentence since it has a greater percentage of keywords matching with the question.

The *NE Matching* scores each candidate answer depending on the number of NEs occurring in both the sentence containing it and the question. For instance, given the question *"When did Pope Julius II hire Michelangelo to paint the Sistine Chapel?"*, with *"Pope Julius II"*, *"Michelangelo"* and *"Sistine Chapel"* as NEs, the procedure assigns the greatest score to the candidate answer *"between 1536 and 1541"*, since contained into a sentence like *"between 1536 and 1541, Pope Julius II commissioned Michelangelo to paint the giant fresco The Last Judgment behind the Sistine Chapel altar."* with all those NEs.

The *Proximity with Gap - PwG* scores each candidate answer in relation to the distribution of question terms that are present in the sentence containing it. This procedure is inspired by dynamic programming algorithms typically used to align sequences. In particular, given the text containing a candidate answer to a question, the procedure inspects all the neighbor terms of the candidate answer and, in case of a match with a question term, it increments the partial score with a default value equal to 1, whereas, in case of a gap, it decrements that score with a value calculated by using the series described in Eq. 1, where X is the set of terms of the sentence placed on the left/right of the candidate answer. The index i is set back to 2 every time a match is encountered.

$$\sum_{i=2}^{|X|} \frac{1}{2^i} \tag{1}$$

The procedure looks at both sides of the candidate answer, keeping the calculation of the partial scores distinct. At the end, the partial scores calculated moving on both sides are summed and normalized to form the final score. For instance, given the question *"Who hired Michelangelo to paint the Sistine Chapel's ceiling?"*, the procedure calculates a greater score for the candidate answer *"Pope Julius II"* in the sentence *"In 1508, Pope Julius II hired Michelangelo to paint the ceiling of the Sistine Chapel"* with respect to the same answer in the sentence *"Pope Julius II, after 5 years from the beginning of his reign, hired Michelangelo to paint the ceiling of the Sistine Chapel, that later become the most famous piece of artwork of the artist."*. Indeed, the matching terms around the candidate answer are less sparse in the first sentence than in the second one.

The *Proximity of Highlighted Text to LAT - PHTL* scores each candidate answer depending on its average distance, in terms of characters, with all the terms in the sentence containing it that match the query terms, named *highlighted terms*. For instance, given the question *"Who painted the Sistine Chapel?"*, this procedure assigns a greater score to the candidate answer *"Michelangelo"* in the sentence *"In 1508 Michelangelo painted the Sistine Chapel, at the request of Pope Julius II"* with respect to the same answer in the sentence *"In 1508 Michelangelo went back to Rome and received the request from Pope Julius II to paint the Sistine Chapel"*. Indeed, the first sentence shows closer matching terms to the candidate answer than the second one.

Finally, the *Answer Selection procedure* identifies as right answer the one having the highest score. The final score is calculated as the normalized algebraic sum of the scores attributed by each and every of the aforementioned *Scoring filters*. It's worth noting that, in this sum, all the scores produced by the *Scoring filtering procedures* have the same weight equal to 1.

4   Experimental Evaluation

The QA pipeline is implemented as a RESTful API, as described in [6] and an ad-hoc tool named Carya has been used to test it. In more detail, the corpus used for testing consists of 16 documents in Italian crawled from the Internet and specific to a cultural heritage theme of the city of Naples. The test suite consists of approximately 920 questions, split according to their LAT and stored in csv files. The list encompasses *Artifact, Date, Description, Location*, and *Person*.

Many different configurations of the QA pipeline have been used for the test. The first one is named *Old Pipeline* and represents the first version of the pipeline described in [6]. It includes *Answer Type Filtering* as *Cutting Filtering* and *NE Matching* and *Keywords Overlapping* as *Scoring Filtering*. Next, this configuration is enriched with the remaining filtering procedures here defined, by adding first a procedure at a time, then in pairs of two, and finally all together.

Measurements for both performance and accuracy are carried out. The response time is still under 5 s, and on average under 3. Table 1 shows the *Accuracy@1* for each new filtering procedure added, and their combinations.

Table 1. Accuracy@1 for each filtering procedure and their combinations

	Date	Location	Description	Artifact	Person
Old Pipeline	60.38	84.70	-	62.21	82.61
DNEF	60.38	84.70	97.06	62.21	84.35
PwG	78.46	83.98	97.06	62.21	86.95
PHTL	73.08	79.71	97.06	76.96	79.13
DNEF + PwG	78.46	83.98	97.06	62.21	86.96
DNEF + PHTL	73.08	79.71	97.06	76.96	81.74
PwG + PHTL	76.54	81.85	97.06	80.18	84.35
DNEF + PwG + PHTL	76.54	81.85	97.06	80.18	84.35

PwG contributes the most with a boost on average of 5.42%, with the greatest improvement when the question type is *Date* (18.08%). *PHTL* instead generates an average improvement of 4.74% with a peak of 14.75% when the question type is *Artifact*. *DNEF* improves the number of correct answers in the first position of 1.74% (*Person*), and also the number of correct answers in second and third positions, with an increment of 0.87% always when the question type is *Person*.

Based on these results, a new configuration, named *New Pipeline*, has been manually defined that applies filtering procedures depending on the question type. So, for example, it does not apply *PHTL* when the question type is *Person*.

On top of this configuration, another one, named *New Pipeline + SWR*, has been set, which adds the *Stop Word Removal* procedure to the pipeline. The three configurations *Old Pipeline*, *New Pipeline*, and *New Pipeline + SWR* are compared and the results are listed in Table 2.

Table 2. Accuracy@1 for different configurations of the QA pipeline.

Transcriber	Date		Location		Description		Artifact		Person	
	Size	*Acc@1*	*Size*	*Acc@1*	*Size*	*Acc@1*	*Size*	*Acc@1*	*Size*	*Acc@1*
Old Pipeline	260	60.38	281	84.70	34	-	217	62.21	115	82.61
New Pipeline		77.31		83.98		97.06		80.18		88.69
New Pipeline + SWR		78.85		92.88		97.06		80.18		88.69

It is worth noting that the *Old Pipeline* configuration cannot be applied to questions whose type is *Description*, since the previous version of the pipeline did not address the question type *Description*. The *New Pipeline* configuration has a positive impact on three question types, showing an average improvement of 10.06% with a peak of 17.97% (*Artifact*). For *Location*, the Accuracy@1 drops of 0.72%. The *New Pipeline + SWR* configuration shows an average improvement of 2.61% with a minimum of 1.54% (*Date*) till an 8.9% (*Location*). This configuration boosts the system performance, achieving an average improvement, in comparison to the *Old Pipeline*, of 12.67% with a minimum of 6.08% (*Artifact*) and a maximum of 18.47% (*Date*). It is worth highlighting that the question type *Description* is not taken into account in computing the average improvements since the filtering procedures are not used for it.

5   Conclusion

The paper presented a Question Answering pipeline for Italian and based on a corpus of documents pertaining a closed domain. The novelty of this pipeline essentially relies on its capacity of: (i) analyzing natural language questions in Italian by using lexical features; (ii) handling both factoid and description answer types and, depending on them, filtering contextual stop words from the questions; (iii) scoring and selecting candidate answers with respect to their type in order to determine the best one.

The proposed pipeline has been subject to an evaluation of its performance by using an Italian corpus of 16 documents crawled from the Internet and specific to a cultural heritage theme of the city of Naples. The response time is under 3 s on average, whereas the percentage of correct answers as first response is largely over 70%. These results showed the positive impact of the new procedures for *Question Processing* and *Answer Processing* on the whole pipeline.

Although the results are encouraging, there is still space for improvements. In particular, further developments will regard issues about the usage of: (i) syntactic features, by exploiting the structure of a sentence, to improve the relevance of the terms to be included in the query and the selection of answer candidates; (ii) semantic features, by exploiting Wordnet, to automatically expand the query, improve the quality of *answer filtering* and automatically build a set of stop words per LAT; (iii) a machine learning approach to adaptively combine the scores produced by the *scoring filtering* procedures and determine the best answer.

References

1. Amato, F., Moscato, F.: Exploiting cloud and workflow patterns for the analysis of composite cloud services. Future Gener. Comput. Syst. **67**, 255–265 (2017)
2. Amato, F., Moscato, F.: Pattern-based orchestration and automatic verification of composite cloud services. Comput. Electr. Eng. **56**, 842–853 (2016)

3. Amato, F., Moscato, F.: Model transformations of mapreduce design patterns for automatic development and verification. JPDC (2016)
4. Baudiš, P.: Yodaqa: a modular question answering system pipeline. In: POSTER 2015–19th International Student Conference on Electrical Engineering (2015)
5. Carvalho, G., de Matos, D.M., Rocio, V.: IdSay: question answering for Portuguese, pp. 345–352. Springer, Heidelberg (2009)
6. Damiano, E., Spinelli, R., Esposito, M., De Pietro, G.: Towards a framework for closed-domain question answering in Italian. In: Proceedings Workshop KARE 2016 (2016)
7. Feng, M., Xiang, B., Glass, M.R., Wang, L., Zhou, B.: Applying deep learning to answer selection: a study and an open task. CoRR abs/1508.01585 (2015)
8. Gallagher, S., Zadrozny, W., Shalaby, W., Avadhani, A.: Watsonsim: overview of a question answering engine. arXiv preprint arXiv:1412.0879 (2014)
9. Gondek, D., Lally, A., Kalyanpur, A., Murdock, J.W., Duboué, P.A., Zhang, L., Pan, Y., Qiu, Z., Welty, C.: A framework for merging and ranking of answers in DeepQA. IBM J. Res. Dev. **56**(3.4), 14:1 (2012)
10. Hauswald, J., Laurenzano, M.A., Zhang, Y., Yang, H., Kang, Y., Li, C., Rovinski, A., Khurana, A., Dreslinski, R.G., Mudge, T., Petrucci, V., Tang, L., Mars, J.: Designing future warehouse-scale computers for sirius, an end-to-end voice and vision personal assistant. ACM Trans. Comput. Syst. **34**(1), 2:1–2:32 (2016)
11. Kamdi, R.P., Agrawal, A.J.: Keywords based closed domain question answering system for Indian penal code sections and Indian amendment laws. Int. J. Intell. Syst. Appl. **7**(12), 57–67 (2015)
12. Lally, A., Prager, J.M., McCord, M.C., Boguraev, B.K., Patwardhan, S., Fan, J., Fodor, P., Chu-Carroll, J.: Question analysis: how Watson reads a clue. IBM J. Res. Dev. **56**(3.4), 2:1 (2012)
13. Li, T., Hao, Y., Zhu, X., Zhang, X.: A Chinese question answering system for specific domain, pp. 590–601. Springer International Publishing, Cham (2014)
14. Molino, P., Basile, P., Caputo, A., Lops, P., Semeraro, G.: Exploiting distributional semantic models in question answering. In: 2012 IEEE Sixth International Conference on Semantic Computing (ICSC), pp. 146–153. IEEE (2012)
15. Morales, A., Premtoon, V., Avery, C., Felshin, S., Katz, B.: Learning to answer questions from wikipedia infoboxes. In: Proceedings of the EMNLP 2016, Austin, Texas, USA, 1–4 November 2016, pp. 1930–1935 (2016)
16. Pipitone, A., Tirone, G., Pirrone, R.: QuASIt: a cognitive inspired approach to question answering for the Italian language, pp. 464–476. Springer (2016)
17. Prager, J., Brown, E., Coden, A., Radev, D.: Question-answering by predictive annotation. In: Proceedings of the 23rd ACM SIGIR Conference, SIGIR 2000, pp. 184–191. ACM, New York (2000)
18. Schlaefer, N., Gieselmann, P., Schaaf, T., Waibel, A.: A pattern learning approach to question answering within the ephyra framework. In: International Conference on Text, Speech and Dialogue. pp. 687–694. Springer (2006)
19. Solorio, T., Pérez-Coutino, M., Montes-y Gémez, M., Villasenor-Pineda, L., López-López, A.: A language independent method for question classification. In: Proceedings of Coling 2004, p. 1374 (2004)
20. Vargas-Vera, M., Lytras, M.D.: AQUA: a closed-domain question answering system. Inf. Syst. Manag. **27**(3), 217–225 (2010)
21. Wang, C., Kalyanpur, A., Fan, J., Boguraev, B.K., Gondek, D.: Relation extraction and scoring in DeepQA. IBM J. Res. Dev. **56**(34), 91 (2012)
22. Weis, K.: A case based reasoning approach for answer reranking in question answering. CoRR abs/1503.02917 (2015)

23. Xie, Z., Zeng, Z., Zhou, G., He, T.: Knowledge base question answering based on deep learning models, pp. 300–311. Springer (2016)
24. Yao, X., Van Durme, B., Clark, P.: Automatic coupling of answer extraction and information retrieval. In: Proceedings of ACL Short (2013)
25. Yu, L., Hermann, K.M., Blunsom, P., Pulman, S.: Deep learning for answer sentence selection. arXiv preprint arXiv:1412.1632 (2014)

Towards a Cognitive System
for the Identification of Sleep Disorders

Antonio Coronato and Giovanni Paragliola[✉]

National Research Council (CNR) Institute for High-Performance Computing
and Networking (ICAR), Napoli, Italy
giovanni.paragliola@icar.cnr.it

Abstract. Alzheimer's disease (AD) is the most common type of dementia. Patients with AD may show anomalous behaviors such as sleeping disorders. Due to the increasing attention focused on these kinds of behaviors, activities like monitoring and identification are becoming critical. In order to meet these requirements, we propose a cognitive approach based on a combination of machine learning algorithms and a prior knowledge base for the identification of anomalous behaviors during sleep. The results show an improvement in the identification of sleeping disorders of more than 10%.

1 Introduction

The aging of the world population during the last few decades has highlighted problems related to the increasing incidence of cognitive diseases such as dementia and the consequent issue of how best to assist those patients who suffer from them.

Alzheimer's disease (AD) is the most common type of dementia.

Patients with AD could show symptoms such as sleep disturbances, well-formed visual hallucinations, and muscle rigidity or other Alzheimer movement features.

Alzheimer's disease (AD) is the most common type of dementia. Patients with AD could show symptoms such as sleep disturbance, well-formed visual hallucinations, muscle rigidity or other typical Alzheimer's movement irregularities. In this work we focus on those types of disturbances related to sleep disorders (SD). The attention focused on such disorders has been increasing in the last few decades, with the cost of treatment currently estimated in the range of of 30$ to 35$ billion [1].

Due to this heightened interest, the use of information and communications technology (ICT) for the treatment of SD has been increasing in parallel in the last few years. Typical activities relating to the treatment of SD consist in the monitoring and identification of patient behaviors with a particular emphasis on anomalous behaviors.

The approaches widely used for this purpose are generally based on the monitoring of EEG/EMG signals, although other kinds of parameters have been adopted. An example is provided by Flores et al. [2].

© Springer International Publishing AG 2018
G. De Pietro et al. (eds.), *Intelligent Interactive Multimedia Systems and Services 2017*,
Smart Innovation, Systems and Technologies 76, DOI 10.1007/978-3-319-59480-4_10

In their work the authors use a motion sensor to catch significant body movements. An automatic pattern recognition system has been developed to identify periods of sleep and waking using a piezoelectric generated signal Xie et al. [3] adopt machine-learning algorithms to combine the data coming from an electrocardiograph (ECG) and the saturation of peripheral oxygen (SpO2) sensors in order to detect sleep apnea disorders while Alves de Mesquita et al. [4] present a monitoring system to recognize sleep breathing disorders by means of nasal pressure recording technique.

The acquisition and understanding of biomedical signals, such as movements and heart rate, is a core process in order to achieve satisfactory results.

In this context, the identification of anomalous behaviors is extremely challenging due to both the high variability of biomedical signals and the unpredictable nature of the anomalous behaviors.

For this reason, classic approaches like pattern-matching algorithms, such as the ones presented above, produce a high rate of false positives (FP).

In order to overcome these limitations, improve the accuracy of the classification and reduce the number of FPs, we propose a cognitive approach based on a combination of machine learning algorithms and a prior knowledge base for the identification of anomalous behaviors during sleep.

In order to validate our results, we have used a data-set of the sleeping disorders of a patient in the early stages of Alzheimer's disease.

We have defined three analysis processes, each one with a different kind of machine learning technique: *Fuzzy-Rule*, *Neural Network* (NN) and *Clustering* (C).

For each process we have evaluated the improvement achieved by including a corpus in the classification process. The corpus is defined as the knowledge base of ingested data and is used to manage codified knowledge [5]. It supports the data analysis process to gain an understanding of the data.

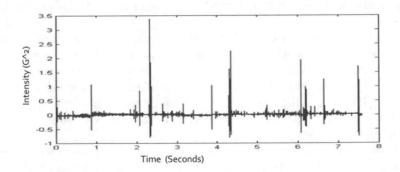

Fig. 1. Magnitude of the acceleration signal

The metric used to validate our process is the *recall*. The *Recall* is defined as the true positive rate and it is calculated as $R = TP/(TP + FP)$, where TPs are true positive.

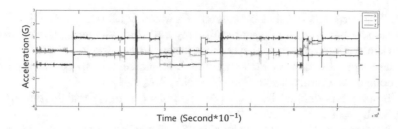

Fig. 2. Raw acceleration signals of the movements of the patient

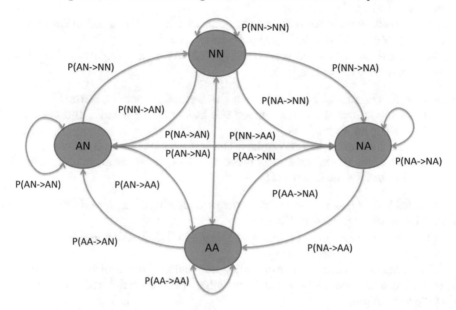

Fig. 3. View of the prior knowledge representation

The cognitive aspect of our system concerns the possibility of continuously improving the corpus by evaluating the results of the classification. This process embodies refined knowledge, in other words, the process of improving the corpus by learning new knowledge.

In this paper we focus on the reduction of the FPs by means of machine learning algorithms and we introduce the first steps to define a refined knowledge process.

The results have demonstrated that the application of the corpus in the classification process shows an increase of the recall of about 10%.

2 Prior Knowledge-Base Representation

In this section we define a knowledge base aimed at describing how a monitored patient switches from a normal to an anomalous behavior during sleep. Following the cognitive computing approach we name this base the knowledge corpus.

corpus. The corpus is defined by means of a Markov Model (MM) [6].

The model has been made by monitoring and then analyzing the movements of a patient with early stage of AD for a period of one week.

It describes how the patient switches from normal sleep to agitated sleep, and is composed of four states S = {NN, NA, AA, AN}.

Each state s is defined by a couple of values (e_i, e_{i-1}) which represent two *events*, with $i = 1....N$, where $e \in E\{N, A\}$.

Each *event* e_i is the result of the classification of a temporal window w_i that describe the patient's movements during sleep; e_i states if the movements are Anomalous (A) or Normal (N).

An anomalous movement is one characterized by a strong agitation (we have faced the problem of how to identify them in [7]).

The model describes the transition probability of moving from the state s_i to s_j, with $s \in S$, $i, j = 1, 2, 3, 4$

- $s_1 = NN$: the model has recognized the last two events as normal
- $s_2 = NA$: the model has recognized the last event as normal and the second to last event as anomalous
- $s_3 = AA$: the model has recognized the last two events as anomalous
- $s_4 = AN$: the model has recognized the last event as anomalous and the second to last event as normal.

A graphical representation of the corpus is shown in Fig. 3 and the transition probabilities are reported in Table 1.

The generic transition $P(s_i -> s_j)$, with $i, j = 1, 2, 3, 4$ defines the probability of moving from the state s_i to s_j.

The model evaluates the probability of moving from a state s_i to another state taking into account both the knowledge of the current prediction and that of the previous one.

Table 1. Transaction probabilities

-	NN	NA	AA	AN
NN	0.67	0	0	0.05
NA	0.05	0	0	0.01
AA	0	0.01	0.1	0
AN	0	0.01	0.05	0

3   Experimental Setup

The system analyzes near real-time data coming from remote sensors with the aim of recognizing anomalous movements that the patient could make while he/she is sleeping.

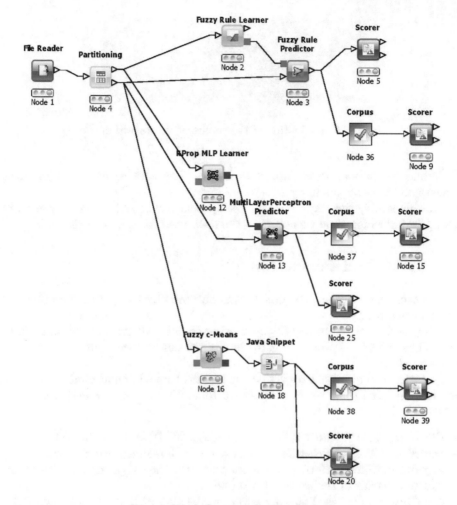

Fig. 4. Work-flows analysis processes

Clinicians can retrieve such semantic information by interacting with the system via the Graphical User Interface (GUI).

In relation to sleeping disorders, whoever is in charge of assisting the patient must set up the sensors before the monitoring commences.

As a first step, the caregiver/clinician must insert a usb/Bluetooth gateway into the pc usb port in order to receive the data stream.

Next, the patient has to put on a body sensor before he/she goes to bed, an operation that can be performed either by the patient or caregiver. When the accelerometer is turned on, it sends the data via Bluetooth.

The caregiver has to run the software application on a laptop and start the monitoring. The system collects and analyzes the gathered data until the monitoring is terminated.

	Recall Without Corpus	Recall With Corpus	Recall Improvemet
Fuzzy-Rules	0,767	0,875	13%
NN	0,70	0,75	7%
Clustering	0,80	0,85	7%

Fig. 5. Evaluation of the results of the algorithms

When the patient wakes up, the caregiver has to turn off the monitoring process and take off the body sensor.

In this study, this experimental set-up was performed repeatedly for one week, during which time we collected the data used for the analysis.

4   Results and Conclusion

In this section we present the first results obtained by applying the corpus to a classification process.

The data used relate to the monitoring of one patient for a period of one week. The aim of the process was to classify the patient's movements as Normal or Anomalous.

In order to achieve this objective, we defined three different kinds of analysis process, applying three techniques: *Fuzzy-Rule, Neural Network* and *Clustering*.

In details:

- the *Fuzzy-Rule* classifier is based on the RecBF-DDA algorithm [8].
- the *Neural-Network* classifier is based on an implementation of the RProp algorithm, which performs a local adaptation of the weight-updates according to the behavior of the error function [9].
- the *Clustering* is based on the fuzzy c-means algorithm is a well-known unsupervised learning technique that can be used to reveal the underlying structure of the data. Fuzzy clustering allows each data point to belong to several clusters, with a degree of membership to each one.

In our scenario the input is a vector of features f_i extracted from each temporal window w_i with $i = 1...N$.

For each window w_i, the extracted features are: the *mean, standard deviation* and the difference $|mean(w_i) - mean(w_{i-1})|$.

The windows w_i have a temporal duration of one minute and are extracted from the row signals (x,y,z) relating to the patient's movements (Fig. 2). For each w_i we calculate the magnitude of the signal: $mag = \sqrt{x^2 + y^2 + z^2}$.

In Fig. 4 it is possible to note the work-flow of the analysis process with the selected algorithms and the corpus. The corpus takes in input the results of the classification of the algorithms and enhances the classification results.

The use of the corpus in this work aims to support the classification process to get a more confident decision when the algorithm is in *doubts* to which class assigns to the window w_i.

From our point of view a *doubt* is a situation in which the difference between the probabilities of assignment w_i to a class is less than a threshold Th, with $Th = 0.10$.

In other words, $|P(Pred_{i+1} = N) - P(Pred_{i+1} = A)| < Th$ where $Pred_{i+1}$ is the predicted class for the temporal window w_{i+1}.

The use of the transition probabilities takes into account the history of patient's behavior in terms of both the last and second to lost classified movement.

In our experiments we wanted to evaluate the improvement that the corpus brings to the classification process with a focus on the reducing of FPs.

We summarize the results in Fig. 5.

As it is possible to note, the three classification processes produce a recall value ranging from {0.70–0.80} Fig. 5. With our approach, we demonstrate that the use of the corpus in the classification gets a value of the recall ranging from {0.75–0.87}.

The improvement of the recall ranges from 7% for the clustering and neural network to 13% for the fuzzy-rule.

We are aware that our study concerns just one patient but we are confident that as the number of patients increases, so the performance will improve.

Finally, it is important to discuss the cognitive aspect of our system.

As we have presented in this work, for each w_i the process generates a hypothesis for the windows concerning the class that it belongs to, {Normal or Anomalous}. Each of these is validated by an expert in order to assess if the system has correctly classified the w_i.

At the conclusion of the monitoring, all the hypotheses are used to tune the probabilities of the corpus, in order to build a process of refined learning which lends to the improvement of the corpus.

As future works, we are planing to increase the number of patients in order to get a huge data-set and investigate how the refining process improve the performance of the classification process.

5   Funding

This work has been funded by the eHealthNet Project.

Acknowledgments. The authors of this paper would like to thanks Giuseppe Trerotola and Raffaele Mattiello for the technical support.

References

1. Hossain, J., Shapiro, C.: The prevalence, cost implications, and management of sleep disorders: an overview. Sleep Breath. **6**(2), 85–102 (2002). doi:10.1007/s11325-002-0085-1
2. Flores, A., Flores, J., Deshpande, H., Picazo, J., Xie, X., Franken, P., Heller, H., Grahn, D., O'Hara, B.: Pattern recognition of sleep in rodents using piezoelectric signals generated by gross body movements. IEEE Trans. Biomed. Eng. **54**(2), 225–233 (2007)
3. Xie, B., Minn, H.: Real-time sleep apnea detection by classifier combination. IEEE Trans. Inf Technol. Biomed. **16**(3), 469–477 (2012)
4. de Mesquita, J.A., dc Melo, P.L.: Respiratory monitoring system based on the nasal pressure technique for the analysis of sleep breathing disorders: reduction of static and dynamic errors, and comparisons with thermistors and pneumotachographs. Rev. Sci. Instrum. **75**(3), 760–767 (2004). doi:10.1063/1.1646734
5. Hurwitz, J., Kaufman, M., Bowles, A.: Cognitive Computing and Big Data Analytics, 1st edn. Wiley Publishing, New York (2015)
6. Welton, N.J., Ades, A.E.: Estimation of Markov chain transition probabilities and rates from fully and partially observed data: uncertainty propagation, evidence synthesis, and model calibration. Med. Decis. Making **25**(6), 633–645 (2005). doi:10.1177/0272989X05282637. pMID: 16282214
7. Coronato, A., Paragliola, G.: An approach for the evaluation of sleeping behaviors disorders in patients with cognitive diseases: a case study (2016)
8. Gabriel, T.R., Berthold, M.R.: Influence of fuzzy norms, other heuristics on "mixed fuzzy rule formation". Int. J. Approximate Reasoning **35**(2), 195–202 (2004). http://www.sciencedirect.com/science/article/pii/S0888613X03001427
9. Riedmiller, M., Braun, H.: A direct adaptive method for faster backpropagation learning: the RPROP algorithm. In: IEEE International Conference on Neural Networks, vol. 1, pp. 586–591 (1993)

An Ensemble Classifiers Approach for Emotion Classification

Mohamed Walid Chaibi[(✉)]

High Institute of Management of Tunis (ISG-Tunis), Tunis, Tunisia
chaibi.walid@gmail.com

Abstract. Decoding the emotional state of a person has a variety of applications. It could be used in human-computer interaction (HCI) or like follow-ups in the therapeutic techniques. Recently, emotion recognition is one of topic that researchers are most interested in and until now, there are several studies relating to the emotion using devices and techniques. To recognize human emotions, various physiological signals have been widely used. In this research, we propose a novel approach for the emotion classification using several physiological signals to classify eight emotions according to the Clynes sentograph protocol of Manfred Clynes. The study has two main objectives. On the one hand a comparative study to choose the best classifiers that addresses the emotion classification problem. And On the other hand to develop an ensemble classifiers approach.

Keywords: Ensemble classifiers · Emotion classification · Physiological signals

1 Introduction

The emotional field of daily life is vast but difficult to define. The study of emotions is a multidisciplinary phenomenon. Affective science which is the scientific study of affect (emotion) was very studied and treated by several researchers. Emotion classification is a contested issue in emotion research and affective science. It was the means by which one emotion is distinguished from another and it allows to identify the feelings of individuals toward specific events. Recently, the analysis of the feelings and the classification by emotion remain an object of ambitious research and until now, there are several studies relating to the emotion using devices and techniques. To recognize human emotions and feelings, various physiological signals have been widely used. In this research, we propose a novel approach for the emotion classification using several physiological signals and the aim consist to achieve a contribution to the work that was done in this field. To do that, we propose an ensemble classifiers method based on the results of eight classifiers. To use and evaluate the effectiveness of statistical learning methods for emotion classification, we need both a training dataset and a testing dataset. In our approach we use a dataset which are generated and tested in many experiences.

© Springer International Publishing AG 2018
G. De Pietro et al. (eds.), *Intelligent Interactive Multimedia Systems and Services 2017*,
Smart Innovation, Systems and Technologies 76, DOI 10.1007/978-3-319-59480-4_11

2   State of the Art

In the literature, there are several emotion theories based on different classification method and many methods for estimating human emotion were proposed.

2.1   Background

Emotion is any conscious experience characterized by a degree of pleasure or displeasure and a mental activity [1]. Some researchers have acknowledged the existence of a limited number of universal fundamental emotions (basic emotion) such as Ekman, Izard.. [2,3] which would then have each an evolutionary function. The more complex emotions would come from a mixture of these basic emotions [4]. The evolutionary current, in psychology of the emotions, draws its origin from the work of Charles Darwin [5]. The theory of the physiologist Walter Cannon shows the relation between the nervous system and the emotions [6]. And the concept of basic emotions appears at Descartes [7]. According to Ekman an emotion is short-lived [2]. Robert Plutchik, an American psychologist, invented the wheel of emotions which defines relationships and combinations between emotions [8]. The concept of emotional granularity is developed by Lisa Feldman Barrett by using the circumplex valence/Arousal (excitation) [9]. The study of music and emotion is a branch of music psychology which consists to understand the psychological relationship between emotion and music [10].

The emotion includes three components: the cognitive component refers to the subjective component of emotional experience. The behavioral component refers to all the expressive manifestations of emotion. And the physiological component based on the set of physiologic concomitants (body temperature, blood pressure, the concentration levels of certain hormones or neurotransmitters..)

2.2   Related Work

Emotion Classification: There are many methods for emotions classification. Conventional methods used mainly audiovisual attributes (speech, facial expressions). Thus, other approaches focus on different physiological information (heart rate, skin conductance, the pupil dilation...) and the signals captured by the brain have been used such as EEG signals. Microblogging is taken as a rich resource of emotion data and social networks has attracted lots of attention of researches for emotion classification [11]. Understanding facial expression is one of the challenging tasks for recognizing a person's emotion state [12,13]. Among the application of emotion classification, the estimation of the speaker's emotion and many studies show that there is a strong correlation between emotional states and glottal features [14]. Also there are a work based on both speech and text [15]. There are other methods more accurate using physiological signals [16–19].

Ensemble Method for Emotion Classification: To improve the classification of emotions, and give more precise results, many researchers are based on combination of several classifiers. Ensemble classifiers also known as ensemble learning are a new concept in machine learning which is a set of classifiers whose individual predictions are combined [20]. In fact, the ensemble classifiers outperform every single classifier by improving the results of prediction and reducing the error rates [21]. There are several methods to generate ensemble classifiers classified into two main families: averaging methods [22] and boosting methods [20]. There are several methods for emotions classification based on ensemble classifiers [23]. Some method treaty the problem of Brain Computer Interface (BCI) [24], other methods deals whith speech [25], written text [26] and facial expression [27–29].

3   Contribution

There are several studies in the literature that treated the problem of emotion classification. Each works focuses on a particular part of the problem and a particular component of emotion. Among the researches there are work based on a particular signal especially EEG signals. Other work deals with a limited number of emotions. While these methods do not give accurate results as the methods using physiological signals. We also noticed that the majority of methods use a single classifier which gives mostly high error rate. Our solution offers an ensemble method for classifying eight emotions (neutral, anger, hate, grief, love, romantic love, joy, and reverence). The proposed method uses several classifiers (J48, BFTree, Kstar, LMT, NBTree, Part, Random Tree and Ridor) and four physiological signals (the GSR, the BVP heart rate, the breathing and the EMG) to reduce the error rate and gives accurate results at the same time.

To study the eight emotion states from the Clynes sentograph protocol, we use a large collection of recorded physiologic signals: the electromyograph (EMG) measures muscle activity, respiration which has a big relation with emotional state, Blood Volume Pressure (BVP) Heart Rate used to measure heart activity and skin conductance (GSR).

3.1   Choice of Classifiers

Before using ensemble classifiers, we need to fix the choice of classifier, choose the best ones uses in emotions classification and we experiment them with our DataSet. First, we recite classifiers that address the emotion classification problem and determine the best ones. Once the selection is done, we experiment them with our DataSet to know which classifier is the best for our ensemble classifiers approach. To determine this later, we use Percentage of Correct Classification (PCC) which is a measure calculates the percentage of correctly classified items and the associated standard deviation from a confusion matrix.

$$PCC = \frac{Instance\ correctly\ classified}{all\ instances} * 100 \tag{1}$$

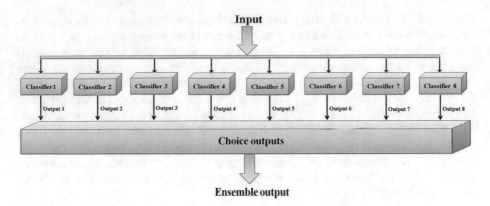

Fig. 1. Concept of ensemble classifiers for emotion classification based on eight classifiers

The ensemble Classifiers combines five classifiers which are Random Tree, J48, NBTree, Part and BFTree and the results were compared to these classifiers and to three other classifiers (Ridor, LMT and Kstar) (Fig. 1).

Random Tree: This algorithm proposed by Breiman [30] combines the concepts of random subspace and bagging. The concept of this algorithm consists to train several decision trees on different subsets of data.

J48: J48 is an open source Java implementation of the C4.5 algorithm [31] in the Weka data mining tool. The concept of J48 consists to identify the attribute that discriminates the various instances most clearly and assign to it the obtained target value.

NBTree: The NBTree is an algorithm presented in the form of tree constructed recursively. It is a hybrid of the Naïve Bayes [32] and the Decision Tree algorithm [31]. It consist to approximate the accuracy of Naive Bayes.

Part: This algorithm have a neural network architecture for clustering high dimensional data in low dimensional subspaces. [33] The key feature of the PART network consist to transmit signal to the cluster neuron only when the similarity measure is sufficiently large.

BFTree: BFtree constructs binary trees which each internal node has exactly two outgoing edges. The resulting tree will be the same but the order in which it is built is different [34]. The best node is the one whose split leads to maximum reduction of impurity.

Ridor: Ripple Down Rule learner (RIDOR) is a direct classification method used to create exceptions. This algorithm consist to construct a set of rule and use an incremental reduced error pruning to find exceptions with the smallest error rate [35].

LMT: Logistic Model Tree (LMT) is a classification model with an supervised training algorithm that combines logistic regression (LR) and decision tree learning [36]. The concept consist to produce a single tree containing binary splits on numeric attributes.

Kstar: Kstar is an instance-based classifier based on the class of training instances similar to it which are determined by a similarity function and an entropy-based distance function [37].

We chose the best classifiers found in the literature that addresses the emotion classification problem and we propose to combine this classifiers to grow their complementarity.

3.2   Algorithm of Ensemble Classifiers for Emotions Classification

```
Multi-Classifier Algorithm
1:  Begin:
2:  i: for each classifier
3     if result of classifier= emotion affected to classifier
4:        results[i]:=1
5:     else
6:        results[i]:=0
7:     end if
8:  end for
9:  NbrTrue:=SUM(results)
10: if NbrTrue=0
11:    emotion:=resultof classifier with the highest PCC of all classifiers
12: else if NbrTrue=2
13:    emotion:=result of classifier with the highest PCC of selected classifiers
14: else if NbreTrue=1
15    emotion:=result of classifier
16: end if
17: End
```

The ensemble classifiers takes as input the DataSet and all classifier uses the same input. At first, the ensemble classifiers takes the results of classifiers which correspond to their assignments. If the ensemble classifiers don't have result which corresponds to assignments of classifier, it gives the result of classifier which has the highest PCC over all emotions. And if the ensemble classifiers have two or more results that correspond to assignments of classifiers, it gives the resultof classifier that has the highest PCC of his emotion.

3.3   Concepts of the Approach

After choosing the classifiers, we combine them into the ensemble classifiers which acting as a coordinator between them. All classifier deploy the same input data and each classifier assigns ranks for his classification to each emotion. Each classifier provides a probability which presents a confidence for each class. The approach aims to take the best classification of each classifier for a common purpose which is the classification of emotional states. We measure the percentage of correct classification of each classifier for each emotion. Thus, we have assigned each emotion to classifier that classify better, so the ensemble classifiers approach takes in consideration the result of classifier which is identical to the emotion of his assignment. The ensemble classifiers assigns an emotion to each classifier. The assignment of classifier is for the classifier which classifies better the emotion depending on its performance.

Priority per Emotion: The priority depends on the PCC of the classifier. The classifier that has the highest PCC per emotion will have the highest priority. The presumption in the selection of classifiers is that each classifier is effective on some emotions in the classification of emotions. When one wishes to issue a decision on data, looking at all responses classifiers and that giving highest PCC is selected to assign him a class. By cons, if two or more classifiers of approach that resulted in the emotion with their assignment, the approach chosen the result of classifier that has the highest PCC of his emotion. The idea behind ensemble classifiers is to increase the PCC with the complementarity of classifiers for that classifiers must operate with different structures. First, the result of classifiers is compared with its reliability of its class. Secondly, a comparison of the results is made to designate the class that best represents the data. Finally, the result of the classifier that classifies the best is taken as the final result. The execution order of the classifiers is not involved in this approach. In this approach, it requires that every classifier provide a confidence (probability) associated with each class. The proposed approach assumes that the results of classifier are reliable and accurate for only some class.

Priority Overall (Average): Based on the results of each classifier, the ensemble classifiers takes the result of the one which classifies better all emotions. In fact, the priority is for the classifier which has the highest PCC in all emotions (PCC average). If the results of each classifier do not provide emotion that corresponds to his assignment, the approach takes the result of classifier that has the highest PCC for all emotions. In this case, we do not have results which corresponding to assignments of classifiers. Ensemble classifiers take the result of classifier which has the highest priority as output. The priority depends on the PCC of the classifier. The classifier which has the highest PCC overall emotions will have the highest priority.

4   Experimentation

The data was generated by the MIT Affective Computing Group's research which is a researcher group in affective computing from the Massachusetts Institute of Technology of Cambridge. The measurements were taken once a day, of a single individual which placed comfortably apart, for over twenty days of recordings in an office setting. At each experimental session, approximately 25 min of data was recorded [38].

4.1   Implementation

In order to predict emotion of the subject, many classifiers were applied using the Wekaprogram. In fact WEKA is a data mining system developed by the University of Waikato in New Zealand that implements data mining algorithms (classification, regression, and clustering..) using the JAVA language [39]. The DataSet were input to Weka and eight classifiers were applied using 10fold cross validation. In fact various classifiers are already implemented by Weka. Then we implement the ensemble classifiers by writing a java code in weka which combine and coordinate the process of classification made by the used classifiers. We used the PCC and precision to measure the data correctly classified of the output of each classifier relative to the data. The results of ensemble classifiers were compared to the five classifiers used in our approach (Random Tree, J48, NBTree, Part, BFTree) and three other classifier (Ridor, LMT and Kstar).

Table 1. PCC results for each emotion and for each classifier

	Random tree	J48	NBTree	Part	BFTree	Ensemble classifiers
No-emotion	0.961	0.958	0.955	0.954	0.96	0.976
Anger	0.89	0.908	0.889	0.899	0.89	0.892
Hate	0.95	0.947	0.923	0.941	0.946	0.942
Grief	0.869	0.876	0.86	0.847	0.871	0.892
p-love	0.833	0.869	0.887	0.881	0.837	0.886
r-love	0.896	0.884	0.856	0.824	0.886	0.885
Joy	0.929	0.947	0.955	0.942	0.942	0.963
Reverence	0.868	0.838	0.785	0.809	0.846	0.828
Weighted Avg	0.9	0.904	0.889	0.887	0.898	0.908

The obtained accuracy results especially the PCC of the proposed approach was compared to the PCC of five classifiers used in the ensemble classifiers approach and the results are shown the Table 1. In fact for the majority of emotions, the PCC of ensemble Classifiers compared to the PCC of the five classifiers is higher. Especially the weighted average of PCC is higher for the ensemble classifiers which is 0.908.

Table 2. Precision results for each emotion and for each classifier

	Random tree	J48	NBTree	Part	BFTree	Kstar	Ridor	LMT	Ensemble classifiers
No-emotion	0.959	0.953	0.33	0.937	0.951	0.886	0.902	0.952	0.948
Anger	0.9	0.898	0.858	0.864	0.9	0.848	0.852	0.894	0.15
Hate	0.949	0.948	0.955	0.936	0.945	0.86	0.939	0.946	0.966
Grief	0.869	0.869	0.867	0.87	0.859	0.867	0.854	0.867	0.864
p-love	0.827	0.816	0.793	0.787	0.827	0.727	0.786	0.823	0.816
r-love	0.903	0.936	0.924	0.931	0.919	0.899	0.95	0.921	0.938
Joy	0.923	0.915	0.901	0.896	0.912	0.847	0.875	0.914	0.908
Reverence	0.866	0.899	0.896	0.892	0.867	0.84	0.873	0.865	0.922
Weighted Avg	0.9	0.904	0.891	0.889	0.898	0.847	0.879	0.902	0.91

The precision result of the ensemble classifiers approach was also compared to the five classifiers used (Random Tree, J48, BNTree, Part) and to three other classifiers (Ridor, LMT and Kstar) and the results are shown the Table 2. The table shows that ensemble classifiers give more precision results that the other classifiers for the majority of emotions and especially the weighted average which is 0.91. This two tables show that the proposed ensemble classifiers gives more precise results for the emotions classification. Indeed, on average the PCC and the precision results of ensemble classifiers is higher compared to the other classifiers.

4.2 Discussion

The experimentation presents an ensemble classifiers approach and ultimately, weaker classifiers when joint can create an ensemble system which increases precision. One of the key issues of this approach is how to combine the results of the various classifiers to give the best estimate of the optimal result. The presumption is that each classifier is an expert in some local area of the feature space. Five classifiers have been deployed and each of them was associated with some emotions. Such classifiers provide a degree of certainty for each classification based on the statistics for each emotion. The outputs of each classifiers are combined to produce a total output for the approach. An algorithm that combined the results of classifiers has been utilized. The architecture of the approach is based on parallel combination of classifier which works independently of the other. The result of the ensemble classifiers approach compared to the results of the eight used classifiers were encouraging. The classification accuracy of the ensemble classifiers is 91 More specifically, out of 30141, only 2712 were incorrectly classified. Decision trees used in our experimentation are BFTree, J48, NBTree, LMT and RandomTree. In this experiment, this decision trees showed their performance for the classification of emotions compared to other classifiers. We can conclude that decision trees are effective for classification of our data.

5   Conclusion

In this research, we are interested in the emotion classification problem. This problem is crucial to understanding the human behavior. Our work was inspired by a preliminary study of existing works which are based mainly on the physiological component. We propose a contribution to the works that were done and we aim a comparative study to several emotion classifiers while basing on physiological signals. And we have developed an ensemble classifiers approach for decoding eigth emotional states emotions according to the Clynes sentograph protocol. For that, we chose a dataset named Eight-Emotion Sentics data which provides the physiological signals of individuals in a reliable, professional and scientific way. Then we chose the best classifiers found in the literature that addresses the emotion classification problem and we propose to combine these classifiers to grow their complementarity. The results from this study indicate that our ensemble classifiers approach to the classification of a complex dataset of emotions can be achieved with a high degree of accuracy. And the combination of many classifiers can increase the accuracy rate.

References

1. Schacter, D.: Psychology, 2nd edn, p. 310. Worth Publishers, New York (2011). ISBN
2. Ekman, P.: An argument for basic emotions. Cogn. Emot. **6**(3–4), 169–200 (1992)
3. Izard, C.: Human Emotions. Plenum Press, New York (1977/2001)
4. Ortony, A., Turner, T.J.: What's basic about basic emotions? Psychol. Rev. **97**, 315–331 (1990)
5. Darwin, C., Ekman, P., Prodger, P.: The Expression of the Emotions in Man and Animals. Oxford University Press, Oxford (1998)
6. Cannon, W.B.: The james-lange theory of emotions: A critical examination and an alternative theory. Am. J. Psychol. **39**, 106–124 (1927)
7. Damasio, A., Sutherland, S.: Descartes' error: Emotion, reason and the human brain. Nature **372**, 287–287 (1994)
8. Plutchik, R.: The Emotions. University Press of America, New York (1991)
9. Lisa, F.: Valence focus and arousal focus: Individual differences in the structure of affective experience. J. Pers. Soc. Psychol. **69**(1), 153–166 (1995)
10. Clynes, M.: Sentography: Dynamic forms of communication of emotion and qualities. Comput. Biol. Med. **3**, 119–130 (1972)
11. Li, W., Xu, H.: Text-based emotion classification using emotion cause extraction. Expert Syst. Appl. **41**(4), 1742–1749 (2014)
12. Sheng, H.K., Mohamed, A.: 3D HMM-based facial expression recognition using histogram of oriented optical flow (2015)
13. Hongying, M., Nadia, B., Yangdong, D., Jinkuang, C.: Time-delay neural network for continuous emotional dimension prediction from (2015)
14. Muthusamy, H., Polat, K., Yaacob, S.: Improved emotion recognition using gaussian mixture model and extreme learning machine in speech and glottal signals. Math. Probl. Eng. **2015**, 13 (2015). doi:10.1155/2015/394083. Article ID 394083
15. Bhaskar, J., Sruthi, K., Nedungadi, P.: Hybrid approach for emotion classification of audio conversation based on text and speech mining. Procedia Comput. Sci. **46**, 635–643 (2015)

16. Park, B.J., Jang, E.H., Kim, S.H., Huh, C., Chung, M.A.: The design of fuzzy C-means clustering based neural networks for emotion classification. In IFSA World Congress and NAFIPS Annual Meeting, pp. 413–417 (2013)
17. Zheng, L., Zhu, Y., Peng, Y., Lu, L.: EEG-based emotion classification using deep belief networks. In: IEEE International Conference on Multimedia and Expo (2014)
18. Verma, G.K., Tiwary, U.S.: Multimodal fusion framework: A multiresolution approach for emotion classification and recognition from physiological signals. NeuroImage **102**, 162–172 (2014)
19. Lokannavar, S., Lahane, P., Gangurde, A., Chidre, P.: Emotion recognition using EEG signals. Emotion 4(5) (2015)
20. Rokach, L.: Ensemble-based classifiers. Artif. Intell. Rev. **33**(1), 1–39 (2010)
21. Dietterich, T.G.: Ensemble methods in machine learning. In International Workshop on Multiple Classifier Systems, pp. 1–15 (2000)
22. Opitz, D., Maclin, R.: Popular ensemble methods: An empirical study. J. Artif. Intell. Res. **11**, 169–198 (1999)
23. Vaish, A., Kumari, P.: A comparative study on machine learning algorithms in emotion state recognition using ECG. In: Proceedings of the Second International Conference on Soft Computing for Problem Solving (2014)
24. Mano, L.Y., Giancristofaro, G.T., Faical, B.S., Libralon, G.L., Pessin, G., Gomes, P.H., Ueyama, J.: Exploiting the use of ensemble classifiers to enhance the precision of user's emotion classification. In: International Conference on Engineering Applications Of Neural Networks (INNS) (2015)
25. Sun, Y., Wen, G.: Ensemble softmax regression model for speech emotion recognition. Multimed.Tools Appl. **76**(6), 1–24 (2016)
26. Perikos, I., Hatzilygeroudis, I.: Recognizing emotions in text using ensemble of classifiers. Eng. Appl. Artif. Intell. **51**, 191–201 (2016)
27. Rani, P.I., Muneeswaran, K.: Recognize the facial emotion in video sequences using eye and mouth temporal gabor features. Multim. Tools Appl. **76**(7), 1–24 (2016)
28. Jain, S., Durgesh, M., Ramesh, T.: Facial expression recognition using variants of LBP and classifier fusion. In: Proceedings of International Conference on ICT for Sustainable Development, pp. 725–732 (2016)
29. Neoh, S.C., Mistry, K., Zhang, L., Lim, C.P., Fielding, B.: A micro-GA embedded PSO feature selection approach to intelligent facial emotion recognition (2016)
30. Breiman, L.: Random forests. Mach. Learn. **45**, 5–32 (2001)
31. Quinlan, J.R.: C4.5: Programs for Machine Learning. Morgan Kaufmann Publishers, San Francisco (1993)
32. Langley, P., Iba, W., Thompson, K.: An analysis of bayesian classifiers. In: National Conference on Artificial Intelligence, pp. 223–228 (1992)
33. Cao, Y., Wu, J.: Projective ART for clustering data sets in high dimensional spaces. Neural Netw. **15**(1), 105–120 (2002)
34. Haijian, S.: Best-first decision tree learning. Hamilton, NZ (2007)
35. Gaines, B.R., Compton, J.P.: Induction of ripple-down rules applied to modeling large databases. Intell. Inf. Syst. **5**(3), 211–228 (1995)
36. Landwehr, N., Hall, M., Frank, E.: Logistic model trees. Mach. Learn. **59**, 161 (2005)
37. Cleary, J.G., Trigg, L.E.: K*: An instance-based learner using an entropic distance measure. In: 12th International Conference on Machine Learning (1995)
38. Healey, J., Picard, R.W.: Eight-emotion sentics data (2002). http://affect.media.mit.edu. Retrieved 12 Aug 2016
39. http://www.cs.waikato.ac.nz/ml/weka/

Sign Languages Recognition
Based on Neural Network Architecture

Manuele Palmeri[2]($\boxtimes$), Filippo Vella[1], Ignazio Infantino[1],
and Salvatore Gaglio[1,2]

[1] Italian National Research Council of Italy, ICAR-CNR, Palermo, Italy
{filippo.vella,ignazio.infantino}@pa.icar.cnr.it
[2] DIID, University of Palermo, Palermo, Italy
manuele.palmeri@community.unipa.it, salvatore.gaglio@unipa.it

Abstract. In the last years, many steps forward have been made in speech and natural languages recognition and were developed many virtual assistants such as Apple's Siri, Google Now and Microsoft Cortana. Unfortunately, not everyone can use voice to communicate to other people and digital devices. Our system is a first step for extending the possibility of using virtual assistants to speech impaired people by providing an artificial sign languages recognition based on neural network architecture.

Keywords: Sign languages · ASL · American sign language · Sign recognition · Kinect · Recurrent neural network · RNN · Deep learning

1 Introduction

Sign languages are complete and complex languages that utilize the movements of the hands combined with facial expressions and postures of the body instead of the normal oral-type communication. They are based on the idea that vision and gestures are the main tools that speech impaired people have to communicate with other people. Sign languages should not be confused with body language, which is an involuntary non-verbal communication form that describes emotions and mood of the speaker and it is not an actual, complete language system.

Although spoken and sign languages have a lot in common, there are also a lot of differences [1]. There are usually no correspondences between signs and spoken languages used in the same geographical area.

Sign speakers can easily communicate with each other using hand gestures. However, communicating with deaf people is still a problem for non-sign-language speakers. Therefore, an automatic sign recognition system is desirable in all the occasions in which a human interpreter is not available. Besides, an artificial system that understands sign languages represents a huge step forward in Human-Computer Interaction (HCI), which is more and more important in our daily lives. This paper propose a system based on Recurrent Neural Networks that understand and classify human body movements, in particular signs

© Springer International Publishing AG 2018
G. De Pietro et al. (eds.), *Intelligent Interactive Multimedia Systems and Services 2017*,
Smart Innovation, Systems and Technologies 76, DOI 10.1007/978-3-319-59480-4_12

of sign languages. Our network utilizes only 3D joints position of human skeleton as described in Sect. 4.4, so it is very fast and light. The joints position was extracted with the RGBD camera Microsoft Kinect v2. More information about this camera and the joints position extraction process can be found in Sect. 4.2.

The paper is organized as follows: In Sect. 2 we discuss related work; In Sect. 3 is present a brief introduction of recurrent neural networks; In Sect. 4 is present the description of our system with all the details about our networks, dataset and results; In Sect. 5 is present the conclusion and future works.

2   Related Works

There are many works and research in sign language recognition. Most of them utilize a computer-vision approach and 3D convolutional neural networks [2], that are able to automate the process of feature construction. Sometimes are also used colored or sensory gloves to increase the accuracy of hand detection [3].

Another interesting method is given by [4], utilizing sub-units to detect sign languages. The results are obtained by comparing a hidden Markov model system with sequential pattern boosting.

In [5] faces the problem to detect signs from a dataset of images (ASLID dataset[1]), using a separate pose estimator and hand detection. This paper evaluates the performance of two deep learning based pose estimation methods, by performing user-independent experiments on the dataset. They also performed transfer learning, obtained results that demonstrate that transfer learning can improve pose estimation accuracy.

We think that our system should not be considered necessarily an alternative to those systems, but joint position systems and computer vision systems can be used together, at different levels, to achieve better results.

3   Recurrent Neural Networks

Recurrent Neural Networks [6] (RNNs) are a particular class of Neural Networks that contains directed cycles between units. With this architecture, neurons are allowed to be connected to themselves. This cycle creates an internal state of the network which allows holding information during a time period. This behavior is inspired on how human brain works: people don't start their thinking from scratch every second. Traditional Neural Networks don't have this type of persistence, so they can't be used in all occasion where temporal dependencies are required.

However, they aren't all that different than a standard neural network. A recurrent neural network can be thought of as a set of multiple copies of the same network, each passing a message to a successor as shown in Fig. 1.

---

[1] American Sign Language Image Dataset: http://vlm1.uta.edu/%7Esrujana/ASLID/ASL_Image_Dataset.html.

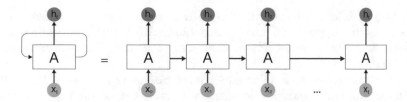

Fig. 1. An unrolled recurrent neural network.

In the last few years, there have been incredible success applying RNNs to a variety of problems such as handwriting recognition [7] and speech recognition [8]. A conventional RNN is constructed by defining the transition function and the output function as:

$$h_t = \phi_h(\boldsymbol{W}^T h_{t-1} + \boldsymbol{U}^T x_t) \tag{1}$$

$$y_t = \phi_o(\boldsymbol{V}^T h_t) \tag{2}$$

where ϕ_h, ϕ_o, x_t, y_t and h_t are respectively a state transition and output function, an input, an output, a hidden state, and $\boldsymbol{W}$, $\boldsymbol{U}$ and $\boldsymbol{V}$ are the transition, input and output matrices.

Unfortunately, as described in depth in [9], traditional RNN has a problem with "long-term dependencies". To solve this issue, we use Long Short Term Memory networks (LSTM) [10], a variant of RNN that was explicitly designed to avoid the long-term dependency problem.

4   Proposed System

4.1   Overview

The aim of the proposed system is to classify and recognize signs from American Sign Language (ASL) using a low-cost RGBD camera, which is Microsoft's Kinect v2. The system uses a Recurrent Neural Network (RNN) Architecture with a dataset created using the Kinect for Windows SDK 2.0². The details about the technology, dataset, and Neural Networks will be explained in the following chapters.

4.2   Microsoft Kinect

Microsoft Kinect v2 is a low-cost RGBD camera released in 2014. The second generation of Kinect sensor offers a higher resolution and a wider field of view compared to the first generation of Kinect.

Microsoft Kinect comes with official Kinect for Windows SDK 2.0, which include a huge set of functionality like human body skeletal tracker. The second

---

² Kinect for Windows SDK 2.0: https://www.microsoft.com/en-us/download/details.aspx?id=44561.

generation of Kinect allows much more tracking accuracy than the first generation, and it supports 25 body joints (20 in Kinect v1). All joints position coordinates are in Camera Space, as described in the Microsoft official documentation[3].

A disadvantage is that Microsoft Kinect cannot track fingers but only the position of the hand and the thumbs, so it could not be used as a standalone to track every sign of ASL. Another disadvantage of Microsoft Kinect is that the tracker sometimes captures unnatural and discontinuous motions when self-occlusions occur, but for simple movements like sign languages is not a big problem. However, this issue could be fixed using multiple Kinect sensors in a workspace or using other approaches such as [11].

In addition to skeleton tracking, Microsoft Kinect is used in research for many other purposes. In [12], for example, it is used as a non-invasive method to detect gaze movements on the monitor.

4.3   Neural Network Details

As mentioned above, we created a Recurrent Neural Network with LSTM layers with the high-level Deep Learning library Keras using TensorFlow back end.

Keras is an open source high-level Python deep neural networks library that runs on top of either TensorFlow or Theano. It was developed as part of the project ONEIROS (Open-ended Neuro-Electronic Intelligent Robot Operating System), and the original author is François Chollet. The library contains numerous implementations of commonly used neural network *building blocks* such as layers, objectives, activation functions, optimizers, and a lot of useful tools to work with image and text data. We decided to use Keras because it is fast and easy to use. As described in the Official Keras documentation[4], some other advantages are modularity, minimalism, and extensibility.

For our network, we used two LSTM layers with 128 hidden units each and a Dense layer with 20 hidden units. To prevent overfitting [13] we also added dropout layers of 0.3. This layer randomly drops units and their connections from the neural network during training and prevents units from co-adapting too much. Each value was found experimentally as described in Sect. 4.4. As activation function, we used a *rectified linear unit* (ReLU) that doesn't face gradient vanishing problem as with sigmoid and *tanh* function [14,15]. Gradient vanishing problem is present when gradient at each step is too small; in that case, the weights does not change enough at each iteration, so it takes many iterations to converge.

Finally, we add a *softmax* layer as output function to ensure that the sum of the components of the output vector is equal to 1. As objective function, we used a *categorical cross entropy* (multiclass logloss) function with *Adam* optimizer with default parameters [16]. The metrics used to judge the performance of

---

[3] MSDN Library - "Coordinate mapping": https://msdn.microsoft.com/it-it/library/dn785530.aspx.

[4] Official Keras documentation: https://keras.io/.

the model was *accuracy*, which calculates the mean accuracy rate across all predictions for multiclass classification problems. A complete visual description of our network is shown in Fig. 2.

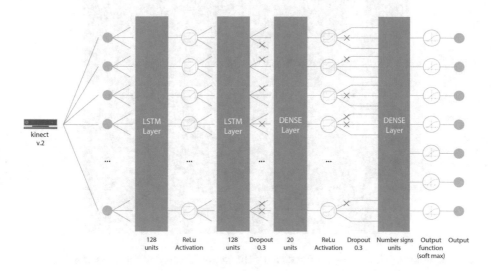

Fig. 2. Visual description of the neural network.

4.4 Dataset Details

The dataset is composed by a set of *samples*. A *sample* is a matrix of fixed length that represents a single performance of a sign movement made by a person. These samples were generated starting from CSV files created with Kinect for Windows SDK 2.0, containing all xyz coordinates of every skeletal joint (for a total of $25 \times 3 = 75$ columns). The number of rows depends on the time interval used for recording (about 30 frames per second).

The CSV files were later imported in a Python script, and we created the matrices of the dataset using two different approaches (see Sect. 4.6 for more details). We also normalized the values in $[0, 1]$, because we are more interested in the *differences* between the frames and not the *absolute* value of the joints position. An example of recording session can be found in Fig. 3.

4.5 Training Details

The parameters of the Neural Network was found experimentally maximizing the accuracy and F_1-score of the test set. First, we iterated the training process to find the best number of LSTM layers. We found that increasing the number of layers we often obtained worst accuracy, as shown in Fig. 4. Thus, we chose to use only two layers of LSTM.

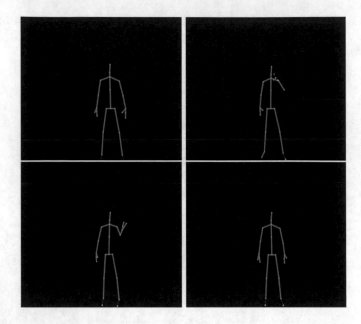

Fig. 3. Four key movements of the sign "thank you" recorded with Microsoft Kinect.

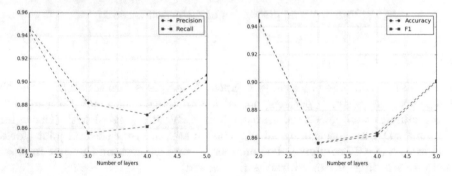

Fig. 4. Variation of matrices with numbers of LSTM layers (using arbitrary 64 units).

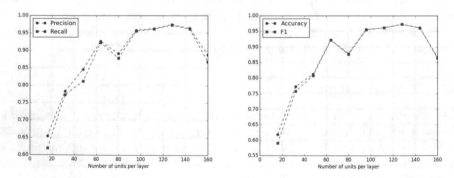

Fig. 5. Variation of metrics with number of units per layer (using two layers of LSTM).

The second parameter is the number of hidden unit per layer. After different tests with various parameters, we found best performances with 112–128 hidden units. In Fig. 5 there is an example of the test.

The last parameter is the dropout rate. We trained the network with different dropout rates, and the best accuracy was in correspondence of 0.3, as shown in Fig. 6.

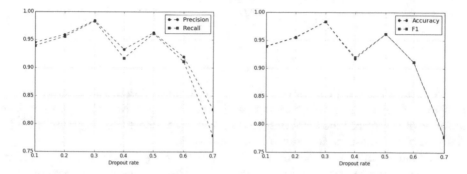

Fig. 6. Variation of metrics with dropout rate (using two layers of LSTM with 128 units).

4.6   Results

For the following results, we used a dataset composed of eight different signs from ASL. The chosen signs are:

1. *Thank you.*
2. *Maybe.*
3. *Wonderful.*
4. *Never.*
5. *Baby.*
6. *Little.*
7. *Rainbow.*
8. *World.*

Some information about these signs can be found in the free online HandSpeak ASL Dictionary[5]. We chose these signs because are all commonly used word and do not contain large fingers movements. Each sign was performed 12 times by nine different people, of various stature and age. So the entire dataset consists of $8 \times 9 \times 12 = 864$ samples. This dataset was divided randomly in train (80%) and test (20%) set.

---

[5] HandSpeak ASL Dictionary: http://www.handspeak.com/word/.

The classification performance is evaluated by different parameters, such as accuracy, precision, recall and F_1-score. They are defined as follow:

$$Acc = \frac{TP + TN}{TP + TN + FP + FN} \qquad (3)$$

$$Prec = \frac{TP}{TP + FP} \qquad (4)$$

$$Rec = \frac{TP}{TP + FN} \qquad (5)$$

$$F_1 = 2\frac{Prec \cdot Rec}{Prec + Rec} \qquad (6)$$

where True Positive (TP) is the number of samples correctly detected, False Positive (FP) is the number of wrong detections, True Negative (TN) is the number of incorrect labels not assigned to a sample, and False Negative (FN) is the number of samples not correctly classified.

We trained and tested our network using two different approaches that differ from how we created the input matrices of the network. The two approaches are:

– **Fixed Length:** we created one matrix per movement. All matrices must have the same number of rows, so we chose a number N greater than the size of the all original matrices, and we interpolated them to have exactly N rows. By doing so, we introduced random noise to the input data. An example of final result can be found in Table 1(a).

– **Sliding Window:** we created a window of fixed size (4–5 times lesser than the length of the matrices). From an original matrix, we created more sub-matrices by sliding that windows with a fixed step. Note that, in general, these sub-matrices can present overlaps. We gave to all sub-matrices the same label of the original matrix.

  By doing so, we have different benefits. First, we do not need to change the size of the matrices. Second, we can have a prediction before the completion of the sign. An example of final result can be found in Table 1(b).

Table 1. Final results with the two approaches.

(a) Fixed Length

Metric	Value
Precision	0.984
Recall	0.983
F_1	0.983
Accuracy	0.983

(b) Sliding Windows

Metric	Value
Precision	0.962
Recall	0.960
F_1	0.960
Accuracy	0.960

5 Conclusion and Future Works

A first study and implementation of a sign languages recognition system has been presented. As shown in the previous tables, both methods used present high accuracy and F_1-score with our test set. The Fixed Length approach has a little more of accuracy but has the disadvantage that each movement has to have the same size.

Those results are very promising; however, in order to use our architecture in a more realistic scenario, a separate system of fingers detection is required. Future development of the system could be to include a lower level fingers movements detector in order to improve the overall quality of classification and to extend the recognition to all the ASL vocabulary.

References

1. Bellugi, U., Fischer, S.: A comparison of sign language and spoken language. Cognition **1**(2), 173–200 (1972)
2. Pigou, L., Dieleman, S., Kindermans, P.-J., Schrauwen, B.: Sign language recognition using convolutional neural networks. In: Workshop at the European Conference on Computer Vision, pp. 572–578. Springer (2014)
3. Oz, C., Leu, M.C.: American sign language word recognition with a sensory glove using artificial neural networks. Eng. Appl. Artif. Intell. **24**(7), 1204–1213 (2011)
4. Cooper, H., Ong, E.-J., Pugeault, N., Bowden, R.: Sign language recognition using sub-units. J. Mach. Learn. Res. **13**, 2205–2231 (2012)
5. Gattupalli, S., Ghaderi, A., Athitsos, V.: Evaluation of deep learning based pose estimation for sign language recognition. In: Proceedings of the 9th ACM International Conference on PErvasive Technologies Related to Assistive Environments, p. 12. ACM (2016)
6. Rumelhart, D.E., Hinton, G.E., Williams, R.J.: Learning representations by back-propagating errors. Cogn. Model. **5**(3), 1 (1988)
7. Graves, A., Liwicki, M., Fernández, S., Bertolami, R., Bunke, H., Schmidhuber, J.: A novel connectionist system for unconstrained handwriting recognition. IEEE Trans. Pattern Anal. Mach. Intell. **31**(5), 855–868 (2009)
8. Sak, H., Senior, A.W., Beaufays, F.: Long short-term memory recurrent neural network architectures for large scale acoustic modeling. In: Interspeech, pp. 338–342 (2014)
9. Bengio, Y., Simard, P., Frasconi, P.: Learning long-term dependencies with gradient descent is difficult. IEEE Trans. Neural Netw. **5**(2), 157–166 (1994)
10. Hochreiter, S., Schmidhuber, J.: Long short-term memory. Neural Comput. **9**(8), 1735–1780 (1997)
11. Park, Y., Moon, S., Suh, I.H.: Tracking human-like natural motion using deep recurrent neural networks (2016). arXiv preprint arXiv:1604.04528
12. Vella, F., Infantino, I., Scardino, G.: Person identification through entropy oriented mean shift clustering of human gaze patterns. Multimed. Tools Appl. **76**(2), 2289–2313 (2017)
13. Srivastava, N., Hinton, G.E., Krizhevsky, A., Sutskever, I., Salakhutdinov, R.: Dropout: A simple way to prevent neural networks from overfitting. J. Mach. Learn. Res. **15**(1), 1929–1958 (2014)

14. Zeiler, M.D., Ranzato, M., Monga, R., Mao, M., Yang, K., Le, Q.V., Nguyen, P., Senior, A., Vanhoucke, V., Dean, J., et al.: On rectified linear units for speech processing. In: 2013 IEEE International Conference on Acoustics, Speech and Signal Processing (ICASSP), pp. 3517–3521. IEEE (2013)
15. Maas, A.L., Hannun, A.Y., Ng, A.Y.: Rectifier nonlinearities improve neural network acoustic models. In: Proceedings of the ICML, vol. 30 (2013)
16. Kingma, D., Ba, J.: Adam: A method for stochastic optimization (2014). arXiv preprint arXiv:1412.6980

Medical Entity and Relation Extraction from Narrative Clinical Records in Italian Language

Crescenzo Diomaiuta, Maria Mercorella(✉), Mario Ciampi,
and Giuseppe De Pietro

National Research Council of Italy, Institute of High Performance Computing
and Networking - ICAR, Via Pietro Castellino 111, 80131 Naples, Italy
{crescenzo.diomaiuta,maria.mercorella,mario.ciampi,
giuseppe.depietro}@icar.cnr.it

Abstract. Applying Natural Language Processing techniques enables
to unlock precious information contained in free text clinical reports. In
this paper, we propose a system able to annotate medical entities in narrative records. Considering that existing NLP systems mainly concern
entity recognition in English language, we propose an NLP pipeline to
manage clinical free text in Italian. The overall architecture includes a
spell checker, sentence detector, word tokenizer, part-of-speech tagger,
dictionary lookup annotator, and parsing rules annotator. Essentially,
it uses a rule-based approach to extract relevant concepts regarding
patient's conditions, administered medications, or performed procedures,
detecting their attributes, negated forms, and relations expressions. The
indexing of the documents allows the user to retrieve relevant information, increasing his/her medical knowledge.

Keywords: Italian natural language processing · Medical entity recognition · Information Extraction · Unstructured medical records · UIMA

1 Introduction

Medical records contain a large amount of clinical data in the form of narrative text written by clinicians. This unstructured part is rich in useful mentions
about patient's significant problems, performed medical procedures, and consumed drugs. The identification of key concepts from these text segments enables
the automated processing of contained clinical data. Information Extraction (IE)
concerns the extraction of predefined types of information from natural language
text. In particular, Named Entity Recognition (NER) is a subfield of IE that aims
at recognizing specific entities, such as diseases, drugs, or names, in free text documents [22]. IE, due to the ability to extract and convert data in a structured
format, can support epidemiology studies, clinical decision, text mining, and
automatic terminology management. Several issues make especially challenging
the processing of clinical narrative text. Indeed, medical reports are usually full

G. De Pietro et al. (eds.), *Intelligent Interactive Multimedia Systems and Services 2017*,
Smart Innovation, Systems and Technologies 76, DOI 10.1007/978-3-319-59480-4_13

of misspellings, abbreviations, typos, and acronyms. Furthermore, one of the main tasks of NLP is to acquire contextual information for an accurate interpretation of the extracted entities. This requires the ability to detect associated attributes, negations, temporality, and relation among entities. There are two main approaches for designing NLP systems: a rule-based one, which requires the definition of dictionaries and a set of rules for matching patterns in the text; and a machine learning approach, which relies on learning algorithms to construct classifiers automatically using annotated examples. Rule-based approaches can be very effective, but require manual effort for writing the rules. Machine learning systems require less human effort, but require large annotated training corpora. Most contemporary NLP systems are hybrid, built from a combination of the two approaches [13]. In this work, we present a Medical Language Processing system able to extract relevant medical information from clinical records written in Italian. We have focused on a rule-based approach, mainly because of the lack of existing medical annotated corpora for the Italian language, able to train a learning algorithm. An NLP system in biomedicine includes two main components: a biomedical background knowledge and a framework to manage the NLP pipeline. The Unified Medical Language System (UMLS) constitutes a suitable biomedical knowledge resource in clinical NLP, bringing together several health standards and biomedical vocabularies of concepts. In general, an NLP pipeline includes the following basic components: sentence detector, word tokenizer, part-of-speech tagger, dictionary look-up annotator, and parsing rules to identify meaningful combinations of tokens [17]. The majority of existing NLP systems consider English as their domain language. To answer to the exigency of managing clinical free text in Italian, we propose an NLP system designed to analyze the content of narrative medical records, extracting from them patient medical problems, interested anatomical parts, performed procedures, dispensed devices, and prescribed medications. In addition to this, the system is capable of recognizing negated expressions, capturing concepts modifiers, and underlining useful relations among the entities. Finally, it allows a user to perform elaborated searches and enables to retrieve clinical information, improving the quality of patient care. The rest of the paper is organized as follows. Section 2 investigates the efforts to apply NLP to clinical text. Section 3 presents the system architecture. Section 4 evaluates the system. Finally, Sect. 5 includes indications for future works.

2   Related Work

Automatic detection of relevant entities in clinical documents, such as mentions about symptoms, examinations, diagnoses, and treatments, increases medical knowledge, providing the clinicians with a quick overview of the patient. There are quite a lot of studies concerning clinical entities recognition in English text, but very few studies on clinical text written in other languages [25]. Actually, several clinical NLP systems have been developed for converting unstructured text to structured data. The cTAKES is a modular system combining rule-based and machine learning techniques, for recognizing medical entities. Its pipeline includes Sentence detector, Tokenizer, Normalizer, Part-of-speech tagger, Shallow parser, NER and negation annotators [24]. The cTAKES showed a high performance in medical concept extraction, making it suitable for successive applications [16,20]. Another widespread clinical NLP system is MedLEE, which aims at generating structured output from patient reports and assigning UMLS codes to relevant clinical information [15,18]. In literature, reviews about information extraction systems from clinical text are available, predominantly built on English. In [23], the authors based their review on several features, such as language, approach, clinical decision support task, and health outcomes, noticing that the majority of the approaches to support clinical decisions have been proposed for processing English free text. The approach presented by Byrd et al. [11] has shown the best results. They aim at identifying Hearth Failure signs and symptoms, through a rule-based NLP system, based on the Unstructured Information Management Architecture (UIMA) framework [5]. Furthermore, an increasing number of NLP research institutes are basing their software development on UIMA specifications [19], which offer a platform for NLP components integration. Moreover, one of the main requirements for developing clinical NLP systems is a suitable biomedical knowledge resource. For this purpose, UMLS offers several vocabularies to help users retrieving information from a wide variety of biomedical information sources [21]. One of the challenges of clinical NER in Italian is that medical terminologies are less extensive for Italian than for English [6]. In biomedicine, NLP offers a powerful instrument for text indexing and document coding. Recognition of relevant medical entities, understanding of their relationships, and, finally, a DB storage, allow powerful information retrieval [12]. In the Italian scenario, there is a small number of studies concerning the extraction of clinical entities from free text. In [10], Attardi et al., considering that Italian corpora annotated with mentions of medical entities are not easily available, have created their own corpus, using a rule-based approach built on regular expressions. They have used the Tanl NER, a statistical sequence labeler, to identify clinical entities, and SVM classifiers, to recognize negations and associations of measures to entities. In other works [8,9], they compensate for the lack of annotated medical Italian resources, creating a silver corpus through a machine translation of an existing one in English. In [7], they propose an unsupervised machine learning methodology for entity and relation extraction, grouping the relations of the same type with the spherical K-means clustering. Finally, Esuli et al. [14] present a solution for extracting a set of concepts of interest from radiological reports, based on a linear-chain CRF learning system, where clauses are the object of tagging. Considering the lack of Italian annotated corpora in biomedicine, we have chosen to base our work on the definition of rich dictionaries to look-up and the implementation of complex parsing rules to match textual patterns of interest. It requires manual efforts, but it can be very effective. Furthermore, annotations deriving from our approach can be employed to train a machine learning system, in the future.

3   System Description

The proposed system is founded on a rule-based approach, which uses manually constructed grammatical rules, built on an NLP pipeline, to extract relevant information from narrative text. The following paragraphs describe preprocessing and processing phases, illustrating technical implementation details.

3.1   Architecture

The implemented system is based on a modular architecture, composed of preprocessing and processing modules, as shown in Fig. 1. The system takes as input the clinical documents, in a PDF format, with information in the form of narrative text. It performs several *pre-processing* operations on each clinical record, obtaining as a result a cleaned and segmented document. In the *processing* operation, the client invokes a web server, by means of Restful APIs, in order to perform the annotation of keywords within the clinical document. Keywords are considered as meaningful words that are extracted from the textual content. The server side consists of a web application that exploits an NLP pipeline connected to a Knowledge Base (KB), which includes several dictionaries, obtained after appropriate *pre-processing* operations. The server sends a response to the client containing the annotation result, structured as an XML file containing the keywords found within the clinical document. In detail, the system invokes the NLP pipeline twice. Firstly, it invokes the pipeline giving as input the cleaned and segmented clinical document. Secondly, it passes to the pipeline a lemmatized document. Lemmatization is a methodical way of converting all the grammatical/inflected forms to the root of the word. Obtaining canonical forms, or lemma, of the words in the clinical record can increase the percentage of matching with words in dictionaries, which mainly contain terms in a base form. The entire process returns two lists of keywords, lemmatized and not, which have to be compared and integrated. The operation of comparison uses a string matching algorithm based on a cosine similarity metric, for measuring the difference between the two sequences. The system integrates the two keywords lists, building a new richer list, containing lemmatized and non-lemmatized entities. Finally, the cleaned/segmented document, the annotated document, and both lists of keywords are stored in a database. In the next subparagraph, we explain the technique used to perform the *pre-processing*. Text pre-processing is an important task and critical step in NLP. It reformats the original text into meaningful units, which contain important linguistic features before performing subsequent text processing strategies. We applied the text *pre-processing* on two different types of documents: health dictionaries and clinical documents.

3.2   Pre-processing

Firstly, this phase aims at creating a suitable Knowledge Base usable by the NLP pipeline. The KB has been derived from the UMLS metathesaurus, considering

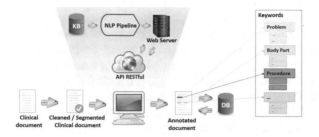

Fig. 1. System architecture.

two types of data files stored in an SQL database: *mrconso*, which contains medical concepts' names and their sources, and *mrsty*, which includes their semantic types. We have selected the following fields of interest, through several queries: CUI (Concept Unique Identifier), a UMLS code for identifying concepts, and STY (Semantic TYpes), a categorization of all concepts. The UMLS information sources, in their Italian translation, used for the creation of the custom dictionaries, are the following: MedDRA, the *Medical Dictionary for Regulatory Activities*; MeSH, the *Medical Subject Headings*; ICPC, the *International Classification of Primary Care*; MTHMST, the *Metathesaurus Minimal Standard Terminology Digestive Endoscopy*. Based on the semantic types and sources of interest, the results of the queries have been integrated in five medical dictionaries. In order to obtain atomic dictionaries and to increase the percentage of matched keywords, a normalization of the dictionaries have been actualized, including: replacement of the accented characters, removal of brackets and their contents, transformation of the terms in lowercase to be case insensitive, and a stop word removal. The resulting custom dictionaries refer to the following semantic types of UMLS: **Problem** includes *Anatomic Abnormality, Disease or Syndrome, Sign or Symptom*, etc.; **Body** includes *Body Part, Organ or Organ Component, Body System*, etc.; **Procedure** includes *Diagnostic Procedure, Health Care Activity, Therapeutic or Preventive Procedure*, etc.; **Device** includes *Medical Device* and *Research Device*; **Medication** includes *Pharmacologic Substance* enriched with drugs dictionary of the *Italian Medicines Agency* (AIFA). Secondly, the system performs a *pre-processing* on the clinical document given in input to the system. It is able to detect errors and suggest corrections respect to a dictionary, using a spell check based on the Symmetric Delete Spelling Correction algorithm (SymSpell) [1]. To allow noise minimization, the spell check operation is not automatic: it lets the user choose whether to replace or not the word gradually found.

3.3 Processing

The *processing* operation, which is the core of the system, refers to the pipeline used to extract relevant entities from clinical documents. It is shown in Fig. 2. The first stage is the ***language identification***, useful to discover the language of the text. Then it actualizes a ***tokenization*** of the text. In particular, the text is broken up into words, phrases, symbols, or other meaningful elements called tokens. At the end of the tokenization process, a **lexical analysis** is implemented. This process marks up a word in a text with a representative part of speech tagging (POS-tag), based on built-in dictionaries and custom dictionaries. Through the built-in dictionaries look-up, the lexical analysis component assigns a grammatical label to the tokens, such as pronoun, verb, noun, adjective, etc. Moreover, via the custom dictionaries look-up, the lexical analysis component identifies *problem, procedure, body part, device,* and *medication* terms inside the text. Based on the lexical analysis tagging, a set of parsing rules has been defined to identify meaningful combinations of tokens in the document.

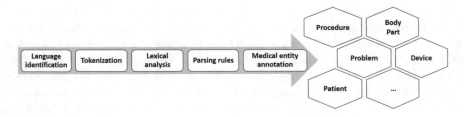

Fig. 2. NLP pipeline.

The **parsing rules** define a sequence of annotations that indicates something of interest inside the text. This stage examines annotations sequence created by preceding stage and selects token combinations that constitute meaningful expressions. Consider the following sentence as an example, "*Si evidenzia una lesione frammentata a livello del menisco mediale*", with its English translation, "*A fragmented lesion of the medial meniscus is appreciable*". It includes a mention about the problem "*lesione frammentata*", and a mention about the body part "*menisco mediale*". In detail, these concepts are fired by the following parsing rules: (i) *Problem = Problem dictionary + Adjective (0 or more) + group [Noun (0 or more) + Adjective (0 or more) + Verb (0 or more)]*; (ii) *Body = group [Body dictionary + Adjective (0 or more) + Adverb (0 or 1) + Verb (0 or 1)]*; (iii) *Body Part/Problem relation = group [Body + Problem]*. The first rule matches the problem *lesion* associated to its qualifier, the adjective *fragmented*. The second one matches the anatomical part. Finally, the third rule is able to connect the previous ones, finding the important relation between the problem and the involved body part. These are just some of the implemented parsing rules. Finally, there is the **medical entity annotation** block, which highlights medical entities in the text.

3.4   Technical Details

For the actualization of the NLP pipeline, the environment IBM Watson Explorer Content Analytics Studio© (ICA) ver. 11.0 has been used, in consideration of its capabilities of analyzing unstructured text [2]. It is based on the Apache UIMA framework. Using the ICA, we have built an NLP pipeline to extract medical entities from free text. ICA allows configuring the following key annotators for text analysis: *Dictionary Lookup annotator*, which matches words from a dictionary with words in the text; Pattern Matcher annotator, which identifies patterns in the text, i.e. sequences of words, by using defined rules. Firstly, the pipeline language detection has been manually set to the Italian language, for removing any language ambiguity. Subsequently, Lexical Analysis parses the structure of the sentence and tags words, looking-up built-in dictionaries and custom dictionaries. During the creation of the custom dictionaries, the ICA studio allows the *generating inflections operation*, useful to add new surface forms and to match the word in different forms. Moreover, thanks to the possibility of adding features to the dictionaries, we have added a column containing the UMLS identifiers (CUI), in order to preserve the concept-code association. Furthermore, ICA Studio provides an interface with the ability to (i) create sophisticated parsing rules and (ii) specify matching criteria for the *Pattern Matcher annotator*. The matching criteria are based on tokens, dictionaries terms, and existing annotations. Subsequently, we have exported the pipeline as a PEAR file (Processing Engine ARchive), which is a fully compliant UIMA annotator, containing descriptor files, compiled classes, configuration files, jar files, and libraries. The PEAR file has been integrated into a Java EE web application, loaded on the Apache Tomcat application server. Furthermore, for ensuring the retrieval of annotated concepts, we have indexed and stored the processed documents, employing MongoDB [3], an open source non-relational database, oriented to the documents, and based on a JSON-style representation. This results in a very fast execution of queries on documents. Moreover, it offers the possibility to achieve a semantic search of information contained within a document, performing a stop word removal and a Snowball stemming [4]. Finally, the system user interface has been designed through the JavaFX framework, which uses Java programming language, to design, test, and deploy rich client applications.

4   System Evaluation

4.1   Use Case

A clinician can employ our pipeline to extract information, increasing his/her medical knowledge and acquiring patient data in a quick and easy way. In detail, our system provides a simple and intuitive user interface, admitting the following operations: (i) simultaneous annotation of one or more documents, (ii) clear displaying of relevant medical information, (iii) semantic search of annotated concepts. The user can select any number of reports to extract entities from them. Then, the system will inform him/her of the success of the operation. Furthermore, the user can display a single annotated report, to obtain a rapid overview of relevant data contained in the unstructured text. In particular, relevant concepts are underlined and a table summarizes recognized medical concepts, their associated UMLS codes, and relations among several entities, such as between a problem and the interested body part, or between a medication and its dosage.

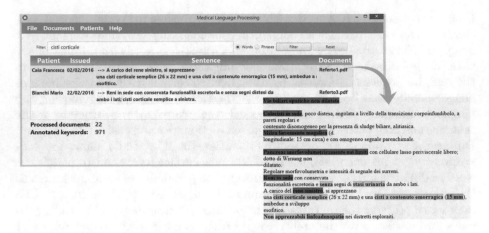

Fig. 3. Semantic search of medical concepts with the related annotated document.

Finally, Fig. 3 shows an operation that can turn to be very useful to a clinician, consisting in the semantic search of a concept among the documents previously annotated. This can support the process and the quality of patient care, due to the availability of a large amount of extracted data that can increase clinician medical knowledge.

Clearly, the knowledge base may be continuously enriched, to increase annotated entities. Furthermore, due to the definition of complex parsing rules, the annotator achieves the following results:

– Capturing negative expressions, thanks to the matching criterion requiring a medical entity and a term coming from a dictionary of common negations;
– Associating qualifiers to medical entities, due to a parsing rule that requires a medical entity combined with a certain type of adjectives or adverbs;
– Obtaining useful temporal expressions, through a parsing rule that matches date triggers of interest and a regular expression able to validate date format;
– Capturing quantifiers information, using a rule that links a medical concept to a regular expression able to match observation results;
– Underlining relations among entities, through a rule that combines entities from the *Problem* and *Body Part* dictionaries in a sentence.

In addition, we have characterized specific rules, not easily generalizable, to recognize the *Patient*, subject of the report, and the *Practitioner*, author of it. Therefore, we have defined dictionaries to find medical entities of interest and parsing rules to better understand their semantic meaning. General errors occur especially because of the complexity of the Italian medical language. To improve the annotations fired by the parsing rules, a procedure of iterative refinement, supervised by a domain expertise, may be adopted.

5   Conclusions and Future Works

This study has been performed on Italian clinical records, written as free text, with the aim of automatically extracting medical entities from them. In particular, we have proposed a system implementing an NLP pipeline, which recognizes keywords of interest in the narrative, extracts meaningful relations, and stores the annotated documents. The indexing of the records allows the retrieval of useful clinical information, increasing the medical knowledge and improving the patient quality of care. The recognition of the entities is mainly based on a dictionaries look-up, while, due to several parsing rules, it is possible to recognize negated mentions of the concepts, measurements associated with the keywords, and relations among them. We aim at improving the knowledge base that constitutes the dictionaries and the rules that have to match meaningful expressions. Moreover, our intent is to employ clinical documents annotated with our rule-based approach for training a machine learning system. Starting from the medical knowledge extracted from unstructured data, several applications can be developed. In the future, our intent is matching stored annotated keywords against a standard medical terminology. We plan to define a tool, based on the implemented system, able to map extracted entities regarding patient's medical conditions with the International Classification of Diseases (ICD). Furthermore, we aim at implementing an application that makes exhaustive and agile the research of information regarding a patient. In particular, we aspire at extracting information from the Patient Summary type of electronic record, a collection of the patient's most significant clinical data. Based on the implemented system, relevant medical entities could be extracted and presented to a clinician, offering a rapid patient overview.

References

1. FAROO spelling correction (2016). http://blog.faroo.com/category/spelling-correction/
2. IBM watson explorer (2016). https://www.ibm.com/us-en/marketplace/content-analytics
3. Mongo database (2016). https://www.mongodb.com/
4. Snowball resources (2016). http://snowball.tartarus.org/
5. UIMA home (2016). https://uima.apache.org/
6. UMLS documentation (2016). https://www.nlm.nih.gov/research/umls/
7. Alicante, A., Corazza, A., Isgrò, F., Silvestri, S.: Unsupervised entity and relation extraction from clinical records in italian. Comput. Biol. Med. **72**, 263–275 (2016)
8. Attardi, G., Cozza, V., Sartiano, D.: Adapting linguistic tools for the analysis of Italian medical records (2014)
9. Attardi, G., Cozza, V., Sartiano, D.: UniPi: Recognition of mentions of disorders in clinical text. In: Proceedings of the 8th International Workshop on Semantic Evaluation, pp. 754–760 (2014)
10. Attardi, G., Cozza, V., Sartiano, D.: Annotation and extraction of relations from Italian medical records. In: IIR (2015)

11. Byrd, R.J., Steinhubl, S.R., Sun, J., Ebadollahi, S., Stewart, W.F.: Automatic identification of heart failure diagnostic criteria, using text analysis of clinical notes from electronic health records. Int. J. Med. Informatics **83**(12), 983–992 (2014)

12. De Bruijn, B., Martin, J.: Getting to the (c)ore of knowledge: mining biomedical literature. Int. J. Med. Informatics **67**(1), 7–18 (2002)

13. Doan, S., Conway, M., Phuong, T.M., Ohno-Machado, L.: Natural language processing in biomedicine: a unified system architecture overview. In: Clinical Bioinformatics, pp. 275–294 (2014)

14. Esuli, A., Marcheggiani, D., Sebastiani, F.: An enhanced CRF's-based system for information extraction from radiology reports. J. Biomed. Inform. **46**(3), 425–435 (2013)

15. Friedman, C., Shagina, L., Lussier, Y., Hripcsak, G.: Automated encoding of clinical documents based on natural language processing. J. Am. Med. Inform. Assoc. **11**(5), 392–402 (2004)

16. Garla, V., Re, V.L., Dorey-Stein, Z., Kidwai, F., Scotch, M., Womack, J., Justice, A., Brandt, C.: The yale cTAKES extensions for document classification: architecture and application. J. Am. Med. Inform. Assoc. **18**(5), 614–620 (2011)

17. Hardeniya, N.: NLTK Essentials. Packt Publishing Ltd. (2015)

18. Johnson, S.B., Bakken, S., Dine, D., Hyun, S., Mendonça, E., Morrison, F., Bright, T., Van Vleck, T., Wrenn, J., Stetson, P.: An electronic health record based on structured narrative. J. Am. Med. Inform. Assoc. **15**(1), 54–64 (2008)

19. Kunze, M., Rösner, D.: UIMA for NLP based researchers workplaces in medical domains. In: Towards Enhanced Interoperability for Large HLT Systems: UIMA for NLP, p. 20 (2008)

20. Lin, C.H., Lai, W.S., Lee, L.H., Tsao, H.M., Liou, D.M.: An entry generation pipeline for converting free-text medical document into clinical document architecture document with entry-level. In: 2014 IEEE-EMBS International Conference on Biomedical and Health Informatics (BHI), pp. 505–508. IEEE (2014)

21. McCray, A.T., Aronson, A.R., Browne, A.C., Rindflesch, T.C., Razi, A., Srinivasan, S.: UMLS knowledge for biomedical language processing. Bull. Med. Libr. Assoc. **81**(2), 184 (1993)

22. Meystre, S.M., Savova, G.K., Kipper-Schuler, K.C., Hurdle, J.F., et al.: Extracting information from textual documents in the electronic health record: a review of recent research. Yearb. Med. Inform. **35**(128), 44 (2008)

23. Reyes-Ortiz, J.A., González-Beltrán, B.A., Gallardo-López, L.: Clinical decision support systems: a survey of NLP-based approaches from unstructured data. In: 2015 26th International Workshop on Database and Expert Systems Applications (DEXA), pp. 163–167. IEEE (2015)

24. Savova, G.K., Masanz, J.J., Ogren, P.V., Zheng, J., Sohn, S., Kipper-Schuler, K.C., Chute, C.G.: Mayo clinical text analysis and knowledge extraction system (cTAKES): architecture, component evaluation and applications. J. Am. Med. Inform. Assoc. **17**(5), 507–513 (2010)

25. Skeppstedt, M., Kvist, M., Nilsson, G.H., Dalianis, H.: Automatic recognition of disorders, findings, pharmaceuticals and body structures from clinical text: an annotation and machine learning study. J. Biomed. Inform. **49**, 148–158 (2014)

Detection of Indoor Actions Through Probabilistic Induction Model

Umberto Maniscalco, Giovanni Pilato, and Filippo Vella[✉]

ICAR, National Research Council of Italy, Via Ugo La Malfa 153, Palermo, Italy
filippo.vella@icar.cnr.it

Abstract. In the present work a system able to classify the indoor action is presented. The data are recorded with multiple kind of sensor collecting the position of the joints of the person in the room, the acceleration recorded on the person wrist and the presence or absence in a specific room. The latent semantic analysis, based on the principal component search, allows to estimate the probability of a given action according the sampled values.

1 Introduction

The possibility to monitor and understand the actions that are carried on in an environment is a key technology to give an aid to people inside their usual environment and proactively help them in the execution of their tasks. Systems can be created to help people with their activity or to provide an help anytime the user requests its work. Systems that clean a room or move heavy loads can be activated for any task that can be done with some difficulty by people in a house. The number of tasks can be increased when people are someway impaired in the perception of the environment or cannot carry out a specific work. This kind of systems has to determine which task is going to be accomplished to provide passive or active help to the user.

Multiple technologies can be employed in Ambient Intelligence technologies [1] and heterogeneous information can be extracted to monitor people in a given environment. Cameras, pressure sensors, motion detectors and wearable technologies in order to perceive and monitor the user's presence and her behaviour in the different spaces of the environment [2,3].

While it is rather easy to collect a significant amount of sensors observations, the great challenge is to properly recognize meaningful patterns in the raw data that can be ascribed to user actions and activities on the environment [4,5]. We aim at identifying the actions that are accomplished and providing labels to the sampled values. We associate a numerical value, that can be thought as the probability, to the set of possible labels. This value is estimated adopting a representation so that the actions are dealt in a similar way to the representation of sentences in a text. In our framework a set of movements is analogue to a sentence while an atomic movement is related to a single word. To extract the atomic movements we used two clustering techniques whose performance

G. De Pietro et al. (eds.), *Intelligent Interactive Multimedia Systems and Services 2017*,
Smart Innovation, Systems and Technologies 76, DOI 10.1007/978-3-319-59480-4_14

are discussed in the Sect. 4. A similar technique has been profitably adopted in [6] for image representation. According to this representation a selection of the principal components is done. As for Latent Semantic Analysis a truncated singular value decomposition is performed. Only the fundamental connections are maintained while the higher order that are linked to sampling noise and accidental cooccurrence are discarded. The maintained relevant connections are the ones the provide an estimation of the labels according the sampled data.

The proposed approach has been tested using the dataset of the SPHERE (Sensor Platform for HEalthcare in Residential Environments) project [7]. The dataset consists in collections of measurements from cameras with an additional channel for depth information, accelerometers and passive environmental sensors. People involved in the acquisition process are asked to perform a set of action in an indoor environment. Their actions have been manually labelled to associate a category to the performed movements.

The paper is organized as follows: the Sect. 2 describes the SPHERE dataset with the recorded data and the processing that has been done to make it processable, discussion about classification through Latent Semantic Analysis is presented in Sect. 3. Section 4 discusses some experimental results, comparing them with a baseline classifier employing a conventional Multi Layer Perceptron Network. Finally conclusions and a discussion on future directions of the work are given the last section.

2   Indoor Actions Data

For the experiments we carried on we took as input the data of the SPHERE dataset. The data are available on line[1] and a detailed description of the dataset is available in [7]. The number of samples is 2787 that have been manually annotated with one of the given labels. The values are acquired by an accelerometer, a color camera (RGB) with depth information (D) and environmental data forming a vector of eighteen values. In particular the value are referred to

- x, y, z: acceleration along the x, y, z axes
- centre_2d_x, y: the coordinated of the center of the bounding box along x and y axes
- bb_2d_br_x, y: The x and y coordinates of the bottom right (br) corner of the 2D bounding box
- bb_2d_tl_x, y: The x and y coordinates of the top left (tl) corner of the 2D bounding box
- centre 3d x, y, z: the x, y and z coordinates for the centre of the 3D bounding box
- bb_3d_brb_x, y, z: the x, y, and z coordinates for the bottom right back (brb) corner of the 3D bounding box
- bb_3d_flt_x, y, z: the x, y, and z coordinates of the front left top (flt) corner of the 3D bounding box.

---

[1] http://irc-sphere.ac.uk/sphere-challenge/home.

Twenty activities labels have been used to annotate the dataset, these labels are. There are three main categories that are *actions*, *positions* and *transitions*.

The actions are all the activities that imply movements and are divided among *ascent stairs*, *descent stairs*, *jump*, *walk with load* and *walk*; The position are referred to a still person and are labelled as *bending*, *kneeling*, *jump*, *lying*, *sitting*, *squatting*, *standing*.

The transition are the intermediate steps between two positions and are: *stand-to-bend*; *kneel-to-stand*; *lie-to-sit*; *sit-to-lie*; *sit-to-stand*; *stand-to-kneel*; *stand-to-sit*; *bend-to-stand*; *turn*.

2.1   Pre-processing

The original data achieved by the project Sensor Platform for Healthcare in Residential Environments (SPHERE) [7], was arranged in ten sections. Each of these sections has a duration between twenty-five and thirty minutes. Data about accelerations are organised in a single file for each section, on the contrary, data regarding the 3D camera are divided into three different files depending on the room where they are taken. Moreover, in a single file, for each section, are stored information about the manually annotated label. In each section are also stored others auxiliary information that we do not take into account in this framework. Thus, we consider the accelerometers and RGB-D data as input for the system of classification introduced in this work and the labels values as the output of the system. Of course, during the training phase both accelerometers and RGB-D and labels can be considered input for the system.

All these series of data are acquired with different sampling frequencies and they can contain holes and not a number values (NaN). Thus, they require a pre-processing before their usage. In particular, a resampling of the data at the same frequency and the elimination of all NaN values should be performed in order to make consistent all the series.

The RGB-D cameras work at a rate of 30 frames per second (fps) that is 30 Hz. On the other side, accelerometers register information with a frequency of 20 Hz. Moreover, the categories of actions are manually annotated and arranged at 1 Hz. Thus, we have three different data series at three different frequencies.

More in details, these data are stored in vectors. The measures achieved by the accelerometers are the values of the accelerations along the x, y and z axes while RGB-D cameras return the coordinates of the center or the limits of the bounding box of the detected human figure.

Both accelerometers and RBG-D cameras also register the time, in seconds, starting from the begin of the sequence.

The first step to process the data is to consider the lowest sampling frequency. All the other series, both the RGB-D values and accelerometers data, must be down sampled to this frequency, as shown in Fig. 1. For each section, the label for a give time instant is used as a reference and the corresponding accelerometers and RGB-D data values are grouped and averaged. When, for a particular time instant, the accelerometers or RGB-D data are missing, a NaN value is assigned

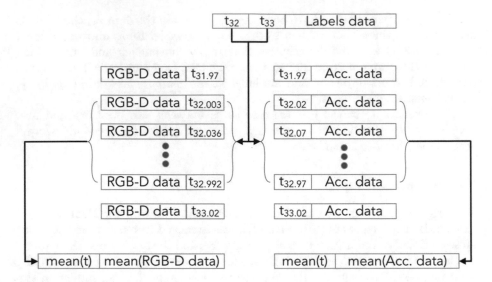

Fig. 1. The labels associated with actions are used as a reference. For each sample the corresponding accelerometers and RGB-D data are grouped and averaged. If there the accelerometers or RGB-D data values are not available, NaN values are assigned.

and will not computed for the temporal average. At the end of the processing we have a single row of data from accelerometers, RGB-D and labels per second.

Input data are also arranged in groups of consecutive values that constitute a time windows of data as shown in Fig. 2. These time windows can have different length maintaining the constraint that the window contains an odd number of seconds. Each time window is referred to the central element and the time of this element identify the correspondent label row.

2.2   Representation of Actions with Vector Quantization

Data collected with multiple sensors and processed as described in the previous sections, are represented with a vector of values that are referred to a given time instant. All the components have continuous values that vary according the position taken by the person in the apartement by all the sensors(accelerometers and passive sensors). We process this data in a way that resembles the representation of words in a text. To make this transposition a clusterization of data is done. We search for the points in the vector space where the represented data are more dense. In this paper we used the k-means [8] and LBG (Linde-Buzo-Gray) algorithm [9] to estimate a set of centroids that are the point in the vector space where the data are aggregated. The number of the centroids is an input parameter for the k-means algorithm while only a number of centroids equal to a power of two number can be selected for LBG algorithm. Once the centroids have been extracted, all the vectors are mapped on the set formed by the collection of the centroids. Given an input vector, the distance between the vector

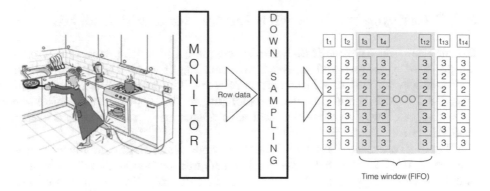

Fig. 2. An example of time window for the input data

and all the centroids vectors is evaluated. The centroid with the least distance from the input vector provides the representation of the input vector. The new representation of the input values, instead of a representation with a dimension equal to the input vector dimensionality will be a vector with a dimensionality equal to the number of centroids where only a component is one, the one corresponding to the nearest centroids, while all the others will be set to zero.

The label associated with the samples, at a given instant, corresponds to a 1-to-n vector in the space formed by the centroids. Since we desire to have a representation that corresponds not only to the single time instant but to a sequence of samples we considered the 1-to-n vector as a word and the sample sequence as the document. If, in consecutive instants, more than one centroid is activated, the corresponding components are set to one. In other words if, during consecutive recordings, the sequence of movements activates the centroids α and β, the components corresponding to α and β centroids are set to one. Furthermore, considering that the temporal information is relevant for our task we also added an information bound to the order in the execution of the movements. We enlarged the representation considering, in a representation analogue to the words, the presence of the bigrams. If N_c is the number of centroids the representation will have a size equal to $N_c + N_c \cdot N_c$ where the first components are bound to the single "words" while the remaining part is bound to the "bigrams". Since the order is important the detection of the centroid α after the centroid β will activate a bigram different from the sequence of centroid β after the centroid α. This representation, that resembles the representation of documents according the present words, is used to associate a label to the detected movement through the Latent Semantic Analysis and the association of a labels to a new detected movement. The number of clusters used with k-means and LBG algorithms and the size of the time windows to evaluate the composition of movements, called *time span*, are varied in the experiments to assess the performance of the system according accuracy. A detailed description is provided in Sect. 4.

3   Classifying Actions with Latent Semantic Analysis

Latent Semantic Analysis (LSA), is a technique based on the vector space paradigm, used to extract and represent the meaning of words through statistical computations applied to a large corpus of texts. The paradigm defines a mapping between words and documents belonging to the corpus into a continuous vector space $\mathcal{S}$, where each word, and each document is associated to a vector in $\mathcal{S}$ [10,11].

Even if Truncated Singular Value Decomposition (TSVD) has been traditionally applied in the text classification and information retrieval fields [12], it can be successfully applied to any kind of dyadic domain, therefore we have applied it to annotation of images [13] and here for the classification/annotation of actions inside an indoor environment. The usual dataset, in a text classification task, is represented as a matrix where each row represents a document according to the words that are bound to the columns.

Let us consider a corpus made of N data, and let M be the number of actions associated to the data. Then let $\mathbf{D}$ be the $N \times M$ matrix whose (i, j)-th entry $[d]_{ij}$ is the value of the j-th feature in the i-th data pattern.

According to the Singular Value Decomposition theorem, $\mathbf{D}$ can be univocally decomposed as $\mathbf{D} = \mathbf{U}\boldsymbol{\Sigma}\mathbf{V}^T$, where $\mathbf{U}$ is a column-orthogonal $N \times M$ matrix, $\mathbf{V}$ is a column-orthogonal $M \times M$ matrix and $\boldsymbol{\Sigma}$ is a $N \times M$ diagonal matrix, whose elements are identified as singular values of $\mathbf{D}$.

$\mathbf{D}$'s singular values can be assumed to be ordered in decreasing manner. Let R be an integer with $0 \leq R < M$, and let $\mathbf{U}_R$ be the $N \times R$ matrix obtained from $\mathbf{U}$ by suppressing the last $M - R$ columns, $\boldsymbol{\Sigma}_R$ the matrix obtained from $\boldsymbol{\Sigma}$ by suppressing the last $N - R$ rows and the last $M - R$ columns and $\mathbf{V}_R$ be the $M \times R$ matrix obtained from $\mathbf{V}$ by suppressing the last $M - R$ columns. Then $\tilde{\mathbf{D}}_R = \mathbf{U}_R\boldsymbol{\Sigma}_R\mathbf{V}_R^T$ is a $N \times M$ matrix of rank R obtained from the matrix $\mathbf{D}$ through Truncated Singular Value Decomposition (TSVD).

Performing the TSVD on $\mathbf{D}$, means to evaluate the best rank R approximation $\tilde{\mathbf{D}}_R = \{[\tilde{m}_R]_{ij}\}$ to $\mathbf{D}$ (among the $M \times N$ matrices) according to the Frobenius distance.

We have built the matrix $\mathbf{D}$ as the composition of two sub-matrices: $\mathbf{F}$ and $\mathbf{A}$, i.e. $\mathbf{D} = [\mathbf{F}|\mathbf{A}]$, where the submatrix $\mathbf{A}$ is associated to the actions to be taken into consideration, while the submatrix $\mathbf{F}$ represents the features that have been considered to represent the timeframes. For clarity, the generic i-th row of $\mathbf{D}$ has the following structure:

$$[f_{i1}, f_{i2}, \cdots, f_{iN}, a_{i(N_f+1)}, a_{i(N_f+2)}, \cdots, a_{i(N_f+K)}] \tag{1}$$

where N_f is the number of columns used to represent the accomplished actions K is the number of action labels, $[f_{i1}, f_{i2}, \cdots, f_{iN}]$ is the vector of the probability distribution of features and $[a_{i(N_f+1)}, a_{i(N_f+2)}, \cdots, a_{i(N_f+K)}]$ is set of the label associated to the features. As a first study, for computational reasons, each element of the submatrix $\mathbf{A}$ has been modified in order to obtain the submatrix $\mathbf{A}_b$ through a binarization process illustrated in the following equation:

$$a_{b(ij)} = \begin{cases} 1 & if \ a_{ij} > T \\ \\ 0 & if \ a_{ij} \leq T \end{cases} \tag{2}$$

where T is a threshold $T \in [0,1]$. For our experiments we have chosen $T = 0.1$. The matrix $\mathbf{D} = [\mathbf{F}|\mathbf{A}]$ is therefore transformed into the matrix $\mathbf{D}_b = [\mathbf{F}|\mathbf{A_b}]$. The TSVD decomposition technique is then applied to the matrix $\mathbf{D}_b$, obtaining the matrix $\widetilde{\mathbf{D}}_b = \mathbf{U}_R \mathbf{\Sigma}_R \mathbf{V}_R^T$.

For clarity we can split the matrix $\mathbf{V}_R^T$ into two submatrices $\mathbf{V}_{R(features)}^T$ and $\mathbf{V}_{R(actions)}^T$ so that $\mathbf{V}_R^T = [\mathbf{V}_{R(features)}^T|\mathbf{V}_{R(actions)}^T]$.

Mapping a new pattern in the space created by TSVD means adding a row $\mathbf{d}^+$ to the matrix $\mathbf{D}_b$ and computing the TSVD over this new extended matrix. However, the last K elements of this new row are unknown, being the target pattern of actions to be computed. Hence we decompose the row $\mathbf{d}^+$ in two parts $\mathbf{d}^+ = \{\mathbf{d}_{features}^+|\mathbf{d}_{actions}^+\}$.

To map this new pattern into the conceptual space we have to compute an approximate solution of the equation $\mathbf{u}^+ \cdot \sqrt{\mathbf{\Sigma}_R} \mathbf{V}_{R(features)}^T = \mathbf{m}^+_{R(features)}$.

Once the vector $\mathbf{u}^+$ has been calculated, it is easy to compute the vector $\mathbf{m}_{R(actions)} = \mathbf{u}^+ \cdot \sqrt{\mathbf{\Sigma}_R} \cdot \mathbf{V}_{R(actions)}^T$, and obtaining the actions which are associable to the $\mathbf{d}_{features}^+$ pattern.

4   Experiments and Results

We have run a set of experiments according to the two possible codings of the data.

In particular 10% of the items of from each dataset has been selected and extracted from the dataset for building a test set. The selection has been done through a random choice. The remaining 90% of items has been used as a training set. Fixed the truncation parameter R, the aforementioned procedure has been run 20 times and the results have been averaged in order to obtaining affordable results.

We made experiments by using values of R, that is the parameter related to the truncation in the singular value decomposition process, ranging from 1 to half the rank of the matrix $\mathbf{D}$. We have chosen a threshold of $T = 0.1$ for the experiments (see Eq. 2). The Table 1 reports the accuracy results for the methodologies that gave the best accuracy results. In Fig. 3 we show the trend of accuracy versus the truncation parameter R of the three approaches (k-means with 10 centroids and time span equal to 9; LBG with 16 centroids and time span equal to 5; k-means with 8 centroids and time span equal to 9) that gave the best accuracy values.

Figure 4 shows the values of the truncation parameter R versus the accuracy over the test set. It is clear that the best results of accuracy are obtained by lower values of R (R less than 4) whatever is the method used.

In the plots shown in Fig. 5a and b we show that the choice of the number of clusters and the choice of the time parameter have a little influence on the accuracy performance.

Table 1. Cluster methodology, number of clusters and time span for the best results of accuracy

Clustering	Num. of clusters	Timespan	R	Precision
k-means	10	9	2	0.9297
LBG	16	5	2	0.9296
k-means	8	9	2	0.9292
k-means	8	5	1	0.9291
k-means	10	5	1	0.9290
k-means	10	7	1	0.9290
k-means	10	11	2	0.9288
k-means	12	9	2	0.9288
k-means	12	11	2	0.9287
k-means	12	11	3	0.9287

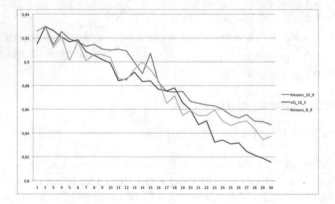

Fig. 3. Trend of accuracy values versus the truncation parameter R for the three approaches that gave the best accuracy values.

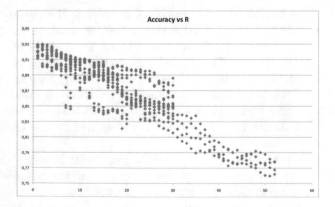

Fig. 4. Average values of accuracy versus the truncation parameter R for all the experiments

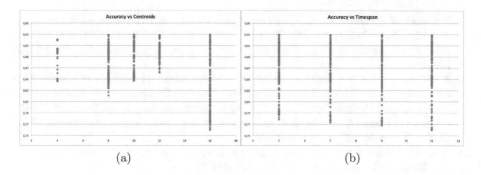

(a)   (b)

Fig. 5. (a) Accuracy versus the number of centroids (b) Accuracy versus the time span size

5   Conclusions and Future Works

A system for the detection of the actions of person in a indoor environment has been presented. The data come from heterogeneous sources and a processing is done to extract the position of the person, the values of the acceleration of her wrist and the room where the person is. The recorded data are mapped on a base selected through two algorithms of clustering. The vectors of the base have been considered as terms enabling the possibility to deal the position as words and the sequence of actions as a sentence. To this representation has been applied the truncated singular value decomposition to analyze the most important components and classify unseen actions. Results are encouraging and such a system can improve the quality of life of many person and, in future, make their home safer.

References

1. Remagnino, P., Foresti, G.L.: Ambient intelligence: a new multidisciplinary paradigm. IEEE Trans. Syst. Man Cybern. Part A Syst. Hum. **35**(1), 1–6 (2005)
2. Bakar, U., Ghayvat, H., Hasanm, S., Mukhopadhyay, S.: Activity and anomaly detection in smart home: a survey. In: Next Generation Sensors and Systems, pp. 191–220. Springer (2016)
3. Aarts, E., Encarnação, J.: True visions: the emergence of ambient intelligence (2006)
4. Castillo, J.C., Carneiro, D., Serrano-Cuerda, J., Novais, P., Fernndez-Caballero, A., Neves, J.: A multi-modal approach for activity classification and fall detection. Int. J. Syst. Sci. **45**(4), 810–824 (2014)
5. Krishnan, N.C., Cook, D.J.: Activity recognition on streaming sensor data. Pervasive Mob. Comput. **10**(Part B), 138–154 (2014)
6. Ardizzone, E., La Cascia, M., Vella, F.: A novel approach to personal photo album representation and management. In: Electronic Imaging 2008, International Society for Optics and Photonics, p. 682007 (2008)

7. Twomey, N., Diethe, T., Kull, M., Song, H., Camplani, M., Hannuna, S., Fafoutis, X., Zhu, N., Woznowski, P., Flach, P., Craddock, I.: The SPHERE challenge: activity recognition with multimodal sensor data. arXiv preprint arXiv:1603.00797 (2016)

8. MacQueen, J., et al.: Some methods for classification and analysis of multivariate observations. In: Proceedings of the Fifth Berkeley Symposium on Mathematical Statistics and Probability, Oakland, CA, USA, vol. 1, no. 14, pp. 281–297 (1967)

9. Linde, Y., Buzo, A., Gray, R.: An algorithm for vector quantizer design. IEEE Trans. Commun. **28**(1), 84–95 (1980)

10. Landauer, T.K., Foltz, P., Laham, D.: An introduction to latent semantic analysis. In: Discourse Processes, vol. 25, pp. 259–284 (1998)

11. Landauer, T., Dumais, T.: A solution to Plato's problem: the latent semantic analysis theory of acquisition, induction and representation of knowledge. Psychol. Rev. **104**(2), 211–240 (1997). http://lsa.colorado.edu/papers/plato/plato.annote.html

12. Agostaro, F., Augello, A., Pilato, G., Vassallo, G., Gaglio, S.: A conversational agent based on a conceptual interpretation of a data driven semantic space. Lecture Notes in Artificial Intelligence, vol. 3673, no. 2, pp. 381–392 (2005)

13. Pilato, G., Vella, F., Vassallo, G., La Cascia, M.: A conceptual probabilistic model for the induction of image semantics. In: 2010 IEEE Fourth International Conference on Semantic Computing (ICSC), pp. 91–96. IEEE (2010)

A ROS Driven Platform for Radiomap Management Optimization in Fingerprinting Based Indoor Positioning

Giovanni Luca Dierna, Alberto Machì, and Sergio Scirè[✉]

Cognitive Robotics and Social Sensing Lab., Institute for High Performance Computing and Networking, National Research Council of Italy (ICAR-CNR),
Via Ugo La Malfa, 153, Palermo, Italy
{giovanniluca.dierna,alberto.machi,sergio.scire}@icar.cnr.it,
http://www.icar.cnr.it

Abstract. An electromagnetic beacons infrastructure is commonly used in positioning applications within buildings where the GPS signal is not present.

Through techniques of multilateration and fingerprinting an average accuracy of about 2 m can be reached, but the accuracy is limited by multiple reflections, obstacles and signal dispersion that make it unreliable analytic field modeling. Dense field sampling field allows a reconstruction more detailed but is costly and clever uneven sampling is appropriate.

This work describes the progress of an interactive robotic platform under development to support field modeling and beacons positioning, through intelligent iterative strategies of data acquisition.

The platform integrates a Matlab-based control system and simulation software: a robot equipped with distance sensors, able to perform autonomous navigation in a known environment, and a data logger module hosted in an Android mobile device, all connected via a ROS framework.

Robot assisted field sampling is here proposed and used to reduce costs of radiomap construction and update. In particular, this technology is suitable for complex environments as museums and exhibitions.

Keywords: Mobile robot · Indoor positioning · ROS · Matlab · SLAM · Mapping · Path planning · Navigation · Path tracking · BLE beacons

1   Introduction

At ICAR-CNR we are testing a minimally invasive indoor positioning system, based on Bluetooth Low Energy (BLE) technology, suitable for accurate positioning in well-defined indoor areas such as expositive buildings. Such a localization system could be able to provide end users with augmented information based upon its position.

This technology could be used in different scenarios: as a virtual guide inside a museum [1], as a safety localization system for children inside a shopping center, or even as an item collector and clues revealer, for entertainment or serious games as (didactic) scavenger hunt games [2].

© Springer International Publishing AG 2018
G. De Pietro et al. (eds.), *Intelligent Interactive Multimedia Systems and Services 2017*,
Smart Innovation, Systems and Technologies 76, DOI 10.1007/978-3-319-59480-4_15

To provide an accurate map of a closed area, with no ambiguity in position and a sufficiently precise positioning, we need to characterize the radiomap resulting from the active elements placed in the area. The position and amount of elements certainly need to be adjusted from a preliminary layout on planimetry, because of interference and overlapping of radio signals (due to walls, open spaces, reflective or opaque surfaces).

To perform this radio signal characterization, in a reproducible and systematic way, we are developing a navigation toolkit for the low-cost Turtlebot2 open source robot platform [3] that uses Simultaneous Localization And Mapping (SLAM) techniques, to localize itself accurately and simultaneously build a map of the environment [4].

Our design incorporates a variety of simple and inexpensive modifications that significantly enhance the Turtlebot2, with the goal of transforming it into an indoor or outdoor service robot capable of performing clearly defined tasks, as described in the next sections, more efficiently and with higher precision than humans can attain. In our scenario, the Turtlebot2 robot performs operations that in any case would be burdensome for a human operator.

2   Indoor Positioning: Problems and Possible Solutions

The positioning techniques allow recognizing location on a map using multilateration compared with known positions.

However, in indoor environments, due to limited accuracy and satellite visibility requirements, GPS is not appropriate for positioning (or self-localization) and localization. For this reason, some authors propose in [5] an indoor positioning methodology relying on fingerprinting techniques and BLE technology, to gently suggest to the user evidence of contextually coherent areas of interest around him.

In radiomap based positioning, emitters of radio frequency signals (beacons) are placed at known positions. The presence of the beacon is recognizable through the measure of characteristics of the perceived radio signal, as the amplitude, the phase, the flight time, the noise.

The absorption of the transmission medium, obstacles, reflections and diffusion of the signal make the signal itself, received at a distance from the source, variable in space and time. The space-time variability of the $k + 3$ dimensional electromagnetic field is representable in a spatial radiomap in which at each point of the space are associated the average values in time and the variance of a characteristic of signal like its perceived intensity.

If the spatial variability of the field overcomes its temporal fluctuation, the vector of perceived beacon signal intensities constitutes a signature (fingerprint) of the point and the ensemble of fingerprints measured in reference points at known positions a radiomap of the location. In methods based on radiomaps, the signature of a test point is then compared with those of the reference points contained in the radiomap, to select most similar candidates to evaluate position by triangulation.

Intrinsic symmetries in the field and reflections produce local minima in the distance function that limit the accuracy of the recognition of geometrically nearest similar landmarks.

Probabilistic methods such as Particle Filter enhance the accuracy of the positioning through an iteration of the distance evaluation process.

Structuring the BLE field around the observer's position determines the obtainable accuracy. Where the variation of the field gradient is low, compared to its temporal variability, the fingerprint is the vector field representative of a large area and not just of the immediate vicinity of a reference point on the map, so the probability filter becomes ineffective in decreasing the positioning uncertainty and even a dense sampling does not provide additional meaningful information.

This effect therefore suggests to iteratively optimizing positioning of the sources after evaluating the effects of field structure on effectiveness of the Particle Filter.

Since repositioning involves re-sampling of the field, it is appropriate to develop a supervised support system that allows interactively sampling, reshaping the field, rechecking the reliability of modeling, estimating the accuracy and providing suggestions for repositioning beacons.

For this purpose, we developed a robotic platform, interacting via Robotic Operating System (ROS), composed by:

1. A control and simulation module built in Matlab [6];
2. A robot with sensors able to navigate independently inside a previously mapped area;
3. An acquisition module hosted in a hand held Android smartphone.

3   The Application Context

The methodology of positioning based on BLE technology aims to provide reliable self-localization inside museum areas of interest and to guide the visitor in its exploration.

Space organization in exhibitions is often hierarchical and presents three levels of coherent organization: the pavilion/hall/small collection level, the expository/window level and the object of interest level. In the first space level, pieces and collections are exposed together for thematic or historical topic; the second space level is constrained by available exposition structures.

In a large part of exhibitions and museums, the first level covers an area which spans tens to hundreds square meters while the second level covers areas from a few to some tens of square meters, the third level a few square meters around the object. Positioning has then to cope with different scales of extent and accuracy.

Table 1 shows the accuracy requirements empirically evaluated.

Table 1. Empirical accuracy requirements

Area	Contiguity	Accuracy
Cultural	Separated by walls or doors	<2 m
Exposition	Constrained by exposition structures	<1.5 m
Objects of interest	Object size dependent	<1 m

In an experiment performed at the Archaeological Museum of Camarina in Sicily (Italy), we placed 31 BLE beacons, in museum roof or walls, at a height of about 2.5 m, in 239 reference points not uniformly spaced to provide full coverage of site walkable areas, with a density of 1 beacon for each 25 m². The time of transmission was set to 100 ms whereas transmit power was set between −59 and −72 dB (at 1 m).

A plausible route for visiting the museum was defined: a complex itinerary and 40 points, along the route, were marked as test points [5].

A raw static reference radiomap was then obtained by averaging RSSI (Received Signal Strength Indication) values for each beacon in reference points and by interpolating values on a grid of one-meter cell size via triangular linear interpolation. Particle Filter was used to estimate the accuracy of position of visitors along the test route. The Cartesian distance between fingerprints in the sample and in reference points (radiomap) was used as the control measure for the Particle Filter.

Figure 1 represents the interpolated map of fingerprint spatial derivative in the museum BLE field. It shows the average value of distances between fingerprint at each grid point and ones in its spatial neighborhood. Where spatial gradient is low, space is poorly structured and Particle Filter will reduce accuracy.

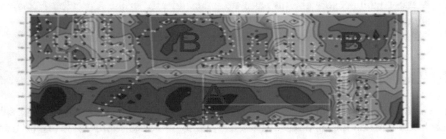

Fig. 1. Interpolated map of fingerprint spatial derivative in the museum BLE field. In the lower part of the site (courtyard, indicated with the letter A) and in largest halls space (indicated with the letter B) the environment is poorly structured and Particle Filter performs worse.

Thanks to dense sampling, accuracy of 2 m was reached (at an average density of about 0.3 samples square meter) and further increased with denser sampling in proximity of singularities of structural elements.

In particle filtering simulation on the advanced track, we obtained an accuracy error of 1.80 m in the average in 10 runs, and a max error of 3.20 m in 90% of cases (Fig. 2).

Unfortunately, dense sampling is in fact a not economic procedure and several weeks were spent in performing beacon position optimization. In [7] Kriz et al. describe a comparable dense sampling experiment where precision lower than 1 m. was obtained with a similar effort and just in a confined poorly structured environment (mainly a corridor). In the following, we describe steps towards smart iterative strategies of not uniform sampling actuated by a robot.

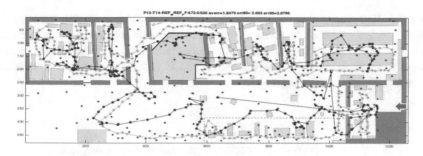

Fig. 2. Monte Carlo Simulation of tracking using Particle Filter with 672 particles. Red dot clouds show waypoints position estimates.

4   The Robotic Platform

The support robotic platform under development is mainly composed by three components: a simulation and control component, a robot equipped with sensors and a mobile BLE receiver and data logger.

The simulation and control component is implemented in Matlab because of its easiness of data representation, programming capabilities, and simplicity of its syntax based on linear algebra. It leverages functions and utilities from the graphics, simulation and robotics toolboxes.

The robot component consists in a Turtlebot2 research robot driven by open source controllers running on an onboard Ubuntu laptop. The data logger is an Android mobile device with a BLE sensor hosting a Java application developed on the Android studio mobile programming environment.

All components use open source packages and interact exchanging messages through the ROS framework.

4.1   ROS

ROS is an open-source middleware that provides libraries and tools to help the software developers create applications for robotics [8, 9].

It is a meta-operating system including the typical characteristics of a real operating system (hardware abstraction, process management, package management, low-level device control) but enriching it with elements of a middleware (provides the infrastructure for communication between processes/different machines), and a framework (utility tools for developing, debugging and simulation).

The Turtlebot2 platform has installed the ROS Indigo Igloo release, running on a laptop equipped with Ubuntu 14.04 LTE Operating System. An open source pure Java ROS implementation compatible with Android s used on the mobile data logger, an LG V490 tablet running Android 4.3 (Jelly Bean).

Finally, the Matlab Robotics System Toolbox provides an interface to communicate with a ROS network, interactively explore robot capabilities, and visualize sensor data between Matlab, Simulink and ROS.

In the ROS nomenclature, a "node" is a process. Each node can communicate, through "topics", with other nodes, using a specific data structure called "message". Messages are sent using the Publisher/Subscriber pattern [10], a process can subscribe to one or many topics to receive the published data. Figure 3 shows the message channels connecting the platform components.

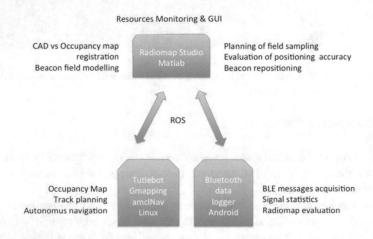

Fig. 3. Activities and roles of the three platform components

4.2   The Robot Component

The Turtlebot2 is a low-cost research robot for robotics applications. It is based on a Yujin's Kobuki electro mechanical base, equipped with proximity sensors, a gyroscope, a depth camera and a mechanical structure able to carry on its own software controller (a notebook) and some payload (the logger tablet in our case).

The open source SDK includes a GUI and application modules for tele-controlled and Autonomous Monte Carlo Localization (AMCL) of a known environment, as well as a module for creating an occupancy grid map in the environment based on range measures estimated from depth camera images (GMapping). The available software modules communicate with external components via ROS.

Sections 6 and 7 describe some details of the GMapping and amclNavigation modules and their use in our platform.

4.3   The Android Logger

The Android logger is an interactive Java mobile app developed and debugged with tools of the Android Studio programming environment. At present, it is devoted to detect and register beacon advertisement messages and to perform simple statistics on them.

The logging procedure is optimized for catching as many messages as possible from the various beacons to allow signal amplitude estimates at frequency of 1 Hz. The logger waits for synchronization commands from the controller, and then it starts iteratively a timed scan of BLE channels and records the contents of the received packets. The scan is performed using the MAX_FREQUENCY_SCAN mode of the Android BLE API to catch as many packets as possible. Detection of up to 3 packets per second from the same source has been experimented.

Statistics of the received messages are evaluated and stored in a raw representation of the radiomap and uploaded to the controller via a dedicated ROS topic.

5   The Control and Simulation Component

The control component implements a simple Integrated Development Environment (IDE) supporting the creation and update of radiomaps. It includes a main GUI for coordinating activities and viewers to display pictorial or vector maps and to manage feature points, tracks, structure elements represented as vectors (Fig. 4).

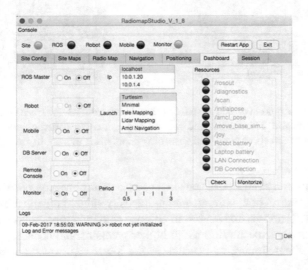

Fig. 4. Resources status monitoring in the Dashboard Tab of the Control Component GUI

The front end GUI contains controls such as buttons, edit fields, menus, toolbars and indicators.

Active panels are devoted to configure or control main activities, namely: Session (for work session management), Dashboard (for node activation, connection settings, active topic verification and process management), Site Configuration (for geometric configuration features and site map resolution), Site Maps (for occupancy grid map management), Radio Map (for radiomap management), Navigation (for setting the initial position of the robot, setting goal, track loading and manual navigation guided by remote control operations) and Positioning (for activating graphical tool to create track and choose measurement points for radiomap building) and Logger (for acquisition data process management and communication with logging data app).

The simulation tool includes, at present, procedures for map interpolation, evaluation of field statistics, feature distance metrics, Monte Carlo a priori probability distributions, and for simulation of Particle Filter based tracking.

Figure 5 shows a simple example of the usage of map interpolation during selection of meaningful reference points for sampling in a site (see CAD map in Fig. 7). After sampling of the field of a beacon in a limited number of reference points distributed pseudo randomly in the corridor area, interpolation (Fig. 5 left) helps to identify new candidate points where site structures is expected to deform field shape. Figure 5 on the right shows the restructured field after resampling.

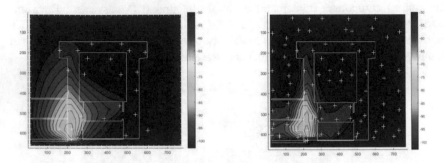

Fig. 5. First approximation of Beacon field (interpolation from 23 reference points sampled + 40 points valuated at site boundaries) and second approximation (interpolation of 85 reference points). Structuring effects of walls and absorbing doors made evident in the left lower quadrant of figure.

Figure 6 shows a simple field statistics used to control beacon positioning. It represents the interpolation of maximum among intensity values perceived from any beacon after the first (left) and second stages. In absence of symmetries, areas with high spatial signal gradients are expected to lead to higher precision and accuracy in positioning. Optimization of positioning of beacons in the lower right quadrant of the figure is

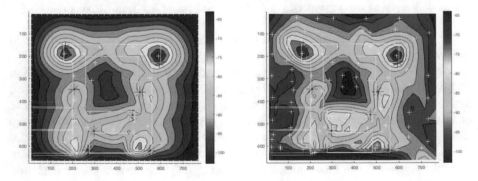

Fig. 6. Left: First approximation of global beacons field (interpolation of maxima of 23 reference points sampled + 40 points valuated at site boundaries). Right: second approximation (interpolation of 85 reference points). Structuring effects of walls and absorbing doors made evident in the left lower quadrant of figure.

probably required, while it is not worthwhile in the center area, just because it is a not accessible one.

6 Mapping Site and Registering Its Occupancy Map

In order to create the navigation map we used the GMapping algorithm. It is a highly efficient algorithm of construction of the map starting from laser scans.

In particular, the algorithm we use is the one that uses a grid-based map (occupancy grid) and the estimation technique known as Rao-Blackwellized particle filters (RBPF). This approach was developed by Grisetti et al. [11, 12] and it constitutes an ameliorative version of the algorithm proposed by Hahnel et al. [13].

The OpenSLAM ROS GMapping package is based on this algorithm. It combines odometry data with laser range finder (or depth image) data, to produce a 2-D occupancy grid map; this map represents the world as seen by the robot itself, in terms of free space and obstacles (e.g. walls, furniture, etc.).

The obtained map is not perfect because of systematic and accidental errors, in particular noise in the odometry and noise in the sensor data. The datasets gathered with robot can in fact present some inconsistency because of odometry error, the maximum range of the sensor and presence of moving objects. These errors can be minimized by carefully guided navigation in the selected environment. It is commonly known that rapid rotations, high speed of the robot, not visibility of points by the laser, people walking along with the robot, in view of its laser (people close to the robots appear in fact as objects in the resulting map), represent noise factors in the creation phase of the map. Furthermore, loop closure is the hardest part: when closing a loop, it is necessary to drive another 5–10 m to get plenty of overlap between the start and end of the loop. In addition, can be present errors to wheel slip, etc.

The main inconvenient that comes from using maps created by the robot itself is that obstacles representation is static, the map cannot be edit or update after its creation. Every time something is moved from its recorded position the map is no more a realistic representation of the environment, so it is necessary to create again the map.

For all these reasons, after acquiring the occupancy grid map, the built map is recorded and eventually modified (edited or updated) through operations managed by the control system (Site Maps tab of console), such as panning, tilting, scaling and clipping, in order to make it compatible with the environment image (for example CAD format) if available. Purpose of this operation is to make compatible the occupancy grid with the initial map of the environment created by human operator (CAD format, etc.).

In Fig. 7 are depicted the reference site image (CAD format) and the occupancy map obtained by robot while navigating open spaces and a few accessible rooms.

Optimization of this process will be the subject of future research, which will consist in the search for methodologies to compute consistent maps and so to obtain an accurate map of robots very close to the reference image.

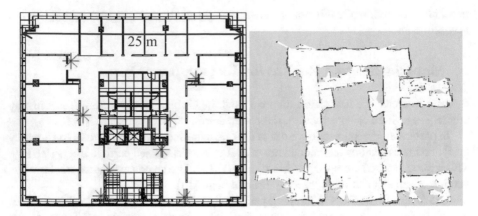

Fig. 7. Reference map of aisle and rooms of II B floor with beacons (on the left) and navigation map (on the right)

7   Navigation and Path Tracking

To build and update radiomaps we need to collect a large amount of data, from conveniently chosen locations in the environment.

Path planning is an interactive procedure used to select a set of locations in the map, where direct the robot to collect data. That procedure helps to estimate the distribution of the field and consequently to evaluate in real time where the accuracy can be improved.

The recursive algorithm for path tracking consists of several steps. The robot needs to have initial pose coordinates assigned, in order to localize itself in the maps, sometimes needs also to execute some calibration revolution, to refine the coordinates and orientation values. After selecting the destination point by clicking on the map, a track path is saved and an iterative routine starts. For each point of the path track, a ROS message containing the destination coordinates is sent, and then the AMCL algorithm begins and obtains a trajectory to be followed if one exist. During the route time, the ROS topic containing the status of the task is checked, to verify if the desired location is reached or not. If no problem occurred during the route and the result is successful, the data acquisition routine on tablet starts and after a predefined time slot stops, storing RSSI values in a relational Database, with location coordinates, otherwise, the algorithm skips to the next point of the track, logging and notifying that an error has happened (Fig. 8).

In our experiment, we placed 8 BLE beacons, in roof or walls, at the height of about 2.4 m, with a density of 1 beacon for each 78 m^2 (the position is shown by the red asterisks on Fig. 7 left). The time of transmission was set to 100 ms while transmit power was set to −72 dB (at 1 m), according to the size of environment. RSSI values were monitored in each reference point at least 10 s.

The beacons were positioned asymmetrically to provide complete coverage of the site. Measurements were taken by a robot holding a tablet at height of 1 m, oriented always in the same direction.

Fig. 8. Turtlebot2 equipped with data logging tablet

8 Conclusions and Future Work

This work describes some features of an interactive robotic platform under development at ICAR-CNR to support BLE field modeling for indoor positioning. The use of intelligent iterative strategies, in dense field sampling and in optimization of beacons positions, is beneficial to improve accuracy of positioning based on fingerprint.

We sketched main features of the three platform components, namely the simulation and control node, the Turtlebot2 robot and finally the data logger, all developed as far as possible reusing open-source software packages. In particular using ROS as the interconnecting framework, GMapping and amclNavigation robotics packages on the robot node, Android BLE scan APIs on the mobile receiver, the robotics plugin and extensions on the Matlab controller.

The usage of the robotics platform is still at an initial stage and a number of problems have to be overcome. Among others:

- ROS is dependent on stability of Wi-Fi connection. The latter can drop in presence of network extenders, because of timeout while switching from one device to another.
- *GMapping* is a self-consistent procedure and registration between its output occupancy map and a site CAD map must be done manually;
- *amclNavigation* is sometimes unable to find a path to some points in the map and a strategy has to be developed to allow robot to skip unreachable points while following a mission track.

References

1. Gallozzi, A., et al.: The MuseBot project: robotics, informatic, and economic strategies for museums. In: Handbook of Research on Emerging Technologies for Digital Preservation and Information Modeling. IGI Global (2017). Chap. 3
2. Georgiadi, N., et al.: A pervasive role-playing game for introducing elementary school students to archaeology. In: Mobile-CH 2016: Workshop on Mobile Access to Cultural Heritage (in Conjunction with ACM Mobile HCI 2016), MobilcHCI 2016 Proceedings of the 18th International Conference on Human-Computer Interaction with Mobile Devices, Florence (Tuscany), Italy, 6–9 September 2016
3. http://www.turtlebot.com
4. Durrant-Whyte, H., Bailey, T.: Simultaneous localization and mapping: part I. IEEE Robot. Autom. Mag. **13**(2), 99–110 (2006)
5. Dierna, G.L., Machì, A.: Towards accurate indoor localization using iBeacons, fingerprinting and particle filtering. In: 2016 International Conference on Indoor Positioning and Indoor Navigation (IPIN), Alcalá de Henares, Spain, 4–7 October 2016
6. http://www.mathworks.com
7. Kriz, P., et al.: Improving indoor localization using bluetooth low energy beacons. Mob. Inf. Syst. **2016** (2016). Hindawi Publishing Corporation, Article ID 2083094, 11 p.
8. http://www.ros.org
9. Quigley, M., et al.: ROS: an open-source robot operating system. In: ICRA Workshop on Open Source Software 2009, vol. 3(3.2) (2009)
10. Patrick Thomas Eugster et al: The many faces of publish/subscribe. ACM Comput. Surv. **35**(2), 114–131 (2003)
11. Grisetti, G., et al.: Improving grid-based SLAM with Rao-Blackwellized particle filters by adaptive proposals and selective resampling. In: Proceedings of the IEEE International Conference on Robotics and Automation (ICRA) (2005)
12. Grisetti, G., et al.: Improved techniques for grid mapping with Rao-Blackwellized particle filters. IEEE Trans. Robot. **23**, 34–46 (2007)
13. Hahnel, D., et al.: Map building with mobile robots in populated environments. In: Proceedings of the IEEE/RSJ International Conference on Intelligent Robot and Systems (IROS) (2002)

Improving Spatial Reasoning by Interacting with a Humanoid Robot

Agnese Augello[1]([✉]), Giuseppe Città[2], Manuel Gentile[2], Ignazio Infantino[1],
Dario La Guardia[2], Adriano Manfré[1], Umberto Maniscalco[1],
Simona Ottaviano[2], Giovanni Pilato[1], Filippo Vella[1], and Mario Allegra[2]

[1] ICAR - National Research Council of Italy,
Via Ugo la Malfa 153, 90146 Palermo, Italy
{agnese.augello,ignazio.infantino,adriano.manfre,umberto.maniscalco,
giovanni.pilato,filippo.vella}@icar.cnr.it
[2] ITD - National Research Council of Italy,
Via Ugo La Malfa 153, 90146 Palermo, Italy
{giuseppe.citta,manuel.gentile,dario.guardia,
simona.ottaviano,mario.allegra}@itd.cnr.it

Abstract. This paper analyzes the connection between spatial reasoning and STEM education from the point of view of embodied theories of cognition. A new learning model based on the use of a humanoid robot is presented with the aim of teaching and learning basic STEM concepts in a fruitful and engaging fashion.

Keywords: Mental rotation · Spatial reasoning · Embodied cognition · STEM · Cognitive architecture · Humanoid robots

1 Introduction

Improve students achievement in the fields of science, technology, engineering and mathematics has long been a stated goal for several national school systems. Several researches show some connections between measures of cognitive skills and long-term economic development not only at a quantity level [1,2] but also at a quality level. The Organisation for Economic Co-operation and Development highlighted the key role of such an aspect revealing that quality in education, observed through specific tests on cognitive skills in mathematics and science predicts and influence the economic outcomes of a country [3]. Some studies confirm this overall position [4,5]. Also UNESCO emphasized the economic relevance of connecting education with fields as engineering and mathematics and described this kind of connection as a foundation for the development of society [6,7].

Within this context, an improvement of science, technology, engineering, and mathematics (STEM) competencies enables young people to cope the complexities of the modern life. Despite of the increasing demand of skilled people in this area, there is a lack of offer. For this reason, various research programs and educational policies in many parts of the world were launched to support

© Springer International Publishing AG 2018
G. De Pietro et al. (eds.), *Intelligent Interactive Multimedia Systems and Services 2017*,
Smart Innovation, Systems and Technologies 76, DOI 10.1007/978-3-319-59480-4_16

a multidisciplinary approach to teaching these topics [8,9]. For example, within european countries area, the InGenious project[1] represents a great joint initiative aimed at strengthening the link between science education and careers in STEM fields. The actions of inGenious aim to improve the image of STEM and STEM careers among young people, encouraging them to think about the wide range of opportunities that working in science provides.

The interest in STEM education is steadily growing and the research in the area, investigating the different involved aspects (e.g. cognitive, social, economic), is therefore committed to checking if and how it is possible to improve the effectiveness of STEM education programs.

Within this context, in the literature, the connection between spatial reasoning and STEM education has been faced by mainly highlighting the mental and abstract aspects of this phenomenon. The research in the field focuses on the mental manipulation of three-dimensional spatial forms within spatial visualization tasks along different STEM disciplines such as biology [10], physics [11] or chemistry [12]. Moreover, this linking has been highlighted referring to the possibility to strongly predicts achievement and attainment in science, technology, engineering, and mathematics starting from the analysis of spatial reasoning [13,14].

The core idea of this paper arises from two main remarks by [15,16] about the relationship between STEM education and spatial reasoning:

– "spatial training can help novices in STEM education because they rely more on decontextualized spatial abilities than experts do" [15];
– effectiveness, durability and transferability should be key features of every spatial reasoning training [16].

Referring to the cognitive aspect of STEM educational paths and the role of some specific technological tools, we aim to investigate how STEM learning can be affected by spatial reasoning within structured training sessions in primary school. The goal is to set a suitable context where analyze when, why and how training spatial abilities positively affect STEM learning [15].

Moreover, taking into account the theory of embodied cognition [17] we will present a proposal of an experimental setting able to explore how working on the improvement of some skills involved in spatial reasoning and in particular in mental rotation processes. The proposed setting fosters new paths and methodologies for primary school with the aim of teaching and learning basic STEM concepts in a fruitful and engaging fashion by means of a humanoid robot able to guide students in the accomplishment of the learning activities.

The use of humanoid robots in STEM and STEAM education in the light of embodied cognition theory is not deeply investigated. An experiment has been conducted in [38], where a Nao robot has been proposed as assistant to the educational staff of a kindergarten, in playing educational games aimed to geometrical thinking. To the best of our knowledge there are no other STEM

---

[1] Project funded by the European Commission's 7th Framework Programme and supported of major international industry partners http://www.ingenious-science.eu.

environments that employ humanoids in activities aimed at improving spatial reasoning abilities.

In the following sections, we analyse the mental rotation processes in the light of embodied theories of cognition discussing a possible experimental setting involving a humanoid robot.

2   Mental Rotation Process in the Light of Embodied Theories of Cognition

The expression "spatial reasoning" denotes a complex cognitive phenomenon. This complexity is revealed by a lack of a generally accepted definition of spatial reasoning [15] and by the existence of different opposing perspectives that suggests spatial reasoning as a single complex skill or as a multilayered skill that emerges from the relation among different cognitive components [18].

Although the absence of a unique perspective, according to [19], the different points of view agree that spatial reasoning is an ability that "connects a perceived and constructive 3D world". Within this context, spatial reasoning involves abilities as generating, representing, transforming and recalling nonlinguistic information [20], and abilities that concern with the representation of an object, the use of an object, the 2D and 3D relationships among objects and their actual contexts [21].

According to this view, the knowledge of space coordinates emerges as the fundamental of every action. Visuospatial abilities are the basis to perceive and to act, and operate on mental representations according to the spatial coordinates. The set of visuospatial functions allows a wide range of activities, praxisbuilding skills and topographical orientation, locate stimuli in space and evaluate distance, orientation and size.

In this paper, according to [22], we will embrace a definition of spatial reasoning in a broad sense referring to it as *a set of related skills that include symmetrizing, balancing, locating, orienting, decomposing/recomposing, visualizing, scaling comparing, transforming and navigating* [23].

Specifically, we will focus on one main process within these skills are clearly involved: mental rotation.

Mental rotation processes are usually described as shape-matching activities where an actor has to decide whether two objects, at the same time or sequentially presented, and from different angular orientations proposed, are different or identical [24]. The label mental is from a description by Metzler and Shepard [24,25] that describe the process as based on a specific visuo-spatial ability thanks to which a cognizer imagines how a 2D or a 3D object looks like when is rotated.

In a mental rotation process, complex cognitive skills are required "to manipulate the imagines in a dynamic way that is the ability to make the mental image of a given 2D or 3D object turning in space" [24].

In this paper, we suggest considering the process of mental rotation and its relation to STEM education within some basic concepts of two theoretical frameworks belonging to embodied cognition theories area: Enactivism [26] and Sensorimotor Theory [27].

Within this perspectives, each activity of mental rotation of an object, in a learning setting, can be conceived as a process emerging from a continuous interaction among knowers' bodily actions, cognitive processes and environment. In other terms, according to these embodied points of view, a joint action among these components highlights the essential role of motor processes, body actions and the world as crucial actors [28]. Actions and perception of a body are the key elements of the knowledge construction involved in each activity since they are organs for experiencing the world [29]. Such a perspective redefines the concept of cognition in general and, in particular, the features of those elements defined as mental.

Our proposal of new paths and activities for teaching and learning some basic STEM concepts takes into account this framework. It confers a key role to a humanoid robot that acts as an embodied and enacted object able to rotate and to make a rotation of other objects. Such a setting reveals a fundamental theoretical aim of this paper: suggesting that mental rotation is only the last step of a process of knowledge construction where corporeal and actional elements affect, and are affected by, other related components, such as [30]:

- proprioception, sensory information revealing the position of the self and movements;
- kinaesthesia, the particular sensation through which we perceive movements and bodily position;
- working memory and the executive control related to it, the way in which we manage information from the environment to achieve goals in specific tasks during an action.

An educational training that takes into account these aspects seems to be a suitable framework for implementing an effective, durable and easy transferable training of spatial reasoning to improve novices learning in STEM area.

3   The Proposed Learning Model for Mental Rotation Training

The proposed learning model is conceived as an experimental design with the aim at verifying if an embodied training activities performed with a humanoid robot helps to improve the mental rotation abilities in young children aged from 5 to 6.

In order to evaluate the development of the students' spatial abilities, three different tests will be administered before and after the learning activities.

The following measures will be used:

- the Mental Rotation Test (MRT) developed in 1978 by Vandenberg and Kuse [31] according to the basic ideas of Shepard and Metzler's. The MRT is composed by different image-stimulus pairs elaborated in two-dimensional images of three-dimensional objects drawn by a computer. The image is shown on

an oscilloscope and rotated around a vertical axis. This test asks to subject to compare two 3D objects that rotate according to the different axis, Participant must define if the objects are the same image or if they are mirror images. The score is related to correctly and quickly answer that they can distinguish between the mirrored and non-mirrored pairs.
– the Mental Cutting Test (MCT) that consists of 25 items in which the administrator shows a criterion figure which is to be cut with an assumed plane. The student has five alternative answers to choose the correct cross-section [32].
– the Rey Osterrieth complex figure test (ROCF) that is a perceptual visual test, which allows evaluating the visual-spatial skills of a subject that must reproduce a complicated line drawing, using his perceptual organisation and visual working memory [33].

The student will be engaged in a learning path which consists of 8 two-hour meetings which might take place once or twice a week, in order to give the students the indispensable time to interiorize the trained abilities and, at the same time, to allow the learning environment to create a profile for each involved student.

The training sessions between the pre and post measurement will be split into different phases:

– **phase1**: in this phase, the child assumes the postures, with the support of the teacher; for each posture, the robot first takes the reflected posture, then it replicates the child's posture;
– **phase 2**: in this phase the robot assumes the postures as described in the next section, asking the child to replicate the posture according to a mirror function or perspective taking function. The system evaluates the movement of the child to check if he/she performs the task correctly. The sequence that leads from the starting position to the final position is analysed to understand better the causes of potential errors in order to profile the student.
– **phase 3**: according to extracted student's profile a personalised learning path is proposed to the child, in order to train the student's abilities with personalised tasks in order to correct the typical errors he/she produce.

4   Humanoid Cognitive Architecture

The cognitive processes of the humanoid rely on the cognitive architecture depicted in Fig. 1.

The robot learns to execute and coordinate its movements and to act according to the purposes of the learning activities. By process of observation it acquires a dataset of postures, that is then stored in the long term memory (LTM) and used as input in the interaction with the students. The different postures acquired in this phase are also clustered to allow the robot to establish a similarity among them. The postures are arranged in sequences according to the learning activities.

The robot acts according to its motivation, influenced by its *urges* to satisfy physiological, mental, or social demands. The urges that require immediate

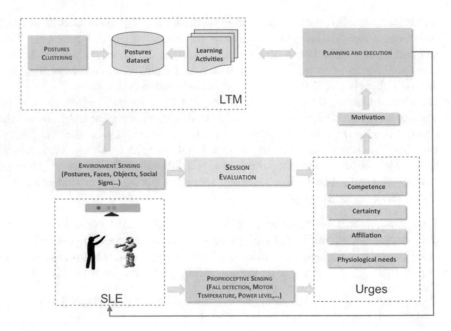

Fig. 1. Humanoid architecture

attention or action determine a motivation to accomplish a correct behaviour. The urges are influenced by different factors, such as the results of evaluation mechanisms [34] (in the specific case the evaluation of the students learning sessions) and the robot internal state, sensed by its somatosensory system [35]. A detailed description of the urges is reported in [34].

During the student's learning sessions, the humanoid chooses from time to time the sequence of postures to produce. Its behaviour it is also driven by what it senses from the external environment. The recognition of a face, the result of the interaction and the observation and evaluation of the postures performed by the students can lead to the accomplishing or a replanning of the activities. For the first experimentations, the humanoid robot NAO [36] will be used.

4.1   Postures Acquisition

During the training, the robot acquires the set of postures defined during the design of the learning activities. A 3D acquisition device (RGBD camera) captures postures performed by a child. Such an approach is not invasive because children have not to wear particular devices, and it is also cheaper with respect to other motion capture solutions. In particular, we use a Microsoft Kinect v1 in our system. It is a horizontal bar equipped with different sensors: an RGB camera, an infrared (IR) emitter and an IR depth sensor, to capture a color image and a depth image, and four microphones for capturing sound.

The captured output is a skeleton, i.e. a list of 3D positions of relevant points of the human body. The open source ROS (Robotics Operating System) framework [37] will be used to process the acquired data.

By mean of the Kinect sensor, it is possible to extract the 15 joints values of the human. Since the NAO size and movements are different from human ones, the skeleton data are transformed in the robots coordinate space. Precisely, the human skeleton values are translated into the corresponding joints robot values. The skeleton markers package of ROS was used to detect the motion of the child captured by the Kinect and tracks the position of arms, head and legs.

The acquired and processed data are stored into the Long Term Memory (LTM) of the robot. Each posture is encoded by a vector of 9 float values representing in radians the angles of the robot's joints. Then a clustering of the acquired postures is obtained by adopting a K-Means approach. The data are then partitioned into K mutually exclusive groups, where the elements in a cluster are close as possible and different from the elements belonging to other clusters, according to a city-block distance.

4.2   Student's Evaluation and Activities Replanning

As discussed before, each activity with the students is composed of three main phases. In the first phase the robot reproduces the postures of the students. The student's postures are acquired with the same methodology discussed in the previous section and therefore translated in the robot's coordinate space and reproduced by the humanoid.

In the second phase, the robot has an active role. It assumes a sequence of postures required by the specific learning activity which are interpreted by the child to assume the right posture as a response. The robot analyses the student posture by evaluating the city block distance between the acquired skeleton and the expected one. If necessary, (for example the child assumes an unknown posture) the most similar posture in the LTM is compared with the acquired skeleton opportunely translated in the robot coordinate space.

According to the evaluation the robot decides if continue with the ongoing activity or if is necessary to replanning it. The replanning leads to the third phase, where a new activity is composed. The robot can also propose again the activity insisting on the postures that generated confusion and mistakes in the students. In this case, the robot introduces a variability in the sequence, choosing other similar postures (i.e. a posture belonging to the same cluster).

5   Conclusion

In this paper, a new learning model to teach and learn basic STEM concepts in a fruitful and engaging fashion is presented. The proposed experimental setting is designed according to the theories of embodied cognition of Enactivism and Sensorimotor theory, to improve the spatial reasoning skills involved in mental rotation processes. The cognitive architecture of a humanoid robot, exploited as an embodied and enacted object able to rotate and to make a rotation of other objects, is presented. Future works will regard the implementation and testing of the proposed STEM approach.

References

1. Sianesi, B., Reenen, J.V.: The returns to education: macroeconomics. J. Econ. Surv. **17**(2), 157–200 (2003)
2. Krueger, A.B., Lindahl, M.: Education for growth: why and for whom? (No. w7591). National Bureau of Economic Research (2000)
3. Hanushek, E.A., Woessmann, L.: The High Cost of Low Educational Performance: The Long-Run Economic Impact of Improving PISA Outcomes. OECD Publishing, Paris Cedex (2010). 2, rue Andre Pascal, F-75775 16, France
4. Hanushek, E.A., Woessmann, L.: Do better schools lead to more growth? Cognitive skills, economic outcomes, and causation. J. Econ. Growth **17**(4), 267–321 (2012)
5. Hanushek, E.A., Woessmann, L.: The role of cognitive skills in economic development. J. Econ. Lit. **46**(3), 607–668 (2008)
6. UNESCO 2010, Engineering: Issues, challenges and opportunities for development,UNESCO, the World Federation of Engineering Organisations, the InternationalCouncil of Academies of Engineering and Technological Sciences, and the International Federation of Consulting Engineers
7. UNESCO 2007, Science, technology and gender: An international report, Science-and Technology for Development Series, UNESCO, Division for Science Policy and Sustainable Development
8. Marginson, S., Tytler, R., Freeman, B., Roberts, K.: STEM: country comparisons: international comparisons of science, technology, engineering and mathematics (STEM) education. Final report, Australian Council of Learned Academies (2013)
9. Kanematsu, H., Barry, D.M.: STEM and ICT Education in Intelligent Environments, vol. 91, pp. 3–198. Springer (2016)
10. Russell-Gebbett, J.: Skills and strategiespupils' approaches to three-dimensional problems in biology. J. Biol. Educ. **19**(4), 293–298 (1985)
11. Kozhevnikov, M., Motes, M.A., Hegarty, M.: Spatial visualization in physics problem solving. Cogn. Sci. **31**(4), 549–579 (2007)
12. Wu, H.K., Shah, P.: Exploring visuospatial thinking in chemistry learning. Sci. Educ. **88**(3), 465–492 (2004)
13. Wai, J., Lubinski, D., Benbow, C.P.: Spatial ability for STEM domains: aligning over 50 years of cumulative psychological knowledge solidifies its importance. J. Educ. Psychol. **101**(4), 817 (2009)
14. Shea, D.L., Lubinski, D., Benbow, C.P.: Importance of assessing spatial ability in intellectually talented young adolescents: a 20-year longitudinal study. J. Educ. Psychol. **93**(3), 604 (2001)
15. Uttal, D.H., Cohen, C.A.: 4 spatial thinking and STEM education: when, why, and how? Psychol. Learn. Motiv. Adv. Res. Theor. **57**, 147 (2012)
16. Uttal, D.H., Meadow, N.G., Tipton, E., Hand, L.L., Alden, A.R., Warren, C., Newcombe, N.S.: The malleability of spatial skills: a meta-analysis of training studies. Psychol. Bull. **139**, 352–402 (2013)
17. Shapiro, L.: Embodied Cognition. Routledge, London (2010)

18. Gersmehl, P.J., Gersmehl, C.A.: Spatial thinking by young children: neurologic evidence for early development and educability. J. Geogr. **106**(5), 181–191 (2007)
19. Nagy-Kondor, R.: Spatial ability: measurement and development. In: Visual-Spatial Ability in STEM Education, pp. 35–58. Springer International Publishing (2017)
20. Linn, M.C., Petersen, A.C.: Emergence and characterization of sex differences in spatial ability: a meta-analysis. Child Dev. **56**, 1479–1498 (1985)
21. Williams, C.B., Gero, J., Lee, Y., Paretti, M.: Exploring spatial reasoning ability and design cognition in undergraduate engineering students. In: ASME 2010 International Design Engineering Technical Conferences and Computers and Information in Engineering Conference, pp. 669–676. American Society of Mechanical Engineers, January 2010
22. Khan, S., Francis, K., Davis, B.: Accumulation of experience in a vast number of cases: enactivism as a fit framework for the study of spatial reasoning in mathematics education. ZDM **47**(2), 269–279 (2015)
23. Bruce, C.D., Moss, J., Sinclair, N., Whiteley, W., Okamoto, Y., McGarvey, L., Davis, B.: Early years spatial reasoning: learning, teaching, and research implications. In: Workshop Presented at the NCTM Research Presession: Linking Research and Practice, Denver, CO. (2013)
24. Shepard, R.N., Metzler, J.: Mental rotation of three-dimensional objects. Science **171**(3972), 701–703 (1971)
25. Metzler, J., Shepard, R.N.: Transformational studies of the internal representation of three-dimensional objects (1974)
26. Di Paolo, E.A., Thompson, E.: The enactive approach. In: Shapiro, L. (ed.) The Routledge Handbook of Embodied Cognition, pp. 68–78. Routledge, New York (2014)
27. Bishop, J.M., Martin, A.O. (eds.): Contemporary Sensorimotor Theory. Springer, Heidelberg (2014)
28. Smith, L.B.: Cognition as a dynamic system: principles from embodiment. Dev. Rev. **25**(3–4), 278–298 (2005)
29. Määttänen, P.: Experience and Embodied Cognition in Pragmatism, vol. 18. Springer, Cham (2015)
30. Crifaci, G., Città, G., Raso, R., Gentile, M., Allegra, M.: Neuroeducation in the light of embodied cognition: an innovative perspective. In: Proceedings of the 2015 International Conference on Education and Modern Educational Technologies (EMET 2015), pp. 21–24 (2015)
31. Vandenberg, S.G., Kuse, A.R.: Mental rotations, a group test of three-dimensional spatial visualization. Percept. Motor Skills **47**(2), 599–604 (1978)
32. CEEB Special Aptitude Test in Spatial Relations, developed by the College Entrance Examination Board, USA (1939)
33. Osterrieth, P.A.: Le test de copie d'une figure complexe. Arch. Psychol. **30**, 206–356 (1944)
34. Augello, A., Infantino, I., Manfrè, A., Pilato, G., Vella, F., Chella, A.: Creation and cognition for humanoid live dancing. Robot. Auton. Syst. **86**, 128–137 (2016)
35. Augello, A., Infantino, I., Maniscalco, U., Pilato, G., Vella, F.: The effects of soft somatosensory system on the execution of robotic tasks. IEEE Robot. Comput. (2017)

36. NAO robot. https://www.ald.softbankrobotics.com/en
37. Robotics Operating System. http://wiki.ros.org/it
38. Keren, G., Ben-David, A., Fridin, M.: Kindergarten assistive robotics (KAR) as a tool for spatial cognition development in pre-school education. In: 2012 IEEE/RSJ International Conference on Intelligent Robots and Systems (IROS). IEEE (2012)

An Artificial Pain Model for a Humanoid Robot

Umberto Maniscalco$^{(\boxtimes)}$ and Ignazio Infantino

Cognitive Robotics and Social Sensing Lab., Istituto di Calcolo e Reti ad Alte
Prestazioni - C.N.R., Via Ugo La Malfa, 153, Palermo, Italy
{umberto.maniscalco,ignazio.infantino}@icar.cnr.it
http://www.icar.cnr.it

Abstract. The primary aim of this paper is to give to a humanoid robot
the sense of pain. That because the pain is one of the ways to which we
protect our body. Thus, we consider that a humanoid robot can take
into account this sense in its behaviours to make them more realistic.
Moreover, sensing the pain the robot can act in a protecting way. The
perception of the pain is just one of the peculiarities, though one of the
most important, of a somatosensory system that is our concluding goal.

Keywords: Soft sensor · Cognitive robotics · Somatosensory system ·
Artificial pain

1 Introduction

The complex system of nerve cells, including thermoreceptors, mechanoreceptors,
chemoreceptors and nociceptors, in the humans, is known as the somatosensory
system [1]. It is a distributed system in which there are three different functional
components:

1. The perceptive function, performed by the sensory receptors spread over all
   body including the skin, epithelial tissues, skeletal muscles, internal organs,
   and also in the cardiovascular system.
2. The transport function, operated by a set of fibres of different kinds each one
   dedicates to carry a specific kind of stimulus.
3. The elaboration function fulfilled by the somatosensory cortex.

Of course, it is a very simplified schema, but it describes the macroscopic
aspect, different function and, above all, this is a useful guideline when we want
to design an artificial somatosensory system for a humanoid robot.

The somatosensory system acts as the interface between you and not-you
and it includes several bodily senses, codes the experiences and classifies them
influencing the human behaviours to protect us from pain, risks or dangers. The
free nerve endings named nociceptors, widely cover the human body, they are
responsible for the perception of the pain and, they bring information from the
body's periphery toward the somatosensory cortex.

The nociceptors transform an external energy (mechanical or thermal) in an
internal electrical energy. Thus, they act like the transducers sensors. The value

© Springer International Publishing AG 2018
G. De Pietro et al. (eds.), *Intelligent Interactive Multimedia Systems and Services 2017*,
Smart Innovation, Systems and Technologies 76, DOI 10.1007/978-3-319-59480-4_17

of the internal electrical energy generate by the nociceptors depends on both the magnitude and the area covered by the external stimulus. As shown in the left side of Fig. 1, the activation function of the nociceptors is proportional to the external stimulus, in this case, the pressure, but it must be strong enough to exceed the threshold. Moreover, the trend of the number of potential action per second is an exponential function of the external stimulus as shown in the right side of the Fig. 1. This function tends to saturate growing the external stimulus.

The nociceptors do not perform the real perception of the pain; they are only the sensors able to localise the source of the pain. The perception of the kind of the pain is performed by identifying type of the fibres involved in the transport function. The perception of the intensity of the pain is determined by the somatosensory cortex.

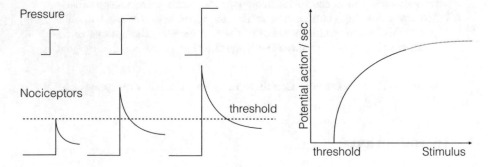

Fig. 1. In the left side, the activation function of the nociceptors proportional to the external stimulus. In the right side, the trend of the number of potential action per second that tends to saturate growing the external stimulus.

In this work we focused on the pain, that is one of the most relevant somatosensory sensation. Humans sense the pain through the nociceptors that transmit action potentials when stimulated by a mechanical, thermal or chemical action, and a system of fibres that drive the stimulus up to the somatosensory cortex [1].

Indeed, this description based on a single direction of the pain, from nociceptors to cortex, hides an important function of modulation of the pain performed by the PAG (Peri-Aqueductal Gray) along the downward path of the pain. Endogenous substances like endorphins or exogenous substances like opiates can operate this modulation function. Thus, the final perception of the pain is always an algebraic sum of these two components.

Following the model of the pain sketched above an artificial pain bio-inspired model is designed for a robot. Starting from signals coming from its hardware sensors as described in Fig. 2, a set of soft sensors [2] increase the semantic level of information. They are mathematical tools able to calculate or estimate quantities that are impossible to measure using real sensors.

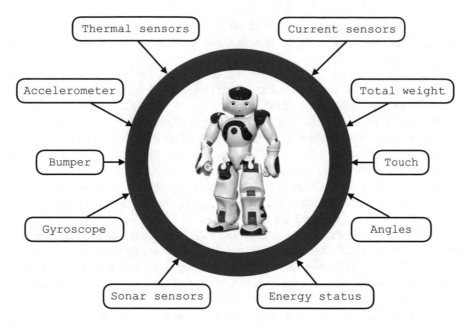

Fig. 2. The humanoid robot NAO and some of the main sensors involved in our soft somatosensory system.

The soft sensors were employed in several real world application by authors and they alway used as stochastic estimators replacing the real sensor [3–10]. The soft sensors used for the pain estimation will be described in depth in the next section.

They exploit a particular activation function, also bio-inspired, to reproduce the somatosensory cortex elaboration function. The modulation mechanism is taken into account in the proposed artificial pain model to copy the biological model.

At the aim to test the model and its implementation, we have chosen to use the humanoid robot NAO of SoftBank Robotics[1]. Indeed, the model and the general framework is independent of such a kind of robot but just applied to it. The organisation of the paper is in four sections included the introduction. Next section outlines the general framework describing the main features of the robot and the parameters take into account for the pain estimation. Section 3 introduces the artificial pain model and the last section reports about the conclusion and some ideas about future works.

2 The General Framework, the Robot and His Sensors

As mentioned above, we have chosen a NAO robot to design, implement and test our artificial somatosensory system and in particular the pain sensation.

---

[1] https://www.ald.softbankrobotics.com/en/cool-robots/nao.

NAO is a humanoid robot equipped with several basic hardware sensors that measure critical parameters as temperature, currents, acceleration, positions, bumps and so on[2].

This work points the attention only on the pain derived from the temperature and the pilot current of the 25 actuators of the robot. Figure 3 sketched all these actuators and reports their name.

The release of NAO chosen in this context uses four different kinds of "stepper motors" as actuators. Each one of these is piloted by its pulse current signal that can range from zero to a C_{max}. Of course, the four different kinds of actuators have four different values of C_{max}.

For each actuator, a thermal sensor measure also the temperature of the junction that can range from zero to a T_{max}. In the case of the temperature, T_{max} is the same for each actuator.

Typically in the operating condition, both the temperature and the pilot current of each motor does not reach the 75% of its maximum value.

Regarding the pilot current, when something or someone blocks one movement of the robot, the motors involved in this movement works in a stressed condition, and its pilot current can exceed the 75% of its C_{max} or even reached C_{max}. Similarly, when the robot performs for a long time movements which involve the same actuator, its temperature can grow over the operating value.

Thus, we consider two different types of pain, the first one linked to the pilot current and the second one linked to the temperature of the junction. Starting from the measures of the temperature and the pilot current, two specific soft sensors estimate the pain caused by these two stimuli. Of course, all 25 actuators and junctions are continually monitored by 25 couple of soft sensors.

"Soft sensor" is just one of the name used to define a set of mathematical tools able to estimating quantities that cannot be measured or, more in general, that are difficult to measure. They are also known as virtual sensors, and they are substantially based on technologies that provide an estimation of measurements learning a mathematical model from data.

In this case, the soft sensors do not perform any forecasting or evaluation, and they use a particular activation function to compute the artificial pain of the robot starting from a basic value of current or temperature measured in one of 25 the actuators.

We use the name *Soft sensor* because they estimate the artificial pain of the robot that is a no measurable quantity using real sensors.

3   The Artificial Model of the Pain

Starting from all above consideration about the human model, the robot characteristic and adopting the appropriate adjustments, we try to replicate the main features described, in an artificial model of the pain for a humanoid robot.

The robot's sensors can measure the temperature and the pilot current in the 25 actuators (see Fig. 3). In this setting, temperature and current values are

---

[2] http://doc.aldebaran.com/2-1/family/nao_dcm/actuator_sensor_names.html.

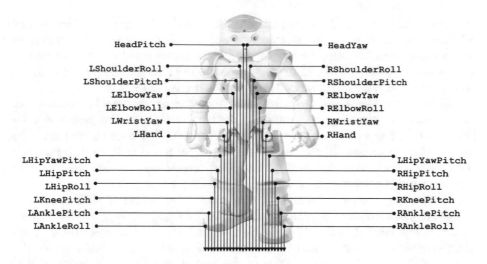

Fig. 3. The humanoid robot NAO and its 25 actuators and thermal sensors (**Head:** HeadPitch, HeadYaw. **Arms:** RShoulderRoll, RShoulderPitch, RElbowYaw, RElbowRoll, RWristYaw, LShoulderRoll, LShoulderPitch, LElbowYaw, LElbowRoll, LWristYaw, LHand. **Legs:** RHipPitch, RHipRoll, RKneePitch, RAnklePitch, RAnkleRoll, LHipYawPitch, LHipPitch, LHipRoll, LKneePitch, LAnklePitch, LAnkleRoll).

considered as the external stimuli and, being measured by different hardware sensors at distinct points of the robot's body we can localise the area and the kind of stimulus. In other words, the position of the hardware sensors guarantees the localisation aspect. Furthermore, the different types of sensors measuring temperature and pilot current identify the kind of external stimulus.

The soft sensors, perform the transduction feature, from external stimulus to internal energy, the inhibition and the modulation aspect, and the final estimation of the pain.

Figure 4 shows the two lines of the artificial pain. The current line on the right and the thermal line on the left. In the both lines the ascendant direction, from hardware sensor to the cortex, and the descendant direction that operates the inhibition or the modulation are visible.

The variable resistor at the end of the ascendant path represents the modulation action coming from of the descendant path. In this artificial model, it is a scalar parameter in the range [0, 1]. The AND port at the origin of the ascendant path describes the inhibition action, and it can be zero (inhibition on) or one (inhibition off). The whole model contemplates that each soft sensor can have different values of modulation and inhibition parameters. Thus, the artificial model of the pain performs the modulation action by two vectors (one for each type of pain) of 25 elements in the range [0, 1], and the inhibition by two binary (also, in this case, one for each type of pain) vectors.

The ascendant direction of the current line and the thermal line of are similar as shown in Fig. 4, but being the variation of the pilot current in the actuators much faster than the temperature in the junctions, their values are sampled at a different frequency. More in details the sampling rate for the pilot current is 100 Hz and the sampling rate for the temperature is 0.2 Hz

To replicate the exponential trend of the number of potential action per second and the character of saturation of this function, in the proposed artificial model of the pain we have considered the sampled pilot current and the temperature analogous to the supply voltage in an RC circuit. In doing so, the pain caused both to a high level of the pilot current and to high values of the temperature, is computed in according to the Eqs. 1 and 2.

The first equations of 1 and 2 are used when the values of pilot current or temperature overcome respectively the values of C_{pain} or T_{pain}. This circumstance is represented in Fig. 4 by the right branch after the comparator symbol. On the contrary, when values of pilot current or temperature remain under the values of C_{pain} or T_{pain} the model uses the second equations of 1 and 2. In this latter case, the computation of the artificial pain follows the left branch after the comparator symbol in the Fig. 4. In the Eqs. 1 and 2, τ_c and τ_t represent the time constant that in the RC circuite is computed as the product $R * C$. The parameter mod represents the scalar in the range $[0, 1]$ that perform the modulation action. The parameter $inhi$ represents the scalar in the range 0 or 1 that perform the inhibition action. In other words, the artificial pain is computed as the charging phase of a capacitor when the external stimulus overcome a critical threshold and as the discharging phase when the external stimulus remains under the threshold.

$$P_{curr}(t) = \begin{cases} P_{Curr} = inhi * mod * (1 - e^{-t/\tau_c}) & C(t) > C_{Pain} \\ P_{Curr} = inhi * mod * e^{-t/\tau_c} & C(t) \leq C_{Pain} \end{cases} \quad (1)$$

$$P_{Temp}(t) = \begin{cases} P_{Temp} = inhi * mod * (1 - e^{-t/\tau_t}) & T(t) > T_{Pain} \\ P_{Temp} = inhi * mod * e^{-t/\tau_t} & T(t) \leq T_{Pain} \end{cases} \quad (2)$$

The following fragment of pseudo code shown how the artificial pain is computed when it is caused by the pilot current. The system monitors the current values and compares it with critical values of pilot current defined as current of pain C_{pain}. Typically, this values is defined as the 76% of C_{max} for each actuator (in the same way is defined the temperature of pain). The timeline of the charging phase and the discharging phase are separately computed, but the timelines of each charging phase and each discharging phase linked together by the inverse equation $t_{discharging} = -\tau * log(P)$ and $t_{charging} = -\tau * log(1 - P)$. In doing so, each charging phase starts where ends its previous discharging phase and vice versa.

```
while ((monitor is on) & (Inhibit is off))
      if (C(t) > Cpain)
           P(t) = Inhi*Mod*(1−exp(−x/Tau));
           y = round(−Tau*log(P(i)));
```

```
        x =x+1;
    else
        P(t) = Inhi*Mod*(exp(-y/Tau));
        x = round(-Tau*log(1-P(i)));
        y = y+1;
    end
t = t+1;
```

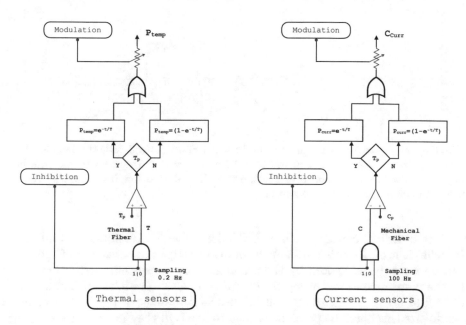

Fig. 4. The two lines of the artificial pain. The current line on the right and the thermal line on the left. In the both lines are reported the two directions of the pain: the ascendant direction from hardware sensor to the cortex and the descendant direction that operates the inhibition or the modulation.

The Fig. 5 reports how works the soft sensor designed to estimate the current in the RShoulderRoll (Right Shoulder Roll) actuator. To do more evident the behaviour of the soft sensor, the critical value of the current of pain, in this circumstance, was established at 50%, instead of the typical value of 76% of the C_{max}.

The horizontal line represents this threshold value beyond which the robot starts to feel the pain. The dotted line indicates the trend of the pain computed by the soft sensor. And the solid line describes the pilot current in the shoulderRoll actuator sampled at 100 Hz.

The trend of the estimated current pain resumes all the characteristic we wish consider in our artificial model of pain for a humanoid robot, besides the already mentioned aspect of localisation, modulation, inhibition and detection of the type of pain. The perception of the pain started when the external stimulus overcame e critical value, the activation function has an exponential character both when the pain grows, and the pain wanes.

Hence, concluding 50 soft sensors completely configurable regarding modulation, inhibition, current and temperature of pain constitute the artificial model of the pain for a humanoid robot.

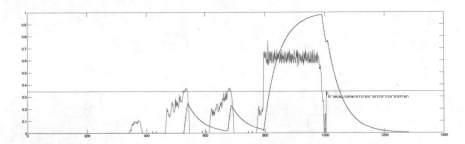

Fig. 5. The figure reports the current in the RShoulderRoll (Right Shoulder Roll) actuator by the solid line. The horizontal line at value 3.5 represents the threshold beyond which the robot starts to feel the pain. The dotted line indicates the trend of the pain computed by the soft sensor.

The described pain system is embedded in a cognitive architecture that drives te behaviour of the robot [13]. As in human beings, the behaviour of the robot can be strongly influenced by the somatosensory subsystem, and in particular by the pain perception. During the execution of a task, or in the planning phase, the robot has to take into account its physiologic wellness [14], by reasoning on the future effects of the possible actions to do: Will they preserve its hardware integrity? Will they assure the requested safety and security during the execution? Moreover, the cognitive system mixes different concurrent subsystems to ensure complex interactions both with the environment and humans [15]. For example, other somatosensory parameters could mitigate pain: physical touches such as caresses, the perception of a relaxing environment (low audio levels of noise, the recognition of familiar faces or objects, and so on). In such context, also the interaction of the somatosensorial mechanisms with the emotional and the motivational subsystems (see for example [11,12]) has a key role in producing human-like behaviours. It is necessary in the case of a robot caring old peoples, interacting and playing with children, cooperating to execute complex tasks with human operators [16].

4   Conclusion and Future Works

In this paper is introduced an original approach to estimate the artificial pain sensation that can be perceived from a humanoid robot by the using of particular soft sensors. We have assumed a bio-inspired model of the pain, and we have adapted this model to a humanoid robot. The principal characteristic of the human model is taken into account and replicate in the artificial model.

In particular, the two lines of the pain, the ascendant and the descendant line, the inhibition or the modulation effect. The decoding of pain type depending on the fibres involved in the transport of the stimulus. And finally the function of activation based on the exponential functions. Functions of this nature get on and off the level similarly to what happens it the humans.

The perception of the pain is just one of the characteristics, though one of the most important, of a somatosensory system that is our concluding goal. Thus, we are working to extend to other stimuli similar consideration faced in this work. More in details, we are considering to involve in the artificial somatosensory system measure regarding accelerations, energy, touch and more in according to Fig. 2, and for each of this measure design a particular soft sensor able to transduce basic values in high-level perception.

References

1. Dubin, A.E., Patapoutian, A.: Nociceptors: the sensors of the pain pathway. J. Clin. Invest. **120**(11), 3760–3772 (2010). doi:10.1172/JCI42843
2. Fortuna, L., Graziani, S., Rizzo, A., Xibilia, M.G.: Soft Sensors for Monitoring and Control of Industrial Processes. Springer Science and Business Media, Heidelberg (2007)
3. Cipolla, E., Maniscalco, U., Rizzo, R., Stabile, D., Vella, F.: Analysis and visualization of meteorological emergencies. J. Ambient Intell. Human. Comput. **8**, 57–68 (2016). ISSN:1868-5137, Springer Berlin Heidelberg
4. Maniscalco, U., Pilato, G., Vassallo, G.: Soft sensor based on E-αNETs. Front. Artif. Intell. Appl. **226**, 172–179 (2010). ISSN:0922-6389
5. Maniscalco, U., Pilato, G.: Multi soft-sensors data fusion in spatial forecasting of environmental parameters. Adv. Math. Comput. Tools Metrol. Test. IX **84**, 252–259 (2012)
6. Maniscalco, U.: Virtual sensors to support the monitoring of cultural heritage damage. In: Biological and Artificial Intelligence Environments, pp. 343–350 (2005)
7. Ciarlini, P., Maniscalco, U.: Mixture of soft sensors for monitoring air ambient parameters. In: Proceedings of the XVIII IMEKO World Congress (2006)
8. Maniscalco, U., Rizzo, R.: Adding a virtual layer in a sensor network to improve measurement reliability. In: Advanced Mathematical and Computational Tools in Metrology and Testing X, pp. 260–264. World Scientific Publishing Co, Singapore (2015)
9. Maniscalco, U., Rizzo, R.: A virtual layer of measure based on soft sensors. J. Ambient Intell. Human. Comput. **8**, 69–78 (2016). ISSN:1868–5137, Springer
10. Ciarlini, P., Maniscalco, U., Regoliosi, G.: Validation of soft sensors in monitoring ambient parameters. Adv. Math. Comput. Tools Metrol. Test. VII **72**, 142 (2006)
11. Infantino, I., Pilato, G., Rizzo, R., Vella, F.: I feel blue: robots and humans sharing color representation for emotional cognitive interaction. In: Biologically Inspired Cognitive Architectures 2012, pp. 161–166 (2013)
12. Augello, A., Infantino, I., Pilato, G., Rizzo, R., Vella, F.: Binding representational spaces of colors and emotions for creativity. Biol. Inspired Cogn. Archit. **5**, 64–71 (2013). Elsevier

13. Infantino, I.: Affective human-humanoid interaction through cognitive architecture. In: The Future of Humanoid Robots - Research and Applications. InTech, pp. 147–156 (2012). ISBN:978-953-307-951-6
14. Infantino, I., Pilato, G., Rizzo, R., Vella, F.: Humanoid introspection: a practical approach. Int. J. Adv. Robot. Syst. **10**, 246 (2013). InTech
15. Infantino, I., Rizzo, R.: An artificial behavioral immune system for cognitive robotics. In: Advances in Artificial Life, ECAL, vol. 12, pp. 1191–1198 (2013)
16. Augello, A., Infantino, I., Pilato, G., Rizzo, R., Vella, F.: Creativity evaluation in a cognitive architecture. Biol. Inspired Cogn. Archit. **11**, 29–37 (2015). Elsevier

Interaction Capabilities of a Robotic Receptionist

Carlo Nuccio[1,2]([✉]), Agnese Augello[1], Salvatore Gaglio[1,2], and Giovanni Pilato[1]

[1] ICAR-CNR, Italian National Research Council Palermo,
Via Ugo La Malfa 153, 90145 Palermo, Italy
agnese.augello@icar.cnr.it, giovanni.pilato@cnr.it
[2] DICGIM, University of Palermo,
Viale Delle Scienze Building 6, 90128 Palermo, Italy
carlo.nuccio@community.unipa.it, salvatore.gaglio@unipa.it

Abstract. A system aimed at facilitating the interaction between a human user and an humanoid robot is presented. The system is suited to answer questions about laboratories activities, people involved, projects, research themes and collaborations among employees. The task is accomplished by the HermiT reasoner invoked by a speech recognition module. The system is capable of navigating a specific ontology making inference on it. The presented system is part of a broader social robot framework whose goal is to give the user a fulfilling social interaction experience, driven by the perception of the robot internal state and involving intuitive and computational creativity capabilities.

Keywords: Humanoid robot · Human-robot interaction · Ontology

1 Introduction

The last years have seen the explosion of social robotics, i.e. the development of autonomous robots capable of interacting with human beings according to social rules and following specific conventions suited for the role that robots are playing.

This has been possible thanks to the last advances in this research area, making it possible to equip robots with more and more advanced capabilities, improving, in particular, their expressive and communication skills.

They can interact in different ways, recognizing faces, gestures, expressions [10]. This led to an increased interest and investment on human-robot interaction (HRI) [3] and on all possible fields of application of robots for the society [8,9].

At present, social robots are assuming an increasingly important role in society and it is already possible to imagine that in a short time they will be ever more integrated within organizations and houses. The first robot concierge [5] and doormans in hotels have been proposed. Other examples of effective robots are socially assistive robots [9], customers assistants [11], house collaborators [2].

In order to execute these tasks, robots must be able to understand both verbal and non-verbal communication of the users, with the aim to properly answer, to provide information and acting in a suitable manner in response to a

© Springer International Publishing AG 2018
G. De Pietro et al. (eds.), *Intelligent Interactive Multimedia Systems and Services 2017*,
Smart Innovation, Systems and Technologies 76, DOI 10.1007/978-3-319-59480-4_18

specific request. An essential aspect of HRI research concerns the formalization of the cognitive abilities of these virtual companions [4,6].

The definition of ontologies allows the robot to adequately represent a set of concepts in order to understand and react to the users requests [1,14]. As an example, Kobayashi et al. [7] aligned the Japanese Wikipedia Ontology with an ontology of Robot Actions enabling a Nao humanoid robot [15] to perform related actions to dialogue topics.

In this work, we focus on the knowledge formalization and reasoning abilities of a receptionist robot in a scientific lab environment. The presented system is part of a broader social robot framework that we are currently setting-up, whose goal is to give the user a fulfilling social interaction experience, by means of capabilities of finding associations among dialogue topics as well as computational creativity and humorous features embedded in the robot. Moreover, in the proposed framework the interaction will be strongly influenced by the perception of pleasant and unpleasant sensations, by means of somatosensory system modelled in the robot [16].

As a case of study, in this paper, we describe the definition of a robot receptionist for a research laboratory. The robot is able to welcome visitors, answer to information request about the laboratory personnel, reacting to the verbal communication in an appropriate manner. In the work we focus on the formalization of a verbal interaction module. The module exploits QiChat [12], the conversational module of the Pepper humanoid robot [13], strengthening it with a proper reasoner. The reasoner queries an ontology to get access to the knowledge about the laboratory according to the ongoing dialogue, providing from time to time the requested information. If the answer is not present in the ontology, the robot looks for it on the Internet. It also exploits the data coming from the robot's sensors to determine the occurrence of events. This allows the robot to pro actively drive the conversation adapting it to the different situations.

2 System Overview

The proposed system is aimed at facilitating the interaction between a human user and a Pepper robot through a natural language interaction, using QiChat. An overview of the overall system is shown in Fig. 1. The system is suited to answer questions about laboratories activities, people involved, projects, research themes and collaborations among employees. Information is stored in an ontology, which is exploited both for describing the domain and for making inferences, by using the HermiT reasoner [21].

Axioms are created by using ROWL [17], a Desktop Protégé plugin which transforms SWRL (Semantic Web Rule Language) [19] rules in OWL (Web Ontology Language) axioms where OWL [20] is a Semantic Web language designed to represent complex knowledge about individuals, classes, relations between them and properties. SWRL extends the set of OWL axioms to include Horn rules. It enables Horn rules to be combined with an OWL knowledge base.

The interaction system is based on pattern recognition: the questions asked to the robot are compared with the rules on the QiChat module; if a matching with the rules is found, a *NAOqi* method, running on a Python web service, is called. The method infers the most appropriate answer in the ontology by using the SPARQL (SPARQL Protocol and RDF Query Language) language and manages the result.

If the answer is not present in the ontology, further research on the internet by using Wikipedia API for Python is conducted. The answer found by Wikipedia API is sent to the QiChat module as an *NAOqi event*. As an example, *Dialog/NotSpeaking*20 is an event caught when a human has not talked for 20 s. For an event $e : eventName$, a variable \$eventName is also available. In this system, we are sending an answer in the variable \$eventName. The event $e : eventName$ will activate robot output. The Pepper Robot will answer with the sentence corresponding to user input or event.

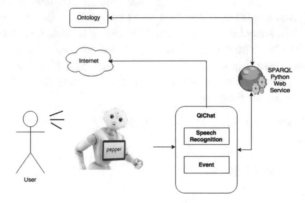

Fig. 1. Architecture and system components

3   Technologies

In the following subsections, we illustrate the technologies used for developing the first version of our robotic receptionist: the Pepper robot, the QiChat module, the HermiT Reasoner, and the developed ICAR Ontology.

3.1   Pepper Robot

Pepper is a humanoid robot produced by Aldebaran Robotics and SoftBank [13] as day-to-day companion. It is able to perceive emotions and as consequence adapt his behaviour to the mood of his interlocutor. It is able to perceive four main emotions: joy, sad, angry and surprise. This is possible by means of Emotion Engine, which provides to the robot to empathize with the interlocutor by analyzing the voice of the speaker and the expressions on his face.

Pepper is also equipped with a tablet, which can be used to express his personal emotions or to help human users to make a choice. The tablet interface can be used to play videos, images, open web pages or to handle wifi connections.

The robot can move in all directions by means of three multidirectional wheels at a speed of 3 km/h. It is equipped with 20 engines that allow it to move the head, arms and back. It has 12 h battery life. Pepper possesses two sonars, six laser sensors and three obstacle detectors placed in his legs.

Pepper's API allows to build and use an ontology representing the knowledge of the robot, by using RDF and OWL formalisms.

3.2 QiChat

The interaction with Pepper Robot has been implemented using conversation module of Pepper: QiChat [12]. The dialogue mechanism of QiChat is based on a list of rules written to handle the flow of the dialogue between the human and the robot. A set of *Topics* constitutes the dialogue mechanism. A *Topic* is a script box containing a *Rule*. These rules can be of two types: User rules and Proposal rules. A User rule associates a specific user input to possible robot output. A Proposal rule needs to be triggered with a function, i.e. ^*nextProposal* actives the first proposal in the *Topic* which has the *Focus*. The *Focus* is provided to the Topic containing the last triggered rule, except if it has the ^*noStay* property. An example is reported below:

```
topic: ~introduction ()
language: enu

u:(Hi) Hello ^nextProposal

proposal: how is your day going?
  u1:(Good, thanks) Cool!
  u1:(I'm Tired) Oh, you should take a nap
```

In this case, the conversation could be:

```
U: Hi
Robot: Hello how is your day going?
U: Good, thanks
Robot: Cool!
```

Another critical component is *Concept*, which is a list of words and/or expressions that refer to an idea. There are two types of *Concepts*: static, which cannot be modified at run time, and dynamic ones, which must be modified at run time. We used static *Concepts* because they are usable in different *Topics*. Every time a user asks the Pepper for a question, a pattern matching mechanism recognizes the message and the Rule is triggered. When this happens, the Robot output is executed, i.e. something is said or some movement is executed, and the *Focus* is set to the *Topic* containing the rules. A human input rule can contain ~ *conceptName*, which allows catching any word included in a *Concept*. The Robot output allows catching one or several words in the Human input and reuses it within the Robot output directly following. It allows to store it in a variable and can be reused after in the interaction.

In QiChat, a *NAOqi event* can be used as input rule. As an example, the robot output can be triggered by a movement. For example, the event *PostureChanged* active the Robot Output when the robot has changed his posture.

QiChat allows to call functions and program logic. In this case, we used the instruction ^*call(MyService.myMethod*($1)), which invokes a *NAOqi* method, defined in a Python web service and handles the result. The evaluation of the call is always completed at the beginning of the sentence.

The focus of QiChat is the pattern recognition mechanism. We used keywords to extend input rules. In this manner, we have one rule, which can be matched by multiple sentences. We used two kinds of keywords: _*, which matches anything, and *Concepts*. Keywords can be saved into variables. An example is reported below:

```
topic: ~QueryOntology()
language: enu

concept:(lab) ["Cognitive Robotics and Social Sensing" "Cognitive System"]
concept:(people) ["Giovanni Pilato" "Agnese Augello"]

u:(Which are the research activities at _*) $resOrganization = $1
                    ^call(Sparql.countResGroup($resOrganization))
    c1:( _*) There are $1 main research groups, concerning
                ^call(Sparql.researchOrganization($resOrganization))
        c2:(_*) $1

u:(Do you know some research themes in which _~lab lab is involved)
            $researchLab = $1 ^call(Sparql.researchLab($researchLab))
    c1:(_*) $researchLab lab is involved in $1
    c1:(None) I don't know research themes for this lab.

u:(Is someone in _~lab lab who working on _~themes)
                    ^call(Sparql.someoneInLabWhoWorkOn($1, $2))
    c1:(_*) The answer is $1

u:(Is there someone who works with _~people) $workWith = $1
                        ^call(Sparql.workWith($workWith))
    c1:(_*) $workWith and $1 work together
    c1:(None) No one work with $workWith
```

In this case, i.e., the user input could be "Is someone in Cognitive Robotics and Social Sensing lab who working on Conceptual Spaces and Geometric Representation of Knowledge?". In this example we used two *Concepts* keywords: "Cognitive Robotics and Social Sensing", which is stored in variable $1, and "Conceptual Spaces and Geometric Representation of Knowledge", which is stored in variable $2. The ^*call(Sparql.someoneInLabWhoWorkOn*($1, $2)) uses the two variables to find a result in the ontology and handles it in a subrule **c1**. In this case, we have only one subrule, which will match any result and that will activate the Robot output. The result is stored in variable $1. The answer will be "The answer is Giovanni Pilato".

3.3 HermiT Reasoner

HermiT is reasoner for ontologies written in Web Ontology Language (OWL). HermiT can determine ontology consistency, identify subsumption relationships between classes, and much more. HermiT is based on a "hypertableau" [22] calculus which provides much more practical reasoning than any previously known algorithm. HermiT can classify some ontologies which had previously proven too complex for any available system to handle; furthermore, it can handle SWRL (Semantic Web Rule Language) rules.

3.4 The ICAR Ontology

We created the "ICAR" OWL ontology. A small portion of the created ontology is illustrated in Fig. 2, where light ovals represent Classes, and dark ovals represent instances of the classes. In the following, a description of the main concepts is reported for clarity.

ResearchInstitution is the class of research institution at which performing research activities. ICAR is an instance of this class representing the *Istituto di Calcolo e Reti ad Alte Prestazioni del Consiglio Nazionale delle Ricerche* (ICAR-CNR). *ResearchGroup* represents the classes of research groups. *CognitiveRoboticsAndSocialSensingLab* is a specialization of *ResearchGroup*. The instances of this class are groups which conduct research in Cognitive Robotics and Social Sensing. An instance of *CognitiveRoboticsAndSocialSensingLab* is *CognitiveRoboticsAndSocialSensingLabICAR*, which represents the group of *ICAR*. *ResearchTopic* represents the classes of research topics. *CognitiveComputing* is a specialization of *ResearchTopic*. *CognitiveRoboticsAndSocialSensing* is a specialization of *CognitiveComputing*. The instances of this class are research topics in Cognitive Robotics and Social Sensing. An instance of *CognitiveRoboticsAndSocialSensingLab* is *ConceptualSpacesAndGeometricRepresentationOfKnowledge*, which represents Conceptual Spaces and Geometric Representation of Knowledge Topic.

AcademicStaff represents the class of the academic employees. *PostdocStaff*, *ResearchStaff*, *TechnicianStaff* are specializations of *AcademicStaff*. We also represent people working at the *ResearchInstitution* by specifying their main skills and tasks. For example, *GiovanniPilato* is a *ResearchStaff* who is employed at the *ICAR*, and who is an expert about *ConceptualSpacesAndGeometricRepresentationOfKnowledge*.

We used ROWL Protégé plugin [17] which accepts rules as input and adds them as OWL axioms to a given ontology. The following rule can be used to characterize all individuals who work together:

$$involvedInResearchGroup(?x, ?p) \widehat{\ } involvedInResearchGroup(?y, ?p) \rightarrow workWith(?x, ?y)$$

This means that if exists a research group p and x is involved in p, and y is involved in p, so x and y work together.

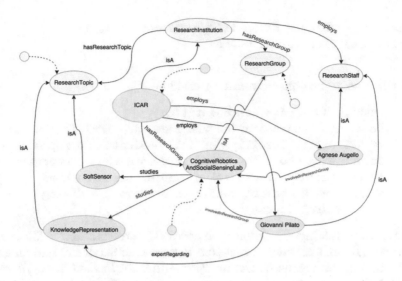

Fig. 2. A detail of the ICAR ontology

4 Human - Robot Interaction

In the following subsections we illustrate the interaction process between a human being and our system. An example of interaction is also reported.

4.1 Interaction Between Pepper Robot and Ontology

The Pepper Robot is able to interact with the OWL ontology by means of a Python web service which integrates SPARQL language. This module contains a function for each question. Each function receives keywords of the question as parameters.

The web service uses *rdflib* package [18], which uses RDF graph, which is a set of RDF triples. RDF Graphs support the Python *in* operator, as well as iteration and some operations like union, difference and intersection.

The module is invoked using QiChat *^call(Sparql.method(params))* function. It uses SPARQL to executes a query on the ontology. If there isn't an answer in ontology, the search is extended to the Internet. Otherwise, the result is sent to QiChat module, which enables the Robot Output.

4.2 Interaction Between Pepper Robot and Internet

The module is triggered by receiving input parameters by QiChat module. We used Wikipedia Python API to find an answer on internet. After receiving the parameters, a test on the language used by the robot is performed. Therefore, we will get the article summary for the that keyword. Once processed the request, the result is sent to the QiChat module as an event, which, as we have seen, can activate the robot output. An example is reported below:

```
u:(what is _*) $key = $1 ^call(Sparql.whatIs($key))
   c1:(None) $searchWiki = $key
   c1:(_*) $1
```

In this case, a possible conversation could be:

```
Human: what is Knowledge Representation
Robot: Knowledge representation and reasoning (KR) is the field
       of artificial intelligence (AI) dedicated to representing
       information about the world in a form that a computer
       system can utilize to solve complex tasks such as
       diagnosing a medical condition or having a dialog
       in a natural language. [...]
```

In this case, user input is "what is Knowledge Representation?". The keyword "Knowledge Representation" is sent to a Python Web Service to find an answer in the ontology by means of the function *call(Sparql.whatIs($key))*. In this case, there is not an answer in the ontology and the *call* function returns *None*. Therefore, the key is sent to the Internet module. The result of the Internet module is sent to QiChat as an event, as below:

```
u:(e:wiki) $wiki
```

In this case, for the event *e:wiki*, the result is contained in variable *$wiki*.

4.3 Example

An of interaction exploiting the ontology is given below.

```
Human: Hi

Robot: Hi , my name is Nao. Welcome to ICAR-CNR. I can give you informations
       about research activities , research group , people involved and more.
       Can I help you  ?

Human: Which are the research activities at ICAR?

Robot: There are four main research groups , concerning Knowledge and Data,
       Bioinformatics, Distributed systems and IOT, Cognitive Computing  ,

Human: Is there a lab who working on Cognitive Computing?

Robot: Yes , Cognitive Robotics and Social Sensing lab is working on Cognitive
       Computing

Human: Do you know some research themes in which Cognitive Robotics and Social
       Sensing lab is involved

Robot: Cognitive Robotics and Social Sensing lab is involved in Conceptual
       Spaces and Geometric Representation of Knowledge  ,
```

Human: Is someone in Cognitive Robotics and Social Sensing lab who working on
       Conceptual Spaces and Geometric Representation of Knowledge?

Robot: The answer is Agnese Augello and Giovanni Pilato

Human: Is there someone who works with Giovanni Pilato?

Robot: Agnese Augello and Giovanni Pilato work together

Human: Ok thanks

Robot: Thank you , Bye .

Furthermore, an example of interaction with Internet is given below.

Human: what is Cognitive Computing?

Robot: Cognitive computing (CC) describes technology platforms that, broadly
       speaking, are based on the scientific disciplines of Artificial
       Intelligence and Signal Processing. These platforms encompass machine
       learning, reasoning, natural language processing, speech and vision,
       human-computer interaction, dialog and narrative generation and more .

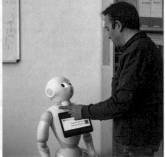

Fig. 3. Some moments of interaction with Pepper at the CRSS lab

5 Conclusion and Future Developments

A first study and implementation of a social robot has been presented. The
robot is part of a broader social robotic platform we are currently developing at
ICAR-CNR. The presented system has been implemented on a Pepper Robot
and makes use of an ad hoc ontology and an HermiT reasoner to make inference.
The system is capable of interacting in natural language with the user and it
can look for information also on the Internet. Future development of the system
regard embedding intuitive, associative skills as well as humorous capabilities to
give the human users a fulfilling interaction experience.

References

1. Sorce, S., Augello, A., Santangelo, A., Pilato, G., Gentile, A., Genco, A., Gaglio, S.: A multimodal guide for the augmented campus. In: Proceedings of the 35th Annual ACM SIGUCCS Fall Conference, pp. 325–331. ACM (2007)
2. Holthaus, P., Leichsenring, C., Bernotat, J., Richter, V., Pohling, M., Carlmeyer, B., Engelmann, K.F.: How to address smart homes with a social robot? a multimodal corpus of user interactions with an intelligent environment. In: International Conference on Language Resources and Evaluation (2016)
3. Sheridan, T.B.: Human robot interaction status and challenge. Hum. Factor J. Hum. Factor Ergon. Soc. **58**(4), 525–532 (2016)
4. Aly, A., Griffiths, S., Stramandinoli, F.: Metrics and benchmarks in human-robot interaction: recent advances in cognitive robotics. Cogn. Syst. Res. **43**, 313–323 (2016)
5. Statt, N.: Hilton and IBM Built a Watson-Powered Concierge Robot. The Verge, National Park (2016)
6. Mutlu, B., Roy, N., Sabanovic, S.: Cognitive human-robot interaction. In: Springer Handbook of Robotics, pp. 1907–1934. Springer (2016)
7. Kobayashi, S., Tamagawa, S., Morita, T., Yamaguchi, T.: Intelligent humanoid robot with Japanese Wikipedia ontology and robot action ontology. In: 2011 6th ACM/IEEE International Conference on Human-Robot Interaction (HRI), pp. 417–424. IEEE (2011)
8. Breazeal, C., Dautenhahn, K., Kanda, T.: Social robotics. In: Springer Handbook of Robotics, pp. 1935–1972. Springer (2016)
9. Mataric, M.J., Scassellati, B.: Socially assistive robotics. In: Springer Handbook of Robotics, pp. 1973–1994. Springer (2016)
10. Yang, L., Cheng, H., Hao, J., Ji, Y., Kuang, Y.: A survey on media interaction in social robotics. In: Pacific Rim Conference on Multimedia, pp. 181–190. Springer (2015)
11. Demirel, E., Gur, K.D., Erdem, E.: Human-robot interaction in a shopping mall: a CNL approach. In: International Workshop on Controlled Natural Language, pp. 111–122. Springer (2016)
12. QiChat. http://doc.aldebaran.com/2-4/naoqi/interaction/dialog/dialog.html
13. Pepper robot. https://www.ald.softbankrobotics.com/en/cool-robots/pepper
14. Pilato, G., Augello, A., Vassallo, G., Gaglio, S.: Sub-symbolic semantic layer in cyc for intuitive chat-bots. In: Proceedings of ICSC 2007 International Conference on Semantic Computing, Irvine CA, United States, pp. 121–128, 17–19 September 2007
15. Nao robot. https://www.ald.softbankrobotics.com/en/cool-robots/nao
16. Augello, A., Infantino, I., Maniscalco, U., Pilato, G., Vella, F.: Creation and cognition for humanoid live dancing. Robot. Auton. Syst. **86**, 128–137 (2016). Elsevier
17. Sarker, M.K., Carral, D., Krisnadhi, A.A., Hitzler, P.: Modeling OWL with rules: the ROWL protege plugin. In: International Semantic Web Conference (2016)
18. RDFLib: https://rdflib.readthedocs.io/en/stable/
19. SWRL. https://www.w3.org/Submission/SWRL/
20. OWL. https://www.w3.org/OWL/
21. Reasoner, H.: http://www.hermit-reasoner.com
22. Motik, B., Shearer, R., Horrocks, I.: Hypertableau reasoning for description logics. J. Artif. Intell. Res. **36**, 165–228 (2009)

Artificial Pleasure and Pain Antagonism Mechanism in a Social Robot

Antonello Galipó[2(✉)], Ignazio Infantino[1], Umberto Maniscalco[1], and Salvatore Gaglio[1,2]

[1] ICAR - National Research Council of Italy,
Viale Delle Scienze - Edificio 11, 90128 Palermo, Italy
`ignazio.infantino@cnr.it, gaglio@unipa.it`
[2] DIID, University of Palermo, Viale Delle Scienze, 90100 Palermo, Italy
`antonello.galipo@community.unipa.it`

Abstract. The goal of the work is to build some Python modules that allow the Nao robot to emulate a somatosensorial system similar to the human one. Assuming it can perceive some feelings similar to the ones recognized by the human system, it will be possible to make it react appropriately to the external stimuli. The idea is to have a group of software sensors working simultaneously, providing some feedback to show how the robot is feeling at a particular time. It will be able to feel articular pain and stress, to perceive people in his surroundings (and in a future work to react according to the knowledge of them with face recognition), feel pleasure by recognizing caresses on his head and respond to the surrounding noise level.

Keywords: Soft sensors · Artificial somatosensorial systems · Cognitive robotics

1 Introduction

The behavior of the living beings depends on many basic biological mechanisms based on the processing of opposite effects. Among them, pain and pleasure have a high relevance on the general well-being of an organism, and in human biology opioid and dopamine systems regulate their mutually inhibitory effects [8]. Such biological mechanisms are useful also in artificial agents, and various researchers have tried to simulate them in cognitive architectures or sensorial systems [4] with the purpose to link elementary sensations with high-level behavioral parameters directly influenced by emotion, intention, mood, and so on.

In living beings, the interactions of an organism's nervous system with the external environment produce effects that the organism can predict and are voluntary whereas the interactions of the nervous system with what is inside the body give rise to involuntary and unpredictable effects [1]. Among the various strategies for gaining a significant control of the not voluntary effects on their behavior, humans try to acquire the awareness of their internal states. Such

G. De Pietro et al. (eds.), *Intelligent Interactive Multimedia Systems and Services 2017*,
Smart Innovation, Systems and Technologies 76, DOI 10.1007/978-3-319-59480-4_19

biological mechanism seems useful also in robotics to define complex behavior in social interactions and unstructured environment. Parisi in [10] proposes an internal robotics, which in addition to the external robotics (i.e. standard processing of environment sensing) describes interesting and complex scenarios aiming to *humanize* a robot ([5,6]). Current research focuses on robots capable of showing emotions (see for example [3]) or intentions, by simulating facial expressions, voice modulation, or proper body postures. But any internal model of robot emotion or affective system also has to consider the influence of the degree of wellness of the body. Our idea is to build such subsystem, to explore how to capture the internal sensations of the robot, and experimenting simple mechanisms of pleasure and pain antagonism. We use an approach based on soft-sensors (see for example [9]) allowing the robot to perceive its relevant physiological data (such as current, temperature, audio perception, touch sensing, and so on) and transform them in pain or pleasure values.

The paper has the following structure: next section describes the robot sensors involved in our system; then we exploit the corresponding soft sensors, and finally we report some experiments during the human-robot interactive session.

2 Self-sensing of the Humanoid Body

The proposed artificial somatosensorial system extends the perceptive capabilities of the robotic platform named NAO[1]. Such robot has a robust software platform called *naoqi* that allows the user to interact both with the real robot or a simulated one. In particular, we use Python SDK v2.1.4 for the Nao robot V4 and V5. In order to briefly introduce the robot software architecture in the following, we give some details about the main modules used.

ALProxy is the core of the work. It enables the programmer to ask the system for services about every needed area of work, in our case we used ALMemory, ALTextToSpeech, ALSoundDevice, and DCM.

ALMemory provides access to the memory of the robot. It allows to read values from memory, write and update values into the memory and register to specific events or value changes. It was used to get the data from the robot sensors.

ALSoundDevice provides access to the robot microphones, allowing us to work on what the robot is hearing. It can operate some automatic calculations on the energy of the perceived sounds, which enabled us to make a simple noise level detector for the robot to react.

The DCM (Device Communication Manager) is in charge of the communication with almost every electronic device in the robot (boards, sensors, actuators ...), excepting sound (in or out) and cameras. In our case, DCM was used to activate the sonar transmitters and receivers to get information about the robot surroundings.

Moreover during the experiments, the robot explicits his sensations by simple vocal reactions. *ALTextToSpeech* allows the programmer to make the robot say

---

[1] https://community.ald.softbankrobotics.com.

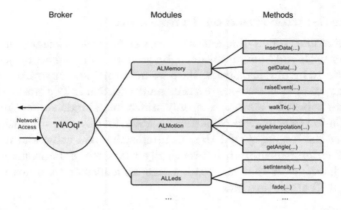

Fig. 1. Architecture of the Naoqi framework.

phrases. It was used to provide some vocal feedback. An overview of the whole Naoqi architecture can be seen in Fig. 1.

3   The Soft Sensors

We choose to consider some sensorial values corresponding in a vague sense to muscular fatigue and suffering, the perception of the presence in proximity to humans, the skin sensibility to external touches, the perception of the environmental sound level. Then the implemented soft sensors are the following:

– Pain Detector based on the current flowing in the joints
– Pain Detector based on the temperature of the joints
– Social Space Detector
– Caress Detector
– Noise Detector

Each sensor will update a memory value at its frequency of work, providing information about the situation. All these values will be read and processed by a module, that will output the overall result.

3.1   Pain Detector Based on Current

This sensor emulates the human sensorial system part that informs us about the pain that is occurring in certain joints caused by some stress on them. We want to monitor the current flowing through the joints motors, process its values and see if the result is above a certain pain threshold. A similar pain computation is reported in [2]. In the present work we use instantaneous values from the soft sensors instead of using a threshold. In this way we can have a more detailed idea of the evolution of the pain. The Python implementation consists of a PainCurrentDetector class that scans all the joints at the frequency of 10 Hz, computes the pain value and updates it in memory. Here we can see how we use of ALProxy to get ALMemory, to read and write variable values to the memory of the robot, and ALTextToSpeech to make Nao say the supplied text.

3.2 Pain Detector Based on Temperature

This sensor emulates in a different way the way we feel articular pain. This time the analogy with the human pain can be seen as pain caused by joint overuse and not strictly by physical stress. Here we monitor the temperature of the joint motors and see if it is above a certain pain threshold. The math is the same of the current based detector; the only difference stands in the input values. The Python implementation consists of a PainHotDetector class that scans all the joints at the frequency of 10 Hz, computes the pain value and updates it in memory. The implementation is practically the same of PainCurrentDetector. The only difference is that the read values from memory are not the Current ones but the Temperature ones.

3.3 Social Space Detector

This sensor emulates our feeling of comfort zones. The idea is based on the social model made by Edward T. Hall [7], who divides our space perception into three range areas:

- Public space where all the people are
- Private space closer than public. We can start to perceive some discomfort if the people in it are not friendly
- Intimate space closer than private. We allow very few people in it.

The Python implementation of this process consists of a SocialSpaceDetector class, which works by taking the separate sonar values of the left and right receivers, picks the minimum value (corresponding to the closest object) and computes an anxiety function, which expresses the decreasing comfort of the robot. This function has an exponential behaviour for rising and falling the anxiety value at different speeds, according to the distance perceived. The value of this function is stored in memory. The whole process works at the frequency of 10 Hz.

3.4 Caress Detector

This sensor allows the robot to *feel pleasure* upon receiving some caresses on his head. A sequence of touches on the three tactile head sensors identifies a caress. The module distinguishes a normal caress, a backward touch or simply a hand placed on its skull. The hypothesis is that the contact lasts a brief time (i.e. like a patting). In fact, a prolonged touch period can be a cause of discomfort (the user is, for example, tilting the robot head by grabbing it from above). The Python implementation consists of a headCaresseDetector class that scans the head tactile sensors. When the front sensor is touched, it starts the process of caress forward detection. If a human starts to touch robot head from the rear, the detection of a caress backward is activated. When we have all three sensors touched at the same time, we start to count time, and if the touching lasts more than one seconds the robot will feel bad about it. If a caress is detected, the satisfaction grow up to a given fixed positive value, then will decay over time until it reaches 0.

3.5 Sensorial Monitoring

A simple Python module (named sensorial switching) allows the developer to enable only the soft sensors that he needs in an easy way, without having to take care of all the things to be done under the hood. The switcher takes a list of names as input and enables the corresponding sensors to monitor, and modules will run in parallel. Finally, a suitable module computes the general wellness parameter of the robot taking into account pain and pleasure detected with different weight. In the following experimental part, we consider equal weights.

4 A Working Example

We will now present a hypothetic situation in which the discussed sensors apply. Tha Nao robot is sitting quietly and relaxed. The user walks in front of the robot, who recognises him and stands up, waving at him. The robot will invite the user to replicate some movements, for example, it will invite the user to put his arms repeatedly up and down. The user will not do it. Instead, he will walk straight to the robot and block one arm. The robot will initially feel uncomfortable with the user getting closer, and will certainly feel articular pain, inducing a demotivation. The user will now stop holding Nao's arm but will keep staying close to the robot. A few seconds later the user will caress the head of the robot, motivating it and lowering the discomfort felt due to the close distance. After a while he will place his hand for a long time on the robot's head, making it feel uncomfortable, and after a while he will caresse it again. The user will now go away, and the robot will keep moving his arms until it is tired. In the folllowing, we report the graphics related to each soft sensor and the global pain-satifaction funtion.

Figure 2 depicts a graphical evolution of the Pain Detector based on current of a joint involved in the action. During the reported session, the robot moves his arms freely, then the user blocks one of them twice, resulting in a spike in the current values approximately at 5 and 12 s (corresponding to the abscissas a bit after 0 and a bit after 100 in the graph). Without other stimuli, the pain decays over time.

Similar at previous, Fig. 3 depicts a graphical evolution of pain based on the temperature of the same joint.

Figure 4 shows the effect of detected free space in front of the robot using its sonars. The objects, the humans within 2.5 m are not identified. If the space near the robot is occupied, a discomfort occurs

Figure 5 shows the effect of human caresses over the head robot. In particular, the user caresses the robot, then holds his hand over the robot's head, then touches it again. Above is depicted all the sequence. The second graph shows a zoom of the same sequence to appreciate the three touches in the first part of it.

Figure 6 depicts the satisfaction (pleasure) values derived from caresses for the previous explained sequence.

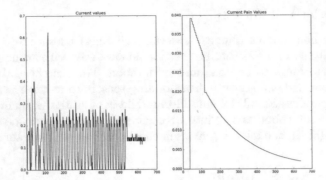

Fig. 2. The robot moves his arms freely, then the user blocks one of them twice, resulting in a spike in the current values (on the left). The pain (values on the right) decays over time. Notice that the second current peak is too brief to cause any pain increase.

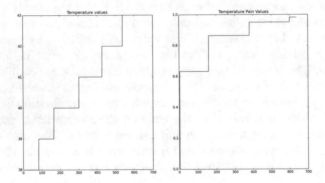

Fig. 3. As the robot gets tired, the temperature of the joint rises (left values) and the pain rises with it (right values).

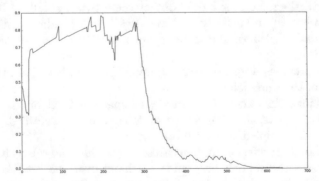

Fig. 4. The social space feeling: the initial value is not zero because the user will stand at a certain distance from the robot in order to be detected. The user gets closer, interacts with the robot, then goes away. The discomfort of the robot goes to zero after some time the user is gone. Note that the noise is caused by the movement of the arms in front of the sonars, altering the values.

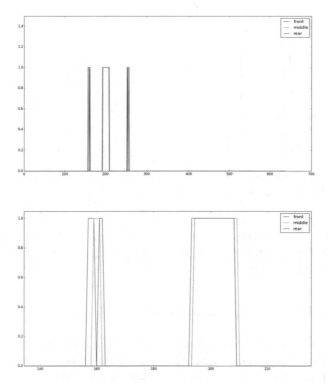

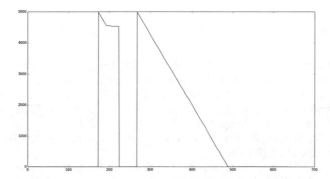

Fig. 5. The user caresses the robot, then holds his hand over the robot's head, then caresses it again. Above is depicted all the sequence. The second graph shows a zoom of the same sequence to appreciate the three touches in the first part of it.eps

Fig. 6. The satisfaction values for the caress increase, then spike to zero due to the hand holding, then rise again because of the second caress and decay over time.

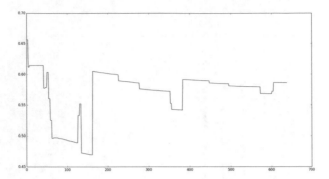

Fig. 7. A naive calculation of the overall pain felt by the robot. It's calculated as a simple average of the non zero pain values.

Finally, Fig. 7 reports the overall pain felt by the robot for the given sequence. In particular, it is possible to distinguish the positive effects of human touches on the general level of pain.

5   Future Work and Conclusions

We are experimenting the somatosensorial system with a cognitive architecture that takes into account robot basic demands. In particular, we will study the effect of the proposed system on the motivation and mood of the robot, and how it influences the execution of a task with human presence. Moreover, it will be interesting to investigate on the human reactions to an acting robot equipped with the somatosensorial system, and the difference with a standard robot. The proposed approach is independent of the robotic platform, and somatosensorial soft sensors can be developed on other entities (lasers, infrared sensors, and so on).

References

1. Acerbi, A., Parisi, D.: The evolution of pain. In: European Conference on Artificial Life, pp. 816–824. Springer, Heidelberg, Berlin, September 2007
2. Augello, A., Infantino, I., Maniscalco, U., Pilato, G., Vella, F.: The effects of soft somatosensory system on the execution of robotic tasks. In: IEEE International conference on Robotic Computing 2017, Taiwan (2017)
3. Breazeal, C., Brooks, R.: Robot emotion: a functional perspective. In: Who Needs Emotions, pp. 271-310 (2005)
4. Dodd, W., Gutierrez, R.: The role of episodic memory and emotion in a cognitive robot. In: IEEE International Workshop on Robot and Human Interactive Communication, ROMAN 2005, pp. 692–697, August 2005
5. Duffy, B.R.: Anthropomorphism and the social robot. Robot. Auton. Syst. **42**(3), 177–190 (2003)

6. Gray, K., Wegner, D.M.: Feeling robots and human zombies: mind perception and the uncanny valley. Cognition **125**(1), 125–130 (2012)
7. Hall, E.T.: The Silent Language, vol. 3. Doubleday, New York (1959)
8. Leknes, S., Tracey, I.: A common neurobiology for pain and pleasure. Nat. Rev. Neurosci. **9**(4), 314–320 (2008)
9. Maniscalco, U., Rizzo, R.: A virtual layer of measure based on soft sensors. J. Ambient Intell. Human. Comput. **8**, 1–10 (2016)
10. Parisi, D.: Internal robotics. Connect. Sci. **16**(4), 325–338 (2004). doi:10.1080/09540090412331314768

Move Your Mind: Creative Dancing Humanoids as Support to STEAM Activities

Giuseppe Città[1]([✉]), Sylvester Arnab[2], Agnese Augello[3], Manuel Gentile[1],
Sebastian Idelsohn Zielonka[4], Dirk Ifenthaler[5], Ignazio Infantino[3],
Dario La Guardia[1], Adriano Manfrè[3], and Mario Allegra[1]

[1] ITD - National Research Council of Italy, Via Ugo La Malfa 153,
90146 Palermo, Italy
{giuseppe.citta, manuel.gentile,dario.laguardia,
mario.allegra}@itd.cnr.it
[2] Disruptive Media Learning Lab, Coventry University, Coventry, UK
aa8110@coventry.ac.uk
[3] ICAR - National Research Council of Italy, Via Ugo la Malfa 153,
90146 Palermo, Italy
{agnese.augello,ignazio.infantino,adriano.manfre}@icar.cnr.it
[4] Eurecat Technology Center, eHealth Unit,
Av. Diagonal, 177, 08018 Barcelona, Spain
sebastian.idelsohn@eurecat.org
[5] Department of Educational Science, School of Social Sciences,
University of Mannheim, A5,6, 68131 Mannheim, BW, Germany
ifenthaler@bwl.uni-mannheim.de

Abstract. Educational activities based on dance can support interest in comprehension of concepts from maths, geometry, physics, bio-mechanics and computational thinking. In this work, we discuss a possible use of a dancing humanoid robot as an innovative technology to support and enhance STEAM learning activities.

Keywords: Cognitive architecture · Embodied Cognition · Enactivism · STEAM · Humanoid robots · Computational creativity

1 Introduction

Several studies show how STEM education, a teaching paradigm based on a multidisciplinary interaction between Science, Technology, Engineering and Mathematics allows the improving convergent skills in students curricula [1,3]. At the same time, there is a growing interest in investigating the role of Arts as a glue between these disciplines to improve also divergent skills. This leads to the growth of a new interdisciplinary paradigm called STEAM that, in recent research [4–6], seems to be an approach to increase the efficiency of learning and motivation for STEM concepts. According to these study, the inclusion of liberal arts and humanities in STEM curricula can be described as not a simple addition of an A in a STEM area but as an integration of Arts in STEM education with the

© Springer International Publishing AG 2018
G. De Pietro et al. (eds.), *Intelligent Interactive Multimedia Systems and Services 2017*,
Smart Innovation, Systems and Technologies 76, DOI 10.1007/978-3-319-59480-4_20

aim of creating an expanded transdisciplinary perspective [4, 7]. Such a perspective rejects the idea that Art is creative but not logical and scientific disciplines are logical but not creative. Art in its several variations (painting, sculpture, theater, dance, etc.) can represent different ways to interact with STEM for creating and expanding the mentioned transdisciplinary area. Among these variations, dance, if it is analyzed within an embodied cognitive paradigm [8], represents an excellent perspective for analyzing teaching/learning processes and activities involving STEM disciplines [9–12]. Namely, according to the Embodied Cognition approaches and in particular to Enactivism [13] and sensorimotor theory [14], that define intentional movements as a primary modality of thought [15], the sensorimotor interactions of the body with the physical world can be read as key elements of the process of cognition, learning and knowledge construction. Taking into account this point of view, dance can be described as a social practice which, relating to movements - primary modalities of thought - and integrating physical and motor aspects with aesthetic, active, communicative and social elements, allows investigations on the different skills. Moreover, it helps to unleash experimental performers' creativity, allowing it to spread and also facilitate the transfer of behavioral outcomes into everyday contexts and experiences. Learning activities (LAs) based on dance, therefore, emerge as suitable in the understanding of concepts from maths, geometry, physics, biomechanics and computational thinking (e.g. symmetry, reflection, acceleration, motion, equilibrium, self-similarity, progression patterns, compositional patterns, iteration) [9, 16, 17], suggesting an original dance-based STEAM approach.

In this work, we describe MoveYourMind (MYMi), a proposal for a STEAM Smart Learning Environment within which learning activities are conceived as educational practices where dance performances involve teachers, students, and technologies working together to implement creative ways to learn and teach STEAM notions at different levels.

2 Proposed Approach

Although within the STEM research field, several educational experiences of teaching and learning scientific concepts through dance can be reported, each experience can be described as an individual attempt to account for some concepts of a specific scientific discipline. There is a common intrinsic shortcoming that emerges within the context of this kind of studies: they appear as unrelated educational experiences of teaching and learning that exploit dance as a mere educational instrument.

MYMi proposal takes into account these several unconnected educational experiences of teaching and learning scientific concepts through dance and aims to create a learning framework able to connect these experiences within a unique educational framework. In this context, MYMi, using new technological integrations [18], proposes the realization of new tech-based and dance-based activities that exploit dance compositions as embodied STEAM concepts and foster the creation of abstract concepts from concrete activities [11]. Specifically, the created learning framework aims to be the theoretical/methodological/educational background for connecting in a whole coherent perspective educational the dance-based experiences that have explored:

– standard concepts as circles, squares, triangles, rectangles, lines and line segments, angles, exemplifying points [19];
– the classification of polygons according to the number of sides [19];
– geometrical concepts as shapes, patterns, angles, symmetry, reflection, rotation, rescaling [10];
   platonic solids, duality, self-similarity, regularity [20];
– the relationships between Math and Dance for teaching and learning fractions [21];
– the relationships between Math and Dance for teaching and learning progression patterns and matrices [22];
– the possible relationship between Physics and Dance to teach and learn concepts as equilibrium, motion, rotation, acceleration, mass of a body [12];
– modalities for creatively improving logical and computational thinking through computer programming [23–25].
– modalities for teaching and learning key computational concepts as identifying phases and orders within different tasks (sequencing), reiterate procedures to reach different goals (iteration), combine different parts in a whole (modularization), divide an activity into a concatenation of sub-activities or recognize parts (decomposition), fixing mistakes (debugging) [26].

An important assumption of the MYMi approach is that the integration of specific enabling technologies supports knowledge construction and educational processes within a dance-based STEAM (science, technology, engineering, arts, and mathematics) curriculum. Such an integration brings out the MYMi Smart Learning Environment (SLE) as the physical realization of the MYMi Learning Framework. This SLE, equipped with some key technological features [27], aims to involve students in learning paths within which the different possible connections between dance and technology are the essential cornerstones. As regards to these connections, the MYMi approach takes into account some crucial points such as the behavioral investigation about body perception and action recognition carried out through the use of motion-capture technology [28,29], the 3D human body motion tracking based on inertial sensing to record the trajectories of movements during learning activities and reconstruct the kinematics of a dancing body and its parts [30,31], the reconstructions with different modalities of representations and tools - e.g. a Serious Game context [32] - fitting for the MYMi specific educational goals, the corporeal interaction with the Smart Learning Environment during learning activity by means of the constant learners relation with a humanoid dancing robot with an active role. In the next section the key role of the humanoid robot will be examined in depth.

2.1 Dancing Humanoid Robots as Catalysts of Knowledge Construction

Within the MYMi perspective, the interaction with humanoid robots can enhance STEAM learning activities. A humanoid robot can act as a smart and amusing learning assistant and can help the students in performing some specific and ad-hoc developed learning activities. The knowledge management and reasoning mechanisms modeled in the robot can guide students in their learning experiences by acting as a captivating cultural mediator and fostering their creativity.

The body and its movements have a key role in the proposed STEAM setting, either in the choice of "dance" as the artistic discipline and also in the physical embodiment of the chosen technology. The robot is able to exploit its physicalness to perform dancing choreographies designed for the learning activities. It is able of reproducing real dancer movements and it is also able to creatively compose elementary movements by following the typical design steps of a choreographer. The behavior of the robot and the performed choreographies depends on the specific LA and the ongoing interaction with the students. The different typologies of LAs require behaviors of various complexity for the robot, from the reproduction of basilar movements to the performing of more complex choreographies and also the automatic generation of novel movements and choreographies. This last ability, obtained using a computational creativity approach, is exploited for introducing some variations in the LAs, with the result of stimulating the students and helping them in identifying possible mistakes. Besides, students can also deepen the cognitive model of the robot and exploit the coding interface of the robot to train the robot to perform specific activities or simulate some human behavior. The design and implementation of the robot behavior, therefore, represents itself a powerful component of LAs. In fact, some LAs can be based on the study and the use of the cognitive architecture on which the robot is based on, with the aim to train the robot to perform specific activities or simulate some human behavior.

According to this perspective, the robot is exploited not as a passive technological tool, but it has an active role in the STEAM learning setting.

3   MYMi STEAM Activities

In this section, we discuss possible activities for the proposed STEAM framework. The activities are aimed to review and practice mathematical concepts previously introduced in a third-grade class in a primary school. The concepts and their related vocabulary have been already introduced to the students. The learning purpose is the improvement of what learned in class and the student's abilities of observation and analysis of forms (passive observation of the concepts), expressing verbally and with the body what is observed and also the answers to the teacher questions (active expression of the concepts). Tree samples of ad hoc designed dance-based STEAM learning activities are proposed below. For each LA, at least the following key information is reported:

- The topic of the activity;
- The learning outcome of the activity (goal);
- The role and the number of participants;
- The description/implementation of the LA;

- The cognitive implication of the LA
- The didactic implication of the LA

All this information will be text based and will be used as a starting point to formal describe the learning activities. All the activities in the proposed framework require the exploitation of cross-sectional competencies from dancers, teachers in different fields (e,g math, geometry, physics and PE) and researchers during the design, implementation and assessment phases.

3.1   Sample Activity 1: PERMUDANCE

- Topic: Permutations (Math)
- Goal: arranging all the members of a set into some sequences or orders
- Participants: Teacher, a small group of students (2–3) and the robot.
- Description of the learning activity: The teacher proposes to students to play a role in the PERMUDANCE game. Both student and robot are placed in a row (side by side). Each position is marked by a number in ascending order. Students are the coaches of the robot and have to teach it a sequence of rhythmic movements to allow each member of a group to occupy, shifting rotation, the position number 1. Within this context, students collaborate to create and plan the sequence of movements and steps that robot has to learn. Then they execute the performance repeatedly to test the sequence of movement they created. At the end of the activity an avatar of the robot in a mobile platform suggests some exercises recalling the real situation (e.g. students have to find possible combinations/permutations rejecting repetitions).
- Cognitive implications: PERMUDANCE as learning activity involves memorization processes. Referring to the codification of sequences of movements in to be remembered movement tasks it involves both working memory (sequencing online) and long-term memory (sequences stored through corporeal patterns).
- Didactic implications: PERMUDANCE as learning activity allows students to acquire the ability to arrange all the members of a set into some sequence or order. Moreover, it can be adapted to different target groups modifying the number of participants and the complexity of spatial relationships.

3.2   Sample Activity 2: ORIGIN OF SYMMETRY

- Topic: Symmetry
- Goal: Ability to symmetrically/asymmetrically reproduce movements and actions and transfer (generalize) symmetrical and asymmetrical relationships to different contexts.
- Participants: Teacher, students of a classroom, robot/avatar
- Implementation: Imitation, Role-Changing, Creative configurations (3 Steps).
- Description of the learning activity:
  • Step 1 Imitation: The teacher involves students in a learning game in which the robot plays the role of tutor/coach. Introducing the robot as a dancing coach, the teacher asks students to place themselves in front

of the coach. After checking the accuracy of students disposition teacher reveals the goal of the game: each one is the mirrored image of the robot and has to exactly duplicate its actions to create several mirrors of the dancer coach.

- Step 2 Role-Changing: The teacher organizes different groups of students that have to create and design n-movements. They will teach robot these movements within an imitation game similar to Step 1 activity.
- Step 3 Creative Configurations: The robot, sequencing the learned movements by similarity, will propose to students a new dance. In this step, students will have to recognize the different movements created by each group and evaluate the accuracy of the performance.

At the end of each step, within a virtual learning environment an avatar of the coach, monitored by teacher, suggests some exercises and activities recalling the real situations (e.g. students have to distinguish symmetrical and asymmetrical movements, symmetrical in opposite/same direction and in/out of phase).

- Cognitive implications: ORIGIN OF SYMMETRY as learning activity involves motor skills related to posture and visual-motor imagery (spatial and kinaesthetic). Referring to the imitation step, synchronizing with movements of other people stimulates attentional processes. Moreover, this activity affects long-term memory that stores information as non-verbal grammar or movement vocabulary that can be employed in contexts other than dance practices.
- Didactic implications: ORIGIN OF SYMMETRY as learning activity allows students to acquire the ability to recognize symmetrical and asymmetrical relationships starting from a real situation. Moreover, it can be adapted to different level of complexity: e.g. within the exercises suggested by the avatar, modifying the number of different objects in symmetrical/asymmetrical relation or extending/generalizing the concept of symmetry to other connected element of other branch of knowledge (e.g. the logical concept of same as).

3.3   Sample Activity 3: BODY-SCRATCHING

- Topic: Scratch (programming).
- Goal: Learn to program in a virtual or a mixed reality to convert a simple sequence of movements into a creative embodied goal-directed action.
- Participants: Teacher, students, avatar/robot.
- Implementation phases: Body-scratching learning activity can be implemented through several levels of complexity and across different levels of generalization starting from two possible points: (1) from abstract STEAM concepts to real situations and (2) from real situations to abstract STEAM concepts.
- Description of the learning activity:
  - Starting point 1 From abstract STEAM concepts to real situations - Students learn to program by scratch in a virtual learning environment where the avatars actions need to be created and programmed. Within this event-driven programming context, using a mobile platform or a PC, students have to select from a database movements and dance steps fitting

with a specific proposed rhythm. After movements and dance steps selection, with teachers support, students learn to create a regular sequence of movements having a starting and an ending point (choreography). Each choreography will be embodied in a real learning context by the humanoid robot. During this phase, the teacher can highlight how the scratch activities are translated into concrete movements and how programming a choreography allows implementing actions with a specific goal.

- Starting point 2 From real situations to abstract STEAM concepts - Students equipped with specific technological tools (e.g. inertial sensors) learn to program an agent (the avatar) in real-time by means of their body. They implement an innovative body-scratching with the aim of creating a combination of rhythmic movements within a dance context. During this activity, the teacher will manage the avatar ability to recreate, within a virtual environment, students movements combination highlighting regular repetitions and possible errors emerging from the designing process. Moreover, teacher stresses the importance of movements iteration and rhythmic consistency in a choreography creation activity and emphasizes how real movements can be translated into a combination of scratch processes.

At the end of the body-scratching activity, the different choreographies can be implemented in a real situation by the robot.

- Cognitive implications: BODY-SCRATCHING as learning activity can be described as an embodied and enactive participatory sense-making activity and involves cognitive skills connected to creativity. It specifically involves long-term memory that is stimulated thanks to the learning of sequences coded in corporeal patterns. Moreover, it stimulates logical skills as inference and generalization and allows to evaluate heuristics processes within a computational learning activity.
- Didactic implications: BODY SCRATCHING as learning activity allows students to acquire knowledge connected to STEM disciplines and skills belonging to logical reasoning, especially the ability to generalize programming concepts starting from real situations (inductive and abductive reasoning) - POINT 2 - and the ability to reason involving inference (deduction) POINT 1.

4   Abilities of the Humanoid Dancer

A cognitive architecture [2] underlies the autonomous cognitive and creative processes of the artificial agent and supports the STEAM activities. The main modules of the architecture allow the robot to perceive the stimuli of the external environment, to understand its internal state, to learn through experience, to execute and coordinate its movements and to create new dancing choreographies.

The robot learns the basilar movements, postures and gestures designed in the learning activities. The robot behaviors are organized in modules, each one having a proper learning goal. The modules can be re-used during the definition of other LAs, with consequent advantages regarding the extensibility of the system.

The robot can also learn associations between music patterns and suitable motion sequences through an internal mechanism of evaluation, as in a dance school. The dancing repertoire is stored into the long-term memory (LTM) and used as input by the planning and execution module.

During the planning and execution phase, the humanoid dancer estimates the best sequence of movements to produce, choosing time to time the best movement according to the previously executed movement and the perceived sound.

Moreover, the robot is able to create new dancing steps in real time, by means of an artificial creative process, relying on the combination of an interactive genetic algorithm and a Hidden Markov Model (HMM).

The main parameters of the HMM are the states, the possible observations and two stochastic processes governing the system through a Transition Matrix (TM) and an Emission Matrix (EM). Briefly, in our model [2] the states correspond to the possible movements that the robot can perform whereas the observations are the perceived music.

The Transition Matrix of the HMM represents the correlation between the different movement; it has a key role in the composition of the movements, and it is built to select those that suit better each other. The Emission Matrix is used to link the movements to the rhythm perceived by the system; it is set up to select some movements rather than others in the presence of specific musical patterns. The HMM is employed by the robot to estimate the best motion sequence associated with the perceived music. The robot also employs an interactive genetic algorithm to learn different dancing styles under the guide of a human teacher producing original and harmonized movements in relations with the perceived music.

5 Conclusion and Future Works

In this work, an innovative proposal for a STEAM Smart Learning Environment, called MYMi, has been discussed. The assumption is that the introduction of art in teaching processes of STEM disciplines can increase the efficiency of learning, improving at the same time students' motivation and engagement.

The aim is to exploit body movements and dance in the definition of the learning activities since it is possible to obtain several connections with STEM concepts. We also propose the use of a humanoid robot as an amusing assistant in the accomplishment of the activities. Future works will regard the effective implementation and experimentation of the proposed SME.

References

1. Kanematsu, H., Barry, D.M.: STEM and ICT Education in Intelligent Environments, vol. 91, pp. 3–198. Springer (2016)
2. Augello, A., Infantino, I., Manfrè, A., Pilato, G., Vella, F., Chella, A.: Creation and cognition for humanoid live dancing. Robot. Auton. Syst. **86**, 128–137 (2016). Elsevier

3. Cotabish, A., Dailey, D., Robinson, A., Hughes, G.: The effects of a STEM intervention on elementary students' science knowledge and skills. Sch. Sci. Math. **113**(5), 215–226 (2013)
4. Land, M.H.: Full STEAM ahead: the benefits of integrating the arts into STEM. Procedia Comput. Sci. **20**, 547–552 (2013)
5. Madden, M.E., Baxter, M., Beauchamp, H., Bouchard, K., Habermas, D., Huff, M., Plague, G.: Rethinking STEM education: an interdisciplinary STEAM curriculum. Procedia Comput. Sci. **20**, 541–546 (2013)
6. DeSimone, C.: The necessity of including the arts in STEM. In: Integrated STEM Education Conference (ISEC 2014), pp. 1–5. IEEE, March 2014
7. Spector, J.M.: Education, training, competencies, curricula and technology. In: Ge, X., Ifenthaler, D., Spector, J.M. (eds.) Emerging Technologies for STEAM Education, pp. 3–14. Springer International Publishing (2015)
8. Shapiro, L.: Embodied Cognition. Routledge, London (2010)
9. Schaffer, K., Stern, E., Kim, S.: Math Dance. MoveSpeakSpin, Santa Cruz (2001)
10. Wasilewska, K.: Mathematics in the world of dance. In: Proceedings of Bridges 2012: Mathematics, Music, Art, Architecture, Culture, pp. 453–456 (2012)
11. Schaffer, K.: Math and dance windmills and tilings and things. In: Bridges Proceedings, pp. 619–622 (2012)
12. Capocchiani, V., Lorenzi, M., Michelini, M., Rossi, A.M., Stefanel, A.: Physics in dance and dance to represent physical processes. J. Appl. Math. **4**(4), 71–84 (2011)
13. Di Paolo, E.A., Thompson, E.: The enactive approach. In: The Routledge Handbook of Embodied Cognition, pp. 68–78 (2014)
14. Bishop, J.M., Martin, A.O. (eds.): Contemporary Sensorimotor Theory. Springer, Heidelberg (2014)
15. Smith, L.B., Thelen, E.: Development as a dynamic system. Trends Cogn. Sci. **7**(8), 343–348 (2003)
16. Leonard, A.E., Daily, S.B.: Computational and Embodied Arts Research in Middle School Education. Voke (2013)
17. Wilson, M., Kwon, Y.H.: The role of biomechanics in understanding dance movement: a review. J. Dance Med. Sci. **12**(3), 109–116 (2008)
18. Parrish, M.: Technology in dance education. In: International Handbook of Research in Arts Education, pp. 1381–1397. Springer, Netherlands (2007)
19. Moore, C., Linder, S.M.: Using dance to deepen student understanding of geometry. J. Dance Educ. **12**(3), 104–108 (2012)
20. Parsley, J., Soriano, C.T.: Understanding geometry in the dance studio. J. Math. Arts **3**(1), 11–18 (2009)
21. Watson, A.: Engaging senses in learning. Aust. Senior Math. J. **19**(1), 16–23 (2005)
22. Mui, W.L.: Connections between contra dancing and mathematics. J. Math. Arts **4**(1), 13–20 (2010)
23. Hamner, E., Cross, J.: Arts & Bots: techniques for distributing a STEAM robotics program through K-12 classrooms. In: Proceedings of the Third IEEE Integrated STEM Education Conference, Princeton, NJ, USA, March 2013
24. Oh, J., Lee, J., Kim, J.: Focus on 6th graders science in elementary school. In: Multimedia and Ubiquitous Engineering, pp. 493–501. Springer, Netherlands (2013)
25. Yanco, H.A., Kim, H.J., Martin, F.G., Silka, L.: Artbotics: combining art and robotics to broaden participation in computing. In: AAAI Spring Symposium: Semantic Scientific Knowledge Integration, p. 192, March 2007
26. Soh, L.K., Shell, D.F.: Integrating computational creativity exercises into classes (2015)

27. Augello, A., Infantino, I., Manfrè, A., Pilato, G., Vella, F., Gentile, M., Città, G., Crifaci, G., Raso, R., Allegra, M.: A personal intelligent coach for smart embodied learning environments. In: Intelligent Interactive Multimedia Systems and Services 2016, pp. 629–636. Springer International Publishing (2016)
28. Hove, M.J., Keller, P.E.: Spatiotemporal relations and movement trajectories in visuomotor synchronization. Music Percept. Interdisc. J. **28**(1), 15–26 (2010)
29. Neri, P., Luu, J.Y., Levi, D.M.: Meaningful interactions can enhance visual discrimination of human agents. Nat. Neurosci. **9**(9), 1186–1192 (2006)
30. Sevdalis, V., Keller, P.E.: Captured by motion: dance, action understanding, and social cognition. Brain Cogn. **77**(2), 231–236 (2011)
31. Loula, F., Prasad, S., Harber, K., Shiffrar, M.: Recognizing people from their movement. J. Exp. Psychol. Hum. Percept. Perform. **31**(1), 210 (2005)
32. De Gloria, A., Bellotti, F., Berta, R.: Serious games for education and training. Int. J. Serious Games **1**(1) (2014)

A Recommender System for Multimedia Art Collections

Flora Amato[1,2]([⊠]), Vincenzo Moscato[1,2], Antonio Picariello[1,2],
and Giancarlo Sperlí[1]

[1] DIETI - Department of Electrical Engineering and Information Technology,
University of Naples "Federico II", Naples, Italy
{flora.amato,vmoscato,picus,giancarlo.sperli}@unina.it
[2] CINI (Consorzio Nazionale Interuniversitario per L'Informatica) ITEM Lab,
Naples, Italy

Abstract. In this paper we present a novel *user-centered* recommendation approach for multimedia art collections. In particular, *preferences* (usually coded in the shape of items' metadata), *opinions* (textual comments to which it is possible to associate a particular sentiment), *behavior* (in the majority of cases logs of past items' observations and actions made by users in the environment), and *feedbacks* (usually expressed in the form of ratings) are considered and integrated together with *items' features* and *context information* within a general and unique recommendation framework that can support an intelligent browsing of any multimedia repository. Preliminary experiments show the utility of the proposed strategy to perform different browsing tasks.

1 Introduction

In the last decade, *Cultural Heritage Information Management* has been one of the most important research area that mainly took advantage by the advances of multimedia technologies.

As well known, this field involves the development of tools and applications for the processing, storage and retrieval of cultural digital contents, coming from distributed and heterogeneous data sources such as digital libraries and archives of cultural foundations (often in the shape of virtual museums), multimedia art collections (Europeana), web encyclopedias and social media networks (Wikipedia, Flickr, Picasa and Panoramio) and so on.

Indeed, such repositories in many cases include a large amount of multimedia contents with rich and different metadata, and one of the most interesting open research challenge is to provide recommendation techniques able to support in an effective and efficient way the intelligent browning of such collections, exploiting at the same time (low-level) features and (high-level) metadata description (together with the attached semantics) of cultural objects and trying to match user needs and preferences [1,2].

In other terms, the use of recommendation technologies can radically change the mode of fruition of multimedia art collections, from merely providing static

© Springer International Publishing AG 2018
G. De Pietro et al. (eds.), *Intelligent Interactive Multimedia Systems and Services 2017*,
Smart Innovation, Systems and Technologies 76, DOI 10.1007/978-3-319-59480-4_21

information of collections to personalized services to various visitors worldwide, in a way suiting visitors' personal characteristics, goals, tasks and behaviors together with information about the current *context*. In particular, personalization using recommender systems enables changing "the museum monologue" into "a user-centered information dialog" between the museum and its visitors. This interactive dialog can occurs not only in the real museum (on-site), but also in the *virtual museum* (on-line) usually accessible by the museum web site and realized through different technologies (e.g. rich Internet applications, 3D game engines, etc.).

Thus, museums are increasingly experimenting with and implementing more and more personalized and interactive services on their own web sites, and all over the world, the number of museum web site visits is growing fast [3]. Moreover, information from museum digital archive is often enriched and integrated with contents from social media network and other multimedia repositories, generating a large amount of data that poses new problems for the related management. Visitors spend more and more time on the museum web sites, e.g. to discover interesting artworks, prepare a museum tour, or learn knowledge about artworks, usually in relation to a (possible) physical museum visit. This brings a great challenge for museums to provide a personalized and extended museum experience for visitors in an "immersive" museum environment, which includes both the virtual museum and the real one.

Generally, recommender systems help people in retrieving information that match their preferences or needs by recommending products or services from a large number of candidates, and support people in making decisions in various contexts: what items to buy, which photo or movie to watch [1], which music to listen, what travels to do [4], who they can invite to their social network, or even which artwork could be interesting within an art collection [5,6] just to make some examples.

Formally, a recommender system deals with a set of *users* $U = \{u_1 \ldots, u_m\}$ and a set of *items* $O = \{o_1, \ldots, o_n\}$. For each pair (u_i, o_j), a recommender can compute a *score* (or a *rank*) $r_{i,j}$ that measures the expected interest of user u_i in item o_j (or the expected utility of item o_j for user u_i), using a *knowledge base* [7] and a *ranking* algorithm that generally could consider different combinations of the following characteristics: (i) user preferences and past behavior, (ii) preferences and behavior of the user community, (iii) items' features and how they can match user preferences, (iv) user feedbacks, (v) context information and how recommendations can change together with the context (see [8] for more details). From the architectural point of view, the last generation of recommender systems is usually composed by one or more of the following components [4].

- A *pre-filtering* module that selects for each user u_i a subset $O_i^c \subset O$ containing items that are good candidates to be recommended; such items usually match user preferences and needs.
- A *ranking* module that assigns w.r.t. user u_i a rank $r_{i,j}$ to each candidate item o_j in O_i^c using the well-known recommendation techniques (i.e., *content-based, collaborative filtering* and *hybrid* approaches) that can exploit in several ways

items' features and users' preferences, feedbacks (in the majority of cases in terms of *ratings*) moods (by opinion or sentiment analysis) and behavior [9].
– A *post-filtering* module that dynamically excludes, for each user u_i, some items from the recommendations' list; in this way, a new set $O_i^f \subseteq O_i^c$ is obtained on the base of user feedbacks and other contextual information (such as data coming from the interactions between the user and the application) [10]. Eventually, depending on applications, recommended items can be arranged in *groups* according to additional constraints [4].

In this paper, we propose a novel *user-centered* approach that provides *cultural* multimedia recommendations on the base of several aspects related to users [11,12]. In particular, *preferences* (usually coded in the shape of items' metadata), *opinions* (textual comments to which it is possible to associate a particular sentiment), *behavior* (in the majority of cases logs of past items' observations and actions made by users in the environment), *feedbacks* (usually expressed in the form of ratings) - are considered and integrated together with *items' features* and *context information* within a general and unique recommendation framework that can support the intelligent browsing of any multimedia art collection.

The paper is organized as follows. Section 2 describes the proposed strategy for recommendation and provides a recommender system overview with some implementation details. Section 3 reports a functional overview of the proposed recommender system. Section 4 reports preliminary experimental results and Sect. 5 gives some concluding remarks and discusses future works.

2   The Multimedia Recommendation Strategy

The basic idea behind our proposal is that when a user is browsing a particular multimedia art collection, the recommender system: (i) determines a set of useful *candidate* items on the base of user actual needs and preferences (*pre-filtering stage*); (ii) opportunely assigns to these items a rank, previously computed exploiting items' intrinsic features, users' past behaviors, and also using users' opinions and feedbacks (*ranking stage*); (iii) dynamically, when a user "selects" as interesting one or more of the candidate items, determines the list of most suitable items (*post-filtering stage*), also considering other context information in the shape of a set constraints on items' features.

In our approach, items to be recommended are multimedia data (i.e. texts, images, videos, audios) related to specific artworks (e.g. paintings, sculptures or other kinds of artifacts that can be exhibited within a museum). As in the majority of multimedia systems, items are described at two different levels:

– from an "high-level" perspective, by one or more set of *metadata* that do not depend on the type of multimedia object [1];
– from a "low-level" perspective, by a set of *features* that are different for each kind of multimedia object.

---

[1] In the Cultural Heritage domain different harvesting sets of metadata and possibly domain taxonomies or ontologies can be considered.

To cope with the heterogeneity of multimedia data, our recommender system is constituted by a *multichannel browser* [13]. Each single channel is then specialized to recommend multimedia items of a given type. Recommended items of different types can be eventually grouped in a unique "multimedia presentation".

In the following, we describe the recommendation strategy provided by each single browsing channel.

2.1   Pre-filtering Stage Using User Preferences

In the *pre-filtering* stage, our aim is to select for a given user u_h a subset $O_h^c \subset O$ containing items that are good "candidates" to be recommended: such items usually have to match some (static) user preferences and (dynamic) needs from the "high-level" perspective.

Each item subjected to recommendation may be represented in different and heterogeneous feature spaces. For instance, a photo may be described by a set of metadata as title, set of tags, description, etc. The first step consists in clustering together "similar" items, where the similarity should consider all (or subsets of) the different spaces of features.

To this goal, we employ *high-order star-structured co-clustering* techniques - that some of the authors have adopted in previous work [6] - to address the problem of heterogeneous data filtering, where a user is represented as a set of vectors in the same feature spaces describing the items.

2.2   Ranking Stage Using User Behavior and Items Similarity

The main goal of this stage is to automatically rank the set of items O embedding in a collaborative learning context (user preferences are represented modeling the choice process in recommender system considering users' *browsing behaviors*) their *intrinsic features* (those on the top of which it is possible to introduce a *similarity* notion).

In particular, we use a novel technique that some of the authors have proposed in previous works [1,14] to provide useful recommendations during the browsing of multimedia collections.

Our basic idea is to assume that when an item o_i is chosen after an item o_j in the same user *browsing session* (and both the explored items have been positively rated or have captured attention of users for an adequate time), this event means that o_i "is voting" for o_j. Similarly, the fact that an item o_i is "very similar" in terms of some intrinsic features to o_j can also be interpreted as o_j "recommending" o_i (and viceversa).

Thus, we are able to model a browsing system for the set of items O as a labeled graph (G, l), where: (i) $G = (O, E)$ is a *directed graph*; (ii) $l : E \rightarrow \{pattern, sim\} \times R^+$ is a *labeling function* that associates each edge in $E \subseteq O \times O$ with a pair (t, w), where t is the type of the edge which can assume two enumerative values (*pattern* and *similarity*) and w is the weight of the edge.

We list two different cases:

1. a *pattern label* for an edge (o_j, o_i) denotes the fact that an item o_i was chosen immediately after an item o_j and, in this case, the weight w_j^i is the number of times o_i was chosen immediately after o_j;
2. a *similarity label* for an edge (o_j, o_i) denotes the fact that an item o_i is similar to o_j and, in this case, the weight w_j^i is the "similarity" between the two items. Thus, a link from o_j to o_i indicates that part of the importance of o_j is transferred to o_i.

Given an item $o_i \in O$, its *recommendation grade* $\rho(o_i)$ is defined as follows:

$$\rho(o_i) = \sum_{o_j \in P_G(o_i)} \hat{w}_{ij} \cdot \rho(o_j) \tag{1}$$

where $P_G(o_i) = \{o_j \in O | (o_j, o_i) \in E\}$ is the set of predecessors of o_i in G, and $\hat{w}_{ij}$ is the normalized weight of the edge from o_j to o_i.

We note that for each $o_j \in O$ $\sum_{o_i \in S_G(o_j)} \hat{w}_{ij} = 1$ must hold, where $S_G(o_j) = \{o_i \in O | (o_j, o_i) \in E\}$ is the set of successors of o_j in G.

In [14], it has been shown that the ranking vector $R = [\rho(o_1) \ldots \rho(o_n)]^T$ of all the items can be computed as the solution to the equation $R = C \cdot R$, where $C = \{\hat{w}_{ij}\}$ is an ad-hoc matrix that defines how the importance of each item is transferred to other items.

The matrix can be seen as a linear combination of:

- a *local browsing matrix* $A_h = \{a_{ij}^h\}$ for each user u_h, where its generic element a_{ij}^l is defined as the ratio of the number of times item o_i has been chosen by user u_h immediately after o_j to the number of times any item in O has been chosen by u_h immediately after o_j;
- a *global browsing matrix* $A = \{a_{ij}\}$, where its generic element a_{ij} is defined as the ratio of the number of times item o_i has been chosen by any user immediately after o_j to the number of times any item in O has been chosen immediately after o_j;
- a *similarity matrix* $B = \{b_{ij}\}$ such that b_{ij} denotes the similarity between two items o_i and o_j (a semantic relatedness [15,16] based on a set of taxonomies using some high-level features values [13], eventually combined with a low-level features comparison using Windsurf multimedia libraries [17]).

The successive step is to compute *customized* rankings for each individual user. In this case, we can rewrite previous equation considering the ranking for each user as $R_h = C \cdot R_h$, where R_h is the vector of preference grades, customized for a user u_h considering only items in the related O_h^c.

We note that solving the discussed equation corresponds to finding the stationary vector of C, i.e., the eigenvector with eigenvalue equal to 1. In [1], it has been demonstrated that C, under certain assumptions and transformations, is a real square matrix having positive elements, with a unique largest real eigenvalue and the corresponding eigenvector has strictly positive components. In such conditions, the equation can be solved using the *Power Method* algorithm.

Rank can be finally refined using user attached sentiments and ratings (see [4] for more details).

2.3 Post-Filtering Stage Using Context Information

We have introduced a *post-filtering* method for generating the final set of "real" candidates for recommendation using *context* information. The context is represented by means of the well-known *key-value* model [18] using as dimensions some of the different feature spaces related to items.

In our system, context features can be expressed either directly using some *target items* (e.g. objects that have positively captured user attention) or specifying the related values in the shape of *constraints* that recommended items have to satisfy.

Assume that a user u_h is currently interested in a target item o_j. We can define the set of candidate recommendations as follows:

$$O_{h,j}^f = \bigcup_{k=1}^{M} \{o_i \in O_h^c \,|\, a_{ij}^k > 0\} \cup \{o_i \in NNQ(o_j, O_h^c)\} \tag{2}$$

The set of candidates includes the items that have been accessed by at least one user within k steps from o_j, with k between 1 and M, and the items that are most similar to o_j according to the results of a *Nearest Neighbor Query* ($NNQ(o_j, O_h^c)$) functionality[2]. The ranked list of recommendations is then generated by ranking the items in $O_{h,j}^f$, for each item o_j selected as interesting by user u_h, using the ranking vector R_h thus obtaining the final set O_h^f.

Finally, for each user all the items that do not respect possible context constraints are removed from the final list.

3 System Overview

Figure 1 describes at a glance an overview of the proposed system in terms of its main components, which we are describing in the following.

Data to be recommended are retrieved by a *Wrapper* component that is composed by several modules. The *Data Crawler and Schema Mapper* is responsible of: (*i*) periodically accessing to the items' repositories (e.g., museum digital archives, DBpedia, Europeana, Flickr, Panoramio, etc.), (*ii*) extracting for each item all the *features* (e.g., metadata, possible ratings and reviews, etc.) and other information of interest (e.g. user preferences, time-stamped items' observations, etc.), (*iii*) performing a mapping w.r.t. a global schema. A part of such information will be then exploited by the *User Logs and Preference Analyzer* and *Opinion Miner* modules to determine the user behavior graphs with the related profile and the users' mood on the different items.

---

[2] Note that a positive element a_{ij}^k of A^k indicates that o_i was accessed exactly k steps after o_j at least once.

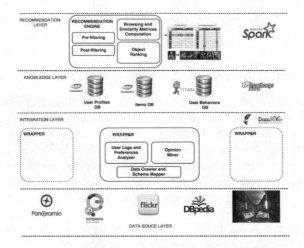

Fig. 1. System overview.

After the wrapping phase, all the information will be stored in the *Knowledge Base* [19] of the system. In particular, it is composed by: (*i*) the *User Behaviour DB* containing a set of graphs related to the users' browsing sessions, (*ii*) the *Items DB* containing items with all the related features, and (*iii*) *User Profiles DB* containing user preferences in terms of items' features.

The *Recommender Engine* provides a set of recommendation facilities for multi-dimensional and interactive browsing of items. Exploiting user preferences, the *Pre-filtering* module selects a set of *candidate* items for recommendation; successively, the *Object Ranking* module assigns a ranking of such candidates exploiting a proper social strategy (that uses the *Users and Similarity Matrices Computation* module and information on users' opinions). Finally, the *Post-filtering* module dynamically selects a subset of candidates, on the base of the item that a user is currently watching and context information.

4   Case Study and Preliminary Experiments

We designed and implemented a first prototype - for the system depicted in Fig. 1 - that builds and manages different multimedia art collections, providing the basic facilities for querying and recommendation, using data from several multimedia repositories. In particular, our recommender system was realized on the top of the Apache Spark engine based on the Hadoop technological stack to meet Big Data issues.

More in details, the log files and multimedia raw data are stored on HDFS while user and object descriptions are managed by Cassandra. Co-clustering, ranking and post-filtering techniques were implemented on the top Spark machine learning and graph analysis libraries and leveraging SPARK SQL facilities; in turn, multimedia similarities were computed using Windsurf library [17].

Wrapping functionalities were implemented using DataRiver[3] solution (based on the MOMIS system).

In addition, the system provides some REST API that can be dynamically invoked by the virtual museum applications to suggest a set of items that can be of interest for a particular user browsing multimedia art collections.

As case study, we test our system to support the intelligent browsing of a virtual museum containing about 1000 digital reproductions of paintings belonging to more than 50 different artistic genres and 250 different authors.

In particular, the virtual world has been realized by the Unity3D framework[4] within a particular game, which goal is to dynamically suggest images of interests for users to maximize his/her satisfaction by the generation of *visiting paths*[5]. In the game, a proper graphic user interface gives in any moment the detailed view of the suggested path (obtained invoking API of our recommender system) on an proper cartography, reporting a preview of artworks and of the related descriptions (see Fig. 2).

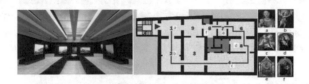

Fig. 2. Virtual museum.

Concerning the preliminary experimental results, we evaluated the impact of the proposed system on users engaged in several *browsing tasks* of multimedia items and compared its performances without any recommendation utility. We performed a *user-centric* evaluation based on *user satisfaction* with respect to assigned activities, evaluating how our recommendations can effectively support browsing tasks of increasing complexity. To such purpose, we asked a group of about 50 people to browse the collection of items and complete several search tasks (20 tasks per user) of different complexity (five tasks for each complexity level), using recommendation facilities. Then, we asked them to browse the same collection without any assistance and complete other 20 tasks of similar complexity.

We subdivided browsing tasks into three categories: (i) **Low Complexity** search tasks (e.g. find at least 10 images related to woman portraits); (ii) **Medium Complexity** search tasks (e.g. find at least 15 images depicting landscape and related to the Romanticism art); (iii) **High Complexity** search tasks (e.g. find at least 35 images with a religious subject related to the baroque art).

---

[3] http://www.datariver.it/it/.

[4] Virtual rooms with paintings are related to specific historical periods.

[5] The suggested paths represent the most easy way for the player to find and access the paintings of interest.

Table 1. Utility of recommendation services in terms of TLX factors

TLX factor	With recommendations	Without recommendations
Mental demand	34	44
Physical demand	32.1	48
Temporal demand	37,4	54
Effort	33	60
Perfomances	63.8	78.4
Frustration	30.1	50

Note that the complexity of a task depends on several factors: the number of items to explore, the type of desired features and the number of additional constraints. The strategy was used to evaluate the results of this experiment is based on the TLX (*NASA Task Load Index factor*) evaluation protocol [20]. TLX is a multi-dimensional rating procedure that provides an overall workload score based on a weighted average of ratings on six sub-scales: mental demand, physical demand, temporal demand, own performance, effort and frustration. Lower TLX scores are better and the average scores are then reported in Table 1.

5   Conclusions and Future Works

Our proposal represents a recommender system supporting cultural heritage applications. We have shown that the recommendations may be computed combining several features of objects, past behavior and preferences of individual users and overall behavior of the entire community of users, and finally, context information. Future works will be devoted to extend the experimental evaluation to a larger multimedia data set, also considering the accuracy performances.

References

1. Albanese, M., d'Acierno, A., Moscato, V., Persia, F., Picariello, A.: A multimedia recommender system. ACM Trans. Internet Techn. **13**(1), 3 (2013)
2. Kabassi, K.: Personalisation systems for cultural tourism. In: Multimedia Services in Intelligent Environments, pp. 101–111. Springer (2013)
3. Karaman, S., Bagdanov, A.D., Landucci, L., D'Amico, G., Ferracani, A., Pezzatini, D., Del Bimbo, A.: Personalized multimedia content delivery on an interactive table by passive observation of museum visitors. Multimedia Tools Appl. **75**(7), 3787–3811 (2016)
4. Colace, F., Santo, M.D., Greco, L., Moscato, V., Picariello, A.: A collaborative user-centered framework for recommending items in online social networks. Comput. Hum. Behav. **51**, 694–704 (2015)
5. Albanese, M., d'Acierno, A., Moscato, V., Persia, F., Picariello, A.: A multimedia semantic recommender system for cultural heritage applications. In: Proceedings of the 5th IEEE International Conference on Semantic Computing (ICSC 2011), Palo Alto, CA, USA, 18–21 September, pp. 403–410 (2011)

6. Bartolini, I., Moscato, V., Pensa, R.G., Penta, A., Picariello, A., Sansone, C., Sapino, M.L.: Recommending multimedia visiting paths in cultural heritage applications. Multimedia Tools Appl. **75**(7), 3813–3842 (2016)
7. Minutolo, A., Esposito, M., De Pietro, G.: A mobile reasoning system for supporting the monitoring of chronic diseases. Springer, Heidelberg, pp. 225–232 (2012)
8. Adomavicius, G., Sankaranarayanan, R., Sen, S., Tuzhilin, A.: Incorporating contextual information in recommender systems using a multidimensional approach. ACM Trans. Inf. Syst. (TOIS) **23**(1), 103–145 (2005)
9. Ricci, F., Rokach, L., Shapira, B., Kantor, P.B. (eds.): Recommender Systems Handbook. Springer (2011)
10. Essmaeel, K., Gallo, L., Damiani, E., De Pietro, G., Dipanda, A.: Comparative evaluation of methods for filtering kinect depth data. Multimedia Tools Appl. **74**(17), 7331–7354 (2015)
11. Brancati, N., Caggianese, G., Frucci, M., Gallo, L., Neroni, P.: Experiencing touchless interaction with augmented content on wearable head-mounted displays in cultural heritage applications. In: Personal and Ubiquitous Computing, pp. 1–15
12. Caggianese, G., Gallo, L., De Pietro, G.: Design and preliminary evaluation of a touchless interface for manipulating virtual heritage artefacts. In: 2014 Tenth International Conference on Signal-Image Technology and Internet-Based Systems (SITIS), pp. 493–500. IEEE (2014)
13. Albanese, M., Chianese, A., d'Acierno, A., Moscato, V., Picariello, A.: A multimedia recommender integrating object features and user behavior. Multimedia Tools Appl. **50**(3), 563–585 (2010)
14. Albanese, M., d'Acierno, A., Moscato, V., Persia, F., Picariello, A.: Modeling recommendation as a social choice problem. In: Proceedings of the 2010 ACM Conference on Recommender Systems, RecSys 2010, Barcelona, Spain, 26–30 September, pp. 329–332 (2010)
15. Amato, F., Mazzeo, A., Moscato, V., Picariello, A.: A framework for semantic interoperability over the cloud. In: Proceedings - 27th International Conference on Advanced Information Networking and Applications Workshops, WAINA 2013, pp. 1259–1264 (2013)
16. Amato, F., Mazzeo, A., Penta, A., Picariello, A.: Using NLP and ontologies for notary document management systems. In: Proceedings - International Workshop on Database and Expert Systems Applications, DEXA, pp. 67–71 (2008)
17. Bartolini, I., Patella, M.: Multimedia queries in digital libraries. In: Data Management in Pervasive Systems, pp. 311–325. Springer (2015)
18. Strang, T., Linnhoff-Popien, C.: A context modeling survey. In: Workshop Proceedings (2004)
19. Colantonio, S., Esposito, M., Martinelli, M., De Pietro, G., Salvetti, O.: A knowledge editing service for multisource data management in remote health monitoring. IEEE Trans. Inf. Technol. Biomed. **16**(6), 1096–1104 (2012)
20. Hart, S.G., Staveland, L.E.: Development of nasa-tlx (task load index): Results of empirical and theoretical research. Adv. Psychol. **52**, 139–183 (1988)

Using Multilayer Perceptron in Computer Security to Improve Intrusion Detection

Flora Amato[✉], Giovanni Cozzolino, Antonino Mazzeo, and Emilio Vivenzio

DIETI - Dipartimento di Ingegneria Elettrica E Tecnologie Dell'Informazione, Università degli studi di Napoli "Federico II", via Claudio 21, 80125 Naples, Italy
{flora.amato,giovanni.cozzolino,mazzeo}@unina.it

Abstract. Nowadays computer and network security has become a major cause of concern for experts community, due to the growing number of devices connected to the network. For this reason, optimizing the performance of systems able to detect intrusions (IDS - Intrusion Detection System) is a goal of common interest. This paper presents a methodology to classify hacking attacks taking advantage of the generalization property of neural networks. In particular, in this work we adopt the multilayer perceptron (MLP) model with the back-propagation algorithm and the sigmoidal activation function. We analyse the results obtained using different configurations for the neural network, varying the number of hidden layers and the number of training epochs in order to obtain a low number of false positives. The obtained results will be presented in terms of type of attacks and training epochs and we will show that the best classification is carried out for DOS and Probe attacks.

Keywords: Network security · Intrusion detection · Multilayer perceptron · Machine learning · Neural networks

1 Introduction

Today computer networks have a widespread distribution and daily people make use of a growing number of network services [13, 14] that are becoming even more pervasive. This has led, dually, to a security problem for devices connected to a network. In order to find out attacks against information systems, there are many tools (hardware and/or software), called IDS (*Intrusion Detection System*) [1], designed to protect the accessibility of systems, the integrity and confidentiality of data.

Based on the location in a network, IDS can be categorized into two groups:

- **NIDS**: Network IDS. They analyse the packets transmitted through the network looking for their *"signatures"* (set of conditions) and comparing them with those stored in a database [2]
- **HIDS**: Host-Based IDS. They operate directly on a machine detecting intrusions by monitoring the operating system through its log, the file system and hard drives, etc.

© Springer International Publishing AG 2018
G. De Pietro et al. (eds.), *Intelligent Interactive Multimedia Systems and Services 2017*,
Smart Innovation, Systems and Technologies 76, DOI 10.1007/978-3-319-59480-4_22

There are two main techniques of analysis adopted by a NIDS [2]:

- **Pattern Matching Based**: It determines intrusions by comparing an activity with known signatures. It has a *low* false positive rate but it does not allow to recognize new kinds of attack.
- **Anomaly Detection Based**: It determines intrusions by looking for anomalies in network traffic. It has a relatively *high* rate of false positives, but allows to detect new kinds of attack, not yet stored in the database.

Existing IDS, like *Snort*, a very popular and open-source network intrusion detection system, present a limitation related to the detection of new attacks, because the detection mechanism is based on *"signatures"*. In fact, they are not able to acquire new knowledge unless the system administrator does not update the definitions, just like anti-viruses. So, if an unknown attack occurs, although it may slightly differ from another one stored in definitions database, the IDS will not be able to identify it.

Therefore, we exploit the advantages of the generalization property, typical of neural networks. We apply the perceptron algorithm in order to classify a generic attack through a predictor function, combining a set of weights with a feature vector. A such type of network is able to identify both known and unknown intrusion and, if properly trained with a series of examples, to reduce false alarms.

In literature many projects have exploited artificial intelligence approaches [19, 20] (decision trees, Bayesian classifiers, multilayer perceptron, etc.) to mitigate generics intrusion detection risks or single class of anomalies. We can read them in [3–7].

2   The Multilayer Networks

Multilayer networks have been introduced to cope with the limitations of single-layer perceptrons. Minsky and Papert demonstrated in [8] how a simple exclusive OR (XOR), which is a classification problem but not linearly separable, could not be solved by a single-layer perceptron network. Therefore, it is possible to consider more levels (also called *layers*) of neurons connected in cascade.

The multi-layer networks are composed of:

- **Input layer:** composed of n nodes, without any processing capacity, which send the inputs to subsequent layers;
- **Hidden layer** (one or more): composed of neural elements whose calculations are input to subsequent neural units;
- **Output layer:** composed of m nodes, whose calculations are the actual outputs of the neural network.

In case of a competitive learning, the output is selected through the *Winner-Takes-All* computational principle, according to which only the neuron with the greatest "activity" will remain active, while the other neurons will be inactive. Graphically, input data can be represented on a plane and each layer draws a straight line inside. The intersections of these lines generate decision regions. This is a limit to be considered, because the inclusion of too many layers can create too many regions, which means that the perceptron loses the ability to generalize, but it specializes on the set of training data samples. This phenomenon is called **overfitting**. Dually the **underfitting** problem exists, where the network has a number of neurons unable to learn. The use of a preventive mechanisms, such as cross-validation and the early stopping, can avoid falling into similar excesses (Fig. 1).

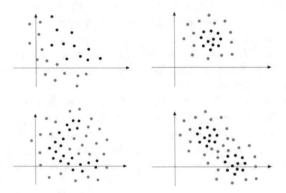

Fig. 1. Examples of not linearly separable data

Unlike to the single perceptron, supposing some hypotheses about activation functions of the individual element, it is possible to approximate any continuous function on a compact set and then to solve the problems of classification [11, 12] of not linearly separable sets. This reasoning agrees to Kolmogorov theorem (1957). It state that with three layers it is possible to implement any continuous function in $[0, 1]^n$, where in the first layer we place the n input elements, in the intermediate layer $(2n + 1)$ elements, and in last layer we place the m elements (equal to the number of elements of the co-domain space $\mathbb{R}^m$).

2.1   Training a Neural Network

The modeling of a multilayer neural network leads to two main problems [9]. We have to:

– **Select the architecture**, in other words, the number of layers and neurons that each layer should possess;
– **Train the network**, namely to determine the appropriate weights of each neuron and its threshold;

Typically, if we fix architecture, the training problem can be seen as the ability of the system to produce the outputs according to our needs. This is equivalent to minimizing the error between the desired error and the obtained output. The most used is the squared error:

$$E_i(w) = \frac{1}{2}||D_i - O_i||^2$$

where, for simplicity, D_i indicates the desired output of the generic i-th neuron (in place of y_i), and O_i indicates the obtained output (in place of $y(x_i, w)$, depending on the weight and input).

Usually, to compute and find this error, we follow two heuristic methods [9], because often the statistical theories are not adequate. The first method is **structural stabilization:** it consists in gradually growing, during the training, the number of neural elements (whose set is called *training set*). It is estimated, initially, the error of this network on the training set and on a different set, said *validation set*. Then we can select the network that produces the minimum error on the latter. Once trained, the network will be evaluated using a third set said *test set*. The second method, known as **regularization**, consists of adding penalty to the error, with the effect of restricting the choice set of weights w.

3   The Dataset KDD '99 and Features Description

We aim training a neural network to make it able to predict and distinguish malicious connections from not malicious ones (normal connections). To train our network we choose a (publicly available) labelled dataset for IDS, KDD '99, subset of DARPA (agency of U.S. Department of Defense) dataset [10]. It was created by acquiring nine weeks of raw TCP dump data from a LAN, simulating attacks on a typical military environment, like U.S. Air Force LAN. The connections are a sequence of TCP packets and each record consists of about 100 bytes.

The attacks fall into four main categories:

- **DOS:** denial-of-service, e.g. syn flood;
- **R2L:** unauthorized access from a remote machine, e.g. guessing password;
- **U2R:** unauthorized access to local superuser (root) privileges, e.g., various "buffer overflow" attacks;
- **Probe:** surveillance and other probing, e.g., port scanning.

For our data analysis, we use an open-source platform, widely used in the field of "data mining" and "machine learning": KNIME (Konstanz Information Miner). It means the aggregation of nodes makes it possible to "pre-process data, in other words do extractions, transformation and loading, modeling, analyzing, and displaying data".

For simplicity, during our experimental session, due to the complexity of KDD dataset (about 500.000 records), we use only a tenth part for training activities discussion. We manipulate the dataset features, originally not organized in a tabular representation, in order to change the format in ARFF type, useful to process it in KNIME environment with the components of Weka (automatic learning software developed by the University of Waikato in New Zealand). In Table 1 we show a classification of attacks according to their type.

Table 1. Classification of attacks

Type of attack	Attack
DOS	Back, Land, Neptune, Pod, Smurf, Teardrop
U2R	Ipsweep, Nmap, Portsweep, Satan
R2L	Bueroverow, Perl, Loadmodule, Rootkit
Probe	Ftpwrite, Imap, GuessPasswd, Phf, Multihop, Warezmaster, Warezclient

3.1 Preprocessing and Features Selection

Given the huge number of features of KDD dataset, we make a selection of the essential attributes. Moreover, the attributes have different types: continuous, discrete and symbolic, each with its own resolution and range of variation. We can convert symbolic attributes into numerical (attributes like *protocol_type*, *service*, *flag*) and normalize the other attributes between 0 and 1. In Fig. 2 we show the features selection meta-node.

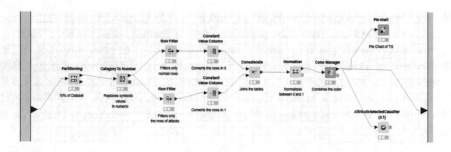

Fig. 2. Block diagram of the meta-node for the selection of the essential attributes

We can observe the following blocks:

- **ARFF Reader:** it reads the file containing the samples;
- **Partitioning:** it partitions the table considering only 10% of the dataset;
- **Category to Number:** it takes symbolic attributes and converts them to numeric;
- **Row Filter:** it filters rows of non-malicious connections by marking them with a 0 and rows of malicious connections by marking them with 1;
- **Concatenate:** it combines the changed tables;
- **Normalizer:** it normalizes between 0 and 1 the attribute values, dividing the value of each attribute to its maximum;
- **Color Manager:** it assigns a colour to the Normal class (0) and one at Attack class (1);
- **AttributeSelectedClassifier (v3.7):** it carries out the selection of the most discriminating attributes, based on various algorithms that we will show;

– **Pie Chart:** it displays a pie chart of the attributes of the training set. We can observe that, in agreement to what was said, there are more attacks that normal connections (Fig. 3);

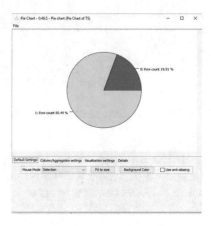

Fig. 3. Pie chart of the training set

To select the discriminating attributes, we tested different configurations for the AttributeSelectedClassifier block, also making use of an external tool, Weka Explorer, to better display the outcomes. No substantial differences were noted: both search algorithms on 41 attributes will select 11: *protocol_type, service, flag, src_bytes, dst_bytes, land, wrong_fragment, root_shell, count, diff_srv_rate, dst_host_same_src_port_rate.*

We considered the two following search algorithms:

– **GreadyStepwise:** it basically uses a Hill-Climbing algorithm. It returns a number of essential features equal to 11 and it is preferred because it employs a lower computational time (equal to 28 s).
– **BestFirst:** it is similar to GreadyStepwise but with the use of backtracking. It returns the same number of features (11) but with a higher computational time (32 s) and that's why it was decided to discard it.

In addition, we apply the ranker in order to obtain a consistency on features choices. From a list of attributes classified by an evaluator, it sorts them in descending vote and still get a consistent choice. All these blocks, for practical reasons, are encapsulated in a single meta-node: *Preprocessing.*

4   Evaluation of the Network with the Entire Dataset and Analysis

In this phase, we test the chosen configuration on the entire data set using the block diagram shown in Fig. 4. We test it also on the individual types of attacks (and not) to provide some statistical utility.

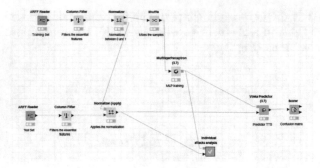

Fig. 4. Test workflow of the entire dataset

The scheme presents all the previous blocks, except the Normalizer Apply, to apply the same normalization derived from the Normalizer block applied to the test set. The out-coming classification system is characterized by an error of about 1.3% with a 98.7% accuracy (Fig. 5).

Correct classified: 487.804	Wrong classified: 6.217
Accuracy: 98,742 %	Error: 1,258 %
Cohen's kappa (κ) 0,979	

Fig. 5. Output of the final scorer

We also test the system on a dataset consisting of individual types of attacks (and same training set), through the scheme shown in Fig. 6:

We observe that, for each category of attack, there is a "Value Filter" block that filters the rows based on the type of attack. Then we use the blocks that perform the prediction on the basis of the training of the previous block. It is possible to see the error rate, in other words the number of correctly classified (or not) of our studies in the following Table 2:

Table 2. Error rate for individual types

Type of attack	Error rate	Correctly classified	Not correctly classified
DOS	0.639%	388.958	2500
Probe	41.953%	2.384	1.723
U2R	100%	0	52
R2L	99.911%	1	1.123
Normal	0.84%	96.461	817
All	1.258%	487.804	6.217

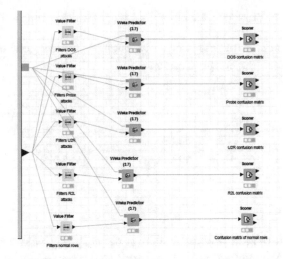

Fig. 6. Meta-node for the analysis of individual attacks

So, with given data set and training configuration, our neural network is able to correctly distinguish between malicious and non-malicious connections. However, through malicious connections, the system classify with more precision DOS and Probe attacks. This behaviour is justified by the insufficiency of instances of the other two categories of attacks, so the network is not able to learn properly (U2R has only 52 instances, while R2L slightly exceeds thousands). This result are consistent with with other studies (e.g. [7]). On the basis of these observations, we have tried to train the network with the largest training time, choosing a training time equal to 500. As expected, the network goes into over-training, as is evident from the outcome shown in Fig. 7 of the scorer downstream of the entire dataset evaluation:

Fig. 7. Output of the final scorer with training time equal to 500

while the analysis for the individual types of attacks, summarized in Table 3, shows lower performances of the classification process, and justify our previous choice.

Table 3. Error rate for individual types (training time = 500)

Type of attack	Error rate	Correctly classified	Not correctly classified
DOS	72.367%	108.170	283.288
Probe	10.251%	3.686	421
U2R	100%	0	52
R2L	100%	0	1.124
Normal	0.711%	96.586	692
All	57.807%	208.442	285.579

5   Conclusion

In this work we have seen that MLP neural networks are well suited to classification problems. We adopted a MLP network to develop a methodology to classify hacking attacks adopting the multilayer perceptron (MLP) model with the
back-propagation algorithm and the sigmoidal activation function. We analyse
the results obtained using different configurations for the neural network, varying the number of hidden layers and the number of training epochs in order
to obtain a low number of false positives. Through this methodology, after the
analysis of various configurations and the evaluation of their pros and cons, we
achieved a high classification rate and a low error rate.

Although they require time and a good knowledge to be trained, they are
able to quickly detect old attacks as well as new ones. Another advantage of this
model is its scalability: it isn't needed to retrain the entire network after adding
a new type of attack, but only the set of layers that have the new attack as
input.

Future work are focused on removing the classification errors, trying new
types of attacks and changing other parameters such as learning rate and momentum, comparing the results with other machine learning models, and the application of proposed methodology to different domains [15–18].

References

1. Intrusion detection system. Wikipedia.it. https://it.wikipedia.org/wiki/Intrusion_
   detection_system
2. Network intrusion detection system. Wikipedia.it. https://it.wikipedia.org/wiki/
   Network_intrusion_detection_system
3. Przemysaw, K., Zbigniew, K.: Adaptation of the neural network-based IDS to new
   attacks detection, Warsaw University of Technology
4. Laheeb, M.I., Dujan, T.B.: A comparison study for intrusion database. J. Eng. Sci.
   Technol. **8**(1), 107–119 (2013)
5. Heba, E.I., Sherif, M.B., Mohamed, A.S.: Adaptive layered approach using
   machine. Int. J. Comput. Appl. (0975–8887) 56(7) (2012)
6. Alfantookh, A.A.: DoS Attacks Intelligent Detection using Neural Networks. King
   Saud University, Arabia Saudita (2005)

7. Barapatre, P., Tarapore, N.: Training MLP Neural Network to Reduce False Alerts in IDS, Pune, India
8. Minsky, M., Papert, S.A.: Perceptrons: An Introduction to Computational Geometry. The MIT Press, Cambridge (1969)
9. Grippo, L., Sciandrone, M.: Metodi di ottimizzazione per le reti neurali, Roma, Italia
10. University Of California, 28 10 1999. http://kdd.ics.uci.edu/databases/kddcup99/kddcup99.html
11. Amato, F., Barbareschi, M., Casola, V., Mazzeo, A.: An FPGA-based smart classifier for decision support systems. Stud. Comput. Intell. **511**, 289–299 (2014)
12. Amato F., Barbareschi M., Casola V., Mazzeo A., Romano S.: Towards automatic generation of hardware classifiers, Lecture Notes in Computer Science (including subseries Lecture Notes in Artificial Intelligence and Lecture Notes in Bioinformatics), 8286 LNCS (PART 2), pp. 125–132 (2013)
13. Moscato, F.: Model driven engineering and verification of composite cloud services in MetaMORP(h)OSY. In: Proceedings - 2014 International Conference on Intelligent Networking and Collaborative Systems, IEEE INCoS 2014, art. no. 7057162, pp. 635–640 (2014)
14. Aversa, R., Di Martino, B., Moscato, F.: Critical systems verification in MetaMORP(h)OSY Lecture, Notes in Computer Science (including subseries Lecture Notes in Artificial Intelligence and Lecture Notes in Bioinformatics), 8696 LNCS, pp. 119–129 (2014)
15. Minutolo, A., Esposito, M., De Pietro, G.: Development and customization of individualized mobile healthcare applications. In: 2012 IEEE 3rd International Conference on Cognitive Infocommunications (CogInfoCom), pp. 321–326. IEEE (2012)
16. Sannino, G., De Pietro, G.: An evolved ehealth monitoring system for a nuclear medicine department. In: Developments in E-systems Engineering (DeSE). IEEE (2011)
17. Cuomo, S., De Pietro, G., Farina, R., Galletti, A., Sannino, G.: A revised scheme for real time ecg signal denoising based on recursive filtering. Biomed. Sign. Process. Control **27**, 134–144 (2016)
18. Coronato, A., De Pietro, G., Sannino, G.: Middleware services for pervasive monitoring elderly and ill people in smart environments. In: 2010 Seventh International Conference on Information Technology: New Generations (ITNG). IEEE (2010)
19. Colace, F., De Santo, M., Greco, L.: A probabilistic approach to tweets' sentiment classification. In: Proceedings - 2013 Humaine Association Conference on Affective Computing and Intelligent Interaction, ACII 2013, art. no. 6681404, pp. 37–42 (2013)
20. Colace, F., Foggia, P., Percannella, G.: A probabilistic framework for TV-news stories detection and classification. In: IEEE International Conference on Multimedia and Expo, ICME 2005, art. no. 1521680, pp. 1350–1353 (2005)

WiFiNS: A Smart Method to Improve Positioning Systems Combining WiFi and INS Techniques

Walter Balzano[✉], Mattia Formisano, and Luca Gaudino

Università degli studi di Napoli Federico II, Naples, Italy
walter.balzano@unina.it,
mattia.formisano.3d@gmail.com,
gaudinoluca18@gmail.com

Abstract. Nowadays, positioning systems and localization techniques have an outstanding application and prominence to reality. In recent times many studies have been concentrated on two localization systems: Inertial Navigation System (INS) and Wi-Fi positioning system (WiPs/WFPS). INS is not much used mainly because of its low reliability on cheaper devices. WFPS is, instead, a recently developed system that presents an appreciable ductility to many environments and a respectable reliability. Techniques and algorithms have been studied to improve precision in localization exploiting more systems to provide a better result. These are taken into consideration by the proposed work which suggests a logic to obtain an as accurate as possible method to retrieve a reliable individuation of devices exploiting and combining WFPS and INS (Inertial Navigation System) in a Wi-Fi provided space. The data obtained are two distances matrices and two errors array referring to the approximate distance among devices in the space selected. To reach the purpose, the arrays of errors are for first normalized and then Wi-Fi and INS matrices are symmetrized taking in more consideration devices with less margin of error, these last are then merged in order to obtain a more accurate distances matrix.

Keywords: INS · Inertial · WiPs · WFPS · Wi-Fi · GPS · Localization system

1 Introduction

With the advent of new technologies it has been necessary to develop device localization systems in a space. Two of the most famous localization systems are Global Positioning System (GPS) and Inertial Navigation System (INS).

GPS is a network of about 30 satellites orbiting around the Earth at an altitude of 20,000 km that retrieves the position through the trilateration technique. The US government originally developed the system for military purposes, but with the technological revolution the system has been adapted to be used by everyone, as matter of fact almost every device provided of computational unit have got a localization system that supports GPS [1].

© Springer International Publishing AG 2018
G. De Pietro et al. (eds.), *Intelligent Interactive Multimedia Systems and Services 2017*,
Smart Innovation, Systems and Technologies 76, DOI 10.1007/978-3-319-59480-4_23

INS is a navigation aid that use accelerometer, gyroscope and magnetometer to calculate continuously the position using the previously determined position; this technique is also known as dead reckoning. On the base of this, many other sub-techniques have been developed, such as: Pedestrian dead reckoning [11], and automotive navigation. At the start, an INS knows its velocity and position in the space (Provided from an external source such GPS), then it calculates the updated velocity and position, using data retrieved from its motion sensor. The main advantage of using an INS is that it does not require external references to determine its position, orientation, or its speed once navigation started but it is subject to navigation drift that decreases its accuracy every time a new velocity and position are calculated.

In the past this process of technology creation was more complex and the cost of new devices was not affordable for everyone; today instead, obtaining a device equipped with localization chip, Wi-Fi card and many other features is almost inexpensive [2].

Analyzing the poor GPS behavior in closed environments [9], there was necessity to find a way for locating devices in this kind of environment, then a positioning system has been studied for Wi-Fi: the Wi-Fi positioning system (WiPs/WFPS). This system can exploit one of the existing localization techniques. The most employed techniques are RSSI [3] and Fingerprinting [4, 5]. A different approach has been used for IMU (Inertial Measurement Unit) which is always been employed for INS, but it was used in a few environments. In recent times instead, many different IMUs have been developed, concurrently new techniques have been studied [6].

2 Related Works

To explain in detail how the study operates, it is necessary to focus on the reliability of the two selected methods that provide the distances among devices. The considered methods are affected both by errors; indeed, in last years, they have become subject of interest. Concerning the INS, since it was developed, errors in localization's prediction were a well-known problem. In [7], Filyashkin and Yatskivsky studied a system to predict the evolution of the errors in INS, to improve its accuracy. Many studies have been done, in recent times, also on the possibility of improving the WFPS [8]. In this work, it is meant to find a way to merge data given by the above-mentioned localization methods, to obtain a more accurate localization than those obtained using only one of them. In similar works, authors show ideas to exploit two localization system: PDR (Pedestrian dead reckoning) and WFPS based on RSS. In [10], the systems have been fused to improve the localization of a device moving through an indoor path. The work in literature [11] exploits the two methods to ensure a better localization of a device in an indoor space. Differently from these papers, this one would show a method to decrease the errors in localization, considering more devices over a space and exploiting their inertial and Wi-Fi distances matrix.

The proposed work is composed of five different steps:

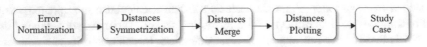

Fig. 1. Matrix building steps

3 Error Normalization

The first step is a normalization process on the array of errors needed for ensuring the working on data contained in a range with an upper and lower limit; this is important to avoid enormous differences among the values of arrays of errors.

In order to reach the goal, it is necessary to determine an upper-lower limited range to identify values, selected in {0, 1} interval.

There are different techniques on data normalization useful to obtain values within the range chosen; after a study it is deemed appropriate using the Feature scaling. It is a method used to standardize the range of independent variables, arrays and variously sort of data structures. This is proved by the large amount of works realized thanks to this technique, as seen in [11, 12, 17].

Feature Scaling is divided in three macro-categories that offers various methods to normalize raw data.

The first methodology is "Rescaling", the simplest method to rescale 'features' to {0, 1} or {− 1, 1} range. It is used as technique in [15]. Unfortunately, it is not usable in this work because the biggest value presented by an array of error would be identified by the upper limit of the interval, and the lowest by the lower limit. This is a big issue, because the lowest value of an array (the value which presented the largest error) is not considered in the weighted averages used to retrieve the distances matrix.

The second method known as "Standardization" [14] retrieves value through the use of variance: it is a sturdy technique but it is not applicable to this work because many tests on variables are needed to use it, while the only data available in this work, to obtain a normalization, are the two non-symmetrized distances matrix.

The technique taken in exam and then exploited is "Scaling to unit length" [13], a method thought for vectors: in order to obtain them in a normalized form, it is necessary to calculate the Euclidean norm [12] for the analyzed array, and then, divide the array for the norm. Assuming x as non-normalized array and $||x||$ as Euclidean norm for x, the used formula is: $normalized\_x = x/||x||$.

The chosen method has to be refined for the study needs. The first improvement concerns the necessity of the whole array of errors system to be normalized following a generic norm applicable to both arrays. Each array has values that could be widely different from values present in another one, consequently the norm calculated for the first array would be different from the one calculated for the second one, and this difference would cause an inconsistency of values throughout the steps of this work. To fix this issue, it has been chosen to unite values from both arrays in a third one and then calculate the norm on it.

The Euclidean norm has been calculated with the formula: $||x|| = \sqrt{(\sum_{i=1}^{n} x_i^2)}$.

Therefore, the Scaling to unit length formula has been applied, but once done, it has been noticed that values returned in *normalized_x* are closest to 1 with higher values and closest to 0 with lower ones. In the array of errors presented, instead, a short value indicates a better accuracy of information.

$$normalized_x = 1 - \left(\frac{x}{||x||}\right). \tag{1}$$

To achieve this purpose, *normalized_x* has been reversed and the result is (1). In this way it is possible to convert higher values into values closest to 0 and lower ones into values closest to 1.

4 Distances Matrix Symmetrization

At this point, as a matter of interest, the observation of distances matrices has revealed a discrepancy, for example, between the distance from the A point to the B point and its reverse: they are not the same. This situation does not allow individuating distances among distinct devices since it is not possible for a device to be at a different distance from another. For this reason, it has been looking for a way trying to solve the problem through the symmetrization of matrices. In literature, there are many studies that use this technique for similar scenarios, in [18] it is shown a sturdy method that finds its application in the matrix symmetrization problem.

$$X[m, i] = \frac{R[m, i] * N[m] + R[i, m] * N[i]}{N[m] + N[i]}. \tag{2}$$

Further analysis led to the *DISY* algorithm that, using the matrix of errors and the array of errors, through a weighted average, retrieves a symmetrical matrix.

Algorithm 1. Distances Symmetrization (DISY)

Input:
        $R / E / N \rightarrow$
Raw Matrix of Distances / Array of Errors /
                    Normalized Array of Errors
Output: X $\rightarrow$ Symmetric Matrix
 1: /* initialize output matrix and other variables */
 2: $T \leftarrow -1$ // Output matrix initialized to -1
 3: $n, z, i \leftarrow$ matrix length
 4: **for** $x \leftarrow 1$ to n **do**
 5:   $m = pos\_min(E)$ //position of min in E array
 6:   $E[m] = max(E) + 1$ //increase min to not select it again
 7:   **for** $i \leftarrow 1$ to n **do**
 8:     **if** $i = m$ **then**
 9:       $X[m, i] = 0$ //it is a diagonal point
10:     **elif** $X[m, i] = -1$ **then**
11:       $X[m, i] = (R[m, i] * N[m] + R[i, m] * N[i])/(N[m] + N[i])$
12:       $X[m, i] = CSV(E[m], X[m, i])$
13: **return** X

Procedure 1. Consistency Symmetric Value (CSV)

Input: $E / V \rightarrow$ Error of Value to Check / Value to Check
Output: C $\rightarrow$ Checked Result
 1: $C = V$
 2: **if** $(V > V + E)$ **then**
 3:   $C = V + E$
 4: **elif** $(V < V - E)$ **then**
 5:   $C = V - E$
 6: **return** C

In (2) is indicated the weighted average between the product of a distance with his error and its transposed value with transposed error value. The distance in $R[m, i]$ is multiplied for its normalized error ($N[m]$); at the same time, also the distance in $R[i, m]$ is multiplied for its normalized error ($N[i]$); the two results are summed and then divided for $N[m] + N[i]$. At the end of this calculations *DISY* could build the symmetric matrix, but there are cases where some calculated distance value is greater than the raw distance value added to his error or lower than the raw distance value subtracted to his error. *CSV* checks if this is the case; if it is, it considers as correct the value the raw value added to his error or raw value subtracted to his error according to case.

5   Distances Matrix Merge

Once obtained the symmetric matrices of Wi-Fi and Inertial surveys, it is calculated the merge matrix using the two matrices and their errors vector with the purpose of ensuring a more correct localization in the space considered. Algorithm *DIME* shows the complete building process of this new matrix.

Algorithm 2. Distances Merge (DIME)

Input:	W / I	→ Triangular Wi − Fi Matrix / Triangualar Inertial Matrix
	Ew / Ei	→ Wi − Fi Errors Array / Inertial Errors Array
	Nw / Ni	→

Normalized Wi −
Fi Errors Array /   Normalized Inertial Errors Array

Output: R   → Symmetrical Result Matrix

1: /* initialize output matrix and other variables */
2: $R \leftarrow -1$ // Output matrix initialized to -1
3: $Ea \leftarrow (Ew + Ei)/2$ // Average errors array
4: $n \leftarrow$ matrix length
5: **for** $x \leftarrow 1$ to n **do**   // n = matrix size
6:    $m = pos\_min(Ea)$ //position of min in Ea array
7:    $err = [Ei[m], Ew[m]]$ //array that contains Inertial and Wi-Fi errors cycle
8:    $Pmin = pos\_min(err)$ //position of min in "err" array
9:    $Ea[m] = max(Ea) + 1$ //increase min to not select it again
10:    **for** $i \leftarrow 1$ to n **do**
11:       **if** $i = m$ **then**
12:          $R[m, i] = 0$ //it is a diagonal point
13:       **elif** $R[m, i] = -1$ **then**
14:          $R[m, i] = (I[m, i] * Ni[m] + W[m, i] * Nw[m])/(Ni[m] + Nw[m])$
15:          $R[m, i] = CMV(Pmin, err, R[m, i])$
16:          $R[i, m] = R[m, i]$
17: **return** R

Procedure 2. Consistency Merge Value (CMV)

Input:
                  Pmin / err / V →
Position of Lower Error / Array with Inertial and WiFi Error merged / Value to Check

Output: C   → Checked Result

1: $C = V$
2: **if** $(V > V + err[Pmin])$ **then**
3:    $C = V + err[Pmin]$ //follow description
4: **elif** $(V < V - err[Pmin])$ **then**
5:    $C = V - err[Pmin]$ //follow description
6: **return** C

The logic used to obtain the symmetric merge matrix is to calculate first the rows and the columns of devices with lower average error. In this way, it is possible to fix in the space the most reliable distances and then build on them the remaining part gradually less reliable. The average error array is obtained through a mathematical media between Wi-Fi and Inertial errors arrays. In this way, it is possible to get an overall view, though approximate, of error that affects each device. Thanks to this array it is conceivable a better understanding of which device is affected on average by lower error and so fix for first in the space distances calculated from that device and then move on to next.

To obtain the merged symmetric matrix, this formula is used: $(I[m, i] * Ni[m] + W[m, i] * Nw[m])/(Ni[m] + Nw[m])$.

This last is a specialization of (2) it fills the Symmetrical Result Matrix R by calculating the average between Wi-Fi and Inertial distances, giving more importance to detection system with a lower error in that instance. After that, *CMV* procedure checks if calculated distance is greater or lower than the distance calculated with more precise detection system added or subtracted to error device of same detection system. At the end, *DIME* fills the symmetric part too to prevent it from being recalculated the distance starting from a device having a higher error.

6 Tools

Each algorithm created in this work, has been developed in *MATLAB*, using built-in library and functions (e.g. *Norm*).

The algorithm used to plot all the graphs of distances is instead developed in *Phyton*, specifically exploiting two libraries: *Numphy* and *Networkx*. The first one is used to save the matrix at start; the second one is used to represent the matrix as a graph object. To plot the graph obtained, the algorithm transforms the *networkx* graph into a *graphviz* graph; this operation is important because through *Neato* program the graph visualization tool graphviz can plot a graph of distances respecting edge length. In case that the distances are not substantial, the *graphviz's Neato* program builds the graph by changing the least as possible number of distances. It is studied to generate graphs with a lot of nodes, almost 80.000, so it can be used to represent a huge number of nodes without having complexity issues. Other methods for graph drawing are saw in [16].

7 Study Case

In this phase, a study case is proposed to show the effective improvements in a system above specified, starting from two raw distances matrix and then arriving to a result matrix. For a proper representation of the whole work, as sample, the starting matrices chosen are 7×7.

Step 1 Raw Matrices and Array of errors: In this initial step, the system obtains two raw 7×7 matrices (The Inertial and Wi-Fi Raw Matrix, Tables 1 and 2) and two arrays that contain the error in distance (Table 3), associated to each cell of both matrices. Both matrices and arrays are affected by inconsistence, except the array of errors average (composed by the average between both arrays), that will help to find a starting point to upcoming phases (Fig. 1).

Table 1. Raw Wi-Fi distances matrix. (In Meters)

Pts.	A	B	C	D	E	F	G
A	0,00	16,33	15,20	11,90	5,33	12,59	9,57
B	18,21	0,00	7,23	18,95	11,25	25,23	7,80
C	15,50	11,80	0,00	10,10	9,90	17,20	14,90
D	9,50	14,80	14,90	0,00	2,50	7,80	17,50
E	9,80	17,20	9,55	6,99	0,00	11,78	15,24
F	8,55	25,84	19,75	2,50	12,56	0,00	20,50
G	11,80	7,90	18,20	21,04	13,29	21,90	0,00

Table 2. Raw inertial distances matrix. (In Meters)

Pts.	A	B	C	D	E	F	G
A	0,00	11,22	14,50	10,50	7,90	15,00	13,70
B	17,90	0,00	4,60	6,10	20,30	19,80	4,60
C	22,10	3,50	0,00	15,90	13,84	20,22	18,30
D	15,62	14,66	15,98	0,00	9,26	4,55	23,81
E	7,50	12,25	12,90	8,20	0,00	7,80	11,90
F	4,56	27,16	20,15	2,21	13,29	0,00	15,87
G	13,50	5,98	19,12	20,59	16,54	21,90	0,00

Table 3. Array of errors. (In Meters)

	A	B	C	D	E	F	G
WiFi	2	4	2	5	4	6	2
Ins	5	7	10	7	2	9	5
AVG	3,50	5,50	6,00	6,00	3,00	7,50	3,50

Step 2 Normalization: The data have to be now standardized. Previously it has been discussed the adopted methodology to normalize the arrays according to the necessity of the problem. In this case, normalization problem is solved implementing (1) in an algorithm developed in *MATLAB*, exploiting the built-in Norm function (Table 4).

Table 4. Normalized array of errors.

Pts.	A	B	C	D	E	F	G
WiFi	0,90	0,81	0,90	0,76	0,81	0,71	0,90
Ins	0,76	0,67	0,52	0,67	0,90	0,57	0,76

Step 3 Distances Matrix Symmetrization: The Array of errors normalized form allows to refine also the raw data associated to the given matrices. The arrays retrieved from Step 2 will be used to symmetrize matrices. To symmetrize matrices, (2) has been exploited to create *DISY* algorithm. For the case needs.

Table 5. Symmetric Wi-Fi distances matrix. (In Meters)

Pts.	A	B	C	D	E	F	G
A	0,00	17,22	15,35	10,80	7,33	10,81	10,69
B	17,22	0,00	9,80	16,94	14,23	25,52	7,85
C	15,35	9,80	0,00	12,10	9,73	18,32	16,55
D	10,80	16,94	12,10	0,00	4,81	5,24	19,42
E	7,33	14,23	9,73	4,81	0,00	12,15	14,21
F	10,81	25,52	18,32	5,24	12,15	0,00	21,28
G	10,69	7,85	16,55	19,42	14,21	21,28	0,00

Table 6. Symmetric inertial distances matrix. (In Meters)

Pts.	A	B	C	D	E	F	G
A	0,00	14,34	17,59	12,89	7,68	10,53	13,60
B	14,34	0,00	4,12	10,38	14,25	23,20	5,34
C	17,59	4,12	0,00	15,94	13,24	20,18	18,79
D	12,89	10,38	15,94	0,00	8,65	3,47	22,09
E	7,68	14,25	13,24	8,65	0,00	9,80	13,90
F	10,53	23,20	20,18	3,47	9,80	0,00	19,32
G	13,60	5,34	18,79	22,09	13,90	19,32	0,00

Tables 5 and 6 show the results of the Experiment on matrices. The distances matrix has got consistent values that is utilised to calculate the last matrix.

Table 7. Symmetrical result matrix. (In Meters)

Pts.	A	B	C	D	E	F	G
A	0,00	15,90	16,37	11,76	7,52	10,68	12,02
B	15,90	0,00	7,23	13,98	14,24	24,47	6,70
C	16,37	7,23	0,00	13,51	11,59	19,00	17,57
D	11,76	13,98	13,51	0,00	6,84	4,41	20,64
E	7,52	14,24	11,59	6,84	0,00	10,91	14,05
F	10,68	24,47	19,00	4,41	10,91	0,00	20,38
G	12,02	6,70	17,57	20,64	14,05	20,38	0,00

Step 4 Distances Matrix Merge: The symmetric matrices are now used to obtain the conclusive distances matrix which presents a more accurate representation of values than the one of the two symmetrized matrices.

Table 7 shows data obtained by applying *DIME* algorithm. As it is possible to notice, this last matrix is extremely more reliable even accounting for errors. In example taken the point C of each matrix and array:

It presents 0,90 as Wi-Fi error, 0,50 as Inertial error (Table 4).

Looking at its values in Tables 5 and 6, and then in 7, it is trivial to deduce that values referred to C in the final Matrix are mostly similar to that in Table 5 (Wi-Fi). This is because the error associated to Wi-Fi detection is lower than the Inertial one.

Step 5 Plotting: Now, all data is obtained: to prove that there is an effective improvement using data calculated in the final matrix, a representation of all this distances matrices is needed. Through the already quoted Neato and GraphWiz is possible to represent distances matrices as Graph.

Fig. 2. Inertial distances matrix graph

Fig. 3. WiFi distances matrix graph

Fig. 4. Merged distances matrix graph

Fig. 5. Distances matrix compared graph

Figures 2, 3, 4 and 5 contain the graph associated to matrices in Tables 5, 6 and 7.

After the single representation, the three resulting graphs have been overlapped in order to provide an all-around view to the effective amendment brought by the graph in Fig. 4. By observing Fig. 5, it is possible to notice that merged matrix nodes are located in positions that keep count of both Wi-Fi and Inertial values and their errors.

8 System Performance

The first section of the algorithm consists substantially in various assignation all constants or asymptotically equivalent to $\Theta(n)$, considering n as arrays and matrices length; the value normalization could be a $\Theta(n)$ too because numerator and denominator values of normalization formula could be calculated at the same array of errors access. The core section of algorithm consists in three pairs of *for* nested in two, both having $\Theta(n)$ computational complexity, raising it from $\Theta(n)$ to $\Theta(n^2)$.

9   Conclusions and Future Work

DIME algorithm represents a simple way to obtain, starting from two inaccurate distances matrices and their errors array, a third more accurate matrix. To acquire a visual feedback, the N*eato* program can be used; however, it does not realize a highly accurate graph because of data inconsistency it changes the least possible distances data. Therefore, the algorithm could be improved so that when there is a data inconsistency it modifies the graph adapting the distances calculated with greater error. Recent studies developed car2car systems [19] useful to exchange messages; exploiting this work it would be possible to exchange also positions, building a car network [20], perhaps taking advantage of the cloud [21, 22] to save data [23, 24].

References

1. Balzano, W., Del Sorbo, M.R., Del Prete, D.: SoCar: a Social car2car framework to refine routes information based on road events and GPS. In: 2015 IEEE International Conference on Computer and Information Technology; Ubiquitous Computing and Communications; Dependable, Autonomic and Secure Computing; Pervasive Intelligence and Computing, (CIT/IUCC/DASC/PICOM). IEEE (2015)
2. Cullen, G., Curran, K., Santos, J.: Cooperatively extending the range of indoor localisation. In: 24th IET Irish Signals and Systems Conference (ISSC 2013), pp. 1–8 (2013)
3. Dong, Q., Dargie, W.: Evaluation of the reliability of RSSI for indoor localization. In: 2012 International Conference on Wireless Communications in Underground and Confined Areas, ICWCUCA 2012 (2012)
4. Vaupel, T., et al.: Wi-Fi positioning: system considerations and device calibration. In: 2010 International Conference on Indoor Positioning and Indoor Navigation (IPIN). IEEE (2010)
5. Balzano, W., Murano, A., Vitale, F.: WiFACT–wireless fingerprinting automated continuous training. In: 2016 30th International Conference on Advanced Information Networking and Applications Workshops (WAINA). IEEE (2016)
6. Wis, M., Colomina, I.: Dynamic dependent IMU stochastic modeling for enhanced INS/GNSS navigation. In: 2010 5th ESA Workshop on Satellite Navigation Technologies and European Workshop on GNSS Signals and Signal Processing (NAVITEC), Noordwijk, pp. 1–5 (2010)
7. Filyashkin, N.K., Yatskivsky, V.S.: Prediction of inertial navigation system error dynamics in INS/GPS system. In: 2013 IEEE 2nd International Conference on Actual Problems of Unmanned Air Vehicles Developments Proceedings (APUAVD). IEEE (2013)
8. Liao, X.-Y., Hu, K., Yu, M.: Research on improvement to WiFi fingerprint location algorithm. In: 10th International Conference on Wireless Communications, Networking and Mobile Computing (WiCOM 2014), Beijing, pp. 648–652 (2014)
9. Ho, C.C., Lee, R.: Real-time indoor positioning system based on RFID Heron-bilateration location estimation and IMU inertial-navigation location estimation. In: 2015 IEEE 39th Annual conference on Computer Software and Applications Conference (COMPSAC), vol. 3 (2015)
10. Chen, L.H., Wu, E.H.K., Jin, M.H., Chen, G.H.: Intelligent fusion of Wi-Fi and inertial sensor-based positioning systems for indoor pedestrian navigation. IEEE Sens. J. **14**(11), 4034–4042 (2014)

11. Hu, K., Liao, X.-Y., Yu, M.: Research on indoor localization method based on PDR and Wi-Fi, pp. 653–656 (2014)
12. Abdi, H., Edelman, B., Valentin, D., Dowling, W.J.: Experimental Design and Analysis for Psychology. Oxford University Press, Oxford (2009)
13. Abdi, H., Williams, L.J.: Normalizing data. In: Salkind, N.J., Dougherty, D.M., Frey, B., (eds.) Encyclopedia of Research Design, pp. 935–938. Sage, Thousand Oaks (2010)
14. Stolcke, A., Kajarekar, S., Ferrer, L.: Nonparametric feature normalization for SVM-based speaker verification. In: 2008 IEEE International Conference on Acoustics, Speech and Signal Processing, Las Vegas, NV, pp. 1577–1580 (2008)
15. Bardak, B., Tan, M.: Prediction of influenza outbreaks by integrating Wikipedia article access logs and Google flu trend data. In: 2015 IEEE 15th International Conference on Bioinformatics and Bioengineering (BIBE), Belgrade, pp. 2–3 (2015)
16. Kamada, T., Kawai, S.: An algorithm for drawing general undirected graphs. Inf. Process. Lett. 31(1), 7–15 (1989)
17. Li, D., Zhang, B., Li, C.: A feature-scaling-based k-nearest neighbor algorithm for indoor positioning systems. IEEE Internet Things J. 3(4), 590–597 (2016)
18. Uçar, B.: Heuristics for a matrix symmetrization problem. In: International Conference on Parallel Processing and Applied Mathematics. Springer, Berlin, Heidelberg (2007)
19. Balzano, W., Murano, A., Vitale, F.: V2 V-EN–Vehicle-2-Vehicle elastic network. Procedia Comput. Sci. 98, 497–502 (2016)
20. Balzano, W., et al.: A logic-based clustering approach for cooperative traffic control systems. In: International Conference on P2P, Parallel, Grid, Cloud and Internet Computing. Springer International Publishing (2016)
21. Amato, F., Moscato, F.: Pattern-based orchestration and automatic verification of composite cloud services. Comput. Electr. Eng. 56, 842–853 (2016)
22. Amato, F., Moscato, F.: Exploiting cloud and workflow patterns for the analysis of composite cloud services. Future Gener. Comput. Syst. 67, 255–265 (2017)
23. Amato, F., Moscato, F.: Model transformations of MapReduce Design Patterns for automatic development and verification. J. Parallel Distrib. Comput. (2016). doi:10.1016/j.jpdc.2016.12.017
24. Amato, F., Moscato, F.: A model driven approach to data privacy verification in e-health systems. Trans. Data Priv. 8(3), 273–296 (2015)

PAM-SAD: Ubiquitous Car Parking Availability Model Based on V2V and Smartphone Activity Detection

Walter Balzano$^{(\boxtimes)}$ and Fabio Vitale

Dipartimento di Ingegneria Elettrica e Tecnologie dell'Informazione,
Università degli Studi di Napoli Federico II, Naples, Italy
walter.balzano@unina.it, fvitale86@gmail.com

Abstract. GNSS based systems (like GPS or GLONASS) for outdoor localization are nowadays widespread in common smartphones. Vehicle-2-vehicle technology allows vehicles to communicate via wireless to share any kind of information and detect mutual distances via RSS-to-distance evaluation. Smartphones can be used to detect when an user switches between car driving, walking and several other kind of activities.

In this paper we discuss a novel methodology to discover nearest available street-parking spot using a smart combination of V2V, GNSS systems and smartphone-driven activity detection. In order to achieve system ubiquitousness across large areas (beyond the size of a single city), while still keeping low computation requirements, for localization purposes we only consider a fragment of the whole network of parked cars. Additionally we consider recognition of newly available parking clusters (for instance, in case of a fair or other kind of events) and disqualification of previously available spots (in case of work in progress or permanent street modifications). Smartphone activity detection is therefore used in tandem with V2V to mark new parking availability or occupancy based on user driving or walking away or toward the vehicle. When the user stops the car and starts walking, the vehicle is informed by the user smartphone of the context switch, and shares this positional information with nearby cars; when the user then goes back to his car and starts driving, the vehicle informs the local network of the newly available spot.

Keywords: Parking · V2V · Activity detection · LBS

1 Introduction

Finding a valid spot in roadside parking can nowadays be an issue. Car number keeps growing faster than public parking lots, and is therefore more difficult to find a valid parking which is also nearby desired destination. Moreover, finding a spot sometimes requires user to drive in the same area over and over waiting for a place to become available, which in turn increases stress and pollution levels.

In this paper we present PAM-SAD, which is a new model for ubiquitous car parking availability detection, specifically designed to help drivers rapidly

© Springer International Publishing AG 2018
G. De Pietro et al. (eds.), *Intelligent Interactive Multimedia Systems and Services 2017*,
Smart Innovation, Systems and Technologies 76, DOI 10.1007/978-3-319-59480-4_24

find the most appropriate parking slot available, using a smart combination of user activity detection via smartphone accelerometers, GPS for localization and a Vehicle2Vehicle network for communication and cloud storage. When the user parks his car and leaves by foot, the system marks the spot as occupied, while getting back to the car and driving away marks the spot as free for other users of the system. All the communication happens on the V2V network: when a new car looking for parking enters an area, it connects to the nearby car cluster and asks for available places, allowing user to drive directly to the most appropriate roadside parking slot. At the same time, when a car leaves, the local cluster is informed of the new condition for any subsequent user. This methodology requires little computation power in order to work, and it's therefore possible to use small and cheap on-board units (OBU) on vehicles, since they are only used for communication and storage purposes over V2V network.

2   Related Works

Location related problems, activity detection and applications of vehicle-2-vehicle technology are having a lot of attention in the current literature. In this section we are going to report some works which are closely related to our project, subdivided in three macrocategories: human activity recognition, vehicle 2 vehicle technology and parking issues and solutions.

Human activity recognition (HAR) has many interesting applications. In [1] authors present an interesting survey on recent advancements in activity recognition using smartphone sensors and data mining. In [2], a step detection algorithm based on sensors is described. It uses a signal filtering algorithm in order to reduce noise, detecting steps with an accuracy up to 98.6%.

HAR problems have several applications, one of which is improving health and fitness of users [3,4], like in [5] which uses smartphone sensors in order to calulate calories burnt and improve user health.

Vehicle2Vehicle (V2V) technology has several interesting practical applications. Starting with security, in *WiFiHonk* [6] is described a system which tries to improve pedestrian safety regarding car traffic, while in [7] and *iBump* [8] it is possible to find two possible applications related to car accidents and bumps, one for parked vehicles and one for car rentals and sharing services. Considering other possible uses of V2V technology, in *ScudWare* [9] authors present a complex system which allows communication between entities in a smart vehicle space, while in [10] authors consider using cars and several other kind of devices as sensors and/or for processing power and storage.

Parking is often an issue in modern cities due to the heavy traffic and little parking lots availability. In big parking lots it is easy to forget where one parked. Several solutions to this problem have been proposed in the latest years: [11] allows users to use their own smartphone to retrieve their car in a large crowded parking lot using QR codes to identify spots, while [12] proposes the usage of smartphones in order to take control of augmented reality controlled RC vehicles called Augmented-Cars. It's a complete software and hardware extendable solution, usable in a number of different applications. Regarding parking availability, several system have been proposed. Some consider using visual detection, like [13] which proposes a parking vacancy system using visual features extracted from parking spots, also considering light intensity changes, which are one of the main issues with visual-based systems, while in [14] the usage of fisheye optics is expoited. These cameras are deployed on vehicles and aid user in finding the nearest parking spot. Several systems have been also proposed with regards to parking reservation: in *Parkingain* [15] authors present a methodology which allows users to reserve parking slots via smartphone application. A similar consideration is applied in *iSCAPS* [16], where authors proposed an interesting innovative parking system which ensures parking lot availability for customers, allowing parking reservation and payment using smartphone application and NFC technology. Finally, in *DiG-Park* [17], authors consider the usage of a combination of V2V and a DGP class problem in order to find empty spots in a big parking lot, allowing user navigation to the most appropriate slot available.

Outline

The rest of the paper is organized as follows: first of all we analyze each methodology used by our project, then broadly explain PAM-SAD model, its validation [18–20], and lastly some conclusions and possible future evolutions.

3   Proposed System

In this preliminary section we introduce the concepts which are then used in our model.

For localization purposes, there are nowaday plenty of methodologies available. Most common are Global Navigation Satellite Systems (GNSS) for outdoor localization and path recording [21], which is widespread and commonly available in modern smartphones, like GPS and GLONASS. On the other hand, for indoor localization, several system have been proposed in the last years, like Wireless Positioning Systems [22] (WPS) which exploits wireless sensors and transmitters to determine user position in closed spaces, or Inertial Navigation Systems (INS) which use a combination of several sensors like accelerometers and gyroscopes in order to determine user movement in space based on a previously determined position (dead-reckoning).

There are several smartphone apps able to detect user activity. Most use available sensors in order to detect if user is walking (steps), running (faster steps) or bike-riding (circular movement), in order to improve user health and wellness. There are, however, several other ways to detect user activity beyond detecting leg movement via smartphone sensors. It's possible to exploit user position with regards to nearby devices, for example we can detect if user is inside his own car via bluetooth connectivity, or detect if it is at home or at work via nearby wireless spots recognition. Moreover, it's possible to user personal devices

Table 1. User activity detection based on smartphone position and engine status.

Car engine	SmartPhone	
	Inside car	Outside car
On	Driving activity: marks new available spot on switching to this status	Car theft or user forgot smartphone
OFF	Activity switch	Walking activity: marks new occupied spot on switching to this status

like smartwatches and smartbands to recognise even more complex activities like swimming or other sports like tennis or soccer.

Vehicle-2-Vehicle (V2V) systems are networks in which vehicles are the communicating nodes, able to provide spatial and temporal informations to nearby units regarding safety warnings and traffic [23]. Vehicles using this kind of systems are equipped with an On Board Unit (OBU) which enables communication and may offer several kind of services like localization [24]. OBUs are able to communicate up to 1000 m in range.

3.1 Parking Availability Model

PAM-SAD model is based on the simultaneous usage of these technologies. We use an user smartphone app to detect whether he is walking or driving.

In a common scenario, a user gets to his car by feet, drives toward a new destination, then parks his car in a valid location and walks away. We can therefore recognise 2 different activities: user walking toward or away from his car and user driving. These activities can be discerned using one or more of the following methods:

- **smartphone sensors:** user steps are visible while walking, but not while driving (see Fig. 2, red line);
- **user speed:** a walking pace is slower than a driving one;
- **user position in space:** a user walking in a pedestrian zone or indoor, a car driving on the highway;
- **devices proximity:** using bluetooth or wi-fi we can detect if a user is inside his car, can be used in combination with engine status (Table 1);
- **devices relative location:** using GPS on both the car and the smartphone, we can detect if they are in the same place or far apart;
- **car engine status:** if the engine is on, the user is probably in the car (see Fig. 2, green dashed line);
- **car movement via OBU GPS:** it's possible to detect car movement using its OBU GPS receiver (see Fig. 2, blue dotted line). Since GPS is inaccurate for lower speeds and inside narrow urban canyons, it's possible to use a two-thresholds system for incremental movement detection. Borrowed from video

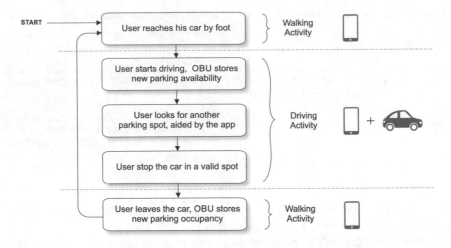

Fig. 1. Main PAM-SAD flow using smartphone GPS/sensors and car engine status for user activity detection.

shots analysis, it works by defining two different thresholds: below the lower one, no movement is detected and the car is considered stationary. Smaller movements (between the two thresholds) are cumulated until their sum grows over the higher threshold, at which point it is considered that the car moved by a significant amount. If the vehicle halts for some time (below the lower threshold), the accumulator is reset. Any movement above the higher threshold is automatically considered movement;

- **car movement via wheels monitoring:** it's also possible to detect car movement based on wheels rotation speed. The OBU is connected to the car control unit, and it's therefore possible to use this informations for movement detection (this can easily be combined with car engine status monitoring for improved reliability).

We exploit this contextual differences to determine in which step of the model the user is at any time.

Let's analyze a common scenario in steps:

1. A user walks toward his car, which is parked roadside (Fig. 1, step 1);
2. When the user leaves with his car, the vehicles network is informed of a new spot availability (Fig. 1, step 2). This information is then made available to any other car entering the same area via V2V;
3. The user drives toward his destination, then looks for a free parking spot (Fig. 1, step 3). The vehicle OBU interrogates the nearby cluster and, if needed, their adiacents in order to find the nearest available place. It's also possible for the user to decide his own destination, in order to find the best parking place, which in turn is the one nearest to his final destination;
4. The user parks the car (Fig. 1, step 4) in a free spot guided by the OBU navigation system;

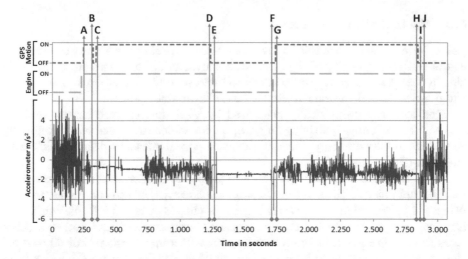

Fig. 2. Accelerometer values, engine status and GPS movement recorded by user smartphone and car OBU over 50 min of hybrid walking/driving activity.

5. The user leaves the car: the place is marked as occupied (Fig. 1, step 5) and the information is broadcasted to the V2V network;
6. The whole cycle is repeated when the user gets back to his car.

In Fig. 2 it's possible to see a recording sample of 50 min of hybrid walking and driving activities.

First of all we have the user which walks toward his car which is parked roadside. The user reaches his car after ~250 s, starts the engine and drives away (point A). At this point the system detects an activity switch, from walking to driving, which marks the parking slot as available. This information is broadcasted to the local V2V network, as explained in step 1 of the previous section.

Between points B and C there is a small gap in gps movement due to a red traffic light, which brought the car to a halt. The gap lasts for ~25 s, after which the car starts moving again. During this time the engine stays on.

At point D we have a second stop in car movement, as indicated by the GPS. The engine is halted (point E) but the user does not gets off his car, as the smartphone sensors do not register any steps pattern. After ~8 min, the engine is started again (point F) and the car leaves the temporary parking (point G). There is no activity switching in any of these situations because no walking has been detected.

At point H the car halts again, as indicated by the GPS receiver, in a valid parking spot. This time, after shutting down the car engine (point I) the user gets off the car (point J). The smartphone sensors register several steps, and therefore the system detects a new activity switch, from driving to walking, which marks the previously available slot as occupied. This information is broadcasted to the local V2V network, as explained in step 5 of the previous section.

3.2   System Ubiquitousness

Since we consider using this model for on-street parking, we need to carefully evaluate the complexity of the calculations involved. Since there are many vehicles in a common city, it is mandatory to build several smaller clusters to reduce the queries needed in order to find a proper available spot.

Clusters are automatically generated by V2V devices with nearby vehicles (up to 1 Km in range), striving to group nodes which are closer one to another, in order to reduce energy consumption and communication delays which may occur in distant nodes.

Each car OBU may be part of more than one cluster in order to not have any discontinuity in the service. Clusters are however able to communicate each other via information relay: when a car in a cluster A needs information about an area covered by cluster B, a car nearby both clusters may act as a relay to recover information in the interesting area. It's important to note, however, that it is very possible to have clusters which are isolated from the rest of the network: for instance, a parking lot in a highway service area may not have any other cluster nearby.

Vehicles belonging to each cluster form a sort of "cloud storage" in which each one knows which places around are available. If a valid spot is not available in a single cluster, borderline vehicles are informed via the V2V network and can check for availability in their other clusters, effectively acting as relays. This operation is repeated as needed, up to the maximum parking distance decided by the user. Finally, it is important to note that each car can only be part of a limited number of clusters, as each one requires a variable amount of storage which is a limited resource in embedded devices without internet connectivity.

4   Conclusions and Future Work

In this paper we presented PAM-SAD, which allows ubiquitous parking spot detection in modern cities using a combination of activity detection using smartphones and V2V technology for communication between vehicles. It allows a user to find the best possible available parking spot with regards to his destination using a smartphone app which works in conjunction with the car OBU, and automatically marks places as available or occupied based on user activity. Future work may include finding better ways to detect car positions in narrow spaces, in order to overcome issues related to inaccuracy of GPS systems. It may be worth using an Inertial Navigation System (INS) or Wireless Positioning System (WPS) in assistance to the satellite-based localization, or by exploiting V2V car connections using a Distance Geometry Problem (DGP) class algorithm. This system may also be useful for future integration with automatic parking cars technology.

References

1. Xing, S., Tong, H., Ji, P.: Activity recognition with smartphone sensors. Tsinghua Sci. Technol. **19**(3), 235–249 (2014)
2. Ryu, U., Ahn, K., Kim, E., Kim, M., Kim, B., Woo, S., Chang, Y.: Adaptive step detection algorithm for wireless smart step counter. In: 2013 International Conference on Information Science and Applications (ICISA), pp. 1–4, June 2013
3. Amato, F., Moscato, F.: A model driven approach to data privacy verification in e-health systems. Trans. Data Priv. **8**(3), 273–296 (2015)
4. Amato, F., De Pietro, G., Esposito, M., Mazzocca, N.: An integrated framework for securing semi-structured health records. Knowl.-Based Syst. **79**, 99–117 (2015)
5. Higgins, J.P.: Smartphone applications for patients' health and fitness. Am. J. Med. **129**(1), 11–19 (2016)
6. Dhondge, K., Song, S., Choi, B.-Y., Park, H.:. WiFiHonk: smartphone-based beacon stuffed wifi car2x-communication system for vulnerable road user safety. In: 2014 IEEE 79th Vehicular Technology Conference (VTC Spring), pp. 1–5. IEEE (2014)
7. Ebert, A., Feld, S., Dorfmeister, F.: Segmented and directional impact detection for parked vehicles using mobile devices. In: 2016 International Conference on Systems, Signals and Image Processing (IWSSIP), pp. 1–4. IEEE (2016)
8. Aloul, F., Zualkernan, I., Abu-Salma, R., Al-Ali, H., Al-Merri, M.: iBump: smartphone application to detect car accidents. In: 2014 International Conference on Industrial Automation, Information and Communications Technology (IAICT), pp. 52–56. IEEE (2014)
9. Zhaohui, W., Qing, W., Cheng, H., Pan, G., Zhao, M., Sun, J.: ScudWare: a semantic and adaptive middleware platform for smart vehicle space. IEEE Trans. Intell. Transp. Syst. **8**(1), 121–132 (2007)
10. Altintas, O., Dressler, F., Hagenauer, F., Matsumoto, M., Sepulcre, M., Sommery, C.: Making cars a main ict resource in smart cities. In: 2015 IEEE Conference on Computer Communications Workshops (INFOCOM WKSHPS), pp. 582–587. IEEE (2015)
11. Li, J., An, Y., Fei, R., Wang, H.: Smartphone based car-searching system for large parking lot. In: 2016 IEEE 11th Conference on Industrial Electronics and Applications (ICIEA), pp. 1994–1998. IEEE (2016)
12. Alepis, E., Sakelliou, A.: Augmented car: a low-cost augmented reality rc car using the capabilities of a smartphone. In: 2016 7th International Conference on Information, Intelligence, Systems & Applications (IISA), pp. 1–7. IEEE (2016)
13. Jermsurawong, J., Ahsan, M.U., Haidar, A., Dong, H., Mavridis, N.: Car parking vacancy detection and its application in 24-hour statistical analysis. In: 2012 10th International Conference on Frontiers of Information Technology (FIT), pp. 84–90. IEEE (2012)
14. Houben, S., Komar, M., Hohm, A., Luke, S., Neuhausen, M., Schlipsing, M.: On-vehicle video-based parking lot recognition with fisheye optics. In: 2013 16th International IEEE Conference on Intelligent Transportation Systems-(ITSC), pp. 7–12. IEEE (2013)
15. Sauras-Perez, P., Gil, A., Taiber, J.: ParkinGain: toward a smart parking application with value-added services integration. In: 2014 International Conference on Connected Vehicles and Expo (ICCVE), pp. 144–148. IEEE (2014)

16. Ang, J.T., Chin, S.W., Chin, J.H., Choo, Z.X., Chang, Y.M.: iSCAPS-innovative smart car park system integrated with NFC technology and e-Valet function. In: 2013 World Congress on Computer and Information Technology (WCCIT), pp. 1–6. IEEE (2013)
17. Balzano, W., Vitale, F.: Dig-park: a smart park availability searching method using v2v/v2i and dgp-class problem. In: International workshop on Big Data Processing in Online Social Network (BOSON). IEEE (2017)
18. Amato, F., Moscato, F.: Exploiting cloud and workflow patterns for the analysis of composite cloud services. Future Gener. Comput. Syst. **67**, 255–265 (2017)
19. Amato, F., Moscato, F.: Model transformations of mapreduce design patterns for automatic development and verification. J. Parallel Distrib. Comput. (2016)
20. Amato, F., Moscato, F.: Pattern-based orchestration and automatic verification of composite cloud services. Comput. Electr. Eng. **56**, 842–853 (2016)
21. Balzano, W., Del Sorbo, M.R.: SeTra: a smart framework for gps trajectories' segmentation. In: 2014 International Conference on Intelligent Networking and Collaborative Systems (INCoS), pp. 362–368. IEEE (2014)
22. Balzano, W., Murano, A., Vitale, F.: WiFACT-wireless fingerprinting automated continuous training. In: 2016 30th International Conference on Advanced Information Networking and Applications Workshops (WAINA), pp. 75–80. IEEE (2016)
23. Balzano, W., Del Sorbo, M.R., Del Prete, D.: SoCar: a social car2car framework to refine routes information based on road events and GPS. In: 2015 IEEE International Conference on Computer and Information Technology; Ubiquitous Computing and Communications; Dependable, Autonomic and Secure Computing; Pervasive Intelligence and Computing (CIT/IUCC/DASC/PICOM), pp. 743–748. IEEE (2015)
24. Balzano, W., Murano, A., Vitale, F.: V2v-en-vehicle-2-vehicle elastic network. Procedia Comput. Sci. **98**, 497–502 (2016)

A Composite Methodology for Supporting Early-Detection of Handwriting Dysgraphia via Big Data Analysis Techniques

Pierluigi D'Antrassi[1], Iolanda Perrone[2], Alfredo Cuzzocrea[1,3(✉)], and Agostino Accardo[1]

[1] DIA Department, University of Trieste, Trieste, Italy
pierluigi.dantrassi@phd.units.it,
alfredo.cuzzocrea@dia.units.it, accardo@units.it
[2] DCD Department, ULSS 7, Treviso, Italy
iolanda.perrone@gmail.com
[3] ICAR-CNR, Rende, Italy

Abstract. Handwriting difficulties represent a common cause of under-achievement in children's education and low self-esteem in daily life. Since proper handwriting teaching methods can reduce dysgraphia problems, the evaluation of these methods represents an important task. In this paper a methodology to compare visual and spatio-temporal teaching methods is proposed and applied in order to assess the influence of different teaching approaches on handwriting performance, via big data analysis techniques. Data was collected from children in their final years of primary school, when cursive writing skills have typically been mastered. Qualitative and kinematic parameters were considered: the former were calculated by means of quality checklists, whereas the latter were automatically extracted from digitizing tablet acquisitions. Results showed significant differences in pupils' handwriting depending on the teaching method applied.

1 Introduction

Handwriting is a complex task that involves cognitive, linguistic, motor and perceptual components [1, 2]. Dysgraphia, a difficulty in handwriting performance, is quite frequent being between 12% and 33% of school-age children producing dysfluent and illegible writing, compromising self-expression, communication skills and the possibility to record their own ideas [3–5]. These difficulties can affect not only the child's school performance, but also his/her self-esteem and everyday life in the future [6]. In order to circumvent handwriting problems, dysgraphic students can derive great benefits from using assistive technologies, such as word processor and note-taking software. Indeed, the ability to produce a result in the same time as classmates, that can be edited, spell-checked, read, and presented to the teacher can increase motivation and encourage writing. Although assistive technologies can help children with handwriting problems to keep up easier with other students in the classroom, explicit handwriting teaching is still a fundamental matter [7].

G. De Pietro et al. (eds.), *Intelligent Interactive Multimedia Systems and Services 2017*,
Smart Innovation, Systems and Technologies 76, DOI 10.1007/978-3-319-59480-4_25

Explicit and supplemental handwriting instructions are important elements in preventing writing difficulties in the primary grades: students receiving correct handwriting instructions are able to improve accuracy and fluency of their handwriting through the grades [8]. Thus, a proper handwriting teaching method can reduce dysgraphia problems. Unfortunately, there is no single method of teaching handwriting in schools, and no objective study of dysgraphia prevention.

Several models of handwriting production have been proposed. The earlier models considered phonological skills [9] or visual and graphomotor skills [10–13]. Later, a more complex neuropsychological model (Fig. 1) was developed considering handwriting as an information flow developing from the central processing to the peripheral components of movement execution [14–16].

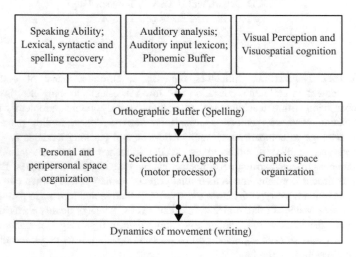

Fig. 1. Neuromotor model of handwriting generation process

This process can require speaking ability, auditory analysis and visual perception. The first is mainly required in the processing of spontaneous production of a text; the second in writing from dictation and the last is primary involved in the copying process. Whatever the starting point, the retrieved information is sent to an orthographic buffer that stores a graphemic representation for conversion into specific letter shapes [17, 18].

Motor processes play a role in the model below the Orthographic Buffer (Spelling) module. From this level, the model converts graphemes into allographs, and identifies the motor sequence required for their realization. The final stage is the dynamics of movement as a result of the allographic motor pattern processor, the graphic spatio-temporal constraints (e.g. writing size and speed; writing a letter inside a row without crossing borders or colliding with other letters; the starting and ending points of letters), and the muscular adjustment due to personal and peripersonal space organization (e.g. posture and handgrip) [15]. These motor levels are directly connected to handwriting production or difficulties like dysgraphia.

Graphemes are classified in letterform families that share geometric features (the hook of the c, the roof of the r, etc.) and their placement on the row (main body, ascenders and descenders). Each geometric element starts from a specific point in the row space; it is linked to the next one with a specific time and space sequencing; and it ends in a specific point in the space that allows an efficient connection with the next grapheme.

The difficulties in dysgraphic subjects in writing production are mainly related to an incorrect biomechanical control [19–21] producing wrong kinematics of movement and scarce legibility. Handwriting legibility or quality considers several parameters concerning quality and pertinence of letter formation, spatial placement (alignment and spacing), sizing of the written elements [22, 23] and features related to handwriting production such as seating posture [24, 25] and pencil handgrip [26, 27]. All of these parameters can describe different aspects of the handwriting process model (Fig. 1) and can be used for handwriting analysis [28]. The kinematics of handwriting also provides important information for diagnosis and treatment of dysgraphia [29].

The traditional handwriting teaching method provides a visual, and usually static approach in letterform learning. The teacher requires children to observe the letterform from blackboard or book, to remember it, and to reproduce it on their copybook. The movements required to achieve the formation of letters from a static picture could be assisted by some symbols: e.g. a dot illustrates the starting point, an arrow indicates the direction to follow when writing the letter, and a cross represents the end point. In cases a letter or a numeral contains two or more individual components (where the pencil/pen must leave the page) numbers are used to indicate which component is to be written first, second and so forth.

An alternative handwriting teaching method is the Terzi's approach [30] that was developed for blind people in the first half of the 1900s, and later adapted for sighted students. In this method, for each letter, the child learns the specific sequence of movements within time and space constraints, following specific phases [31]. Terzi's approach classifies alphabet letters in subsets of Euclidean geometrical elements (straight lines, curves and rotations) suitably integrated in the correct spatio-temporal sequence. The sequence is at first performed with blindfolded eyes and bare feet in order to focus the subject's attention on his/her own body feedback (kinesthetic, proprioceptive and tactile information) without visual distractions. Through different constructional and representative phases, a spatial image of the letter is generated. The child plans the correct starting point, the sequence of the geometrical elements and the correct endpoint of the letter. In the next phase, a motor-spatial image representation of the letter is generated, determining the correct graphomotor scheme: the blindfolded child represents the graphic symbol on the wall, following it with large and fluid movements of the arm, wrist, and hand with (and successively without) teacher's help. In the final step the internal spatial-motor image is enriched by the visual representation of the motion: the graphomotor pattern is reproduced (with open eyes) at first on large sheets with a brush and paint till, with decreasing size, on the appropriately-sized ruled paper.

The general theoretical model used by Terzi's approach is the Embodied cognition model [32, 33] in which cognition is hypothesized as a product of the experience that the body has and of the perception of the external space through which it moves and interacts. Therefore, conscious and organized movements of the body in space and time are tools to facilitate learning and cognition. In the case of handwriting, proprioceptive and kinesthetic information from the body (personal space) and from its interaction with peri-personal space allow for the realization of the grapheme.

In the literature many ways to analyze kinematics of handwriting acquired by digital tablet have been proposed [34, 35] but an objective methodology that takes into account also qualitative characteristics of writing does not still proposed.

The aim of this study is to identify a methodology able to compare objectively two different teaching methods, the traditional and the Terzi's methods, in terms of kinematic and qualitative parameters in order to assess their influence on handwriting performance of students in the last grade of primary school. To this end, a *big data analysis* approach has been defined.

2 Methodology

2.1 Participants

Two groups of 20 students each, at the end of the final year of primary school, participated in the study. For each group, 7 students were boys and 13 students were girls. Their ages ranged between 10–11, and they were Italian mother-tongue, right-handed, belonging to the same area with medium socioeconomic status, and with no handwriting problems or cognitive and motor impairments. Informed consent was obtained from parents before the tests.

Each group was made up of one whole class which followed a different teaching handwriting method: the "Don Milani" primary school of Cernusco sul Naviglio (labelled CE) followed the Ida Terzi method [30], and the primary school of Pioltello (labeled PI) instead used the traditional teaching method.

The handwriting samples were taken at the end of the final year of primary school when cursive skills have been achieved and consolidated.

2.2 Apparatus

The Wacom Intuos 3 Tablet with the Intuos Ink Pen, and 5th grade ruled paper were used for the data acquisition. The digitizing tablet has a spatial resolution of 5 μm and samples horizontal and vertical pen displacement at 200 Hz. The Intuos Ink Pen enables simultaneous collection of both digital and hardcopy data, reproducing a normal pen and paper context.

2.3 Procedure

Kinematic and qualitative handwriting evaluations were based on two tests that require adequate linguistic competences and cursive writing skill. These tests consist in writing as accurately as possible (the A test) and as fast as possible (the F test) the following Italian sentence: *In pochi giorni il bruco diventò una bellissima farfalla che svolazzava sui prati in cerca di margherite e qualche quadrifoglio* (meaning "In a few days the caterpillar became a beautiful butterfly fluttering on the lawns in search of some daisies and clover"). This sentence is not important for the meaning but for its structure: in fact it contains all the letters of the Italian alphabet and several phonological rules as the most difficult connection between letters (b-r, v-e, c-h, q-u, g-l, g-h) and some repetitive graphomotor patterns (s-s, z-z, l-l).

2.4   Data Analysis

In order to evaluate the differences between the two teaching methods, qualitative and kinematic parameters were separately calculated for each test.

Hand-motor performance quantification was undertaken with special regard to the basic writing elements: strokes and components [36]. A proprietary MATLAB program [37] was used to perform this analysis. Components were identified as the written tracts between two consecutive pen lifts. Strokes were identified as segments between points of minimal curvilinear velocity, as suggested by the bell-shaped velocity profile theory [38].

In order to provide information on the level of automation and fluency achieved by a child, a series of kinematic and static parameters [39] were calculated and analysed for each test on the digital data. In particular, the duration, length, mean and peak of curvilinear, horizontal and vertical velocities evaluated for the whole written work, as well as for the components and strokes were calculated. Pen lift duration, number of components, strokes and letters were computed too.

Qualitative analysis was performed, on the hardcopy data, through a manual approach, i.e. visual analysis of the written product, as proposed by a recent study [28]. In this study, starting from the model (Fig. 1), an evaluation scale for qualitative analysis (Table 1) is defined. Indeed, each model's area (A_n) was subdivided into subareas (S_p) and characterized by several parameters (E_p). Each E_p was defined as the number of errors for the specific parameter to represent the different aspects of hand-writing quality. The set of these parameters represents the criteria for the quality evaluation.

The hierarchy in the evaluation scale allows the identification, in case of dysgraphia, of the specific deficit as well as of the area to which the specific deficit is related. In this approach, for each parameter, the number of errors (E_p) for each criterion produced a normalized score ($\hat{E}_p$) calculated by the ratio between the E_p and the maximum number of the possible errors (E_{pmax}):

$$\hat{E}_p = E_p / E_{pmax} .\tag{1}$$

Thus, normalized scores assume a dimensionless value between 0 (no error) and 1 (maximum number of errors). For example, the parameter $\hat{E}_4$ (*right margin un- respected*) represents the ratio between the number of rows in which the right margin is exceeded or the line break is premature, and the total number of rows. Similarly, the parameter $\hat{E}_{11}$ (*confusion of visuospatiallysimilar letters*) is the ratio between the number of wrong letters, and the total number of letters.

Table 1. Evaluation scale for qualitative analysis

Model area	Sub area	Parameter/deficit/criteria	C_{AHP}^{a}
A_1. Personal, Peri-personal spaces organization	S_1. posture	E_1. inefficient posture	0.03
	S_2. handgrip	E_2. inefficient handgrip	0.04
A_2. Graphicspace organization	S_3. sheet graphic space	E_3. alignment left variability	0.02
		E_4. right margin un-respected	0.02
		E_5. irregular line spacing	0.03
	S_4. row graphic space	E_6. irregular word spacing	0.03
		E_7. collision of letters	0.12
		E_8. fluctuations on the line	0.05
		E_9. max variation of letter size	0.05
		E_{10}. wrong letter size	0.08
A_3. Selection of allographs	S_5. allographic recovery	E_{11}. confusion of visuospatially similar letters	0.10
		E_{12}. wrong type of character	0.07
	S_6. grapho-motor patterns	E_{13}. wrong graphomotor pattern	0.12
		E_{14}. dysmetria in letter execution	0.08
		E_{15}. incorrect connection of letters	0.09
		E_{16}. grapheme self-corrections	0.07

[a]Analytic Hierarchy Process (AHP) coefficients

Two of the 16 parameters in the evaluation scale were not evaluated from the hardcopy data. Infact, E_1 (*inefficient posture*) and E_2 (*inefficient handgrip*) error scores were evaluated by means of a checklist, filled in by examining the detailed pictures of pupils takenduring the acquisition phase. The checklist aimed at evaluating the correct posture as shown in the literature regarding the head, neck, trunk and feet [40], fore-hand and wrist [41], writing and non-writing hands [42, 43], fingers [44] during the handwriting process. In particular, participants were photographed to allowevaluation of positions of these elements. $\hat{E}_1$ and $\hat{E}_2$ are equal to the number of elements different from the best posture, normalized by the number of elements in the respective subarea.

Qualitative handwriting error weights (coefficients (C_{AHPp}) in the last column of Table 1) were obtained through the Analytic Hierarchy Process (AHP) method [45]. By means of the analysis of the preferences provided by a group of teachers, engineers, speech therapists, psycho-motility therapists and science education experts about the pairwise comparison of each of the parameters of the evaluation scale, the weight of each parameter on the handwriting quality was found [41]. In particular, experts were asked to decide how much the parameter E_1 influences handwriting quality in com-parison with E_2, then with E_3 and so on. Repeating the pairwise comparison for all parameters resulted in a matrix for each evaluator. Assuming that each of the experts has the same importance associated with each of the criteria, a weighted geometric mean of individual evaluation was performed and entered into a summary matrix. This matrix was the input to the AHP SuperDecision software [46] to decide the weight of each parameter in an optimized priority scale.

Finally, in order to provide a general evaluation of handwriting goodness, we calculated the Total Error Score (TES) for the whole evaluated test, as the sum of the normalized scores ($\hat{E}_p$) weighted with the AHP coefficients (C_{AHPp}) multiplied by 100 (WE_p):

$$WE_p = \hat{E}_p \cdot C_{AHPp} \cdot 100 \ . \tag{2}$$

$$TES = \sum_{p=1}^{16} WE_p \ . \tag{3}$$

Since visual analysis of the written product is an extremely subjective procedure, in order to guarantee an objective evaluation of handwriting goodness, a group of two teachers, two engineers, two speech therapists, three psychologists, and one pedagogist performed the analysis separately. A training program was provided for the evaluation group in order to present the protocol for the quality evaluation with handwriting analysis simulations. The scores for each parameter provided by these experts were averaged and the standard deviation was used as a measure of the inter-rater variability.

For both qualitative and kinematic parameters, the significance of the score differences between the PI and CE groups was evaluated by means of the Wilcoxon test for independent samples. In order to identify the most significant parameters, stepwise regression with forward selection was used in both A and F tests. Thus, only the statistically significant ($p < 0.05$) variables were included in a multilinear model. Furthermore, in order to assess the differences between the two teaching methods, principal component analysis (PCA) was performed on normalized (Z-score) kinematic and qualitative parameters selected by stepwise regression.

3 Results

Mean and standard deviation of measured kinematic and static parameters, evaluated over the complete writing task, are shown in Table 2 for each test and group. Moreover, the p-values of differences between the CE and PI groups are shown.

In "fast mode", both groups spend the same time to write the sentence: even though the PI group is significantly faster when the pen is on the paper than the Terzi group (CE), it spends more time during pen lift and makes a greater number of Components (more pen lifts), taking the same time overall. Conversely, in "accurate mode", CE group completes the test in less time than the PI group. The control group spends significantly more time during pen lift compared to both theCE group and itself in the "fast mode".

Table 3 shows the mean values (±1SD) of kinematic parameters related to components together with the p-values of the difference between groups, both in the A and F tests. Components in the CE group have a longer duration and a greater length than in

Table 2. Mean ± 1SD of Kinematic parameters of Full Track calculated in both A and F test for CE and PI group

Parameter	A Test			F Test		
	CE	PI	p-value	CE	PI	p-value
Whole duration (s)	97 ± 16	128 ± 28,3	<0,0001	74,4 ± 11,5	76 ± 9,4	n.s.
Whole length (cm)	126,3 ± 22,3	145,2 ± 20,5	<0,02	136,5 ± 26,9	150,6 ± 24	n.s.
Mean curvilinear velocity (mm/s)	18,6 ± 4,7	21 ± 6,7	n.s.	25 ± 5,2	32,9 ± 6,2	<0,0002
Mean horizontal velocity (mm/s)	8,9 ± 2,3	10,9 ± 3,8	n.s.	12,4 ± 2,5	17,8 ± 4,1	<0,0001
Mean vertical velocity (mm/s)	13,9 ± 3,9	14,8 ± 4,7	n.s.	18,1 ± 4,3	22,3 ± 4,4	<0,004
Whole pen lift duration (s)	28 ± 9,1	55 ± 17	<0,0001	19,2 ± 6	30 ± 8,5	<0,0002
#Components	57 ± 17	104 ± 22	<0,0001	56 ± 18	97 ± 20	<0,0001
#Strokes	431 ± 42	486 ± 86	n.s.	381 ± 43	366 ± 40	n.s.
Mean pen lift duration (ms)	473 ± 163	498 ± 132	n.s.	334 ± 98	290 ± 87	n.s.
Mean newline duration (s)	1,5 ± 0,7	2,2 ± 1,3	<0,02	1 ± 0,3	1,1 ± 0,3	<0,05
#Components/#Letters	0,5 ± 0,2	1 ± 0,2	<0,0001	0,5 ± 0,2	0,9 ± 0,2	<0,0001
#Strokes/#Letters	4,0 ± 0,3	4,5 ± 0,8	n.s.	3,6 ± 0,4	3,4 ± 0,3	n.s.
#Letters	109 ± 3	108 ± 1	n.s.	107 ± 4	107 ± 3	n.s.

n.s. - not significant.

Table 3. Mean ± 1SD of Kinematic parameters of Components calculated in both A and F test for CE and PI group

Parameter	A Test			F Test		
	CE	PI	p-value	CE	PI	p-value
Mean duration (s)	1,3 ± 0,4	0,7 ± 0,2	<0,0001	1,1 ± 0,3	0,5 ± 0,2	<0,0001
Mean length (mm)	23,7 ± 7,1	14,7 ± 4,5	<0,0003	26,5 ± 8,1	16,2 ± 4,9	<0,0002
Mean curvilinear velocity (mm/s)	16,3 ± 3,6	20,2 ± 6,4	n.s.	22,3 ± 4,3	31,8 ± 6,7	<0,0001
Mean horizontal velocity (mm/s)	8,9 ± 2	11,6 ± 3,8	<0,03	12,7 ± 2,3	18,9 ± 4,3	<0,0001
Mean vertical velocity (mm/s)	10,8 ± 2,6	13,1 ± 4,3	n.s.	14,4 ± 3,3	19,9 ± 4,4	<0,0002
Mean peak curvilinear velocity (mm/s)	37,7 ± 6,6	38,2 ± 9,7	n.s.	50,2 ± 8,2	58,1 ± 9,4	<0,02
Mean peak horizontal velocity (mm/s)	24,3 ± 4,7	27,4 ± 7,2	n.s.	33,7 ± 5,3	43,1 ± 9	<0,002
Mean peak vertical velocity (mm/s)	33 ± 6,5	32,2 ± 8,4	n.s.	43,6 ± 8,6	48,1 ± 8,6	n.s.
Mean horizontal length (mm)	6,3 ± 1,6	4,6 ± 1	<0,002	7,2 ± 1,8	5,3 ± 1,2	<0,003
Mean vertical length (mm)	3,7 ± 0,6	3,5 ± 0,5	n.s.	4,1 ± 0,8	3,8 ± 0,6	n.s.

n.s. - not significant.

the PI group, in both tests. As component length is significantly different in the horizontal direction, and not in the vertical direction, the differences in the whole component length are not in the size of the letter, but only in the number of letters written with a single component. Mean and peak component velocities are significantly greater for the PI group only in the F test. The increase in speed in the fast test, compared to the accurate one, is higher for the PI group than for the CE group. Significant differences in mean and mean peak curvilinear velocities are also present in both vertical and horizontal components.

Table 4 shows the kinematic parameters related to strokes. The mean stroke length is similar across groups and tests. On the other hand, the PI group takes less time than the CE group to make the strokes. Faster performance in making the strokes can be noticed especially in the F test in which the PI group is significantly faster (less mean stroke duration) than the CE group.

Table 4. Mean $\pm$ 1SD of Kinematic parameters of Strokes calculated in both A and F test for CE and PI group

Parameter	A Test			F Test		
	CE	PI	p-value	CE	PI	p-value
Mean duration (ms)	160 ± 13	150 ± 125	<0,03	144 ± 12	126 ± 9	<0,0001
Mean length (mm)	$2,1 \pm 0,5$	$2 \pm 0,5$	n.s.	$2,4 \pm 0,5$	$2,5 \pm 0,5$	n.s.
Mean curvilinear velocity (mm/s)	$17 \pm 4,4$	$19 \pm 6,2$	n.s.	$22,9 \pm 4,9$	$30,4 \pm 6,3$	<0,0004
Mean horizontal velocity (mm/s)	$8,5 \pm 2,2$	$10,5 \pm 3,7$	n.s.	$11,7 \pm 2,4$	$17,4 \pm 4,2$	<0,0001
Mean vertical velocity (mm/s)	$12,4 \pm 3,6$	$13,1 \pm 4,3$	n.s.	$16,4 \pm 4$	$20,2 \pm 4,3$	<0,005
Mean peak curvilinear velocity (mm/s)	$27,1 \pm 6,6$	$27,9 \pm 8,5$	n.s.	$35,6 \pm 6,8$	$43,9 \pm 7,9$	<0,004
Mean peak horizontal velocity (mm/s)	$15,9 \pm 3,8$	$18,5 \pm 6,4$	n.s.	$21,9 \pm 4,1$	$31,1 \pm 7,3$	<0,0001
Mean peak vertical velocity (mm/s)	$23,3 \pm 6,3$	$23,3 \pm 7,3$	n.s.	$30,5 \pm 6,9$	36 ± 7	<0,02
Mean horizontal length (mm)	$4,7 \pm 6,4$	$2,4 \pm 1,8$	n.s.	$3,9 \pm 4,1$	$2,9 \pm 2,3$	n.s.
Mean vertical length (mm)	$1,3 \pm 0,3$	$1,4 \pm 0,4$	n.s.	$1,6 \pm 0,3$	$1,9 \pm 0,4$	<0,003

n.s. - not significant.

PCA was separately performed on the data acquired in accurate and fast mode, firstly using all kinematic parameters and secondly only those selected by stepwise regression. Figure 2 shows the score plots for the first two principal components.

In the A test, when all the parameters are considered, the PCA shows that the first two components accounted for at least 70.57% of the variance, that increases to 100% taking in account only the most significant parameters detected by stepwise regression (i.e., whole pen lift duration and number of components). In the F test, the number of components together with mean horizontal velocity of components were selected by stepwise regression. The PCA shows that the associated explained variance is 66.64% considering all the parameters, whereas it is100%, considering only the selected ones.

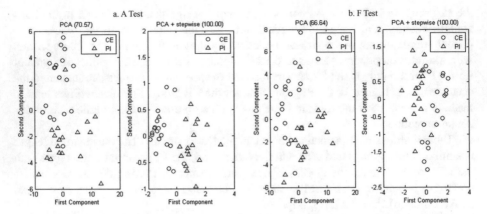

Fig. 2. Loading PCA plot obtained in A (subplot a) and F (subplot b) tests using first all kinematic parameters and then those selected by the stepwise regression. Circle: Terzi's method subjects (CE); Triangle: students of traditional teaching method (PI).

4   Conclusions

The results of this study showed that the methodology we used was able to underline the differences concerning handwriting quality and dynamic movement in pupils' handwriting, depending on the teaching method applied. Students who followed Terzi's approach produced more readable and accurate writing compared to students following the traditional handwriting method. Terzi's teaching method makes students slightly slower with the pen on paper but they end the writing task sooner thanks to a more successful automation of the graphomotor process with fewer pen lifts.

The characteristics of Terzi's teaching method (i.e., embodied perspective) respond to the guidelines of the Italian Ministry of Education and Research (MIUR) regarding the right to education of children with a specific learning disability [47], which also includes handwriting problems. Moreover, the findings of this study highlight some beneficial aspects (i.e. automaticity, fluency and readability) of the aforementioned method, by considering a model of handwriting production, and propose it as a new handwriting teaching method in primary schools for aiding dysgraphia prevention. Future work is mainly oriented to improve our approach via *privacy-preservation functionalities* (e.g., [48]), *performance gain* (e.g., [49]), and *adaptive metaphors* (e.g., [50]).

References

1. Sassoon, R.: Handwriting: A New Perspective. Stanley Thornes, Cheltenham (1990)
2. Maeland, A.E.: Handwriting and perceptual motor skills in clumsy, dysgraphic, and normal children. Percept. Mot. Skills **75**, 1207–1217 (1992)
3. Rubin, N., Henderson, S.E.: Two sides of the same coin: variation in teaching methods and failure to learn to write. Spec. Educ. Forw. Trends **9**, 17–24 (1982)

4. Smits-Engelsman, B.C., Niemeijer, A.S., Van Galen, G.P.: Fine motor deficiencies in children diagnosed as DCD on poor grapho-motor ability. Hum. Mov. Sci. **20**, 161–182 (2001)
5. Karlsdottir, R., Stefansson, T.: Problems in developing functional handwriting. Percept. Mot. Skills **94**, 623–662 (2002)
6. Dunford, C., Missiuna, C., Street, E., Sibert, J.: Children's perceptions of the impact of developmental coordination disorder on activities of daily living. Br. J. Occup. Ther. **68**, 207–214 (2005)
7. Hetzroni, O.E., Shrieber, B.: Word processing as an assistive technology tool for enhancing academic outcomes of students with writing disabilities in the general classroom. J. Learn. Disabil. **37**(2), 143–154 (2004)
8. Graham, S., Harris, K.R., Fink, B.: Is handwriting causally related to learning to write? Treatment of handwriting problems in beginning writers. J. Educ. Psychol. **92**(4), 620–633 (2000)
9. Frith, U.: Beneath the surface of surface dyslexia. Surf. Dyslexia Surf. Dysgraphia **32**, 301–330 (1985)
10. Forster, K.: Accessing the mental lexicon. In: New Approaches to Language Mechanisms. North-Holland, Amsterdam (1976)
11. Coltheart, M.: Lexical access in simple reading tasks (1978)
12. Morton, J., Patterson, K.: A new attempt at an interpretation, or, an attempt at a new interpretation. In: Deep Dyslexia. Routledge, London (1980)
13. Coltheart, M.: Disorders of reading and their implications for models of normal reading. Visible Lang. **15**(3), 245–286 (1981)
14. Margolin, D.I.: The neuropsychology of writing and spelling: semantic, phonological, motor, and perceptual processes. Q. J. Exp. Psychol. **36**(3), 459–489 (1984)
15. Denes, G., Cipollotti, L.: Dislessie e disgrafie acquisite. In: Manuale di Neuropsicologia. Zanichelli, Bologna (1990)
16. Van Galen, G.P.: Handwriting: issues for a psychomotor theory. Hum. Move. Sci. **10**(2–3), 165–191 (1991)
17. Ellis, A.W.: Spelling and writing (and reading and speaking). In: Normality and Pathology in Cognitive Functions. Academic Press, London (1982)
18. Miceli, G., Silveri, M.C., Caramazza, A.: Cognitive analysis of a case of pure dysgraphia. Brain Lang. **25**, 187–196 (1985)
19. Hamstra-Bletz, L., Blote, A.: A longitudinal study on dysgraphic handwriting in primary school. J. Learn. Disabil. **26**, 689–699 (1993)
20. Rosenblum, S., Aloni, T., Josman, N.: Relationships between handwriting performance and organizational abilities among children with and without dysgraphia: a preliminary study. Res. Dev. Disabil. **31**, 502–509 (2010)
21. Smits-Engelsman, B., Van Galen, G.: Dysgraphia in children lasting psychomotor deficiency or transient developmental delay. J. Exp. Child Psychol. **67**, 164–184 (1997)
22. Feder, K., Majnemer, A.: Handwriting development, competency, and intervention. Dev. Med. Child Neurol. **49**, 312–317 (2007)
23. Tseng, M., Chow, S.: Perceptual-motor function of school-age children with slow handwriting speed. Am. J. Occup. Ther. **54**, 83–88 (2000)
24. Pollock, N., Lockhart, J., Blowes, B., Semple, K., Webster, M., Farhat, L., et al.: Handwriting Assessment Protocol. McMaster University, Hamilton (2009)
25. Schneck, C.M., et al.: Prewriting and handwriting skills. In: Occupational Therapy for Children, 6th edn., pp. 555–582 (2010)

26. Schwellnus, H., Carnahan, H., Kushki, A., Polatajko, H., Missiuna, C., Chau, T.: Effect of pencil grasp on the speed and legibility of handwriting in children. Am. J. Occup. Ther. **66** (6), 718–726 (2012)

27. Schwellnus, H., Carnahan, H., Kushki, A., Polatajko, H., Missiuna, C., Chau, T.: Writing forces associated with four pencil grasp patterns in grade 4 children. Am. J. Occup. Ther. **67** (2), 218–227 (2013)

28. Genna, M., D'Antrassi, P., Ajčević, M., Accardo, A.: A new approach for objective evaluation of writing quality. In: 16th Nordic-Baltic Conference on Biomedical Engineering, pp. 32–35. Springer International Publishing (2015)

29. Accardo, A., Chiap, A., Borean, M., Bravar, L., Zoia, S., Carrozzi, M., Scabar, A.: A device for quantitative kinematic analysis of children's handwriting movements. In: 11th Mediterranean Conference on Medical and Biomedical Engineering and Computing, pp. 445–448. Springer, Berlin, Heidelberg (2007)

30. Terzi, I.: Il Metodo spazio-temporale, basi teoriche e guida agli esercizi. Ghedini, Milano (1995)

31. Thomassen, A.J., van Galen, G.P.: Handwriting as a motor task: experimentation, modelling, and simulation. In: Approches to the Study of Motor Control and Learning. Elsevier Science, Amsterdam (1992)

32. Thelen, E.: Time-scale dynamics and the development of an embodied cognition. In: Mind as Motion: Explorations in the Dynamics of Cognition, pp. 69–100 (1995)

33. Iverson, J.M., Thelen, E.: Hand, mouth and brain. The dynamic emergence of speech and gesture. J. Conscious. Stud. **6**(11–12), 19–40 (1999)

34. Tucha, O., Tucha, L., Lange, K.W.: Graphonomics, automaticity and handwriting assessment. Literacy **42**(3), 145–155 (2008)

35. Kushki, A., Schwellnus, H., Ilyas, F., Chau, T.: Changes in kinetics and kinematics of handwriting during a prolonged writing task in children with and without dysgraphia. Res. Dev. Disabil. **32**(3), 1058–1064 (2011)

36. Van Galen, G.P., Weber, J.F.: On-line size control in handwriting demonstrates the continuous nature of motor programs. Acta Psychol. **100**, 195–216 (1998). (Amst)

37. Accardo, A., Genna, M., Borean, M.: Development, maturation and learning influence on handwriting kinematics. Hum. Move. Sci. **32**, 136–146 (2013)

38. Djioua, M., Plamondon, R.: A new algorithm and system for the characterization of handwriting strokes with delta-lognormal parameters. IEEE Trans. Pattern Anal. Mach. Intell. **31**(11), 2060–2072 (2009)

39. Mavrogiorgou, P., Mergl, R., Tigges, P., El Husseini, J., Schröter, A., Juckel, G., et al.: Kinematic analysis of handwriting movements in patients with obsessive-compulsive disorder. J. Neurol. Neurosurg. Psychiatry **70**(5), 605–612 (2001)

40. Alston, J., Taylor, J.: Handwriting: Theory, Research and Practice. Croom Helm, London (1987)

41. Favretto, G., Fiorentini, F.: Ergonomia della formazione. Carocci, Roma (1999)

42. Drew, S.: Movement for writing: practical consideration from an occupational therapist's perspective. Handwriting Today **1**, 55–61 (2000)

43. Guiard, Y.: Asymmetric division of labor in human skilled bimanual action: the kinematic chain as a model. J. Modern Behav. **19**, 486–517 (1987)

44. Thomas, S.: The grip characteristics of pre-schoolers. Handwriting Rev. **11**, 48–56 (1997)

45. Saaty, T.L.: The Analytic Hierarchy Process. McGraw-Hill, New York (1980)

46. Creative Decision Foundation. Super Decision Software for decision making (2012). http://www.superdecisions.com

47. MIUR, "Linee guida per il diritto allo studio degli alunni e degli studenti con disturbi specifici di apprendimento," Ministerial Decree, prot.5669, July 12nd (2011)

48. Cuzzocrea, A.: Privacy and security of big data: current challenges and future research perspectives. In: ACM PSBD 2014, pp. 45–47 (2014)
49. Cuzzocrea, A., Matrangolo, U.: Analytical synopses for approximate query answering in OLAP environments. In: DEXA 2004, pp. 359–370 (2004)
50. Cannataro, M., Cuzzocrea, A., Pugliese, A.: A probabilistic approach to model adaptive hypermedia systems. In: WebDyn 2001 (2001)

SADICO: Self-ADaptIve Approach to the Web Service COmposition

Hajer Nabli[✉], Sihem Cherif, Raoudha Ben Djmeaa, and Ikram Amous Ben Amor

MIRACL, ISIMS, Cité El Ons Route de Tunis Km 10 Sakiet Ezziet, 3021 Sfax, Tunisia
nabli.hajer@yahoo.fr, cherifsihem16@gmail.com,
raoudha.bendjemaa@isimsf.rnu.tn, ikram.amous@isecs.rnu.tn

Abstract. Web service Compositions are rapidly gaining acceptance as a fundamental technology in the web field. They are becoming the cutting edge of communication between different applications all over the web. With the need for the ubiquitous computing and the pervasive use of mobile devices, the context aware web service composition becomes a hot topic. The latter aims at adapting the web service composition behavior according to the user's context, such as his specific working environment, language, type of Internet connection, devices and preferences. Many solutions have been proposed in this area. Nevertheless, the adaptation was carried out only at the run-time and it partially covered the user's general context. In this paper, we proposed a self-adaptive approach to the context-aware web service composition named SADICO for Self ADaptIve to the web service COmposition. Our approach studied a generic context, based on MAPE model (Monitoring, Analysis, Planning, and Execution) to ensure self-adaptation.

Keywords: Web service composition · Context-aware · Self-adaptive · MAPE loop · SABPEL

1 Introduction

Web Services emerged as a major technology for deploying automated interactions between heterogeneous systems. They are autonomous software components widely used in various service oriented applications according to their platform-independent nature (e.g., stock quotes, search engine queries, auction monitoring). The web Service technology allows different applications to be exposed as services via the network and interact with one another through standardized XML-based techniques. These techniques are structured around three major standards: SOAP (Simple Object Access Protocol), WSDL (Web Service Description Language), and UDDI (Universal Description, Discovery, and Integration). These standards provide the building blocks of the Web Service life cycle, such as description, publication, discovery, and interaction.

The increasing interest in web Service composition, the growing number of composite web Services and users' profiles raised new issues in service use. For instance, a composite web Service should be able to deliver to the user an adequate service that fulfills his specific needs and take into consideration his context. In fact, users can access

© Springer International Publishing AG 2018
G. De Pietro et al. (eds.), *Intelligent Interactive Multimedia Systems and Services 2017,*
Smart Innovation, Systems and Technologies 76, DOI 10.1007/978-3-319-59480-4_26

these web Services from various and heterogeneous profiles due to their different interests and preferences. However, a web Service can be accessed from different locations, through a diversity of devices (laptops, mobile devices, PDA, etc.) and network characteristics (Wi-Fi, bandwidth, etc.). Users also want to satisfy their preferences (desired content, layout, etc.) and interests. According to these heterogeneous mobile and changing profiles, adaptation is becoming a major requirement that should be taken into account earlier in the web Service composition. To tackle these adaptation problems, we propose in this paper a self-adaptive approach for web service composition, named SADICO for Self Adaptive to the web service COmposition that allows self-adaptation of web service composition to the used changing context.

This article is organized as follows: Sect. 2 presents the related studies. A detailed explanation follows in Sect. 3 of the general architecture of SADICO focusing on the self-adaptation module in Sect. 4. The evaluation is reported in Sect. 5. Finally, Sect. 6 concludes this article.

2   Related Studies

In this section, we take a look at some research studies about the possibilities of integrating the self-adaptation to the web Service Composition. We provide an overview of some of these studies.

Portchelvi et al. [1] proposed a Goal-Directed Orchestration (GDO) approach that employs an orchestration engine to provide flexibility in responding to the changes in a dynamic service environment. In GDO Process, the user's request for service is given to the engine and the composition request generator generates an abstract and concrete goal tree. The Abstract goal tree is mapped to abstract task tree while the concrete goal tree is mapped to the concrete task tree and thereby a composition plan is generated with abstract service descriptions. If services are not available to construct a composition plan, then, the goal cannot be achieved, which leads to goal failure. This failure is reported back to the composition requestor and an alternate sub-goal for the failed sub-goal is found and given to the composition plan generator. However, the authors need to formalize this goal-directed approach using formal methods.

Jaroucheh et al. [2] introduce an aspect-oriented framework for generating process variants that correspond to contextual changes based on original processes. The proposed framework comprises four main sections: a process model, a context model, an evolution model, and a linkage model. Evolution fragments and evolution primitives are introduced in the evolution model to detect changes. The adaptation of a particular process can be realized at both the process schema level and the process instance level.

Hermosillo [3] described CEVICHE, as a framework that combines Complex Event Processing (CEP) and the Aspect Oriented Programming (AOP) to dynamically support adaptable business processes. The adaptation logic is defined as aspects (reconfiguration component), and the adaptation situations are specified by CEP rules (monitoring component). However, the decision-making is not specified as a component in this framework, but it is integrated into the defined aspects. As we have shown, the study of adaptation at service composition modeling is prolific. Different scopes as well as different mechanisms are defined to achieve such adaptations.

Cheng et al. [4] presented an automatic web service composition method that deals with both input/output compatibility and behavioral constraint compatibility of fuzzy semantic services. First, the user's input and output requirements are modeled as a set of facts and a goal statement in the Horn clauses, respectively. A service composition problem is transformed into a Horn clause logic-reasoning problem. Next, a Fuzzy Predicate Petri Net (FPPN) is applied to model the Horn clause set, and T-invariant technique is used to determine the existence of composite services fulfilling the user's input/output requirements. Then, two algorithms are presented to obtain the composite service satisfying behavioral constraints, as well as to construct an FPPN model that shows the calling order of the selected services.

Madkour et al. [5] proposed a three-phase adaptation approach: firstly, they selected the services suitable for the current context that we recommend to the adaptation process. In the service adaptation phase, they used fuzzy sets represented with linguistic variables and membership degrees to define the user's context and the rules for adopting the policies of implementing a service. Finally, they dealt with the complex requirements of the user by the composition of the obtained adaptable atomic services.

Ordonez et al. [6] present a framework called AUTO which aims at supporting automated composition of convergent services using automated planning for service composition and natural language processing for user request processing. The proposed framework provides an adaptable composite service in four stages: the description of the user's query in natural language, the transformation of user's request to PDDL language, the generation of plans and the implementation of these plans.

Furno et al. [7] presented a design approach based on a semantic model for context representation. It is an extension of the OWL-S ontology aimed at enriching the expressiveness of each section of a typical OWL-S semantic service description, by means of context conditions and adaptation rules. The model proposed by the authors allows to use context-awareness expressions in semantic service descriptions and their adoption during composition.

QoS-aware web service composition problem has been discussed in [8–11]. Liu et al. [8] proposed a two-stage approach for reliable dynamic web service composition. In the first stage, the top K web service composition schemes based on each service's historical QoS values are selected with the proposed algorithm named culture genetic algorithm (CGA). Then, component services in the top K schemes are filtered out and employed as the candidate services for dynamic service composition. Gabrel et al. [9] studied complexity of the QoS-aware Service Selection problem in case of one criteria and proposed a mixed integer program to solve it. Tout et al. [10] proposed AOMD, a novel aspect oriented and model driven approach that defines new grammar to address both adaptability and behavioral conflicts problems, and offers extension for WS-BPEL meta model for high level specification of aspects. Grati et al. [11] proposed a QoS monitoring framework QMoDeSV for composite web services implemented using the BPEL process and deployed in the Cloud environment. The proposed framework is composed of three basic modules to: collect low and high level information, analyze the collected information, and take corrective actions when SLA violations are detected. This framework provides for a monitoring approach that modify neither the server nor the client implementation. In addition, its monitoring approach is based on composition patterns to compute elementary QoS metrics for the composed web service.

A variety of solutions has been proposed in literature that integrate the existing services based on several pieces on information. After the survey of existing research efforts in the area of self-adaptation of web service composition, these requirements can be defined as follows:

- **Context characteristics:** the contextual information in the service composition designed to return results, on the one hand, and adapted to the static requirements of the users and to the dynamic contexts on the other hand.
- **Adaptation type:** three main types have been identified for the implementation of adaptable applications, namely static adaptation (carried out through manual modifications in the source code), dynamic adaptation (modify the service at run-time) and self-adaptation (the system itself detects the conditions needed for the adaptation and executes the modifications).
- **Adaptation mechanism:** To provide the appropriate services and to answer the change contexts, we need to integrate adaptation mechanisms in applications. Among the mechanisms of adaptation, we mention reflection, adaptation policies and variability.
- **Adaptation approach:** Each proposed solution is based on an Adaptation approach: industrial approach, semantic approach, and artificial intelligence approach.

Table 1 summarizes the criteria (context characteristics, adaptation type, adaptation mechanism and approach of adaptation) and the approaches discussed in the previous sections. In summary, we note that most of the approaches do not take into consideration self-adaptation during the composition of web services. For example [1, 2, 5, 6] limited themselves to the static adaptation, while [3, 7–11] provided a dynamic adaptation. We also note that the context of the user proposed by [5] is limited to the user's preferences. As well, we find that the adaptation in [8–11] is limited to QoS parameters. The control of the composition at run-time and the management of events enable to oversee the execution of the composition by checking, for example, the access to services, changes in status, and the exchanges of messages. Based on these works, we find that:

- They do not provide a self-adaptation service composition according to the dynamic and generic context of the user.
- They do not propose tools for the control of web service composition using self-adaptable loop.

Not all of the identified criteria are present in a single architecture, as they focused on different research problems. All surveyed architectures employ a specific context composition model. They address dynamism in the composition, primarily from the perspective of unavailability of selected web services, and deal with the issues of how to replace them with other equally capable web services to perform the desired task. Our approach is different from these approaches in two aspects: (1) the self-adaptive integration of services in a transparent way, without the user's intervention; (2) the respect of contextual information of services appears; and (3) the reasoning of self-adaptation of service compositions through our framework. Then, our solution allows the feedback when an event occurs.

Table 1. Main research challenges and features of automatic web service composition

	Context criteria				Adaptation criteria			
	Context modeling	Context characteristics			Type of adaptation	Adaptation mechanisms	Adaptation time	Adaptation approach
		Application context	Platform context	Infrastructure context				
[1]	-	-	-	Bandwidth	Static	Variability	At runtime	Workflow approach
[2]	EMF	-	Weather	-	Static	Variability	At design time and runtime	-
[3]	XML	Location	Weather	-	Dynamic	AOP	At design time and runtime	Workflow approach
[4]	-	User preferences	-	Memory size	Self adaptable	Policies	At design time	-
[5]	EML	User preferences	-	-	Static	Policies	At design time	Workflow approach
[6]	PDDL	Location	Type of devices	-	Static	-	At runtime	Planning approach
[7]	OWL	Location, user preferences	Type of devices	-	Dynamic	Policies	At runtime	-
[8]	-	-	-	QoS criteria	Dynamic	-	At runtime	Workflow approach
[9]	-	-	-	QoS criteria	Dynamic	-	At runtime	Workflow approach
[10]	-	-	-	QoS criteria	Dynamic	AOP	At runtime	Workflow approach
[11]	-	-	-	QoS criteria	Dynamic	-	At runtime	-
Our approach	**OWL**	**Location, user preferences**	**Type of devices**	**Memory size bandwidth**	**Self-adaptable**	**Reflection**	**At runtime**	**Planning approach**

Our composition framework supports the following features:

1. **The dynamic composition.** Is essential for exploiting the current state of available services and making adaptations based on run time parameters, such as bandwidth and the cost of executing the various services.
2. **Re-composition.** As composite services may be executed in a dynamic environment, the context may change and services may become unavailable. Therefore, it is necessary to have some means of recomposing the service on the fly.
3. **User's interaction.** Whilst service composition is an automated process, it is necessary to help the users provide feedback when they wish to be integrated in the composition process. For example, aside from providing input parameters, users may need to guide the composition by selecting the services and redefining the goals, or guide the failure recovery process.
4. **Automatic service discovery.** Working with a limited domain of services or predefined service types limits the potential of service composition. Moreover, new services, possibly with new capabilities, may become available or existing and change their functionality. Having an automated means of service discovery is therefore an essential feature.

5. **Context monitoring.** For supporting dynamic adaptations, software should be aware of changes in its context. Context-aware applications are based on the acquisition of context (users' preferences or environment), the abstraction, the storage and understanding of the context, and the application behavior based on the recognized context.

3   Architecture of SADICO

According to IBM, the autonomous systems should be a collection of autonomous elements that interact. Autonomous elements are individual components of the system that provide a particular service to the user or to other autonomous elements. The autonomous elements should manage their relationships with their environment (managed system or other autonomous elements) in accordance with established policies of the autonomous system. Therefore, they can relieve the user of responsibility to intervene directly for setting the behavior of the managed system. Indeed, an autonomous or self-adaptive system ensures the adaptation of the composition of web services without any human intervention. In this regard, we propose a new approach Self ADaptIve for web service COmpositon (SADICO). Our proposal is a self-adaptive approach that examines a generic context and ensures the continuous monitoring of the user's context. Our approach can also self-manage without any human intervention using MAPE model. Figure 1 shows an overview of the different components of the proposed architecture. Our self-adaptation module is provided by the following four components: Control, Analysis, Planning and Creation, which are based on the MAPE structure of the Autonomic Computing Architecture [12].

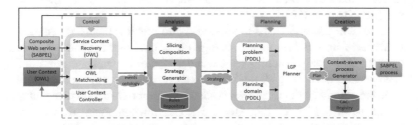

Fig. 1. Overview of the proposed self-adaptive approach

A self-adaptive system alters its own behavior in response to changes in its environment. Indeed, a self-adaptive application helps perform the following functions: (i) the observation of the system environment, (ii) the detection of environment change, (iii) the implementation of the necessary policies to modify application behavior in response to detected changes and (iv) the return of a result adapted to the new used context.

When adapting a process, the composite web service is automatically supplied as an input to the self-adaptive application together with the semantic description of the user's context. The Service Context Recovery module retrieves the context of composite web service from its extended description SABPEL, the service context is represented in the OWL format. OWL files (user and service context) are analyzed and used by OWL Matchmaking module to detect adaptation problems, transform them into events and subsequently build an event ontology. The User's Context Controller module monitors the user's environment and when locating a change, it refreshes and updates the user's context file and remade the matchmaking process to build again the event ontology. Then, the event ontology is exploited for preparing the analysis phase. The Slicing Composition part consists in slicing the web service composition into a set of elementary services. Then, based on the event ontology, the Strategy Generator module reveals the elementary web services that are unsuitable for the new context. Thereafter, based on adaptation rules, the Strategy Generator module provides a strategy that specifies the necessary policies to respond to the current context. The role of the planning component is to find the set of actions to reconfigure the system. To do this, we used the LGP planner [13] that needs two inputs: Planning problem and Planning domain written into the planner language PDDL (we describe it in detail in Sect. 4.3).

Once the planner has defined the action plan, the creation component is called to perform actions on different elementary web services affected by its adaptation to the application. The Context-aware process Generator module uses the CAC-Registry [14] to discover web services and then generates a SABPEL process that can also be an entry for our self-adaptive application if a new context is detected.

4   Self-Adaptation Module

Our self-adaptation module is provided by the four components: Control, Analysis, Planning and Creation, which are based on the MAPE structure of the Autonomic Computing Architecture [12].

The control component is used to supervise and monitor the user's context. It can permanently have information on the change of its environment and transform them later as events. The analysis component decides, from the events carried out by the control component, the coping strategies. Based on strategies supplied by the analysis component, the planning component generates the appropriate action plan that gives the set of changes to ensure the self-adaptation. The creation component respects the previously defined plan and executes the necessary actions to create a SABPEL process [15]. In general, this module helps update and automatically reconfigure the existing compositions according to new contexts.

4.1   Control Component

The control component in our framework is used to gather an instructive and dynamic perspective of the adaptive entity and its environment. This perspective should include every parts of information regarded as useful to take an adaptation decision, including information on the hardware, the user's preferences and location, in addition to the adaptive application and the services themselves. This dynamic perspective is performed by values representing states of the system. Events, coming from the observed elements modify those states.

Precisely, the control component is designed to probe periodically values to generate events and update the view of all the elements of the system. This view is useful to have precise context information and to get a global representation of system's states. Such a context management is a main function in the self-adaptable composition process. Indeed, to be able to adapt an application to the user's context, we must define the necessary means for the detection, interpretation, and context storage.

1. **Detection and Interpretation of context:** This phase ensures the capture of static and dynamic context. To do so, the acquisition of static context information is carried out through a graphical interface, the user entered information that is fixed during the execution of the application such as user's preferences. Simultaneously, sensors listen and automatically picture additional context information about the user's environment, infrastructure and application. These data can be modified at run-time. The context view is implemented using the WildCAT framework [16]. The captured data will then be interpreted in a directly readable format by storage phase.

2. **Context storage:** For each user's session, a context ontology is created and associated with the user. Indeed, this phase is used to store the context detected and interpreted in the context registry, in OWL[1] format. As the user is logged in to the application, his context ontology is refreshed as soon as a context change is detected during the application execution. In this phase, we used Apache Jena[2] that is a Java API that can be used to create and manipulate ontologies.

The user's context groups all the concepts captured or provided by the user of the application. The user's context has three categories: context related to the platform, context related to the infrastructure and context related to the environment of application.

In our framework, the control phase of the MAPE model is allowed also to do comparison mechanisms of context ontologies to extract attributes that are inappropriate to user's context. For example, a composite web service can run on PC ("Hardware_Type" = PC), while the user is connected from a Smartphone ("Hardware_Type" = Smartphone). Then, the control component returns the name of the unsuitable attribute (for example "Hardware_Type") and the value (for example "Smarphone") in order to search for web service running on a "Smartphone". Eventually, these results will be stored in an ontology indicating each time the name of the corresponding event. As the user is already logged on to our application SADICO, the control component monitors and controls its context. We used the WildCat framework [16] to check whether a context switch is detected. Thus, this component always keeps the user's context under control. More clearly, the control component checks if updated values of event are different from previous ones, performs a comparison of two new contexts if it is the case, and finally creates an updated event ontology. Then, the updated event ontology is transferred to the analysis component. The analysis and planning functions

---

[1] http://www.w3.org/TR/owl-features/.

[2] http://jena.apache.org.

of the framework can request values of events, in order for them to get complementary information to execute their tasks when needed.

4.2   Analysis Component

The analysis component in a self-adaptive system is designed to take adaptation decisions. When a change arises in the control component, the control function sends a notice to the analysis component as an event ontology. After receiving a new event, the analysis component has to analyze and decide if an adaptation is needed and to choose an adaptation strategy if needed. Indeed, the analysis component created a strategy when one or more of the dynamic attributes have changed values during the application run-time and the composite service context no longer conforms to that of the user. This means that one or more of the participating services to the composition does not support the new user's context, then a strategy is implemented to correct any faults. The analysis component begins by identifying the failed services. Algorithm 1 symbolically explains how searching for failed (or defective) services works. This function compares the values of attributes of both the user and service context, and if there are attributes affected by a change and that the service cannot fulfil the new context, therefore, the function Compare returns the set of changed attributes (getEvents) that correspond to this service.

```
Algorithm 1: Compare context
Input: File ontoService, File ontoUser
Output: Map <Attribut, Value> getEvents

Function Compare (ontoUser, ontoService)
//Get the attributes of user's context ontology ontoUser,
AttUser = {AttUser1,…,AttUsern}
//Get the attributes of service's context ontology
ontoService, AttServ = {AttServ1,…,AttServn}
//Get common attributes between AttUser and AttServ,
Att = {Att1,…,Attn}
Foreach Atti in Att do
   valUser = Extract the corresponding value from ontoUser
   valServ = Extract the corresponding values from
   ontoService
   If ValUser does not exist in the list ValServ then
      getEvents.put(Atti,valUser)
   End
End
```

Once the defective services are recognized, we continue the process of creating a strategy by extracting, from the events ontology, the corresponding information to triggered events (event name, and attribute value). We used in this step the Semantic Web Rule Language[3] (SWRL), which is a rule language used to increase the amount of knowledge encoded in OWL ontologies. The creation and management of SWRL rule

---

[3] http://www.w3.org/Submission/SWRL/.

bases requires specialized tools that are not usually present in the OWL standard development environments. The authors in [17] proposed a tool called SWRLAPI that provides a rich development environment for working with SWRL rules.

An inference engine based on rules may implement adaptation policies. In our work, they are implemented using the Drools[4] system. The main reason for using Drools in our case is that it is flexible, easy to use and fast in execution.

The last step of the analysis component is the creation of a corresponding strategy. All information collected in the previous three steps is grouped in an XML document. Indeed, to store the strategy parameters, we use an XML representation implemented using the JDOM[5] tool. A strategy is then delivered to the planning component.

4.3 Planning Component

As seen in the previous section, a decision result is represented by a strategy. A strategy defines the new state to reach but it doesn't define how to obtain this state. This is the role of planning phase. To do so, the planner looks for a set of actions that are able to turn the current state of the system into the new state. Therefore, the role of planning component in our framework is to recover the strategies delivered by the analysis component and provides how the monitored application must be adapted to achieve the desired state. Each type of adaptation matches a set of actions that can be used to implement it. The role of the planner is to select the appropriate actions for the current situation. In our case, the adaptation, which consists to remove one or more failed services and add new ones in a composition of services, may be implemented by choosing the new service, connecting it to one or several services of the old composition, after having removed their previous links. In such situation a scheduling policy should be specified by the planner. Indeed some actions are dependent from each other's and they must be executed in a precise order. The result of the planning component is called plan.

The author in [13] proposed a flexible planning model named Lambda Graphplan (LGP). The use of LGP in our case ensures flexible composition with producing flexible plan that integrates control structures. The LGP planner uses the Planning Domain Definition Language (PDDL) to describe the planning issues.

In our approach, the representation of planning problem is carried out automatically. It is based on information provided by the strategy issued by the analysis component. Indeed, as soon as a strategy is present at the planning component, a planning problem will be automatically generated from the data of strategy. The LGP planner is delegated for the treatment of entries, planning problem and planning domain, to produce a satisfactory plan and correct the adaptability problem. (See Fig. 1).

4.4 Creation Component

At this stage, we have a web service composition that is not adapted to the user's context. Indeed, some of the partners' services are not adequate with the user's context that we

---

[4] http://www.jboss.org/drools/.

[5] http://www.jdom.org/.

called defective services. Therefore, the Assembly phase can replace these unsuitable partners' services by other services, not only having the same functionality, but also supporting the updated user's context. To do this, an interaction with the Context Aware Composition Registry (CAC-Registry) [14] is required to discover new services. The discovered services take into account the context information in their description. For this reason, we use Adaptive Web Service Web Service Description Language (AWS-WSDL) [18]. AWS-WSDL is developed to ameliorate the WSDL standard of the W3C and provide a platform for retrieving context information. Contextual information of atomic web services is stored in an ontology. This phase would enable atomic web service discovery mechanisms to find out web services that better match the user's context. These steps are repeated to replace any faulty web service with another service adapted to the new user's context. Algorithm 2 illustrates an abstract process flow of an adaptive service discovering in SADICO. Then, these new services are assembled with the remaining web services and sent to the next phase.

```
Algorithm 2: Discover new services
Input: Failed Service FS, File ontoUser
Output: adaptive service AS

Function Discover (FS,ontoUser)
// connection to the registry
// search for atomic services that have the same
functionality as FS, S = {S1,…,Sn}
Foreach Si in S do
  File ontoService = Extract the corresponding service
  context ontology
  Map <Attribut, Value> getEvents = Compare
  (OntUser,OntService)
  If getEvents isNull then
    // the service Si supports new user's context
    Return Si
    Break;
End
```

Finally, an adapted SABPEL composition to the user's context is created in the last phase. For this composition, we used Eclipse BPEL Designer to model, design and build composite services.

5   Evaluation

As in information retrieval, we use two metrics, Precision and Recall, to evaluate the quality of the results of the selection of composite web services.

$$Precision = \frac{Pertinent\ Web\ Services \cap Returned\ Web\ Services}{Returned\ Web\ Services} \tag{1}$$

Precision is defined by the proportion of the relevant and correctly composite web services of all the composite web services that should be published (as shows in Exp. 1).

$$Recall = \frac{Pertinent\ Web\ Services\ \cap\ Returned\ Web\ Services}{Pertinent\ Web\ Services} \tag{2}$$

Recall is defined by the proportion of the relevant and correctly composite web services of all the existing composite web services that have been published in registries (as shows in Exp. 2).

$$F-measure = \frac{2*(Precision*Recall)}{Precision+Recall} \tag{3}$$

F-measure is a weighted harmonic means of Recall and Precision. We adopt the standard F-measure to combine the precision and recall figures into a single measure of quality.

Figure 2 shows the relation between the performances (recall/precision). The user's context represents the different values for each contextual element observed. To calculate precision and recall, first, we need to determine the threshold so that we can specify the relevance of a composite web service selected to the context of the user. We conduct a series of experiments on a set of users. We note that the service context is divided into five categories (LogicielRessource, MaterielRessource, InfrastructureRessource, Location, Preferences).

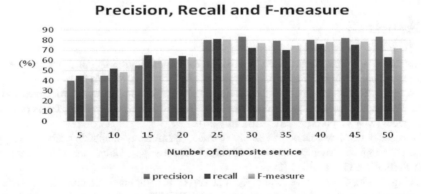

Fig. 2. Evaluation of dynamic adaptive services selection

If the user's relevant composite web service relative to their request and context has a number of context categories equal to three, the number of the user's context attributes is satisfied by the context of the service.

To evaluate the two metrics (precision and recall), we defined a maximum of 15 web services for each user's context. Consequently, we have three user's contexts Ctxt1, Ctxt2 and Ctxt3 with 15 web services at most for each context. This number is calculated after running our framework. To achieve equality between the service context and the context of the user, we used a third ontology. The latter is used to store the attributes of the context of the service that are supported by the context of the user.

The results of the simulations are given in Fig. 2. The precision, recall, and F-measure of the recommendations of the selection process for each of the Ten groups of composite services are given and the average performance of the system for all the requests is drawn. According to these results, the precision, recall and F-measure of the system tend to increase with the number of the user's requests. This is justified by the fact that the number of observations about selections in the system also increases.

6   Conclusion

The proposed work relies on the MAPE structure of the Autonomic Computing Architecture used to adapt dynamically the context-aware applications. The novelty of our work is to entail the adaptation of applications during their execution according to changes in their context. Our proposal consists in providing mechanisms for self-adaptation of composite web services. Thereby, we presented a self-adaptive approach for web service composition named SADICO, which helps manage and dynamically adapt to the environment changes. One of the benefits of our proposed approach is that it enables us to continue monitoring the user's context and react immediately if an inadaptability problem occurs. The solution provided in this work is performed by a case study on tourism services by presenting a variety of choices in order to satisfy the customer's requests and make the application benefit from changing contexts.

References

1. Portchelvi, V., Venkatesan, V.P.: A goal-directed orchestration approach for agile service composition. Int. J. Inf. Technol. Comput. Sci. (IJITCS) 7(3), 60 (2015)
2. Jaroucheh, Z., Liu, X., Smith, S.: Apto: a MDD-based generic framework for context-aware deeply adaptive service-based processes. In: 2010 IEEE International Conference on Web Services (ICWS), pp. 219–226. IEEE, July 2010
3. Hermosillo, G.: Towards creating context-aware dynamically-adaptable business processes using complex event processing. Doctoral dissertation, Université des Sciences et Technologie de Lille-Lille I (2012)
4. Cheng, J., Liu, C., Zhou, M., Zeng, Q., Ylä-Jääski, A.: Automatic composition of semantic web services based on fuzzy predicate petri nets. IEEE Trans. Autom. Sci. Eng. 12(2), 680–689 (2015)
5. Madkour, M., El Ghanami, D., Maach, A., Hasbi, A.: Context-aware service adaptation: an approach based on fuzzy sets and service composition. J. Inf. Sci. Eng. 29(1), 1–16 (2013)
6. Ordonez, A., Alcázar, V., Borrajo, D., Falcarin, P., Corrales, J.C.: An automated user-centered planning framework for decision support in environmental early warnings. In: Ibero-American Conference on Artificial Intelligence, pp. 591–600. Springer, Heidelberg, November 2012
7. Furno, A., Zimeo, E.: Context-aware composition of semantic web services. J. Mob. Netw. Appl. 19(2), 235–248 (2014)
8. Liu, Z.Z., Chu, D.H., Jia, Z.P., Shen, J.Q., Wang, L.: Two-stage approach for reliable dynamic Web service composition. Knowl.-Based Syst. 97, 123–143 (2016)

9. Gabrel, V., Manouvrier, M., Murat, C.: Web services composition: complexity and models. Discrete Appl. Math. **196**, 100–114 (2015)
10. Tout, H., Mourad, A., Talhi, C., Otrok, H.: AOMD approach for context-adaptable and conflict-free web services composition. Comput. Electr. Eng. **44**, 200–217 (2015)
11. Grati, R., Boukadi, K., Ben-Abdallah, H.: A QoS monitoring framework for composite web services in the cloud. In: Proceedings of the Sixth International Conference on Advanced Engineering Computing and Applications in Sciences (Advcomp 2012), September 2012
12. White, S.R., Hanson, J.E., Whalley, I., Chess, D.M., Kephart, J.O.: An architectural approach to autonomic computing. In: ICAC, pp. 2–9, May 2004
13. Martin, C.: Composition flexible par planification automatique. Doctoral dissertation, Université de Grenoble (2012)
14. Cherif, S., Ben Djemaa, R., Amous, I.: Adaptable web service registry for publishing context aware service composition. In: Proceedings of the 17th International Conference on Information Integration and Web-based Applications & Services, p. 62. ACM, December 2015
15. Cherif, S., Djemaa, R.B., Amous, I.: SABPEL: creating self-adaptive business processes. In: 2015 IEEE/ACIS 14th International Conference on Computer and Information Science (ICIS), pp. 619–626. IEEE, June 2015
16. David, P.C., Ledoux, T.: WildCAT: a generic framework for context-aware applications. In: Proceedings of the 3rd International Workshop on Middleware for Pervasive and Ad-hoc Computing, pp. 1–7. ACM, November 2005
17. O'Connor, M.J., Shankar, R.D., Musen, M.A., Das, A.K., Nyulas, C.: The SWRLAPI: a development environment for working with SWRL rules. In: OWLED (2008)
18. El Hog, C., Djemaa, R.B., Amous, I.: A user-aware approach to provide adaptive web services. J. UCS **20**(7), 944–963 (2014)

Autonomous Systems Research Embedded in Teaching

Maria Spichkova[✉] and Milan Simic

RMIT University, Melbourne, Australia
{maria.spichkova,milan.simic}@rmit.edu.au

Abstract. This paper presents how research on autonomous systems is embedded into the curriculum at two departments at the RMIT University, Australia: the School of Engineering and the School of Science. We introduce general structure of our Work Integrated Learning (WIL) modules and the recently completed projects having as a core component research and development of autonomous systems.

1 Introduction

Autonomous systems (AS), i.e. the systems where the autonomy is one of the essential characteristics, have a large application area in robotics, a research field covering both engineering and computer science. The aim is to delegate complex and hazardous tasks from humans to robots: autonomous robots can perform their tasks in the areas where human attendance might be too dangerous, unhealthy, or unprofitable. In many cases autonomous robots might work faster than humans or carry much more heavy load, and their reliability and preciseness might be a big advantage for numerous tasks, cf. e.g., [28].

Development of AS included many research fields, and all of them are equally important to get a reliable AS. Our research on AS covers aspects of path planning, motion control, human-computer interaction (or, more precisely, human-robot interaction), as well as safety aspects, cf. e.g., [8,9]. In this paper, we would like to cover another aspect of our research: embedding of our research activities in teaching, to make our students (not only on at PhD level, but also on final year of Bachelor and Master studies) aware about state of the art within the corresponding research fields. RMIT University aims to create transformative experiences for its students, to get them ready for their next steps in life and work. The 2020 RMIT strategic plan defines the goals that we will pursue together over the following years, and one of the priorities in this plan is *research embedded in teaching and engagement*.

In our earlier work we presented how collaborative industrial project are embedded into engineering curriculum in the School of Science (disciplines *Computer Science and Software Engineering* and *Computer Science and IT*) and the School of Engineering, [26]. The term Work Integrated Learning (WIL) is used to describe a special kind of learning and teaching activities, where an academic learning is taking place in "real life" situations.

© Springer International Publishing AG 2018
G. De Pietro et al. (eds.), *Intelligent Interactive Multimedia Systems and Services 2017*,
Smart Innovation, Systems and Technologies 76, DOI 10.1007/978-3-319-59480-4_27

Outline: Apart from this introductory section, the rest of the paper is organised as follows. Section 2 defines work integrated and project-based learning, as key approaches in modern teaching and learning practices, where Sect. 3 provides a brief overview of the related work on teaching robotics. Sections 4 and 5 introduce how the research on autonomous systems is embedded in teaching within the courses provided by the School of Engineering and the School of Science. In Sect. 6, we discuss our future plans for Bachelors' and Masters' research and development projects within the area of autonomous systems. In Section 7 we summarise the paper and propose directions for future research.

2   Work Integrated and Project-Based Learning

Sedelmaier and Landes [22] provided justification from a pedagogical point of view that project-based learning (PBL) allows learners not only to gain technical knowledge in the related area, but also foster soft skills.

Dagnino [5] presented a method derived from the collaboration between North Carolina State University and ABB. This method brings diverse techniques to simulate an industrial environment for teaching a senior level Software Engineering course. In contrary to the approach of Dagnino, we do not simulate the industry environment for our projects, but use a real one, based on collaboration with local and overseas industrial partners. In addition to that, project based learning is seen as a contributing factor to graduates' better work preparation, cf. [13].

We have already presented a number of collaborative partnerships in engineering education, cf. [20, 23, 26], but in this case we have additional stake holder. It is Government with its grant for the business and innovation. Another good example of Government support for engineering education is in the design of flexible entry pathways to Mechatronics/Robotics profession as previously reported in [25].

3   Teaching Robotics

In this section, we briefly discuss recent approaches on teaching robotics as well as how our approach relates to them. The problem that robotics is often introduced only from the mechanical engineering perspective, was discussed by Kay [14] more than 10 years ago. Since then, many universities extended their Computer Science (CS) curricula to allow students to learn robotics from a CS perspective. An approach on teaching a core CS concept through robotics was also discussed by Magnenat et al. [17] In our work, we aim to cover both, Engineering and Computer Science prospectives.

García-Peñalvo and Colomo-Palacios [12] provided a general overview of recent teaching methods in Engineering. Ruzzenente et al. [21] presented a review of robotics kits for tertiary education, focusing on the modularity, re-usability, versatility and affordability features, i.e. on the toolkits that allow ease of re-use to teach in different curricula.

Blank et al. [1] introduced a programming framework called Pyro, focusing on the pedagogical implications of teaching mobile robotics. This framework provides a set of abstractions that allows students to write platform-independent robot programs.

Galvan et. al. [11] described an approach that follows the "learning by doing" paradigm and utilises LEGO kits to let students experiment with robot kinematic design. Menegatti and Moro introduced an interesting approach [18], where the design of educational robotics, which commence at Master level, was based on the experience developed at secondary school level, using LEGO NXT robot.

Michieletto et al. [19] presented a methodology on teaching robotics with a CS approach in the Master in Computer Engineering. They focused on application of Robot Operating System (ROS), which is becoming a standard *de facto* inside the robotics community.

Recently, distance teaching of robotics using virtual labs becomes more and more popular. For example, Kulich et al. [15] introduced SyRoTek, an e-learning platform for mobile robotics, artificial intelligence and control engineering. SyRoTek provides remote access to a set of fully autonomous mobile robots placed in a restricted area with dynamically reconfigurable obstacles. The platform was applied at the Czech Technical University in Prague, Czech Republic, and at the University of Buenos Aires, Argentina.

Candelas-Herías et al. [4] described the new features included in the Robolab 2 System. The goal was to allow students to use simulation of an industrial robot, and carry out operations with the real robot.

The RMIT University VXLab facilities and their applications for research and teaching were introduced in [2,3,27]. We utilise the VXLab for teaching CS courses, telecontrol, and robotics. For example the student project presented in Sect. 5 was conducted in the VXLab.

4   Research Embedded in Teaching: School of Engineering

In this section, we present a number of examples of autonomous systems developed at the RMIT University, School of Engineering (SoE). Generally, AS can be classified in the following groups:

1. Partial Autonomy (level of autonomy is less than 100%, but more than 0%): Adaptive cruse control, automatic parking, smart and adaptive systems, cf. e.g., [6,10,16,24].
2. Full Autonomy (level of autonomy is 100%): Unmanned Ground Vehicle (UGV), Unmanned Aerial Vehicle (UAV) and other. Level of autonomy is 100%.

The focus of the research is on the systems with the full autonomy, Unmanned Ground Vehicles (UGVs). There are much more intelligent AS to analyse, if we consider other mobile robots, and/or vehicles as Unmanned Aerial Vehicle (UAV), Autonomous Underwater Vehicle, Space Vehicles, or Unmanned Underground Mining Vehicle.

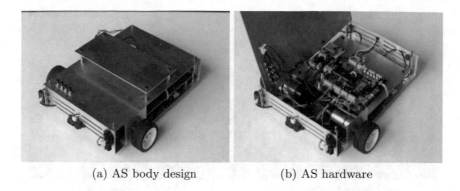

(a) AS body design   (b) AS hardware

Fig. 1. Mobile robot designed third year students at RMIT (SoE)

Most of the subjects in the program of study for the Bachelor of Engineering (Advanced Manufacturing and Mechatronics) (Honours), have a large content of WIL, or Project-Based Learning (PBL) based activities. Two core subjects, delivered in the third year, are Mechatronics Design and Autonomous Systems. In the Mechatronics Design, all skills and knowledge from the Mechatronics area, accumulated in the previous years of study, are integrated in the task of building small mobile robots, as shown in Fig. 1. Robots have to be able to follow certain, predefined paths and to move through the environment escaping obstacles on the way. Various design and deployment platforms are used, like AT Mega 32 for the hardware and C language of the software design. Other platform is based on NI myRIO for the control unit and LabVIEW for the software design environment. This mobile robot uses ultrasonic sensors mounted around the body, to detect environment, and two DC motors powered in differential mode to perform the motion.

In the other core subject, Autonomous Systems, students are dealing with locomotion, vehicle kinematics, autonomous navigation and intelligent path planning and perception. The project, in this subject, is a control system design for mobile robotics competition. Competition in this team based project is on class level. Most of the PBL projects, in the whole program, are team based, but the assessment has large individual component as well. Students are exposed to another design environment, in this case MATLAB/Simulink platform.

In the final year of study, students are performing more sophisticated AS design. Their autonomous systems have to travel from point A to point B, while escaping obstacles, and conduct certain pre-defined tasks, like mining or goods delivery. There is also competition involved, but in this case it is now on national level. Two most popular competitions, in the last few years, were National Instruments Autonomous Robotics Competition (NI ARC) and annual Autonomous Ground Vehicle Competition (AGVC) run by Defence Science and Technology (DST) Group.

Figure 2 presents mobile robots designed by final year Mechatronics students for the NI ARC competition. Figure 2(a) shows mini model of an autonomous

(a) AS for mining   (b) AS for agriculture

Fig. 2. Autonomous robots designed by final year students at the RMIT (SoE)

system that could be used in mining operations. Apart from performing all motion control tasks as a mobile robot, this autonomous system, uses image processing to perform material selection and collection after that. Collected material is then delivered to a target point. Students are using state of the art technology in every stage of design. For example chassis of this AS is designed in CAD and then 3D printed. Figure 2(b) shows agriculture AS application, cf. also [7].

Figure 3 presents an electrical UGV with laptop based control system, used for the research in autonomous systems path planning. Few models of this electrical AS were designed for the DST AGVC. In this case students did not perform whole vehicle design. They used existing golf cart as a platform and designed control system for autonomous operation.

Fig. 3. UGV with control system designed by RMIT (SoE) students of Mechatronics

5   Research Embedded in Teaching: School of Science

The disciplines *Computer Science and Software Engineering* and *Computer Science and (Information Technology) IT* within the School of Science also have a long history of WIL and PBL activities. One of our WIL activities are Software Engineering Projects (SEP), provided as final year courses both on Master and Bachelor levels within the following programs:

– Bachelor of Software Engineering,
– Master of Computer Science,
– Master of IT.

Within SEP, students obtain hands on practical experience in team-based software development, working within a real project environment. The teams consist of 4 – 6 students, where the size of the team depends on the project scope and number of students enrolled into the courses. Allocation to the teams is bases on students' skills and preferences. Students work on the projects within a semester 12 weeks long for approx. 20 – 40 hours/week, depending on the courses they are enrolled in. The number of students enrolled in the corresponding courses grows over last years, e.g., from 87 students in 2013 to 124 students in 2015. In the past years, the projects' focus was on the development of either web/cloud-based systems or iOs/Android apps, which are still the majority of our projects, but in 2016 we also introduced projects with the focus on robotics (which include a research component), cf. Figure 4.

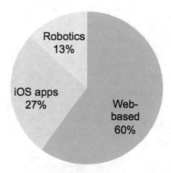

Fig. 4. SEP in semester 2, 2016: types of developed software

In this section, we would like to introduce in more details a project conducted in collaboration with Commonwealth bank (CBA) under support of the Australian Technology Network (ATN). This project was a part of the ATN CBA Robotics Education and Research program, which goals are

– Identify, observe and measure interactions between robots and humans
– Understand the impact and capabilities of robots on the future of work

- Become the leading organisation for STEM (Science, Technology, Engineering and Mathematics)

A humanoid PAL REEM robot was available to the students to conduct the experiments. This robot is

- Equipped with an autonomous navigation system and a touch screen;
- Is capable of roaming through any kind of surroundings;
- Is the first commercial product developed by PAL and the first of its kind in Australia.

Developed by PAL Robotics REEM robot can have many other applications. For this project, RMIT selected the Lab tours use case, where the robot takes guests on tours of our Innovation Labs and answers related questions.

The project had two phases:

- RMIT VXLabs (Melbourne, Australia): Development within a simulated environment provided by a robot software development framework, Robot Operating System (ROS, cf. also Sect. 3). ROS provides interfaces to REEM's sensors and actuators, like motors and speakers. Those interfaces are implemented primary in Python and C++ programming environments.
- CBA Labs (Sydney, Australia): Deployment on a real REEM robot.

To implement the lab tour presented on Fig. 5, students developed corresponding modules on speech, navigation and hand-movements. For the speech recognition, two kinds of processes were used:

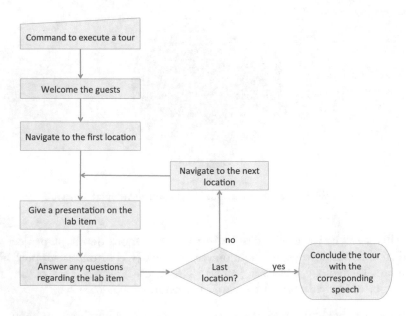

Fig. 5. REEM Lab tour: the process implemented by final year Bachelor and Master students at RMIT (School of Science)

- Commands requiring a prompt reaction (like "stop"): offline speech recogniser Pocketsphinx
- Q&A sessions: online speech recogniser Google Speech API.

The text-to-speech module was implemented using online speech synthesiser IBM Watson API[1] and Acapela[2].

6   Evaluation and Future Plans

In order to achieve quality education, University organises surveys for every subject delivered. Results are extremely encouraging justifying our commitment to project based learning in Autonomous systems teaching, across University. As example survey report is shown in the Fig. 6. We can see that students' engagements and satisfaction are at the highest level.

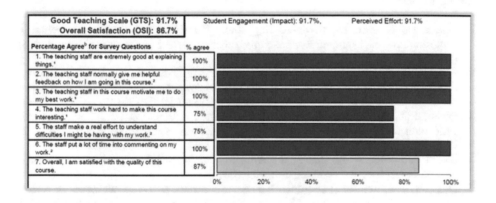

Fig. 6. Example of a survey report

Currently, in Semester 1, 2017, we continue our work on embedding research on robotics into the teaching on Bachelor and Master levels. University is now encouraging more closer cooperation between two Schools in building a collaborative AS team that will consist of students and the staff members with the expertise in this area. The main objective of this new project team is to build a fleet of electrical, autonomous vehicles, as shown in the Fig. 3 that will operate inside the University campuses and transport good and people from a source to destination point inside the University.

---

[1] www.ibm.com/watson/developercloud/text-to-speech.html.

[2] www.acapela-group.com/.

7   Conclusions

This paper presents an overview of our learning and teaching activities: how research on autonomous systems is embedded into the curriculum at the School of Engineering and the School of Science at the RMIT University, Australia. We introduce general structure of our learning modules and the examples of recently completed projects having as a core component research and development of autonomous systems. For example, we are going to conduct the second REEM project in collaboration with CBA, where the developed use case will be extended by new functionalities.

Acknowledgements. We would like to thank the RMIT Alumnus Leroy Clunne-Kiely, Cristhian Betancour, Enrico Ronggowarsito, Luke Payne, and Bijin Idicula for their engagement and hard work within the REEM project.

References

1. Blank, D., Kumar, D., Meeden, L., Yanco, H.: Pyro: a python-based versatile programming environment for teaching robotics. J. Educ. Resour. Comput. **3**(4) (2003)
2. Blech, J.O., Spichkova, M., Peake, I., Schmidt, H.: Cyber-virtual systems: simulation, validation & visualization. In: 9th International Conference on Evaluation of Novel Approaches to Software Engineering (ENASE 2014) (2014)
3. Blech, J.O., Spichkova, M., Peake, I., Schmidt, H.: Visualization, simulation and validation for cyber-virtual systems. In: Evaluation of Novel Approaches to Software Engineering, pp. 140–154. Springer International Publishing (2015)
4. Candelas-Herías, F.A., Bravo, J., Alberto, C., Torres, F.: Flexible virtual and remote laboratory for teaching robotics (2006)
5. Dagnino, A.: Increasing the effectiveness of teaching software engineering: a university and industry partnership. In: 27th Conference on Software Engineering Education and Training (CSEET), pp. 49–54. IEEE Press (2014)
6. Elbanhawi, M., Simic, M.: Examining the use of B-splines in parking assist systems. Appl. Mech. Mater. **490–491**, 1025–1029 (2014)
7. Elbanhawi, M., Simic, M.: Randomised kinodynamic motion planning for an autonomous vehicle in semi-structured agricultural areas. Biosystems Engineering (2014)
8. Elbanhawi, M., Simic, M.: Sampling-based robot motion planning: a review. IEEE Access **2**, 56–77 (2014)
9. Elbanhawi, M., Simic, M., Jazar, R.N.: Continuous path smoothing for car-like robots using B-spline curves. J. Intell. Robot. Syst. **80**(1), 23–56 (2015)
10. Feilkas, M., Hölzl, F., Pfaller, C., Rittmann, S., Schätz, B., Schwitzer, W., Sitou, W., Spichkova, M., Trachtenherz, D.: A Refined Top-Down Methodology for the Development of Automotive Software Systems -The KeylessEntry System Case Study. Technical report TUM-I1103, TU München (2011)
11. Galvan, S., Botturi, D., Castellani, A., Fiorini, P.: Innovative robotics teaching using lego sets. In: Proceedings 2006 IEEE International Conference on Robotics and Automation, pp. 721–726 (2006)
12. García-Peñalvo, F.J., Colomo-Palacios, R.: Innovative teaching methods in engineering. Engineering Education (IJEE) **31**(3), 689–693 (2015)

13. Jollands, M., Jolly, L., Molyneaux, T.: Project-based learning as a contributing factor to graduates? Work readiness. Eng. Educ. **37**(2), 143–154 (2012)
14. Kay, J.S.: Teaching robotics from a computer science perspective. J. Comput. Sci. Coll. **19**(2), 329–336 (2003)
15. Kulich, M., Chudoba, J., Kosnar, K., Krajnik, T., Faigl, J., Preucil, L.: SyRoTek - distance teaching of mobile robotics. IEEE Trans. Educ. **56**(1), 18–23 (2013)
16. Lu, K., Li, Q., Cheng, N.: An autonomous carrier landing system design and simulation for unmanned aerial vehicle. In: Guidance, Navigation and Control Conference (CGNCC), IEEE Chinese, pp. 1352–1356 (2014)
17. Magnenat, S., Shin, J., Riedo, F., Siegwart, R., Ben-Ari, M.: Teaching a core CS concept through robotics. In: Proceedings of the 2014 Conference on Innovation & Technology in Computer Science Rducation, pp. 315–320. ACM (2014)
18. Menegatti, E., Moro, M.: Educational robotics from high-school to master of science. In: Proceedings of International Conference on Simulation, Modeling and Programming for Autonomous Robots, pp. 639–648 (2010)
19. Michieletto, S., Ghidoni, S., Pagello, E., Moro, M., Menegatti, E.: Why teach robotics using ros? Autom. Mobile Robot. Intell. Syst. **8** (2014)
20. Mo, J., Simic, M., Dawson, P.: Collaborative partnership for development of mechatronics engineering education of the future. In: 19th Annual Conference for the Australasian Association for Engineering Education. AAEE (2008)
21. Ruzzenente, M., Koo, M., Nielsen, K., Grespan, L., Fiorini, P.: A review of robotics kits for tertiary education. In: Teaching Robotics Teaching with Robotics: Integrating Robotics in School Curriculum, pp. 153–162 (2012)
22. Sedelmaier, Y., Landes, D.: Active and inductive learning in software engineering education. In: Proceedings of the 37th International Conference on Software Engineering, vol. 2, ICSE 2015, pp. 418–427. IEEE Press (2015)
23. Simic, M.: Courseware design experience. In: Proceedings of the 2006 ICEE 9th International Conference on Engineering Education, pp. 10–15. INEER (2006)
24. Simic, M.: Vehicle and public safety through driver assistance applications. In: Proceedings of the 2nd International Conference Sustainable Automotive Technologies (ICSAT 2010), vol. 490491, pp. 281–288 (2010)
25. Simic, M., et al.: Designing flexible entry pathways to mechatronics/robotics profession. In: Creating Flexible Learning Environments: Proceedings of the 15th Australasian Conference for the Australasian Association for Engineering Education and the 10th Australasian Women in Engineering Forum, p. 348. Australasian Association for Engineering Education (2004)
26. Simic, M., Spichkova, M., Schmidt, H., Peake, I.: Enhancing learning experience by collaborative industrial projects. In: ICEER 2016, pp. 1–8. Western Sydney University (2016)
27. Spichkova, M., Schmidt, H., Peake, I.: From abstract modelling to remote cyber-physical integration/interoperability testing. In: Improving Systems and Software Engineering Conference (2013)
28. Spichkova, M., Simic, M.: Towards formal modelling of autonomous systems. In: Intelligent Interactive Multimedia Systems and Services, pp. 279–288. Springer (2015)

Vehicle Flat Ride Dynamics

Hormoz Marzbani, Dai Voquoc, Reza N. Jazar$^{(\boxtimes)}$,
and Mohammad Fard

School of Engineering, RMIT University, Melbourne, Australia
{hormoz.marzbani,dai.voquoc,reza.jazar,
mohammad.fard}@rmit.edu.au

Abstract. By modelling a four-wheel vehicle as a bicycle, it has been suggested (Olley 1934) that if the radius of gyration, r, in pitch is equal to the multiplication of the distance from the center of gravity of the front, a, and rear, b, wheels of the car ($r^2 = a \cdot b$), the bounce center of the vehicle will be located at one spring and the pitch center on the other spring of a bicycle car model.

Employing the flat ride condition, $r^2 = a \cdot b$, the system of the sprung masses of a vehicle can be considered as two separate mass-spring systems. Therefore, front and rear suspensions may be modelled as two separate one degree of freedom spring-mass system.

This paper mathematically derives the flat ride condition and clarifies analytically how exact is the suggested condition.

Keywords: Flat ride · Smart suspension · Passive suspension optimization · Vehicle dynamics

1 Introduction

Road bump excitation to a car affects the front wheels first and then, with a time lag, it excites the rear wheels. The general recommendation was that the natural frequency of the front suspension should be lower than the rear (Olley 1938). Therefore, the rear part oscillates faster to catch up with the front to eliminate pitch and put the car in bounce before the vibrations die out by damping. Based on this recommendation, scientists tried to find a relation between the front and rear spring ratios to transfer the pitch motions of a car into bounce motion (Best 2002; Crolla and King 2000; Dixon 2008; Odhams and Cebon 2006; Olley 1946; Sharp 2002; Sharp and Pilbeam 1993).

What is known as "Olley's Flat Ride" not considering the other prerequisites can be casted as:

"The front suspension should have around 30% lower ride rate then the rear"

Although measurement of ride has been subjective, those guidelines are still considered valid rules of thumb. A prerequisite for flat ride was the uncoupling condition (Milliken et al. 2002) that bounce and pitch centers of the model locate on the springs. Employing this condition, the front and rear spring systems of the vehicle can be regarded as two separate one degree of freedom systems.

© Springer International Publishing AG 2018

G. De Pietro et al. (eds.), *Intelligent Interactive Multimedia Systems and Services 2017*,
Smart Innovation, Systems and Technologies 76, DOI 10.1007/978-3-319-59480-4_28

In this paper, we use analytical methods to study the flat ride conditions which has been recommended and used by car manufacturers' designers. This paper will prove and provide a mathematical approach for what have been the criteria of design in vehicle dynamic studies.

2 Mathematical Analysis

Consider the beam in Fig. 1, with mass m and mass moment I about the mass center C which is sitting on two springs k_1 and k_2 and models a car in bounce and pitch vibrating motions. The translational coordinate x of C and the rotational coordinate θ are the coordinates that we utilize to measure the kinematics of the beam.

The equations of motion of the system are:

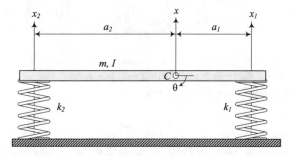

Fig. 1. A beam with mass M and mass moment I, sitting on two springs k_1 and k_2

$$\begin{bmatrix} m & 0 \\ 0 & I \end{bmatrix} \begin{bmatrix} \ddot{x} \\ \ddot{\theta} \end{bmatrix} + \begin{bmatrix} k_1 + k_2 & a_2 k_2 - a_1 k_1 \\ a_2 k_2 - a_1 k_1 & a_2^2 k_2 + a_1^2 k_1 \end{bmatrix} \begin{bmatrix} x \\ \theta \end{bmatrix} = 0 \tag{1}$$

We may also employ the coordinates x_1 and x_2 instead of x and θ to compare the mode shapes of a system. The equations of motion of the system will be:

$$\begin{bmatrix} \dfrac{ma_2^2 + I}{(a_1 + a_2)^2} & \dfrac{ma_1 a_2 - I}{(a_1 + a_2)^2} \\ \dfrac{ma_1 a_2 - I}{(a_1 + a_2)^2} & \dfrac{ma_1^2 + I}{(a_1 + a_2)^2} \end{bmatrix} \begin{bmatrix} \ddot{x}_1 \\ \ddot{x}_2 \end{bmatrix} + \begin{bmatrix} k_1 & 0 \\ 0 & k_2 \end{bmatrix} \begin{bmatrix} x_1 \\ x_2 \end{bmatrix} = 0 \tag{2}$$

We define the following parameters

$$\Omega_1^2 = \frac{k_1}{m}\beta \quad \Omega_2^2 = \frac{k_2}{m}\beta \quad \beta = \frac{l^2}{a_1 a_2} \quad \alpha = \frac{r^2}{a_1 a_2} \quad \gamma = \frac{a_2}{a_1} \quad l = a_1 + a_2 \tag{3}$$

to rewrite the equations as

$$\begin{bmatrix} \alpha + \gamma & 1 - \alpha \\ 1 - \alpha & \alpha + \frac{1}{\gamma} \end{bmatrix} \begin{bmatrix} \ddot{x}_1 \\ \ddot{x}_2 \end{bmatrix} + \begin{bmatrix} \Omega_1^2 & 0 \\ 0 & \Omega_2^2 \end{bmatrix} \begin{bmatrix} x_1 \\ x_2 \end{bmatrix} = 0 \tag{4}$$

Setting

$$\alpha = 1 \tag{5}$$

makes the equation decoupled

$$\begin{bmatrix} 1+\gamma & 0 \\ 0 & 1+\frac{1}{\gamma} \end{bmatrix} \begin{bmatrix} \ddot{x}_1 \\ \ddot{x}_2 \end{bmatrix} + \begin{bmatrix} \Omega_1^2 & 0 \\ 0 & \Omega_2^2 \end{bmatrix} \begin{bmatrix} x_1 \\ x_2 \end{bmatrix} = 0 \tag{6}$$

The natural frequencies ω_i and mode shapes u_i of the system are

$$\omega_1^2 = \frac{1}{\gamma+1}\Omega_1^2 = \frac{l}{a_2}\frac{k_1}{m} \qquad u_1 = \begin{bmatrix} 1 \\ 0 \end{bmatrix} \tag{7}$$

$$\omega_2^2 = \frac{\gamma}{\gamma+1}\Omega_2^2 = \frac{l}{a_1}\frac{k_2}{m} \qquad u_1 = \begin{bmatrix} 0 \\ 1 \end{bmatrix} \tag{8}$$

The nodes of oscillation in the first and second modes are at the rear and front suspensions respectively. Figures 2 and 3 illustrate the mode shapes.

The decoupling condition $\alpha = 1$ yields

$$r^2 = a_1 a_2 \tag{9}$$

It indicates that the pitch radius of gyration, r, must be equal to the multiplication of the distance of the mass canter C from front and rear axles. Therefore, by setting $\alpha = 1$, the nodes of the two modes of vibrations appear to be at the front and rear axles. As a result, the front wheel excitation will not alter the body at the rear axle and vice versa. For such a car, the front and rear parts of the car act independently. Therefore, the decoupling condition $\alpha = 1$ allows us to break the initial two DOF system into two independent one DOF systems as illustrated in Fig. 4, where:

$$m_r = m\frac{a_1}{l} = m\varepsilon \qquad m_f = m\frac{a_2}{l} = m(1-\varepsilon) \qquad \varepsilon = \frac{a_1}{l} \tag{10}$$

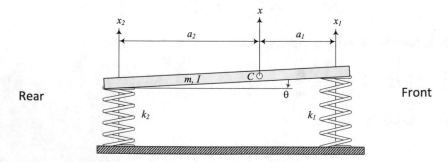

Fig. 2. Mode shape 1, only the front suspension is oscillating and the node is on the rear suspension

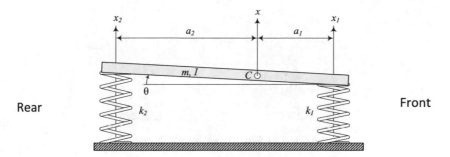

Fig. 3. Mode shape 2, only the rear suspension is oscillating and the node is on the front suspension

The equations of motion of the independent systems will be:

$$m(1 - \varepsilon)\ddot{x}_1 + c_1\dot{x}_1 + k_1 x_1 = k_1 y_1 + c_1\dot{y}_1 \tag{11}$$

$$m\varepsilon\ddot{x}_2 + c_2\dot{x}_2 + k_2 x_2 = k_2 y_2 + c_2\dot{y}_2 \tag{12}$$

The decoupling condition will not necessarily decouple the equations in general. However, if there is no anti-pitch spring or damping between the front and rear suspensions then equations of motion

$$\begin{bmatrix} \alpha + \gamma & 1 - \alpha \\ 1 - \alpha & \alpha + \frac{1}{\gamma} \end{bmatrix} \begin{bmatrix} \ddot{x}_1 \\ \ddot{x}_2 \end{bmatrix} + \begin{bmatrix} 2\xi_1\Omega_1 & 0 \\ 0 & 2\xi_2\Omega_2 \end{bmatrix} \begin{bmatrix} \dot{x}_1 \\ \dot{x}_2 \end{bmatrix} + \begin{bmatrix} \Omega_1^2 & 0 \\ 0 & \Omega_2^2 \end{bmatrix} \begin{bmatrix} x_1 \\ x_2 \end{bmatrix}$$
$$= \begin{bmatrix} 2\xi_1\Omega_1 & 0 \\ 0 & 2\xi_2\Omega_2 \end{bmatrix} \begin{bmatrix} \dot{y}_1 \\ \dot{y}_2 \end{bmatrix} + \begin{bmatrix} \Omega_1^2 & 0 \\ 0 & \Omega_2^2 \end{bmatrix} \begin{bmatrix} y_1 \\ y_2 \end{bmatrix} \tag{13}$$

$$2\xi_1\Omega_1 = \frac{c_1}{m}\beta \qquad 2\xi_2\Omega_2 = \frac{c_2}{m}\beta \tag{14}$$

will be decoupled by $\alpha = 1$.

$$\begin{bmatrix} 1 + \gamma & 0 \\ 0 & 1 + \frac{1}{\gamma} \end{bmatrix} \begin{bmatrix} \ddot{x}_1 \\ \ddot{x}_2 \end{bmatrix} + \begin{bmatrix} \xi_1 & 0 \\ 0 & \xi_2 \end{bmatrix} \begin{bmatrix} \dot{x}_1 \\ \dot{x}_2 \end{bmatrix} + \begin{bmatrix} \Omega_1^2 & 0 \\ 0 & \Omega_2^2 \end{bmatrix} \begin{bmatrix} x_1 \\ x_2 \end{bmatrix}$$
$$= \begin{bmatrix} 2\xi_1\Omega_1 & 0 \\ 0 & 2\xi_2\Omega_2 \end{bmatrix} \begin{bmatrix} \dot{y}_1 \\ \dot{y}_2 \end{bmatrix} + \begin{bmatrix} \Omega_1^2 & 0 \\ 0 & \Omega_2^2 \end{bmatrix} \begin{bmatrix} y_1 \\ y_2 \end{bmatrix} \tag{15}$$

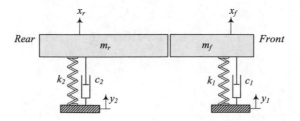

Fig. 4. Car bicycle model after decoupling

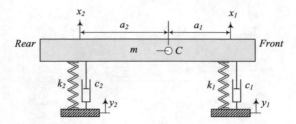

Fig. 5. Car bicycle model with damping.

Figure 5, illustrates the damped model of a bicycle car.

3 Numerical Analysis

To examine the effect of the decoupling condition and having independent front and rear model of a car, let us compare the responses of the model of Fig. 5 and using Eq. (13) for different α.

Consider a car with the given characteristics in Table 1.

Table 1. Specifications of a sample car

m (kg)	420
a_1 (m)	1.4
a_2 (m)	1.47
l (m)	2.87
k_1 (N/m)	10000
k_2 (N/m)	13000
c_1 (Ns/m)	1000
c_2 (Ns/m)	1000
β	4.00238
γ	1.05
Ω_1	95.2947
Ω_2	123.8832
ξ_1	0.05
ξ_2	0.0384

- Case 1: $\alpha > 1$

$$\alpha = 1.2, \ r = 1.571496102\,\text{m}, \ I_y = 1037.232$$

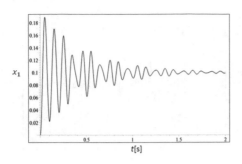

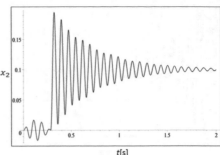

Fig. 6. Oscillations of the front of the car with $\alpha = 1.2$

Fig. 7. Oscillations of the rear of the car with $\alpha = 1.2$

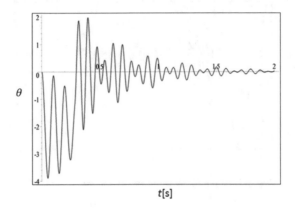

Fig. 8. Angular oscillations of the car with $\alpha = 1.2$

- Case 2: $\alpha = 1$

$$\alpha = 1, \ r = 1.434573107 \, \text{m}, \ I_y = 864.3599996$$

- Case 3: $\alpha < 1$

$$\alpha = 0.8, \ r = 1.283121195 \, \text{m}, \ I_y = 691.4880004$$

Figures 6, 7, 8, 9, 10, 11, 12, 13 and 14 compare and show the effects of uncoupling to the system. Each set of 3 figures are for a value of α which varies from smaller than 1 to 1 and then bigger than 1. Figures 7, 10 and 13 illustrate the oscillations of the rear part of the vehicle after hitting the step. As shown in Figs. 7 and 13, oscillations start with small amplitude, which does not exist in Fig. 10. This delay in the oscillation of the rear of the vehicle is caused by the time lag between the front and the rear wheels hitting the step. This time lag is dependent to the wheelbase of the vehicle and the

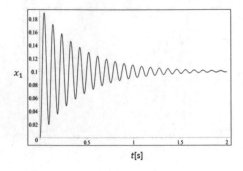

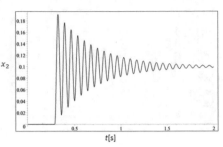

Fig. 9. Oscillations of the front of the car with $\alpha = 1$

Fig. 10. Oscillations of the rear of the car with $\alpha = 1$

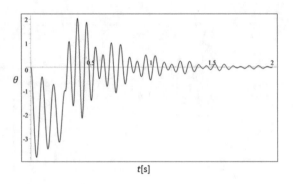

Fig. 11. Angular oscillations of the car with $\alpha = 1$

changes with the traveling speed shown by $\tau = l/v$. Figure 10 shows the oscillation of the rear of a vehicle with $\alpha = 1$, and indicates that in this case the oscillation of the front wheel, which has already started τ seconds ago, does not affect the oscillation of the rear part. That is a result of locating the bounce and pitch centers of the vehicle on the rear and front springs of the vehicle, the condition for uncoupling.

However, considering Figs. 7 and 13 for values $\alpha > 1$ and $\alpha < 1$ respectively, the effect of the front oscillation on the rear, is observable in the form of small amplitude oscillations. The time lag oscillation in Fig. 7 starts off in the form of an upward motion which indicates that the oscillation center of the front wheel is located behind the rear wheel having $\alpha > 1$. The condition of the car is revers for the values of $\alpha < 1$ in Fig. 13.

The same conclusions can be taken by comparing the set of figures which illustrate the pitch motion in the vehicles after hitting the step, Figs. 8, 11 and 14. The angle between the front and rear of the vehicle has been calculated and plotted using the following equation:

$$\theta = \frac{x_2 - x_1}{l} \tag{16}$$

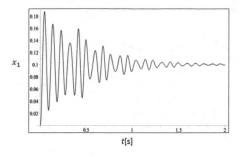

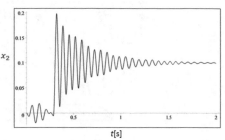

Fig. 12. Oscillations of the front of the car with $\alpha = 0.8$

Fig. 13. Oscillations of the rear of the car with $\alpha = 0.8$

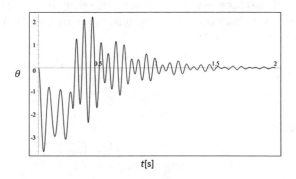

Fig. 14. Angular oscillations of the car with $\alpha = 0.8$

Smaller pitch motion in Fig. 11 for $\alpha = 1$, makes it obvious that an uncoupled system provides a more comfortable ride compared to the change of angle θ, in Figs. 8 and 14.

4 Flat Ride Approximation

The rear suspension must have proper parameters and higher frequency to reach the same amplitude as the front suspension at a reasonable time before the oscillations die out and provide a flat condition. Then the oscillation of the systems dies out before the pitch mode significantly appears again. A near flat ride situation is shown in Fig. 15, for a car going over a unit step.

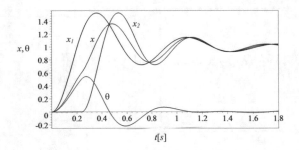

Fig. 15. Response of the front and rear suspensions of a near flat ride car to a unit step.

Solving the equations of motion (23) and (24) for x_1 and x_2 for a unit step input and search for the conditions such that both systems reach their third equal peak amplitude at the same time, as a reasonable catch up time, we find that the third peak of x_1 happens at the time:

$$t_{p_1} = \frac{3\pi\sqrt{1-\varepsilon}}{\omega\sqrt{1-\xi_1^2}} \tag{17}$$

where its displacement is

$$x_1 = 1 + \exp(-3\pi\xi_1/\sqrt{1-\xi_1^2}) \tag{18}$$

The third peak of x_2 happens at

$$t_{p_2} = \tau + \frac{3\pi\sqrt{k/\varepsilon}}{\omega\sqrt{1-\xi^2\xi_1^2}} \tag{19}$$

where its displacement is

$$x_2 = 1 + \exp(-3\pi\xi\xi_1/\sqrt{1-\xi^2\xi_1^2}) \tag{20}$$

The conditions that x_1 and x_2 meet after one and a half oscillations are $t_{p_1} = t_{p_2}$, $x_1 = x_2$. The second equation provides us with damping ratio ξ of

$$\xi = 1 \tag{21}$$

The first equation provides us with spring ratio k:

$$k = \frac{Z_1}{Z_2\tau^2 + Z_3\tau + Z_4} \tag{22}$$

Where,

$$Z_1 = -9\varepsilon\pi^2 m\left(-1 + \xi_1^2\right) \tag{23}$$

$$Z_2 = (\xi^2\xi_1^4 - \xi^2\xi_1^2 + 1 - \xi_1^2)k_1 \tag{24}$$

$$Z_3 = 6\pi m\left(\xi^2\xi_1^2 - \varepsilon\xi^2\xi_1^2 - 1 + \varepsilon\right)\sqrt{-\frac{k_1\left(1 - \xi_1^2\right)}{m(\varepsilon - 1)}} \tag{25}$$

$$Z_4 = -9\pi^2 m(\varepsilon\xi^2\xi_1^2 - \varepsilon + 1 - \xi^2\xi_1^2) \tag{26}$$

Using a set of nominal values,

$$\xi = 1, m = 420, \xi_1 = 0.4, k_1 = 10000 \tag{27}$$

Figure 16 illustrates the plot of k versus τ, for different ε from 0.2 to 0.8.

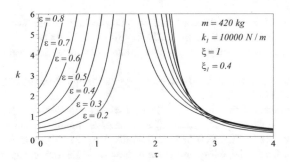

Fig. 16. k vs. τ for different ε

Considering the existing designs of normal street vehicles, only a very small section of the horizontal axis of Fig. 16 should be investigated. The wheelbase for normal street vehicles are not less than 2 m or longer than 3.5 m. Let us assume that the speed of a car which goes over a step and is expected to show a flat ride is between 4 m/s and 20 m/s. Therefore, the practical domain of the time lag between the front and rear wheels would be $0.1 < \tau < 0.875$.

Furthermore, the mass center of street cars are normally in the front half of the wheelbase in order to provide understeer condition. Considering $0.4 < \varepsilon < 0.6$ will cover all street and sports cars.

5   Conclusion

This investigation has shown the analytic condition to have a flat ride in vehicle dynamics. The flat ride condition is dynamic and can be achieved only if an active suspension is implemented. However, the 100 years old suggested condition of $(r^2 = a \cdot b)$, although not ideal, can be a good approximation to be utilized as a passive design.

A series of design curves have been introduced to indicate how an active system must be adjusted to provide frat right for any vehicle and any excitation.

References

Best, A.: Vehicle ride-stages in comprehension. Phys. Technol. **15**(4), 205 (2002)

Crolla, D., King, R.: Olley's "Flat Ride" revisited (2000)

Dixon, J.: The Shock Absorber Handbook. Wiley, New York (2008)

Milliken, W.F., Milliken, D.L., Olley, M.: Chassis Design. Professional Engineering Publishing, London (2002)

Odhams, A., Cebon, D.: An analysis of ride coupling in automobile suspensions. Proc. Inst. Mech. Eng. Part D J. Automobile Eng. **220**(8), 1041–1061 (2006)

Olley, M.: Independent wheel suspension–its whys and wherefores. Soc. Automot. Eng. J. **34**(2), 73–81 (1934)

Olley, M.: National influences on American passenger car design. Proc. Inst. Automobile Eng. **32**(2), 509–572 (1938)

Olley, M.: Road manners of the modern car. Proc. Inst. Automobile Eng. **41**(1), 147–182 (1946)

Sharp, R.: Wheelbase filtering and automobile suspension tuning for minimizing motions in pitch. Proc. Inst. Mech. Eng. Part D J. Automobile Eng. **216**(12), 933–946 (2002)

Sharp, R., Pilbeam, C.: Achievability and value of passive suspension design for minimum pitch response. Veh. Ride Handling **39**, 243–259 (1993)

Autonomous Vehicle Design for Predator Proof Fence Monitoring

Silas Tullah[1], Heinz de Chelard[2], and Milan Simic[1(✉)]

[1] School of Engineering, RMIT University, Melbourne, Australia
s3430412@student.rmit.edu.au, milan.simic@rmit.edu.au
[2] Catchment Health Engineering, Hamilton, Australia
heinz.dechelard@catchmenthealth.com.au

Abstract. Research presented here focuses on developing a new mobile platform, an autonomous cable car, which can be implemented with minimal infrastructure additions, to new and existing reserves, for the reliable fence monitoring. Autonomous cable car solution for infrastructure monitoring is selected after comprehensive investigation in current practices and working environment properties. Inland conservation reserves rely on predator proof fencing in order to separate the conservation reserve from the surrounding environment. The critical infrastructure component, in terms of protecting the animals, and in terms of cost is the fence itself. It requires regular and frequent monitoring in order to maintain its integrity and its effectiveness. Currently used surveillance platforms are often not capable of meeting harsh outdoor requirements.

Keywords: Autonomous vehicle · Fence monitoring · Cable car · Data acquisition

1 Introduction

Predator proof fencing is used to establish a barrier between a conservation reserve and the land that surrounds it, as it is the case at Tiverton, Victoria, Australia. This reserve is protecting and hosting eastern barred bandicoot which is extinct in the wild. It is important to protect them from predation by foxes and feral cats and to facilitate an effective breeding program. Australia has a high rate of mammal extinction in general. Although predator proof fences can be highly effective for protecting endangered species, to remain effective they must be constantly monitored for breaches. Monitoring requires significant manpower and money. Perimeter should be patrolled at least once every 24 h. Tiverton project will enclose approximately 1000 ha with a 15 km long perimeter fence. Man patrolling the fence needs between two and four hours every day, 365 days per year in all weather conditions. As the perimeter of the fence increases so too does the cost of monitoring and maintenance, which decreases the viability of the operation by increasing the overhead costs [1].

It is hoped that this can be overcome through the use of autonomous systems (AS)s. In the initial stages of the project we have conducted few site visits and consultations in order to get better understanding of the real life problems in infrastructure monitoring.

G. De Pietro et al. (eds.), *Intelligent Interactive Multimedia Systems and Services 2017*,
Smart Innovation, Systems and Technologies 76, DOI 10.1007/978-3-319-59480-4_29

Many different possible solutions were comprehensively investigated. Project team has already conducted broad research in Unmanned Ground Vehicles (UGV)s as reported here [2–4]. Various Unmanned Aerial Vehicles (UAV)s are already performing important tasks like surveillance, search and rescue missions as already presented [5]. Multi object detection and tracking applications are implemented and reported [6]. Power requirements were also investigated and novel solutions presented [7, 8]. The decision was finally made to apply autonomous systems, but the question was about the type: UGV, UAVs, or something relatively different.

2   Autonomous System Selection

2.1   UGV Option

UGVs used for precision agriculture should conform to certain design requirements in order to maintain a required level of motion stability. An UGV deployed for precision data collection in a potentially muddy and rocky environment, must be able to comfortably and autonomously drive over skewed and rough terrain, whilst minimising mechanical vibrations and maintaining a level of control appropriate to avoid having the vehicle damage the crop. Article [9] identifies a range of challenges faced by agricultural UGVs and basic requirements for overcoming these challenges. Stable driving characteristics are extremely important in maintaining reasonable data acquisition and therefore system reliability during autonomous operation. In order to minimise effects of mechanical vibrations and environmental variation, researchers, as reported in [9], use a suspension and damping system for the mobile sensor platform. One of the applications of autonomous vehicle in semi-structured agricultural areas is presented here [10], where the optimal path planning was generated.

Given that Tiverton environmental conditions are hard, any UGV deployed there would suffer immensely as a result of the operating environment conditions. Assuming that the vibration noise could be controlled, the machine would still require frequent maintenance and cleaning to account for the accumulation of mud on tyres, body and sensors of the UGV, making it an inappropriate platform for the use in this kind of research project. The issues that prevent UGVs from being fully realised are application specific. An UGV used to patrol the perimeter of a factory, operating on paved roads, in controlled terrain, suffers from very different problems to an UGV intended for deployment in agricultural settings. UGVs operating indoors, or in controlled environments, tend to be successful with respect to their intended mission and are far more reliable than UGVs operating in uncontrolled and rough terrain. Unfortunately, autonomous systems, like commercial unmanned ground vehicles (UGV)s and unmanned aerial vehicles (UAV)s have problems that make them unsuitable for deployment at the tough environments, driving the need for other solutions.

2.2   UAV Option

Over the last few years UAV became an attractive option for many applications [5], including agriculture and environment monitoring. They can carry high resolution video cameras [6], and could also have thermal cameras. Image processing techniques allow 3D maps to be built from captured images, which is useful for surveying land. Drones tend to take either a Fixed Wing (FW) or Quadrotor (QR) form. Quadrotors are well suited to low flying but are generally limited to about 30 min of flight time. Fixed wing crafts are better for aerial surveying, achieving longer flight times. FW crafts are used extensively in agriculture, but they are not currently capable of fully autonomous take-off and landing. Following that, we have focused on the rotary crafts. It is important to note that the utility of certain research varies depending on the geographical area. Western Victoria in Australia, has native Birds of Prey (BoP) that attack UAVs believing that they are prey. This limits places where UAV technology can be applied in a cost effective manner and safe for the native animal.

QRs are the most common form of rotary craft due to their design simplicity. Mechanically, they can be simple with 4 motors and propellers, attached to the X or H frame. Motion is controlled by individually adjusting the speed of each motor. This differs from other forms of rotary crafts, like helicopters, as they do not need to adjust the pitch of the propellers in order to control direction. While this is strength of the QR it is also the source of its two major weaknesses, energetic inefficiency and slow dynamic response time.

QRs tend to suffer similarly to UGVs when operating outdoor. Information they collect becomes less useful as weather conditions deteriorate. If a QR has to operate autonomously, in critical situations, it must be capable of operating under harsh conditions. Weather resistance is a key problem for QRs. There are three major problems preventing the use of UAV technology at Tiverton: battery life, weather conditions, and BoP attacks. The only way to guarantee a solution that effectively and ethically handles BoP attacks, whilst avoiding potential legal issues, is to fly the drones at night. This brings an issue because surveying can not be conducted via traditional methods, as photogrammetry. Commercial solutions are designed for daytime operations.

The advantage of UAVs is that they do not touch the ground and therefore do not need to consider harsh surface conditions, such as loamy soil, but only environmental conditions like rain and wind. All of this means that, the system must stay off of the ground whilst also staying out of the air, to avoid BoP attacks. The system must also avoid implementing significant infrastructure adaptations, such as installing in ground rail system for a monorail.

The solution that this research proposes is a cable car that requires minimal additional infrastructure. The solution differs from traditional cable cars, where the car is attached to the cable and the cable driven, by effectively treating the cables as tracks to drive itself along.

3   Design Stages

3.1   Mechanical Design

The primary function of the system is to autonomously inspect predator proof fence so that rabbit and fox holes, as well as mechanical battering by kangaroos and wallabies can be detected before the perimeter is breached. Detection of fence defects was a subject of a concurrently running subproject. It is performed using video camera and image processing. The whole system has to perform reliably under a wide range of extremely harsh weather conditions. The animals are not deterred by rain, or wind. Our cable car system should handle BoP problem in a safe way, so that no harm, or damages to the birds' life, or habitat could occur.

The mechanical challenges that the system must be able to overcome are: gripping the cables with enough force to remain firmly attached; passing the brackets without dismounting the cables; traversing corners, and climbing inclines, all under extremely harsh conditions. The whole mechanical design of the autonomous cable car prototype was performed in CAD and most subsystems were 3D printed. Initial CAD drawing can be seen from Fig. 1. We can see four wheel clusters and the platform to carry control and data acquisition hardware.

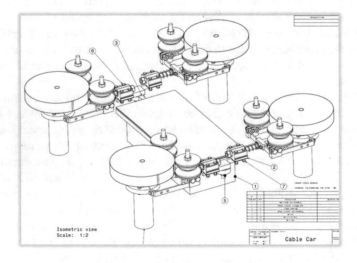

Fig. 1. CAD drawing of initial cable car prototype

To grip the cables, the idle wheels were designed with a groove and placed on a tension spring loaded sliding axle, as shown in Fig. 2. There are two idle wheels per wheel cluster; one placed either side of the driven wheel. The wheel clusters are orientated such that the idle wheels of connected clusters face each other on the inside of the cable with the driven wheel on the outside of the cables.

For the sake of modularity, the fencing and cabling should need minimal modifications. According to that, no rounding has been designed for the corners. A drawing of the projected corner layout is shown in Fig. 3. If the corners were rounded then the distance between the tracks would be constant, and navigating corners would only be a matter of speeding up the motors on one side of the unit by a calculated quantity. However, due to the nature of the existing corners, the track width of the cable car unit must shrink from its natural state which appears on mostly straight parts of the fence. The reduced track width, W_2, i.e. the track width when traversing the

Fig. 2. Assembled wheel cluster, side view

corners, can be calculated as a function of instant corner angle, α, and original track width, W_1, as shown in the Eq. 1:

$$W_2 = W_1 * \sin(\frac{\pi}{2} - \alpha)$$ (1)

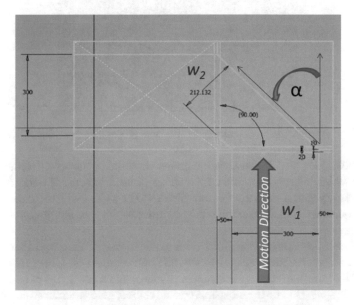

Fig. 3. Bird's eye view showing track corner design

The cable system is designed so that the maximum turn, that the unit would be expected to be capable of crossing, should be 90°, in a transitional layout arrangement of 45% degrees instant corner angle maximum. In that case, the track width was designed to be capable of shrinking by 29.3% (from 300 mm to ~213 mm). When considering that, the wheel clusters must be able to rotate 45° about centre of the linkage arm. While the motor clears the main body to achieve a full 90° turn, the minimum calculated arm length of 80 mm forces the wheel clusters to be staggered. The staggering of the wheel clusters allows the arms to shrink in width whilst also ensuring that at most, one idle wheel is disconnected from the cable at a time, opposed to two for non-staggered clusters.

3.2 Mechatronic Design

Mechatronic systems consist of sensors, actuators and control unit. In our case, four motors are controlled by a National Instruments (NI) Real time Input Output microcontroller myRIO, and powered with Sabertooth 2×12 dual H bridge motor drivers. Figure 4 shows control hardware. Hall effect encoders, with the resolution 1190 pulses per revolution, provide feedback for closed loop control. Four motors allow individual velocity control of wheel clusters through corners, to ensure the system does not become too severely misaligned.

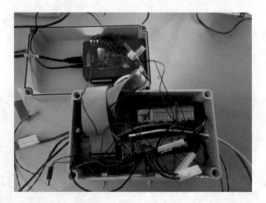

Fig. 4. Autonomous cable car system hardware components

A LS20031 5 Hz GPS Receiver is integrated via serial connection in order to provide the system with relatively accurate GPS readings, with the accuracy range of up to 3 m. This is necessary as it allows a set of coordinates to be paired with the data acquired by the data acquisition (DAQ) system, which in turn enables the highly targeted despatch of repair services.

Our AS has to establish and maintain communication with the control centre, in order to be able to alert the site manager to any attempted breaches. Given that the system needs only to be able to send one compressed image and a few bytes of text at a time, a GSM module appears to be the most appropriate choice, given their cost and the required functionality. A GPRS connection would be used to send a downscaled image along with a standard text message containing the GPS coordinates of the image. Finally 4G technology is now available across the country with the speed of 2 Mbps up to 75 Mbps. Using national communication infrastructure is not an issue and it can easily fulfill our communication, i.e. reporting requirements in the real time. Future solutions may be based on mobile phone application for the image capture, image processing, data storage,

GPS localization, and communication with NI myRIO control unit using wireless local area network (WLAN), and with the control centre using 4G technology.

3.3 Software Design

Individual software subsystems, designed in LabVIEW environment, are combined to create an autonomous system capable of traveling along the fence, collecting the data and reporting in the real time. Block diagram of the software system is shown in Fig. 5. All subsystems are running concurrently in NI myRIO. Since myRIO has a complex structure, which includes microcontroller and field programmable gate array (FPGA), time critical tasks are conducting in the hardware, i.e. in FPGA part of the control system. FPGA is connected to digital and analog inputs and outputs, as well as an accelerometer. Processor part of the system, LabVIEW RT (Real time), has connections to USB communication ports, memory, various indicators and wireless channel running on standard 802.11 b/g/n for wireless local area network (WLAN). Arrow labeled as 4G, shown in figure, is a conceptual solution for the new version of the cable car. A GPS unit was integrated using an existing LabVIEW vi that was downloaded and modified to ensure that the appropriate settings are used for the LS20031 5 Hz GPS Receiver.

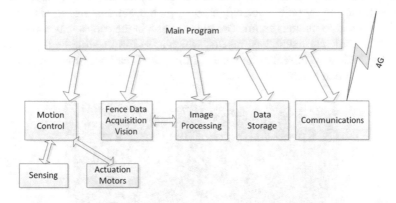

Fig. 5. Autonomous system control software top level block diagram

Actuators control program provides individual closed loop velocity control of the motors, necessary to traverse corners. It uses a Proportional and Integral controller to ensure a reasonable rise time with minimal overshoot and with no settled error. Since our autonomous cable car has distinguishable states of operation we have applied state machines architecture in software design. State machine diagram of the motion control module is shown in Fig. 6.

Image processing, vision and data storage are subject of investigation and development in another subproject which is running concurrently. The findings and achievements will be reported in another publication.

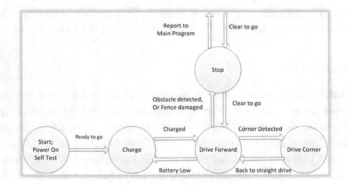

Fig. 6. State machine diagram for motion control module

4   Test Rig Setup and Testing

For the system testing purposes, a small section of the fence, identical to the construction used in the reserve, was erected on RMIT Bundoora East campus, as shown in Fig. 7. It consists of 5 outrigger brackets mounted at 1.8 m, off star pickets driven into the ground 600 mm, and two sub brackets per outrigger bracket. For the initial testing cable was not tensioned to the maximum. The only tension in the cables was resultant of the weight of the cable car itself. Test results, which arise from testing on the current test rig, are only indicative of real system capabilities, with final performance and stability expected to be much better.

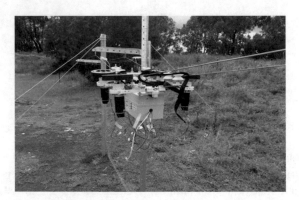

Fig. 7. Cable car on test rig

Our system was tested comprehensively on short straights and partially on long straights and passing sub brackets. The results of initial testing are encouraging. The device handles straight sections well, even with loose cables. It is tolerant to minor problems, such as the idle wheels becoming unseated from the cable as shown in Fig. 8. Idle wheels automatically centre themselves on the cables after becoming dislodged. As in any other mechanical system it is in equilibrium when the vector sum of the forces is equal to zero, as given by Eq. 2.

$$\sum \vec{F} = 0, \quad F_v = F_w = m_g, \quad F_h = F_s = K_s x \quad (2)$$

where F_c represents cable force in the interaction with the wheel, F_h and F_v are horizontal and vertical components of the same force, F_w is the weight of the systems and F_s is the Hooke's force.

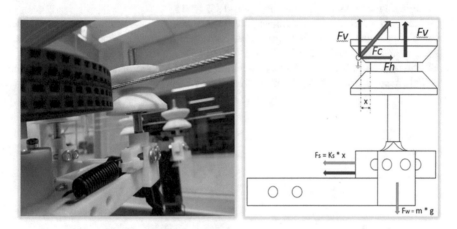

Fig. 8. (a) Unseated idle wheel, (b) Force diagram of unseated wheel

5   Conclusion

Autonomous cable car designed for predator proof fence monitoring project is presented. Prototype vehicle is designed and tested at the University facilities. Initial investigation proves the concept of mechanical and drive train design as one of the optimal solutions for the outdoor harsh environments. While UGV and UAV approaches suffer from various obstructions our novel design overcomes all of those constraints. While this paper presents AV platform's mechanical, hardware and software subsystems, following publications will show applications. Those include fence monitoring DAQ, problems detected, reporting and data storage.

References

1. Doelle, S.: Evaluation of predator-proof fenced biodiversity projects. In: Australian Agricultural and Resource Economics Society 56th Conference, Fremantle (2012)
2. Elbanhawi, M., Simic, M., Jazar, R.: The role of path continuity in lateral vehicle control. Procedia Comput. Sci. **60**, 1289–1298 (2015)
3. Elbanhawai, M., Simic, M., Jazar, R.N.: Continuous path smoothing for car-like robots using B-Spline curves. J. Intell. Robot. Syst. **77**(242), 23–56 (2015)

4. Elbanhawai, M., Simic, M.: Sampling-based robot motion planning: a review. IEEE Access **PP**(99) (2014)
5. Krerngkamjornkit, R., Simic, M.: Human body detection in search and rescue operation conducted by unmanned aerial vehicles. Int. J. Adv. Mater. Res. **655-657**, 1077–1085 (2013)
6. Krerngkamjornkit, R., Simic, M.: Multi object detection and tracking from video file. In: The 2014 International Forum on Materials Processing Technology (IFMPT 2014), Guangzhou, China (2014)
7. Simic, M., et al.: UAV recharging using non-contact wireless power transfer
8. Simic, M., Bil, C., Vojisavljevic, V.: Investigation in wireless power transmission for UAV Charging. Procedia Comput. Sci. **60**, 1846–1855 (2015)
9. Meiser, R.H.V., Šeatović, D., Rotach, T., Hesselbarth, H.: Autonomous unmanned ground vehicle as sensor carrier for agricultural survey tasks. In: International Conference of Agricultural Engineering, Zurich (2014)
10. Elbanhawai, M., Simic, M.: Randomised kinodynamic motion planning for an autonomous vehicle in semi-structured agricultural areas. Biosyst. Eng. **126**, 30–44 (2014)

Sentiment Analysis Method for Tracking Touristics Reviews in Social Media Network

Yasmine Chaabani[1,2]([✉]), Radhia Toujani[1], and Jalel Akaichi[3]

[1] Department of Computer Science, High Institute of Management,
University of Tunis, Tunis, Tunisia
chaabanijasmin@gmail.com
[2] Department of Computer Sciences, College of Adham,
Umm al-Qura University, Mecca, Kingdom of Saudi Arabia
[3] Department of Information Systems, College of Computer Science,
King Khaled University Abha, Abha, Kingdom of Saudi Arabia

Abstract. The touristic sector in Tunisia has declined after the "Arabic Spring". Therefore, the number of comments published by tourists to give their opinions about it has increased. Consequently, this resulted in a high volume of data in the different social networks such as Facebook and Twitter. In this case, the opinion mining plays an important role to more understanding and then ameliorating the situation of tourism in Tunisia. In this paper, the main goal is to select the tourists' viewpoints in Twitter after the revolution. For this reason, we create a sentiment lexicon based on the emoticons and interjections as well as acronyms. We also use a sentiWordnet to build lexical scales for sentiment analysis of different tourist reviews with reference to a travel agency page on Facebook. Then, we propose a method relying on Support Vector Machine (SVM), Maximum entropy and Naive Bayes. Our approach is efficient as it gives encouraging results.

Keywords: Sentiment analysis · Reviews · Medias networks · NP-Complete · Text mining · Machine learning

1 Introduction

In the recent years, the opinion mining has conducted a research to classify the content of reviews in a two-point scale: positive and negative. In an early sentiment analysis approach, such as the topic-based document classification method, a word extracted from the document is used as a feature. Then, in more recent studies, a sentiment lexicon which was constructed from these words is also used as a feature to improve the accuracy of the classification and the vocabulary. In fact, the sentiment lexicon represents an index of emotional words. Therefore, it has the polarity information of the relevant word irrespective of whether it gives a positive sentiment or a negative one.

Since the Tunisian revolution, the number of tourists has decreased. Accordingly, social networks such as Twitter and Facebook became the major source

© Springer International Publishing AG 2018
G. De Pietro et al. (eds.), *Intelligent Interactive Multimedia Systems and Services 2017*,
Smart Innovation, Systems and Technologies 76, DOI 10.1007/978-3-319-59480-4_30

of collecting information and the efficient method of communication between tourists. In this way, social media users may communicate their viewpoints, share their feelings, thoughts and also inform their friends and neighbors about their cities conditions.

In this paper, we explore the application of the text mining techniques for sentiment classification. The latter was performed on reviews through Twitter in order to analyze the Tunisians' behavior after the revolution (from the 10th till the 17nd of July 2016 (One Week)). For this purpose, we chose a random population reviews published in Twitter website about hotels cities conditions in Tunisia. In this study, we analysed the viewpoints of all the different kinds of twitters as males, females, students and so on.

Similarly, we applied a new approach to analyze the tourist's reviews about the different places in Tunisia. Consequently, we created our own dataset using the sentiWordnet to build lexical scales for sentiment analysis. Then, we applied on it three machine learning algorithms: Naive Bayes, Maximum entropy and SVM. In fact, the expected output was to classify hotels into two useful categories, not only for Tunisians who want to know information about hotels in their country, but also for tourists in order to allow them to choose the best hotel and the safest place.

This paper is organized as follows. The state of the art and the background of social networks text mining are described in Sect. 2. In Sect. 3, we discuss the proposed methodology. In Sect. 4, the process of our experiments is described and the results are discussed and evaluated. Finally, in the last section, we present some concluding remarks and propose future works.

1.1   Previous Work

The sentiment can refer to the opinions or emotions which are considered as related concepts. When the sentiment analysis is based on opinions, a distinction is made, for example, between positive and negative ones. Nevertheless, when the sentiment analysis is based on emotions, the distinction is between different kinds of emotions like sadness, fear, joy, love, etc. Previous works indicate that sentiment analysis method are based on lexicons based approaches or machine learning based approaches.

Sentiment Analysis Using Machine Learning Methods. Earlier studies in sentiment analysis domain focused on assigning sentiments to documents [19] using machine learning techniques usually used to train a sentiment classifier and consider frequent terms in the documents. Other studies focused on more specific tasks to find sentiments or semantic orientation of words [12,21]. However, machine learning used the most exploited technique to tackle these relevant problems. The pioneer work on sentiment classification by Pang and Lee (2002) applied naive bayes (NB), maximum entropy (ME) and support vector machines (SVM). Obviously, SVM showed the best performance, while NB was the least precise. Taking into consideration the movies reviews collection from Internet

Movie Database, 82.9 % accuracy was achieved [19]. The work by Dave et al. (2003) used Laplacian smoothing for NB, which enhanced its accuracy to 87% for a product reviews dataset [7]. To improve the precision of NB, Pang and Lee (2004) carried out subjectivity identification as a pre-processing step to sentiment analysis [18]. Boiy et al. (2007) performed experiments using SVM, naive Bayes multinomial and maximum entropy on movie and car brands review. The best accuracy of the study reached 90.25% [4]. Similar reviews of movies dataset used by Annett and Kondrak (2008). Different approaches like SVM, NB, alternating decision tree and lexical (WordNet) based method were used for sentiment analysis and more than 75% accuracy was achieved [2]. Tan and Zhang (2008) compChi Squared Chi Squared Chi Squared are examples of feature selection (Mutual Information, Information Gain and Document Frequency) and learning methods (centroid classifier, K-nearest neighbor, window classifier, Naive Bayes and SVM) applied to extract opinion from Chinese documents. Information Gain was the best technique for sentimental terms extraction, while SVM revealed the best performance for sentiment classification [2]. In a study made by Dasgupta and Ng (2009), a weakly-supervised sentiment classification algorithm was proposed. User feedbacks were provided on the spectral clustering process in an interactive mode to ensure that text is clustered along the sentiment dimension [20]. A survey on travel blogs was done by Ye et al. (2009) based on NB, SVM and the character-based N-gram model. SVM provided the best accuracy in the study [6] with 85.14%. Support Vector Machine was also exploited for sentiment analysis by Paltoglou and Thelwall (2010)) to suggest a combined approach that gave 96.90% accuracy on movie reviews dataset [23]. Xue Bai (2011) experimented with a heuristic search-enhanced Markov blanket model and applied SVM for mining consumer sentiments from online text [17]. Kang et al. (2011) improved Naive Bayes applied to use a restaurant reviews. They obtained 83.6% accuracy on more than 6000 documents [3].

Sentiment Analysis Using Dictionary-Based Approaches. The lexicon of a language forms its vocabulary. The first version and the most well known one is WordNet [10] which is a semantic lexicon where words are grouped into sets of synonyms called Synsets. Another famous example of lexicon is SentiWordNet [9]. This one is a sentiment lexicon that represents an index of sentiment words, and it has the polarity information of the relevant word irrespective of whether it carries a positive sentiment or a negative one. The SentiWordNet is considered as an extension of WordNet. It assigns WordNet Synsets a graded measure with respect to two scales: A positive/negative scale and a objective scale. It is important to notice that the classification is based on Synsets, not on words, because a word can have different meanings [1]. The core application areas of sentiment analysis and opinion mining are Finance [8], reviews [7,19], politics [14], and news [21]. They have a large number of research in the field of sentiment analysis, and in particular sentiment classification. Most of them have focused on classifying large texts, like reviews [11]. Most researchers did not consider these two techniques simultaneously. In this study, we propose lexions method to extract the polarity of each sentence in a document followed by machine learning techniques namely naive bayes and support vector machine.

2   Methodology

In this section, we propose a general architecture (presented in Fig. 1) designed through the 3 following steps.

2.1   Raw Data Collection

This step consists in extracting reviews update, approximately 100 reviews about different hotels in Tunisia posted by Tourists after the revolution.

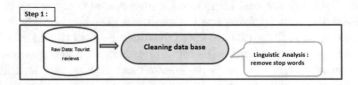

(a) Raw data collection

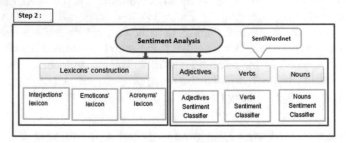

(b) Sentiment Analysis

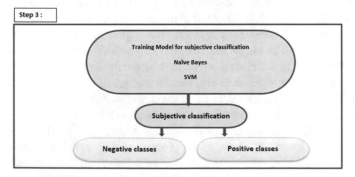

(c) Training Model

Fig. 1. The architecture of our system

Cleaning Data base: removing stop words and non-frequent words.

Document Frequency measures the number of documents in which the feature appears in a dataset. This method removes those features where document frequency is more or less than a predefined threshold. Selecting frequent features will improve the likelihood that the features will be compared with prospective future test cases. The basic assumption [22] is that both rare and common features are either non-informative for sentiment category prediction, or non-impactful to improve classification accuracy. Literature research in [20] shows that this method is the simplest and the most effective for text classification.

2.2 Sentiment Analysis

We define sentiment as a personal positive or negative feeling on the informal language of on-line social networks.

Lexicon Language. For this reason, three types of lexicon were created: lexicon for social acronyms, lexicon for emoticons and lexicon for interjections. The acronyms lexicon (see Table 1) denotes the most used acronyms by Tourists in their reviews about different hotels on the site.

Table 1. Acronyms' lexicon

Acronym	Sentiment
MDR	positive
CB1	positive
DSL	negative
G la N (J'ai la haine)	negative
CU	neural

In Table 2, we collected a lexicon of the most used emoticons on twitter and annotated them manually whether they express a positive or a negative sentiment.

The interjections' lexicon (Table 3) contains the main interjections found particulary in our data set, and generally in tourists reviews site.

Word Sentiment Classifier. For word sentiment classification, we suggested a model. The main idea is to gather a small number of seed words by hand, sorted by polarity into two lists positive and negative. Then, they were developed through adding words obtained from SentiWordNet. We assume that synonyms of positive words are mostly positive, while antonyms are mostly negative, e.g., the positive word 'bon' has synonyms, vertueuse, honorable, juste' and antonyms 'mal, injustes'. Antonyms of negative words are added to the positive

Table 2. The emoticons lexicon

Emoticon	Sentiment
:), :-)	positive
:(, :-(	negative
:/	negative
:'(	negative
;)	positive
:>	positive
:<	negative
:*,:-*	positive
<3	positive
:-x	positive

Table 3. The interjections lexicon

Interjection	Sentiment
Wow, waw	positive
Haha, hihi, hehe, hhh	positive
Oh mon dieu	negative
merci	positive
oui	positive
bien	positive
non	negative
aide	negative
impossible	negative
ay	positive

list, and synonyms are added to the negative ones. The seed lists we select verbs (23 positive and 21 negative) and adjectives (15 positive and 19 negative). Then, we will add nouns later. Since adjectives and verbs are structured differently in SentiWordNet, we obtained synonyms and antonyms for adjectives. But, we got only synonyms for verbs. For each seed word, we extracted, from SentiWord-Net its expansions. Afterwards, we added them back into the appropriate seed lists. Using these expanded lists, we extract an additional cycle of words from SentiWordNet. However, not all synonyms and antonyms could be used because some words have opposite sentiment or are neutral. In addition, some common words such as 'grand', 'fort', 'prendre' and 'obtenir' occurred many times in both positive and negative lists.

2.3   A Training Model and Machine Learning Classification

In this work, we did not consider neutral reviews in our training or testing data. In fact, we used only positive or negative reviews. Another limitation of our work consists in the fact that some reviews express their viewpoints whether they are for or against a specific issue.

Training Model. We started by creating a training dataset model for Supervised learning algorithms; usually requiring handlabeled training data with a specific class (positive or negative).

The training data was post-processed with the following filters:

- Reviews containing both positive and negative emoticons are removed. This removal may happen if a review contains two opinions.
- Rereviews are removed. Indeed, rereviewing is the process of copying another user's review and posting it to another account which usually happens if a user likes another user's review.
- Reviews with ':p' are removed. Because ':p' usually does not imply a negative sentiment.
- Repeated reviews are removed. Occasionally, reviewers return duplicate review. Similar reviews, duplicates are removed to avoid putting extra weight on any particular review.

After post-processing the data, we took the first 50 reviews with positive opinions, and 50 reviews with negative opinions, for a total of 100 reviews.

The test data was manually collected, using the web application. A set of 67 negative reviews and 33 positive reviews were manually marked constructing a rich and varied dataset. which. In our work, we labeled our data with POS if the text expresses a positive sentiment and with NEG if the text conveys negative feeling. Then, we split it into a training set (60% of our preprocessed data), and a testing set (40% of our preprocessed data). The mission consists in determining the training model able to assign the correct class to a new text review. To evaluate the performance of the training model, we classified the testing set with the training model and compare the classified result (predicted labels by the model) with the true labels.

Machine Learning Classification. We tested different classifiers: Naive Bayes, Maximum Entropy, and Support Vector Machines.

Naive Bayes. Naive Bayes is a simple model efficiently applied for text categorization [15]. We use a multinomial Naive Bayes model. Class c^* is assigned to review d, where

$$c^* = argmax_c P_{NB}(c/d)$$

$$P_{NB}(c/d) = \frac{P(c) \sum_{i=1}^{m} P(f/c)^{n_i(d)}}{P(d)} \tag{1}$$

In this formula, f represents a feature and $n_i(d)$ denotes the count of feature f_i found in review d. There are a total of m features. Parameters P(c) and P(f/c) were obtained through maximum likelihood estimates, and add-1 smoothing was used for unseen features.

Maximum Entropy. The idea behind Maximum Entropy models is that one should apply the best uniform models that satisfy a given constraint [16]. MaxEnt models are feature-baseds. Unlike Naive Bayes, MaxEnt makes no independence assumptions for its features. This means we can add features like bigrams and phrases to MaxEnt without worrying about features overlapping. The model is represented by the following equation:

$$P_{ME}(c/d, \lambda) = \frac{exp[\sum_i \lambda_i f_i(c, d)]}{\sum_i exp[\sum_i \lambda_i f_i(c, d)]} \qquad (2)$$

In this formula, c is the class, d represents the review, and corresponds to a weight vector. The weight vectors decide the significance of a feature in the classification process. A higher weight means that the feature is a strong indicator for the class. The weight vector is found by numerical optimization to maximize the conditional probability. We used the Stanford Classifier[1] to perform MaxEnt classification. Theoretically, MaxEnt performed better than Naive Bayes because it handles feature overlap better. However, in practice, Naive Bayes is more efficient [16].

Support Vector Machine. Support Vector Machine is another popular classification technique [5]. In our approach, we used the SVM^{light} [13] software with a linear kernel. We used, as input data, two sets of vectors having size m. Each entry in the vector corresponds to the feature. For example, with a unigram feature extractor, each feature is a single word found in a review. If the feature is present, the value is 1. Otherwise, it is 0.

3   Experiments and Discussion

To evaluate the performance of the performed sentiment classification, we applied the standard classification performance metrics used in previous information retrieval and text classification studies: accuracy and precision.

Accuracy. Accuracy is the most intuitive measure. It represents the overall correctness of the model. It is calculated by dividing the sum of correct classifications, as shown in the following equation:

$$Accuracy = \frac{a + d}{a + b + c + d} \qquad (3)$$

---

[1] The Stanford Classifier can be downloaded from http://nlp.stanford.edu/software/classifier.shtml.

where:

a is the number of terms correctly assigned to this class.

b is the number of terms incorrectly assigned to this class.

c is the number of terms incorrectly rejected to this class.

d is the number of terms correctly rejected to this class.

Precision. Precision measures the exactness of a classifier by calculating, the accuracy of a specificted class:

$$Precision = \frac{a}{a+b} \tag{4}$$

We applied our proposed model on data set collected by crawling one week of public tweets through using the 140dev Twitter framework[2]. Our collection consists of tweets from the 10th till the 17nd of July 2016 (One Week). We explored the usage of unigrams, bigrams and unigrams and bigrams.

Unigrams. The unigram feature extractor is the simplest way to retrieve features from a review. The machine learning algorithms perform better than our keyword baseline. Table 3 reveals that our results are very similar to those provided by Pang and Lee method [19] (Table 4).

Table 4. Comparative study of our methods and Pang and Lee approach of NB, ME and SVM.

Method	NB	ME	SVM
Pang and Lee method	78%	80.9%	79.4%
Our method	80.3%	81.4%	78.5%

Bigrams. We applied bigrams to help reviews express clearly their viewpoints using phrases which contain "bien" or "mal". In our experiments, negation, as an explicit feature with unigrams does not improve accuracy. Although bigrams were more efficient. They were very sparse and the overall accuracy dropped in the case of both Max-Ent and SVM. Even collapsing the individual words to equivalence classes did not help. MaxEnt gave equal probabilities to the positive and negative class for this case, while there was no bigram that tipped the polarity in both directions.

In general using only bigrams as features is not efficient because the feature space is very sparse. Therefore, it is better to combine unigrams and bigrams as features.

Unigrams and Bigrams. Both unigrams and bigrams are used as features. Table 5 presents a comparative study of Unigrams and Bigrams to unigram features. Indeed, accuracy improved for Naive Bayes and Max-Ent. While, our

---

[2] http://140dev.com/free-twitter-api-source-code-library/.

Table 5. Comparative study of using Unigrams and Bigrams or using Unigram only of accuracy values

	NB	ME	SVM
Unigram	80.3%	81.4%	78.5%
Unigrams and Bigrams	81.5%	79.4%	77.2%

Table 6. Comparative study of using Unigrams and Bigrams or using Unigram of precision values

	NB	ME	SVM
Unigram	73.3%	76.2%	81.5%
Unigrams and Bigrams	81.2%	82.4%	83.4%

result show a decline for accuracy value of SVM. For Pang and Lee, there was a decrease in the performance of Naive Bayes and SVM, with an improvement for MaxEnt. Moreover, in Table 6, the Unigrams and Bigrams present the important values of precision for the differents used classifiers: Naive Bayes **NB**, Maximum Entropy **ME**, and Support Vector Mchines **SVM**.

4   Conclusion and Future Works

In this paper, we showed that using emoticons as noisy labels and sentiWordnet to build lexical scales for sentiment classifier to analyze the reviews of tourists is an efficient way to perform a distant supervised learning. Besides, machine learning algorithms (Naive Bayes, Maximum Entropy classification, and Support Vector Machines) achieve high accuracy for classifying sentiment when using this method. The reviews depend on the users' nationality and the period during which they posted them. Either during the Tunisian revolution or after it, we aim to discover the various points of views of tourists about the safety conditions. Finally, we would like to promote our research by tracking changes within people's sentiment on a particular topic, explore the time dependency of our data and analyze their trendy topics dynamically. It will be deeply interesting to involve the temporal feature on such analysis and not to focus solely on the previous posts or discussions.

References

1. Esuli. A., Sebastiani, F: Sentiwordnet: Apublicy available lexical resource for opinion mining. In: Proceedings of the 5th International Conference on Language Resources and Evaluation, Genoa, Italy (2006)
2. Annett, M., Kondrak, G.: A comparison of sentiment analysis techniques: Polarizing movie blogs. In: Advances in Artificial Intelligence, pp. 25–35. Springer (2008)

3. Bai, X.: Predicting consumer sentiments from online text. Decis. Support Syst. **50**(4), 732–742 (2011)
4. Boiy, E., Hens, P., Deschacht, K., Moens, M.: Automatic sentiment analysis in on-line text. In: ELPUB, pp. 349–360 (2007)
5. Cristianini, N., Shawe-Taylor, J.: An Introduction to Support Vector Machines and Other Kernel-based Learning Methods. Cambridge University Press, Cambridge (2000)
6. Dasgupta, S., Ng, V: Topic-wise, sentiment-wise, or otherwise?: Identifying the hidden dimension for unsupervised text classification. In: Proceedings of the 2009 Conference on Empirical Methods in Natural Language Processing, vol. 2, pp. 580–589. Association for Computational Linguistics (2009)
7. Dave, K., Lawrence, S., Pennock, D.M.: Mining the peanut gallery: Opinion extraction and semantic classification of product reviews. In: Proceedings of the 12th International Conference on World Wide Web, pp. 519–528. ACM, New York (2003)
8. Devitt, A., Ahmad, K.: Sentiment polarity identification in financial news: A cohesion-based approach. In: Proceeding of the 45th Annual Meeting Association Computational Linguistics. ACL Press, Prague, Czech Republic (2007)
9. Sebastiani, F., Esuli, A.: Determining term subjectivity and term orientation for opinion mining (2006)
10. Fellbaum, C.: Wordnet: An Electronic Lexical Database. MIT Press, Cambridge (1998)
11. Srinivasaiah, M., Skiena, S., Godbole, N.: Large-scale sentiment analysis for news and blogs. In: International Conference on weblogs and Social Media (ICWSM 2007), Boulder, Colorado, USA (2007)
12. Hatzivassiloglou, V., McKeown, K.R.: Predicting the semantic orientation of adjectives. In: Proceedings of the 35th Annual Meeting of the Association for Computational Linguistics and Eighth Conference of the European Chapter of the Association for Computational Linguistics, pp. 174–181. Association for Computational Linguistics (1997)
13. Joachims, T.: Making large scale SVM learning practical. Technical report, Universität Dortmund (1999)
14. Kim, S.M., Hovy, E.: Identifying and analyzing judgment opinions. In: Proceedings of the Joint Human Language Technology/North American Chapter of the ACL Conference (HLT-NAACL). ACL, Stroudsburg (2006)
15. Manning, C.D., Schütze, H.: Foundations of Statistical Natural Language Processing. MIT Press, Cambridge (1999)
16. Nigam, K., Lafferty, J., McCallum, A.: Using maximum entropy for text classification. In: IJCAI-99 Workshop on Machine Learning for Information Filtering, vol. 1, pp. 61–67 (1999)
17. Paltoglou, G., Thelwall, M.: A study of information retrieval weighting schemes for sentiment analysis. In: Proceedings of the 48th Annual Meeting of the Association for Computational Linguistics, pp. 1386–1395. Association for Computational Linguistics (2010)
18. Pang, B., Lee, L.: A sentimental education: Sentiment analysis using subjectivity summarization based on minimum cuts. In: Proceedings of the 42nd Annual Meeting on Association for Computational Linguistics, p. 271. Association for Computational Linguistics (2004)

19. Pang, B., Lee, L., Vaithyanathan, S.: Thumbs up? Sentiment classification using machine learning techniques. In: Proceedings of the ACL-02 Conference on Empirical Methods in Natural Language Processing, vol. 10, pp. 79–86. Association for Computational Linguistics, Philadelphia (2002)
20. Tan, S., Zhang, J.: An empirical study of sentiment analysis for chinese documents. Expert Syst. Appl. **34**(4), 2622–2629 (2008)
21. Turney, P.D.: Thumbs up or thumbs down? Semantic orientation applied to unsupervised classification of reviews. In: Proceedings of the 40th Annual Meeting on Association for Computational Linguistics, pp. 417–424. Association for Computational Linguistics, Philadelphia, Pennsylvania (2002)
22. Yang, Y., Pedersen, J.O.: A comparative study on feature selection in text categorization. In: ICML, vol. 97, 412–420 (1997)
23. Ye, Q., Zhang, R., Law, Z.: Sentiment classification of online reviews to travel destinations by supervised machine learning approaches. Expert Syst. Appl. **36**(3), 6527–6535 (2009)

Mobility Based Machine Learning Modeling for Event Mining in Social Networks

Radhia Toujani[1]([✉]), Zeineb Dhouioui[1], and Jalel Akaichi[2]

[1] BESTMOD Department, Higher Institute of Management,
University of Tunis, Tunis, Tunisia
toujaniradia@gmail.com, dhouioui.zeineb@hotmail.fr
[2] Department of Information Systems, College of Computer Science, King Khaled
University Abha, Abha, Kingdom of Saudi Arabia
jalel.akaichi@kku.edu.sa

Abstract. Social networks sounds to be a rich source to discover events mobility and analyzing their trends. Hence, the mobility of events refers to the movement of users' opinions, location, velocity and the continuous change over time. Despite the ability of existing methods to deal with the event mobility and evolution. To the best of our knowledge, there is no research able to show the relation between mobility and social interactions. In this work, we associate mobility into event mining issue. We also describe the movement of opinions in social network and we aim at extracting useful information from tweeter posts, especially during the economic and political event "TUNISIA 2020". To achieve this task, we focused on the use of machine learning techniques to analyze tunisian tweeter posts and classify their opinions temporally about this event for each Tunisian region. We introduced decision tree method to model and analyze event mobility and to predict the change of opinions from its spatial and temporal co-occurrence. Therefore, an entropy measure has been proposed based on spatio-temporal attributes as branching attributes. Finally, in order to validate our solution, we used real data and we performed some comparative experiments to show the effectiveness of our method.

Keywords: Social networks · Event detection · Opinion change · Temporal mobility · Spatial mobility · Machine learning · Decision tree · Spatio-temporal attributes · Entropy

1 Introduction

In recent years, social networks have become omnipresent thanks to the increasing propagation of internet-enabled devices, such as personal computers, smart phones, tablets and many other devices that allow users to connect to social networks through the internet services [1]. These new services permit people, all over the world and at any time, to add, update, share and consult massive quantities of new information in real time. These huge amounts of information,

© Springer International Publishing AG 2018
G. De Pietro et al. (eds.), *Intelligent Interactive Multimedia Systems and Services 2017*,
Smart Innovation, Systems and Technologies 76, DOI 10.1007/978-3-319-59480-4_31

added by hundreds of millions of active users [2], are considered as very important source of data for many research fields. Although, while some research groups successfully used information technologies to seek out and understand "what is going on" in the world, they still unable to fully support opinions movement of social network users with a comprehensive spatial viewpoint to satisfy their information needs [3].

In this work, we take Twitter as a representative social networking media to follow the impact of political event on the equitable distribution of the introduced investment plans for each Tunisian region. In fact, we choose this social network since it allows to obtain location information attached to tweets. This is beneficial to enable people to apply this data to produce more relevant location based on real-time results. The goal of our approach is to effectively mining events by geo-location data provided by social network services. The previous mentioned problematic led us to propose an approach based on the use of real time extracted data from social networks to be exploited for the establishment of a system able to recognize, detect and extract different message stream describing the evolution and the effectiveness of "TUNISIA 2020" event. This latter gathers economic partners and investors with political leaders, foreign heads of state and government, as well as numerous international financial institutions and private business dignitaries.

To do so, we suggest a method to model mobility of event by combining machine learning and spatio-temporal information. In spite the existence of many supervised learning algorithms, we choose decision trees since they are considered as an efficient data mining methodology and a powerful solution applied in machine learning and can successfully solve classification problem [4]. The choice of these components namely the testing tree type, the attribute selection measure, the partitioning strategy and the stopping criteria makes the major difference between decision tree algorithms. In this paper, the proposed decision tree is explored to predict the change and the mobility of classification process.

This paper is organized as follows: in the second section, we present an overview of real time event detection techniques. Then, we describe in Sect. 3, our proposed approach for mobility analysis of event mining. Finally, in Sect. 4, we conclude our work and we propose the possible directions for future research.

2   Related Work

Many works could be found for data-driven event extraction approaches, such as [5–8]. The use of data-driven approaches for event extraction is beneficial in the sense that there is no need to expert knowledge or linguistic resources. However, data-driven approaches require large text corpora in order to develop models that approximate linguistic phenomena. Another drawback is that data-driven methods do not deal with text semantic. To overcome this limitation, researchers resorted to knowledge- driven approaches [9] based on patterns that express rules representing expert knowledge. Therefore, various research work in the recent literature relies on hybrid approaches combining data-driven and knowledge-driven approaches [10]. Takeshi et al. [11] elaborated earthquake event detection system relying on the real-time nature of Twitter. They proposed an algorithm to monitor tweets and to detect target events. To detect a target event, they devised a classifier of tweets based on features such as the keywords in a tweet, the number of words, and their context. Subsequently, they produced a probabilistic spatio-temporal model for the target event that can find the center and the trajectory of the event location. Thus, authors considered each Twitter user as a sensor and applied Kalman and particle filtering techniques which are widely used for location estimation in ubiquitous and pervasive computing. The proposed approach gave acceptable results, but it has many drawbacks, like the needed response time due to the rapid spread nature hazard that require a very quick intervention.

To sum up, there are three different methods used to obtain real time information from twitter namely latitude and longitude, Time zone and Content. In [3,12], authors proposed a geo-social event detector method based on latitude and longitude. This method aims at annotating users with geographic coordinates. When the geographic information is acquired from user profile, time zone technique is applied. In [9] the authors used machine learning to identify user location from their posts. They showed that most users enter valid geographic information in the specific location field. Additionally, geographic information can be extracted from contextual posts. Moreover, Cheng's work [13] focused on automatic extraction of location keys word from messages contents. Therefore, from the state of the art, we can conclude that real time event detection methods have some weaknesses. For instance, twitter users have no time zone information in their profile and tweets have no latitude and longitude through tweeter stream API and location ambiguity extracted from contextual posts. Thus, to guarantee the relevance of extracted contextual location, we should focus on the extraction of location from temporally contextual posts and mine the mobility of events using machine learning technique.

3   Decision Tree Algorithm for Mobility of Event Mining (DTAMEM)

In this section, we define and formalize the introduced $DTAMEM$ method. As shown in Fig. 1, the proposed system is based on three fundamental processes. Raw data collection, processing pipeline for temporal event extraction and mobility of event mining process.

3.1   Raw Data Collection

Recently Twitter has released the location service that enables users to publish their tweets with geospatial data as latitude and longitude. Such service encourages research works to focus more on mining the spatio-temporal information on microblogs to get real-time and geo-spatial event information.

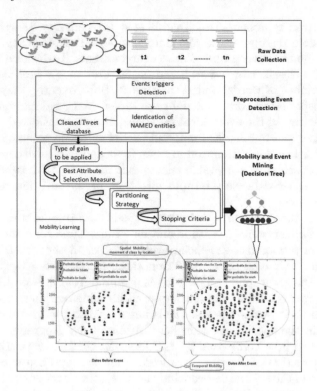

Fig. 1. System targeted architecture

We used the 140dev Twitter framework [14] to extract contextual posts. The input data is a set of temporal twitter post. In fact, a textual post (T_p) is extracted from Twitter acquired at a defined date t. In our case, we deal with extracting events related to 'Tunisia2020' event held on 29–30 November 2016 in Tunis. In addition, tweets about this political event are widely available and provide a source of valuable information that may be useful in the classification process.

Therefore, our approach takes this temporal extracted tweet as input.

3.2 Preprocessing Phase

The language of the given text plays a primordial role in the preprocessing task and the classification one. Therefore, focusing only on one language is necessary to allow data efficient preprocessing, extracting its features and creating the classification model. Subsequently, a translation to English was applied to all the extracted tweets. We used thereafter, a lemmatizer to annotate our un-structured textual data. We apply for our environment the lemmatizer TreeTagger developed by Helmut [15] within the TC project. Based on the annotation result generated by TreeTagger, we browsed text content. We also identified named entities of all text objects, such as DATE, PERSON, LOCATION,...

In fact, in addition to the extracted opinion attribute, these extracted named entities, especially those related to LOCATION and DATE, serve as attributes of each sample. The output of this stage is a cleaned database injected as input to the introduced classification process namely the introduced decision tree method.

3.3   Mobility and Event Mining Process

We aim in this step improving the mobility of opinions through political event. To achieve this purpose, we proposed a widely-used machine learning method namely decision tree. Indeed, the introduced entropy of the proposed method is mainly based on the investigator of three attributes, location, date and extracted opinion. In fact, we focused on the temporal evolution of the opinionated posts for a specific location at a defined date. In our work, we classified opinion users' through the proposed investment plan related to every region during the political event 'TUNISIA2020'. Thus, predicted classes are *profitable* or *unprofitable* plan.

Hence, the proposed classification model is based on both temporal and spatial analysis and it described the movement of event mining at a defined time. The developed decision tree method is with the same representation of a standard decision tree. However, on each tree node, we integrated the type of gain to be applied in order to choose the selection attribute, more especially spatio-temporal attributes.

Information Gain and Attribute Selection. The information gain is very efficient as it reflects the quality of the branching attribute. The attribute-selection part is related to the complexity of the decision tree which is strongly related to the amount of information expressed by the value of a given attribute. In fact, we formalize the idea behind the introduced entropy measure used to calculate the information gain as follows:

We present the input data set in this form:

$$D = (T, A \cup C) \tag{1}$$

where:

- $T = t_1, t_2, ..., t_n$ denotes the set of data samples (temporal extracted tweet),
- $A = a_1, a_2, ..., a_m$ represents the set of condition attributes. In our work, these attributes present spatial information, temporal information and extracted opinions.
- $C = c_1, c, ..., c_j$ is the class label attribute which has q distinct values defining j distinct classes

Table 1 depicts example of sample.

In fact, location and date attributes are extracted during the process of the identification of named entities. Extracted opinions attribute is defined referring to our sentiment analysis method presented in [16]. In addition, the class label shows whether the development plan by Tunisian region is profitable or not. In this case, the calculation of the entropy is as follows:

Table 1. Example of sample data set

Tweet	Location(Spatial attribute)	Date(Temporal attribute)	Extracted opinions	Class
1	North	Before Event	Negative	profitableN
2	South	After Event	Negative	unprofitableS
3	Middle	After Event	Positive	profitableM
4	North	After Event	Positive	profitableN
5	North	After Event	Positive	profitableN
6	North	Before Event	Positive	profitableN
7	South	After Event	Fuzzy	unprofitableS
8	North	Before Event	Negative	profitableN
9	Middle	Before Event	Positive	profitableM

Let s_j be the number of samples of T in class c_i. The entropy needed to classify a given sample is given by

$$entropy(T) = -\sum_{i=1}^{k} p_i \log_2 p_i \tag{2}$$

where p_i is the probability, that an arbitrary sample belongs to label class, is estimated by computing those samples' entropy (k is the number of all samples).

Besides, each attribute a_i has v distinct value $P_1, P_2, ..., P_v$ and can be used to partition T into v subsets $S_1, S_2, ..., S_v$.

Let $s_{i,j}$ be the number of samples of class c_i in a subset S_j, the entropy of attribute a_i is given by

$$E(a_i) = \sum_{j=1}^{v} \sum \frac{s_{1,j} + ... + s_{m,j}}{S} I(s_{1,j}, ..., s_{m,j}) \tag{3}$$

The attribute with the highest information gain, is selected as the branching attribute. For a given subset S_j, we computed the information gain as follows:

$$I(s_{1,j}, ..., s_{m,j}) = -\sum_{i=1}^{m} p_{i,j} \log_2 p_{i,j} \tag{4}$$

where $p_{i,j} = \frac{s_{i,j}}{S_j}$ is the probability that a sample in $_j$ belongs to class c_i. Therefore, information gain of attribute a_i is given by

$$Gain(a_i) = I(s_{1,j}, ..., s_{m,j}) - E(a_i) \tag{5}$$

We computed the information gain of each condition attribute. The attribute with the highest information gain is the most informative and the most discriminating attribute of the given set.

Partitioning Strategy. In classical decision tree, the partitioning strategy consists in partitioning the training set according to all possible attribute values generates one partition per attribute value. However, during the construction of the introduced decision tree, we proposed alternative to partition the training base. This later alternative consists in focusing on spatial and temporal attributes (A_s, A_t with P_v possible value) to partition the training base and construct the node N.

Hence, the training set T composed of e_i sample is partitioned into P_v subset T_v^N such as:

$$\forall e_i \in T^N, if e_i(A_s, A_t) = v_{lk}, then e_i \in T_k^N, 1 < k < P_v \tag{6}$$

where v_{lk} is a value among the set of possible A_s, A_t values.
and

$$T^N = \bigcup_{k=1,\ldots,P_v} T_k^N \tag{7}$$

and

$$\forall x, k = 1, \ldots, P_v, x \neq k, T_x^N \bigcap T_k^N \tag{8}$$

Stopping Criteria. They determine the conditions of stopping the partitioning process. The stopping criterion is generally used to decide whether it is necessary for a training set to continue developing the tree. These criteria can be related to a low number of the example in the considered set or if all examples of the set have the same class, or at least a sufficient number relative to the examples number of the set. The choice of these components (the attribute selection measure, the partitioning strategy, and the stopping criteria) makes the major difference between decision tree algorithms.

The algorithm used in our approach for the mobility of mining event is the C4.5 algorithm [17]. This algorithm has proven its effectiveness and it provides a predictive model represented as a decision tree easily understandable and interpretable.

Consequently, the movement of predicted class by location reveal the spatial mobility analysis of the introduced method. Besides, the change of predicted class over time (before and after the event) prove the temporal mobility.

4   Experimental Results

Here we present the experimental results of our method on both the amelioration of the execution time and to the standard classification performance metrics used in previous information retrieval and text classification studies namely precision and recall. Indeed, the comparative performance was done by referring, on one hand, to another classifier namely Naïve Bayes classifier, on the other hand, to others methods relying on the extraction of spatio-temporal information from tweeter. We used 140dev Twitter framework to extract contextual posts. Our collection consists of tweets from the 15th November till the 6th December 2016 (One week before and one week after the political event TUNISIA 2020).

Figure 2 depicts the specific execution times for the proposed Decision Tree and Naïve Bayes classifiier. Obviously, the gain in execution time using the proposed classifier is very significant and increases polynomially with the sample size. Additionally, the Figure 2, the execution time of our method enhances rather linearly with the sample size and also showing good scaling criteria.

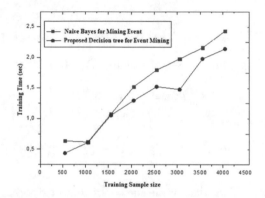

Fig. 2. Comparison in term of execution time

We present a performance comparison as illustrated in Tables 2 and 3 between Naïve Bayes classifier and the proposed Decision Tree classifier (in terms of recall-precision-F-measure).

Table 2. Performance of the introduced decision tree classifier

Class	Precision	Recall	F-measure
ProfitableN	0,93	0,83	0,93
ProfitableM	0,91	0,82	0,91
ProfitableS	0,87	0,81	0,87
UnprofitableN	0,62	0,75	0,62
UnprofitableM	0,59	0,61	0,59
UnprofitableS	0,48	0,40	0,48

The main advantage of the proposed Decision tree for event mining by location is that it allows better classification of all tweets. Using naïve Bayes a tweet expressing profitable class by region is correctly identified with 84% of recall for profitabeN class. However, for the introduced decision tree, the recall is less than that provided by using other classifiers, which means that tweets having

Table 3. Performance of Naïve bayes classifier

Class	Precision	Recall	F-measure
ProfitableN	0,79	0,84	0,79
ProfitableM	0,74	0,83	0,74
ProfitableS	0,72	0,79	0,72
UnprofitableN	0,73	0,71	0,73
UnprofitableM	0,50	0,57	0,50
UnprofitableS	0,45	0,38	0,45

profitable satisfaction opinions through 'TUNISIA2020' is correctly identified with 83% of recall. Indeed, for the developed method, a tweet giving a profitable classification is 93% likely to be correct while profitableN classification of Naïve Bayes approach is 79%. Nevertheless, our method gives a unprofitableN classification is only 62% likely to be correct while the same classification of naïve Bayes is in order of 73%. Compared with naïve Bayes approach, the precision of the proposed decision tree seems good (93% for profitable classification), the precision and recall indicate that the numbers are confusing. This is clearly visible in the F-measure rates (83% for profitable classification and 75% for unprofitable classification).

One possible explanation for these results is that, in this case study, the model built with many more tweets reflects more the satisfaction of the proposed investment plans by regions through 'TUNISIA2020' event than the dissatisfaction reaction, and the test data contains mostly efficient tweets.

Comparison with other methods:
The introduced method is compared with methods relying on cooperating spatio-temporal information into events. We use, on one hand, Hecht et al. method as it is described in [9]. The basic idea of this algorithm is to apply machine learning to identify users location from their posts. On the other hand, the comparative performance focus on Hetch' method which is based on location disclosure behavior.

Figure 3, we give the comparative performance (in terms of Precision/Recall) of Hetch' algorithm [9] and Cheng' method [13] as compared to the performance of the introduced method.

According to the curves above, we notice that the development decision tree gets a higher precision but a lower recall than Hetch' method but finally they present equal values of recall. Thus, we can conclude that the majority of samples are well classified. For the Cheng' algorithm the recall decrease gradually and the precision is relatively high but when we are approaching value 1 in the X axis, the precision falls. Unlike, Hetch' method and the introduced decision tree methods which conserve steady values.

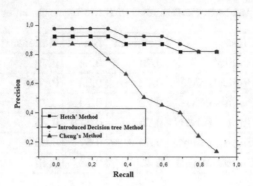

Fig. 3. Comparative performance (in terms Precision/Recall) of the introduced method and Hetch and Cheng methods

Additionally, Fig. 4 shows that the proposed Decision tree shows a potential source of mobility. In fact, we focused on temporal mobility by distinguishing event mining before and after 'TUNISIA 2020' event. Besides, spatial mobility is presented through the extensive movement and change of mining event for each Tunisian region. In Fig. 4, we notice the evolution of profitable class with its various instances, named profitableN denoting profitable investment plan for north of Tunisian, profitableS class for South Tunisian region and profitableM for Middle region.

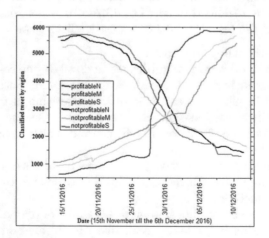

Fig. 4. Mobility of event classification

The suggested mobility analysis method proved the success of 'TUNISIA 2020' event with its ambitious plan with profitable foreign investment in all Tunisian region. Thus, Tunisia succeeded in the 2016 edition of its investment forum. Therefore, trust is built again at Jasmine country, restoring hope to a population seeker of freedom, dignity and shared progress.

5   Conclusion

Decision Tree is one of the successful data mining techniques used in classification or prediction. In this study, decision tree method for mining the mobility of event in social network was discussed. The proposed mobility model was designed to describe the movement of opinion for specific location from tweeter users' location and to demonstrate the change of these opinions over time. Therefore, the introduced entropy measure highlight spatial and temporal attributes related to each detected opinions attribute. Thus, the goal of our approach is to effectively detect and group emerging opinions through political event 'TUNISIA2020' using data provided by Twitter.

By thoroughly experimental analysis, our finding shows that mobility of opinions is highly dependent on both temporal and spatial behavior. Moreover, the preliminary results reveal that our mobility classification model has the potential of event detection and event awareness.

In this work, we deal mainly with the event mobility analysis. Many other challenges could be considered such as the use of specific mobility model namely mobility model with temporal dependency, mobility models with spatial dependency and mobility model with geographic restriction during our classification process. The long term aim of our research is to identify the political preferences and predict electoral process results based on this analysis.

References

1. Kaymak, U., Hogenboom, F., Frasincar, F., Jong, F.D.: An overview of event extraction from text. In: Workshop on Detection, Representation, and Exploitation of Events in the Semantic Web (DeRiVE 2011), CEUR Workshop Proceedings at Tenth International Semantic Web Conference, (ISWC 2011), CEURWS.org (2011)
2. Hogenboom, F., IJntema, W., Sangers, J., Frasincar, F.: A lexico-semantic pattern language for learning ontology instances from text. J. Web Sem. **15**, 37–50 (2012)
3. Lee, R., Sumiya, K.: Measuring geographical regularities of crowd behaviors for twitter-based geo-social event detection. In: Proceedings of the 2nd ACM SIGSPATIAL International Workshop on Location Based Social Networks, CEUR Workshop Proceedings, CEURWS.org. San Jose, California, vol. 779 (2010)
4. Benferhat, S., Amor, N.B., Elouedi, Z.: Decision trees as possibilistic classifiers. Int. J. Approximate Reasoning **48**(3), 784–807 (2008)
5. Chen, L., Zhou, D., He, Y.: A simple bayesian modelling approach to event extraction from twitter. In: Proceedings of the 52nd Annual Meeting of the Association for Computational Linguistics, vol. 2, pp. 700–705, June 2014. Association for Computational Linguistics, Baltimore, Maryland (2014)
6. Okamoto, M.M., Kikuchi, M.: Discovering volatile events in your neighborhood: Local-area topic extraction from blog entries, vol. 5839. Springer (2009)

7. Xiang, L., Chen, X., Liu, M., Liu, Y., Yang, Q.: Extracting key entities and significant events from online daily news. In: 9th International Conference on Intelligent Data Engineering and Automated Learning - IDEAL 2008, South Korea, pp. 201–209, 2–5 November 2008

8. Zhang, Y., Lei, Z., Wu, L.-D., Liu, Y.-C.: A system for detecting and tracking internet news event, vol. 3767. Springer (2005)

9. Suh Hecht, B., Hong, L., Chi, E.H.: Tweets from justin bieber's heart: The dynamics of the "location" field in user profiles. In: Proceedings of the ACM Conference on Human Factors in Computing Systems, South Korea, 2–5 November 2011

10. Lavie, A., Sagae, K., MacWhinney, B.: Combining rule-based and data-driven techniques for grammatical relation extraction in spoken language. In: Proceedings of the Eighth International Workshop in Parsing, pp. pp. 153–162 (2003)

11. Okazaki, M., Sakaki, T., Matsuo, Y.: Earthquake shakes twitter users: Real-time event detection by social sensors. In: Proceedings of the 19th International Conference on World Wide Web, pp. pp. 851–860. ACM, New York (2010)

12. Wakamiya, S., Lee, R., Sumiya, K.: Discovery of unusual regional social activities using geo-tagged microblogs. In: Proceedings of the 19th International Conference on World Wide Web, pp. 1–29 (2011)

13. Caverlee, J., Cheng, Z., Lee, K.: You are where you tweet: A content based approach to geo-locating twitter users. In: Proceedings of the 19th ACM International Conference on Information and knowledge management, Toronto, Canada, pp. 759–768 (2011)

14. http://140dev.com/free-twitter-api-source-codelibrary. Accessed Dec 2015

15. Helmut.: TC project (1994). http://www.cis.uni-muenchen.de/schmid/tools/TreeTagger/. Accessed 19 Apr 2015

16. Dhouioui, R., Toujani, Z., Akaichi, J.: Sentiment analysis in social networks using machine learning and metaheuristic. In: Proceedings of the 11th Edition of the Metaheuristics International Conference (MIC 2015), Agadir, Morocco, 7–10th June 2015

17. Quinlan, J.: C4.5: Programs for Machine Learning. Morgan Kaufmann, San Mateo (1993). CA edition

Ant Colony Optimization Approach for Optimizing Irrigation System Layout: Case of Gravity and Collective Network

Sahar Marouane[1]([✉]), Fahad Alahmari[2], and Jalel Akaichi[2]

[1] High Institute of Management, University of Tunis, Tunis, Tunisia
sahar.marouane@gmail.com
[2] College of Computer Science, King Khaled University Abha, Abha, Saudi Arabia
fahad@kku.edu.sa, j.akaichi@gmail.com

Abstract. Irrigation is the artifcial employment of water to the plants which is used to assist in the growing of agricultural crops. There are several methods of irrigation that differ in how the water is distributed between fields. In fact irrigation systems can be classified into two main categories: gravity irrigation and pressurized irrigation. The allocation of water to the fields is done either collectively or individually. Whatever the used irrigation technique, the goal is to have a well-designed irrigation system. This research applies the metaheuristic method of ant colony optimization (ACO) to design an optimal irrigation layout. The proposed approach uses ACO rules to generate the possible links between fields which distribute water to farmers. And the algorithm ant system was applied to find the optimal link.

Keywords: Ant colony optimization · Irrigation system layout · Gravity irrigation · Collective irrigation

1 Introduction

Irrigation is necessary to gainful crop production in arid climates and it is practiced in all environments by devote water when the plant needs it, and therefore maximize production. The agricultural sector is the largest user of water resources; on demand irrigation systems played an important role in the distribution of water to farmers. The distribution of the irrigation water is defined as the combination of a set of processes and social techniques used to delivery water from the source to the plant. Actually, irrigation sector is heavily impacted by insufficiency of water and lessening may effect an universal decrease of food. Thus, enhancements in agricultural water management are required. Optimization of irrigation systems and improvement of water resources are considered as crucial responses to address water scarcity and provide a water saving. Several researchers have developed optimization models for planning and management irrigation systems. Ant colony optimization (ACO) was introduced for solving

© Springer International Publishing AG 2018
G. De Pietro et al. (eds.), *Intelligent Interactive Multimedia Systems and Services 2017*,
Smart Innovation, Systems and Technologies 76, DOI 10.1007/978-3-319-59480-4_32

hard combinatorial optimization problems and it is used to optimize the irrigation systems [1].

We start with a general introduction. This article is structured into three sections. The first section, Ant Colony Optimization, is devoted to remind some basic concepts about colony optimization and to present some related works. The second section, Contribution, present the problem of irrigation network and the proposed solution. The last section, Experimentation, describe the experiments carried to validate our proposals. Finally, a conclusion which summarize all the work presented in this paper and propose future works.

2   Ant Colony Optimization

2.1   Background

Ants have inspired many methods and techniques. The most studied and successfully applied is the ACO. In fact, ACO algorithms is a class of metaheuristics. It takes inspiration from the foraging behavior of a few ant species [1]. In 1990, Deneubourg and colleagues have demonstrated in their experiences that ants can find food sources using the shortest path between it and their nest [2]. In fact ant deposits a chemical substance called pheromone as a communication medium to inform other ants to follow the same path. The pheromone trails assist as numerical information which the ants used to find shortest paths and probabilistically construct solutions. The pheromone is characterized by a gradual evaporation. The amount of this substance presented on the shortest path is slightly greater. Therefore, the trail having a higher concentration of pheromone is more attractive. This concept was mainly used to solve routing problem and the Traveling Salesman Problem (TSP) play an important role in ACO research. The first ACO algorithm, called ant system is proposed in Dorigo's doctoral dissertation [1,3]. It was first tested on the TSP, as well as many other ACO algorithms which are Elitist Ant System (EAS) [3], AS rank [4], Max-Min Ant System (MMAS) [5], Ant Colony System (ACS) [6], AntQ [7] and ACS-3opt [6]). ACO algorithms can be used to solve a variety of combinatorial optimization problems (routing (sequential ordering problem [8]), assignment(quadratic assignment problem (QAP) [9]), scheduling (car sequencing problem [10]) and subset problems (maximum clique problem [11])... Also, it can be used very well to optimize irrigation system.

2.2   State of the Art

Recently irrigation has been a central element in agriculture. Whatever the irrigation technique used, a well-designed irrigation system reduces water loss and improves water use efficiency. For that many network designs which are considered in many regions, propose optimal settings. Optimizing the irrigation network consists to optimize the layout (minimize the length of pipeline and the number of interconnected elements), the diameters of pipe, reservoir Operation...

Many researchers have been developed to optimize irrigation systems. The design of pressurized irrigation systems is expensive and several studies have treated this problem especially collective pressurized irrigation networks [12]. González et al. propose a new approach of management pressurized irrigation networks using multi-objective genetic algorithms [13]. In 2016 Carríon et al. developed a tool to optimize the process of groundwater abstraction [14]. In 2015, García et al. propose an energy optimization models for irrigation electricity tariff [15]. In 2015, the research of Anamika consist to optimize the water use in irrigation of summer rice through drip irrigation [16]. In 2015 Carríon et al. develop a tool to optimize the management of water for center pivot systems [17] The application of GA include the lengthy computation time. ACO is currently a new alternative in the design of irrigation systems and research observed that with ACO we obtain better results than other approaches. ACO algorithm was applied by Marino and Morales for the first time in 1999 to design a water distribution irrigation network using multiple objective Ant-Q (MOAQ) [18]. And in 2015 Qin Tu et al. applied Ant Cycle System for the design of small scale irrigation systems [19]. Recently Holger et al. conceive a framework for computationally efficient optimal crop and water allocation using ACO [20]. In their study, Kumar et al. use ACO to optimize the reservoir operation by classifying the reservoir volume into several class intervals [21]. The proposed approach was tested in a multi-purpose reservoir system located in India (Hirakud reservoir). Group Steiner Problem (GSP) is a class of problems in combinatorial optimization which include complex real-world problem such as the design of a minimal length irrigation network. Nguyen et al. propose a new based ACO to solve this problem [22]. The solutions proposed to apply ACO for irrigation system indicate that ACO outperforms the other heuristic methods.

Maybe due to the availability of traditional methods, the application of ACO in the optimization of gravity irrigation systems and the collective network is neglected and few contributions are found. Though much attention has been paid to the optimization of pressurized irrigation systems which aims at the least cost of components (involves the reservoir, pipe and pump engine..) or the minimal energy consumption while the optimization of layout precisely the network length is the easiest method to optimize the costs. In fact the optimization of the layout allows to optimize the number of components of irrigation system thus optimize their costs. For this reason we choose to optimize the gravity irrigation system in collective perimeter which focuses on the routing of water in different pipelines which connecting fields.

3 Contribution

3.1 Problem Definition

Irrigation is an important factor in increasing yields. In this paper, we propose a novel method to optimize gravity irrigation network in a collective perimeter. In fact, this method was based in a combination of two kinds of irrigation systems. In this study we conceive an irrigation network in a collective perimeter that includes a set of irrigators. This network must distribute water between different agricultural fields. The proposed method allows to obtain an optimal network route that passes through the wholes fields while respecting the constraint of slope values between fields. The novelty in this approach compared to the existing works is to combine and optimize two types of irrigation systems at the same time which is not proposed in the literature. And also the use of the slope values as a crucial parameter to control the choice of the fields. We are interested in minimizing the network length. To do this, we apply ACO and we suggest AS to minimize the costs associated with such infrastructure. In fact, finding optimal irrigation network is considered as a combinatorial optimization problem which can belong to the class of routing problem if we are interested to find the optimal network length. As we have already mentioned, ACO is a population-based metaheuristic that can be used to find approximate solutions to difficult optimization problems. Indeed, ACO algorithm for its natural advantages, ever since the time it was proposed in the field of combinatorial optimization, especially in the study of routing and networks problems, has attracted much success. For these reasons we chose ACO and specifically AS to solve the irrigation network problem.

Gravity Irrigation: The gravity irrigation consists to irrigate crops by let the water flow by gravity. The cost of building a gravity system is lower because there is no or little consumption of energy. The slope of the land is one of the important parameters of water flow especially the gravity flow. In fact, with too steep slopes, the work is very expensive. Contrarily, on flat land, flood risk and water accumulations are very important. So minimum slope value are more desirable to ensure continuity of water flow.

Collective Irrigation: The presence of small farmers, the existence of groundwater resources and the possibility of their collective use is the result of consolidation of irrigators around a collective use.

3.2   Formal Representation

The collective irrigation network can be represented as a graph easily covered by ants. In this network, we want to find the shortest possible length pipeline through a given set of irrigated fields, visiting each field once. We can transform this network by a complete weighted graph G=(N;A) with N is the set of $n=$ | N | nodes which present the set of fields, A is the set of arcs connecting the totality of nodes. Each arc $(i;j) \in A$ is attribute a value d_{ij} which represents the distance between fields i and j and a value s_{ij} which represents the value of slope between fields i and j. The aim of this work is to find a minimum length hamiltonian path of the network. To ensure that some movement from field to another are undesirable due to the value of slope $(s_{ij} \neq s_{ji})$, we use a boolean matrix $x(i,j) \in \{0,1\}$ intrepted as follows:

- x(i,j)=1 if we move from field i to field j
- x(i,j)= 0, otherwise

Formally the objective of our contribution can be written by the following objective function:

$$min \sum_{i=1}^{n} \sum_{j=1}^{n} x(i,j)d(i,j) \quad \forall i \neq j$$

Subject to

$$\sum_{i=1}^{n} x(i,j) = 1 \quad \forall j \in n$$

$$\sum_{j=1}^{n} x(i,j) = 1 \quad \forall i \in n$$

$$s_{ij} < s_{ik} \quad \forall k \in n, k \neq j$$

The first constraint indicates that each field must be entered exactly once. The second constraint indicate that each field must be departed exactly once (each field there is a departure to exactly one other filed). The last constraint indicates that we must choose the field with minimum slope value from the unvisited fields (the smaller slope must be before the greater slope).

3.3 Algorithm

Modifications, that are proposed with the AS algorithm, are used to solve the traditional TSP in order to permit an optimum length for the gravity irrigation network.

```
Algorithm  AS for Gravity Irrigation Network
1:  Initialization: place randomly each ant on a field
    and initialize the pheromone quantity
2:  for t=1 to t=tmax do
3     for each ant k do
4:      Construct a network Nk(t) with the transition rule
5:      Calculate the length Lk(t) of the network
6:    end for
7:  N+ is the best network found and L+ the corresponding length
8:  Update the pheromone trails
9:  end for
10: Return N+ and  L+
```

At each iteration of the algorithm, a set of agent called artificial ants search for good solutions. To apply AS, the optimization problem is transformed into the problem of finding the best path on a weighted graph. The algorithm has two main phases, the construction of the solution and the pheromone update.

Solution Construction: At initialization each arc linking two fields have a definite quantity of pheromone τ_0. At each iteration all the ants are placed arbitrarily in the fields which are the starting point. The fourth step of the algorithm is to build the network. In fact, every ant builds a network Nk at the iteration (t) by adding new unvisited field to the partial solution under construction a whole network. To choose the next field, we use the pheromone trails and the heuristic information. The pheromone is a set of parameters associated with graph especially edges whose value are modified at runtime by the ants. The fifth step of the algorithm consists to calculate the network length Lk (pipeline length) by summing the edge's weights of the constructed network.

Transition Rule: At each construction step (fourth step in the algorithm), ant k applies a probabilistic transition rule (1) to decide which field visited next. The probability with which ant k, currently at field i, choose to go to field j is:

$$P_{ij}^k(t) = \frac{[\tau_{ij}(t)]^\alpha . [\eta_{ij}]^\beta}{\sum_{I \epsilon J_i^k} [\tau_{ij}(t)]^\alpha . [\eta_{ij}]^\beta} \tag{1}$$

where:

- τ_{ij} is the pheromone trails related to the desirability of visiting field j directly after i
- η_{ij} is the heuristic information or visibility of arc *(i, j)*
- J_i^k is the set of fields not yet visited by ant k while at field i
- α and β are parameters weighting

Pheromone Trails and Heuristic Information: At the first iteration, the pheromone trails are fixed at τ_0 and then it will be updating. As we mentioned, every time we will move to a next field we must go to the one having the lower slope. The visibility η_{ij} (heuristic information) (2) of an edge is related to the value of the slope. Lower value of slope are more desirable. Therefore, the visibility of an edge is taken as the inverse of the slope value of this edge:

$$\eta_{ij} = \frac{1}{s_{ij}} \tag{2}$$

Update Pheromone Trails: After each iteration we determine the shortest path and the pheromone trails are updated. This last step involves pheromone evaporation (3) and the update of the pheromone trails (4).

Pheromone evaporation is implemented by:

$$\tau_{ij} = (1 - \rho).\tau_{ij} \qquad \forall (i,j) \in N \tag{3}$$

After evaporation, all ants deposit pheromone on the arcs they have crossed in their tour:

$$\tau_{ij} = \tau_{ij} + \sum_{k=1}^{m} \Delta \tau_{ij}^k \qquad \forall (i,j) \in N \tag{4}$$

where $\Delta\tau_{ij}^{k}$ is the amount of pheromone ant k deposits on the arcs it has visited (5). It is defined as follows:

$$\Delta\tau_{ij}^{k} = \frac{1}{L^k} \tag{5}$$

where L^k is the length of the network N^k constructed by ant k.

4   Experimentation

The data have been obtained from the Regional Commissary for Agricultural Development of Bizerte (RCAD). The data is composed from a two-dimensional coordinate of 10 fields in irrigated perimeter and a matrix of slopes contain different values of slope between all pairs of fields. To calculate the distances between fields, we use the Euclidean distance. For the parameters values, we choose a good parameters found in an experimental study for the TSP [23]. In this study, MATLAB was used to implement the algorithm. For the parameters: $\alpha = 1$, $\rho = 0.5$, the number of ant m is equal to the number of field (n = 10). In fact, each ant are placed in a field. For the parameter β we tested four different values ($\beta = 2$, $\beta = 3$, $\beta = 4$ and $\beta = 5$). And the number of iteration = 1000.

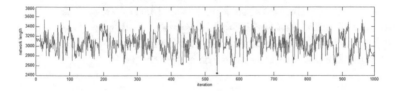

Fig. 1. Running of simulation for $\beta = 2$

For $\beta = 2$, the algorithm find the shortest network with length 2447 at the 535th iteration. The running of search is shown in Fig. 1

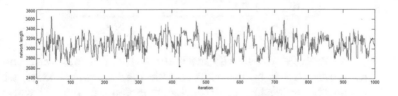

Fig. 2. Running of simulation for $\beta = 3$

For $\beta = 3$, the algorithm find the shortest network with length 2630 at the 420th iteration. The running of search is shown in Fig. 2

For $\beta = 4$, the algorithm find the shortest network with length 2928 at the 328th iteration. The running of search is shown in Fig. 3

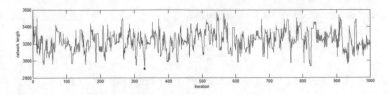

Fig. 3. Running of simulation for $\beta = 4$

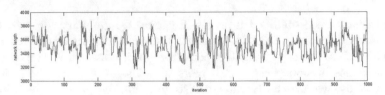

Fig. 4. Running of simulation for $\beta = 5$

For $\beta = 5$, the algorithm find the shortest network with length 3144 at the 340th iteration. The running of search is shown in Fig. 4

After some tests with different parameter values we observed that: If we increase the β value the network length will increase thereafter small β value are more desirable for gravity irrigation system (Table 1). The best solution is found with a length 2447 meters after 535 iterations and with a value of $\beta=2$. The plot representation of this network is shown in Fig. 5

Table 1. Summary of results of AS simulations

β	Iteration	Minimum length value in meters
2	535	2447
3	420	2630
4	328	2928
5	340	3144

This paper proposes a novel ACO method for irrigation network. The innovation of this paper that it addresses a specific method of irrigation which is gravity and collective irrigation and use the slope value to control water flow. However, due to the lack of the application of ACO in the gravity and collective irrigation systems and the availability of traditional methods, we propose an ant algorithm to design gravity irrigation network in a collective perimeter. Although this work seems to be a work that has no improvements on the previous works. On the other hand, our method is more suitable to optimize pipeline length and the results outperform other heuristic because ACO are designed especially to solve routing and network problems by finding the shortest path such as the

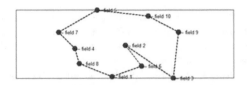

Fig. 5. Plot representation network having minimum length

natural behavior of real ants. The cost of building a gravity network based on the proposed approach is lower compared to the other works because there is no or little consumption of energy such as electric energy related to pumping engines. Thus, the proposed method reduces the cost of infrastructure.

5   Conclusion

The gravity irrigation causes a problem for distributing water to several fields in a collective perimeter; therefore we presented in this paper a new adaptation work of ACO to solve this issue. Indeed, we propose a novel ACO method to optimize gravity irrigation network in a collective perimeter. Our main goal was to minimize the pipeline length. To do so, we proposed modifications to the AS algorithm in order to allow an optimum design for the gravity irrigation network. This algorithm contained two essential steps: The first step represented the solution construct, where every ant builds a network and determine the shortest network length. The second step was the pheromone update. The results of our study proved the feasibility of our algorithm. Indeed, through several experimentations and various tests, we got an optimal pipeline length. As future research, we can improve our proposed approach by considering other parameters, for example we can optimize the number of sprinkler in case of sprinklers irrigation and even use an AS to manage irrigation scheduling. Moreover, we could apply another ACO algorithms (e.g. MMAS, ACS, etc.) to optimize the gravity irrigation network and then compare their performances.

References

1. Dorigo, M., Stützle, T.: Ant Colony Optimization. MIT Press Cambridge, London (2004)
2. Deneubourg, J.L., Aron, S., Goss, S., Pasteels, J.M.: The self-organizing exploratory pattern of the argentine ant. J. Insect Behav. **3**(2), 159–168 (1990)
3. Dorigo, M.: Optimization, learning and natural algorithms (in Italian). Ph.D. Thesis, Department of Electronics and Polytechnic of Milan, Italy (1992)
4. Bullnheimer, B., Strauss, C.: A new rank based version of the ant system-A computational study. In: Adaptive Information Systems and Modelling in Economics and Management Science (1997)
5. Stützle, T., Hoos, H.: Max-Min ant system. Future Gener. Comput. Syst. **16**(9), 889–914 (2000)

6. Dorigo, M., Gambardella, L.: Ant colony system: a cooperative learning approach to the traveling salesman problem. IEEE Trans. Evol. Comput. **1**(1), 53–66 (1997)
7. Dorigo, M., Gambardella, L.: Ant-Q: A reinforcement learning approach to the traveling salesman problem (1997)
8. Gambardella, L., Dorigo, M.: Has-sop: Hybrid ant system for the sequential ordering problem. Technical Report IDSIA 11–97 (2000)
9. Maniezzo, V., Colorni, A.: The ant system applied to the quadratic assignment problem. IEEE Trans. Knowl. Data Eng. **11**(5), 769–778 (1999)
10. Solnon, C.: Combining two pheromone structures for solving the car sequencing problem with ant colony optimization (2008)
11. Fenet, S., Solnon, C.: Searching for maximum cliques with ant colony optimization (2003)
12. Zapata, N., Playan, E., Lecina, S.: From on-farm solid-set sprinkler irrigation design to collective irrigation network design in windy areas. Agric. Water Manag. **87**(2), 187–199 (2007)
13. González, P.M., Poyato, C., Díaz, R.: Optimization of irrigation scheduling using soil water balance and genetic algorithms. Water Resour. Manage. **30**(8), 2815–2830 (2016)
14. Carríon, F., Sanchez-Vizcaino, J., Moreno, M.: Optimization of groundwater abstraction system and distribution pipe in pressurized irrigation systems for minimum cost. Irrig. Sci. **34**(2), 145–159 (2016)
15. García, F., Montesinos, P., Díaz, J.: Energy cost optimization in pressurized irrigation networks. Irrig. Sci. **34**(1), 1–13 (2015)
16. Sonit, A., Hemlata, K.: Optimization of water use in summer rice through drip irrigation. J. Soil Water Conserv. **14**(2), 157–159 (2015)
17. Izquiel, A., Carríon, P., Moreno, M.A.: Optimal reservoir capacity for centre pivot irrigation water supply Maize cultivation in Spain. Biosyst. Eng. **135**, 61–72 (2015)
18. Mariano, C.E., Morales, E.: A multiple objective ant-Q algorithms for the design of water distribution irrigation network (1999)
19. Tu, Q., Li, H., Wang, X., Chen, C.: Ant colony optimization for the design of small scale irrigation systems. Water Resour. Manage. **29**(7), 2323–2339 (2015)
20. Duc, C.H.N., Holger, R.M., Graeme, C.D., James, C.A.: Framework for computationally efficient optimal crop and water allocation using ant colony optimization. Environ. Model. Softw. **76**, 37–53 (2016). Elsevier
21. Kumar, D.N., Reddy, M.J.: Ant colony optimization for multi-purpose reservoir operation. Water Resour. Manage. **20**, 879–898 (2006). Elsevier
22. Nguyen, T.D., Do, P.T.: An ant colony optimization algorithm for solving group steiner problem. In: IEEE Fifth International Conference Communications and Electronics (ICCE), pp. 244–249 (2014)
23. Dorigo, M., Maniezzo, V., Colorni, A.: An investigation of some properties of an ant algorithm. In: Appeard in Proceeding of the Parallel Problem Solving from Nature Conference, Brussels, Belguim. Elsevier (1992)

Query Recommendation Systems Based on the Exploration of OLAP and SOLAP Data Cubes

Olfa Layouni[1(✉)], Assawer Zekri[1], Marwa Massaâbi[1], and Jalel Akaichi[2]

[1] BESTMOD Laboratory, Institut Supérieur de Gestion de Tunis,
Université de Tunis, Tunis, Tunisia
layouni.olfa89@gmail.com, assawer.zekri@gmail.com, massaabi.marwa@gmail.com
[2] College of Computer Science, King Khalid University, Abha, Saudi Arabia
jalel.akaichi@kku.edu.sa

Abstract. Business Intelligence systems refer to technologies and tools responsible for collecting, storing and analyzing data to improve decision-making. In BI systems, users interact with data warehouse by formulating and launching sequences of queries aimed at exploring multidimensional data cubes. However, the volumes of data stored in a data warehouse can be very large and diversified. So, a big amount of irrelevant information returned as results to the user could make the data exploration process inefficient. That's why, it's necessary to help the user by guiding him in his exploration. In fact, query recommendation systems play a major role in reducing the effort of decision-makers to find the most interesting information. Several works dealing with query recommendation systems were presented in the last few years. This paper aims at providing a comprehensive review of literature on a query recommendation based on the exploration of data cubes. A benchmarking study of query recommendation methods is proposed. Several evaluation criteria are used to identify the existence of new investigations and future researches.

Keywords: Query recommendation systems · Business intelligence systems · Data · Analysis · Cube · Data warehouse

1 Introduction

Business Intelligence (BI) system represents the tools that are used to collect, store and analyze data in order to make the best decision [4–6]. The BI system is realized by applying two different steps. The first step is the Extract, Transform and Load data. The ETL tools are responsible for extracting data from different heterogeneous sources, providing the integration and data cleansing according to a target schema or data structure, loading and storing data in a data warehouse. The second step is to analyze data by using an analysis server such as: OLAP or Spatial OLAP server. It is a rapid and flexible way for analysts to navigate, explore and analyze the large amount of data stored in

© Springer International Publishing AG 2018
G. De Pietro et al. (eds.), *Intelligent Interactive Multimedia Systems and Services 2017*,
Smart Innovation, Systems and Technologies 76, DOI 10.1007/978-3-319-59480-4_33

the data warehouse. Indeed, the user can make analysis reports by using: some reporting tools, dashboards, navigation and statistical tools. These tools offer capabilities to explore data and support the analysis process. To analyze data, users interactively navigate a data cube by launching sequences of OLAP or SOLAP queries over a traditional or spatial data warehouse, respectively. The problem appeared when the user may have no idea of what the forthcoming query should be. As a solution and to help the user in his navigation, we need a recommendation system.

The remainder of this paper is organized as follows: Sect. 2 introduces the concepts of recommendation in data warehouse systems. Section 3 presents an overview of several different approaches presented in the field of query recommendation based on the exploration of data cube. Section 4 presents a comparative study that provides a general, comparative view of the different approaches that have been presented. Section 5 concludes the paper.

2   Recommendation System

Recommendation system defined as a system that gives the possibility to generate recommendations of items like books, movies, music, queries, etc.; and products that might interest users [9,16]; those recommendations give the possibility to help the user by guiding him to find relevant information. The current generation of recommendation system is usually categorized into a content-based method, a collaborative method and a hybrid method [1,13,17].

In various studies [17,21–23], we find that the authors described the characteristics of the general algorithm of a recommender system for the exploration of data. These characteristics are the inputs, the outputs and the recommendation steps.

The inputs of the algorithm can be a log of sessions of queries, a schema or an instance of the relational or multidimensional database, a current session that contains the queries launched by the current user and a profile or the behavior of the user.

The outputs of the algorithm can be a query, a set of ordered queries or a set of tuples that can be similar or interested in the current user.

In reality, an algorithm of recommendation is decomposed into three steps. The first step consists in choosing an approach for evaluating the used scores. In fact, in this step we can choose one of the categories of recommendations: a content-based, a collaborative and a hybrid method. The second step is the filter; this step consists in selecting the candidates' recommendations. The last step is the guide; this step consists in ordering the candidates' recommendations.

3   OLAP Query Recommendation Approaches

This section presents a thorough survey on the proposed approaches in the domain of query recommendation for helping users to explore data. Those approaches can be classified into two categories, the first category exploits the

OLAP data cube and so does the second category which exploits the Spatial OLAP data cube.

3.1   Methods Exploiting OLAP Data Cube

Data warehouse stores large volumes of consolidation and historized multidimensional data, to be explored and analyzed by various users. In fact, the user interacts with the data warehouse by launching sequences of OLAP queries aimed at exploring the multidimensional data cube. Since the volume of information to explore can be very huge and diversified, it is necessary to help the user to face this problem by guiding him: by proposing an OLAP query recommendation system in his data cube. In the literature, we can distinguish two different ways to explore OLAP data cube. The first way exploits the profile and so does the second way with the log of queries.

Using the Profile. A lot of researches recommend OLAP queries in the exploration of a data warehouse by exploiting the profile.

The works of Sarawagi et al. in [21–23] were proposed to help the user in his exploration of the OLAP data cube based on the atomization. For this reason, the authors proposed four different operators: DIFF, EXCEP, RELAX and INFORM. Those operators allow the return as results of all sets of tuples that can explain the anomalies detected through the different operators, in fact those operators give the possibility to recommend one or more queries. We find that the proposed algorithms are a recommendation method based on the content. Adding to that, we discover that some proposed operators execute the results obtained after launching the current query and other operators execute only the current query.

The recommendation method proposed by Bellatreche et al. in [7,8] treats the problem of OLAP query personalization. In this method, the authors took into account the particularities of OLAP queries. The proposed method gives the possibility to secure two principal objectives; the first is to compute the utility of query and the second is to display the best query to a user by taken into account not only his preferences but also his visualization constraints. The proposed method is composed in three different steps. The first step consists in using the user profile. In this step, authors proposed two functions Perso and MaxSubset which were used to compute the best subsets of references from the current query and to satisfy the proposed constraint. The second step consists in searching elements firstly by comparing between the stored preferences and a query; secondly by selecting an order set of references for a specific query. The third step, the authors proposed to build a personalization query by using the best references, sorted and recommended them in an ascending order. We deduce that this approach is based on the content method. In fact, this method doesn't take into consideration the previous queries launched in the cube and the sequencing of queries launched by the current user.

The recommendation method proposed by Jerbi in [11] treats the problem of OLAP analysis personalization within data warehouses. In fact, the user must launch several queries in order to obtain a result that can be similar or close to his preferences. The proposed method gives the possibility to improve the current query by using the preferences of the user, and recommends the best query for him. The first step in this method is to analyze OLAP data. An OLAP analysis is modeled through a graph where nodes represent the analysis contexts and edges represent the user operations. The second step is to build a model for user preferences on the multidimensional schema and values. The last step is to propose a framework including two personalization processes. The first process denoting OLAP query personalization depends on contextual preferences stored in a user profile. In this process, two phases must be performed: the selection and the integration of preferences. The second process is recommendation queries. In this process, two types of personalization have to be performed: a personalization of an explicit or a dynamic type. Moreover, the proposed recommendation framework supports recommendation scenarios: assisting the user in a query composition and suggesting the forthcoming and alternative queries. The system recommends a set of queries by comparing the user preferences and alternative queries. Consequently, this framework is based on the content method. Also, this method doesn't take into consideration the sequencing of queries launched by the current user; it takes only the last launched query and the current session.

Using the Log of Queries. A lot of studies recommend queries in the exploration of OLAP data cube by exploiting the log of queries.

The recommendation method proposed by Sapia in [18–20] handles the problem that a user has no idea about the forthcoming query to request. Therefore, the author proposed a method to predict the next OLAP query in order to help him during the rest of the current session. The author proposed a method based on a probabilistic model: the Markov model. This model is used in order to return the similarity between two consecutive queries and predict the probability of occurrence for each prototype. In this method, the author proposed to build for each query in the log of sessions and in the current session a prototype corresponding to it. Then, he suggests to use a distance method to compare between those prototypes. Finally, he proposes to recommend a query to the current user which contains the highest probability of appearance between the prototype and the launched query. We note that this method is a collaborative recommendation method, which uses a statistical model the Markov model. In addition, we remark that this method takes into account the sequencing of queries and the previous queries.

The method proposed by Giacometti et al. in [10] gives the possibility to recommend for the current user the discoveries detected in the previous sessions saved in the log with the same unexpected data as the current session. This approach consists in two different steps. The first step is based on analyzing the query log to discover pairs of cells at various levels of detail for which the measure values differ significantly. The second step is based on analyzing a current query to detect if a particular pair of cells for which the measure values differ significantly can be related to what is discovered in the log. This approach is composed of two parts: the processing of the log and the computation of the recommendations. We find that this method is based on the results and the queries for each session in the log of recommending queries. Besides, it is a collaborative method.

The method proposed by Marcel et al. in [14] gives the possibility to recommend query to the user. The authors proposed a method for computing an intensional answer to an OLAP query by using the previous queries in the current session launched by the current user. The proposed method takes into account the intensional query answers. For this reason, it satisfied the three criteria proposed for the intensional answer: purity, completeness and dependency. So, we find that this method can be classified as mixed, partial and dependent. We describe the proposed method as following. The first step is to compute the extensional answer after the current user launching a query over the cube OLAP. In this method, authors proposed to use an expected cube for storing a model of the user expected values according to the data saved in the extensional answer. In the second step, for each query belonging to the current session, four steps must be done: execute the query over the cube, predict the expected extensional answer, improve the quality of the estimated values in the expected cube and build an intensional answer. We find that this method was suggested for generating recommendations OLAP queries in the context of the collaborative exploration of data cubes and it is based on the extensional and intensional answers. Furthermore, this method doesn't take into consideration the spatial queries and it is based only on the current session.

The method proposed by Aufaure et al. in [3] treats the problem of finding big numbers of possibilities of aggregations and selections that can be operated, all those possibilities may make the user experience disorientating and frustrating. In fact, authors proposed an approach for predicting and recommending the most likely next query to the current user. The proposed approach used a probabilistic user behavior model, which can be built by analyzing previous OLAP sessions and exploiting a query similarity metric. To this end, the authors proposed to analyze the query logs of a user, then, to cluster queries by using both a similarity metric and a Markov-based model based on the user behavior. Finally, by using this model, it is possible to recommend the next query for the current user. We find that this method doesn't only take into consideration sessions in the log and the sequencing of queries but also it takes into consideration queries launched by the current user in the past and the current query. This method is content-based method.

3.2 Methods Exploiting Spatial OLAP Data Cube

By using the new technologies such as: PDA, GPS, RFID, etc., we store different types of data. 80% of the obtained data represent spatial or location components. Spatial data warehouse have been used for storing and manipulating spatial data components [15]. Several spatial data types such as: point, surface, multipolygon, etc.; can be used to represent the spatial extent of real-world objects, which are collected for using the new technologies, in order to store the location or the movement of a real-world object like: car, train, ambulance, etc.

Spatial data types have a set of operations, which can be realized in order to represent spatial characteristics such as: topological, direction and metric distance [15]. A spatial data warehouse is made of multidimensional spatial model by integrating spatial measures and dimensions in order to take into account spatial components. In order to analyze and explore a spatial data warehouse, users need a SOLAP system. The SOLAP system obtained after the combination of Geographic Information Systems (GIS) with OLAP tools and operations. To navigate in the spatial data cube the current user launches a sequence of SOLAP queries over a spatial data warehouse. The problem appeared when the current user may have no idea of what the forthcoming SOLAP queries should be for this purpose we need a SOLAP queries recommendation system. In the literature, we find two methods for recommending SOLAP queries, the first one is proposed by [12,13] and the second one is proposed by [2].

The method proposed by Layouni et al. in [12,13] gives the possibility to recommend SOLAP queries to the current user. The proposed approach consists of the three following steps. The first step consists in computing all the generalized sessions of SOLAP queries of the log. In this step, they proposed a new similarity measure in order to compare between SOLAP queries by taking into account spatial relationships: topological, direction and metric. Also, they propose to use the method of TF-IDF for extracting the spatial measures and spatial dimensions in a launched query. Besides, to do the last classification of SOLAP queries, they decide to choose the Hierarchical Ascendant Classification. The second step is the filter which consists in predicting the candidates SOLAP queries by computing the most similar sessions to the generalized current sessions and searching the set of candidates SOLAP queries. The last step is the guide that consists in ordering the candidates SOLAP queries. We find that this method was suggested for generating recommendations SOLAP queries in the context of the collaborative exploration of spatial data cubes. And it is based on the text query and not the results obtained.

The method proposed by Aissi et al. in [2] recommend a set of SOLAP queries for the current user, this set contains only five queries. This approach takes into account the specific characteristics of spatial data. But the proposed approach for recommending SOLAP queries have some disadvantages. The proposed algorithm eliminates all the old queries in the log. It takes into account only recent queries in the log. In order to recommend a set of queries, the proposed approach detects preferences of the current user and compares between queries by applying a spatio-semantic similarity measure.

4   Comparing Query Recommendation Approaches and Discussion

The following section presents a comparative study that provides a general comparative view of the different approaches that have been presented and discussed in the field of methods proposed for recommending queries. The different models are compared according to these criteria.

Objectives of the works: The works proposed by [7,8,11] have two objectives: the personalization and the recommendation queries in the exploration of the data warehouse but the works proposed by [3,10,12–14,18–23] have as objective the recommendation queries.

Recommendation SOLAP vs non-SOLAP queries: The works proposed by [3,7,8,10,11,14,18–23] recommend queries in the exploration of the data warehouse so they exploit an OLAP data cube. The work proposed by [12,13] recommend queries in the exploration of the spatial data warehouse so the authors in this method explore a SOLAP data cube.

Inputs of the algorithm of recommendation: The works proposed by [7, 8,11,21–23] exploit the profile, but the works proposed by [2,3,10,12–14,18–20] exploit the log. We find that the methods proposed by [7,8,11,21–23] take as inputs of the algorithm the profile of the current user. Also, the methods proposed by [3,10,12–14,18–20] take as inputs of the algorithm the log of sessions of queries. Indeed, we find that the inputs of the algorithm can be: a schema, an instance, the current query, the current session, the previous sessions and visualization constraints. In fact, we remark that a schema was used as input in the algorithms proposed by [10–14,18–23]; an instance was used as input in the algorithms proposed by [7,8,10–14,18–23]; the current query was used as input in the algorithms proposed by [2,3,7,8,10–14,18–23]; the current session was used as input in the algorithms proposed by [2,3,10–14,18–20]; the previous sessions were used as input in the algorithms proposed by [2,3,10,12,13,18–20] and visualization constraints were used as input only in the algorithm proposed by [7,8].

Output of the algorithm of recommendation: Comparing the different proposed methods, we find that the output of the methods proposed by [3,11–14,18–20] is a query, those proposed by [2,7,8,10,12–14] the output is a set of queries and only the method proposed by [21–23] the output is a set of tuples.

Category of recommendation proposed system: We find that the methods proposed by [3,7,8,11,21–23] are content-based methods and the methods proposed by [2,10,12–14,18–20] are collaborative methods.

Filter step: We find that, in the filter step of the proposed algorithm, the method proposed by [2,3,7,8,11–14,21–23] gives the possibility to select the candidate's recommendation queries. Besides, for computing the candidate's recommendations the methods proposed by [14,21–23] apply the maximum entropy theory; also the methods proposed by [7,8,10,11] use a graphic model; as well the methods proposed by [3,18–20] apply the Markov model and the methods proposed by [2,12–14] apply distance model.

Guiding step: The guiding step was applied in the methods proposed by [11–13,18–20].

Intension and extension approach: We also find that the methods proposed by [2,3,7,8,11–14,18–20] applied to queries; only the method proposed by [10, 14,21–23] applied to the results obtained after launching a query.

Table 1. Comparative studies between the different approaches.

Proposed method		Proposed by								
		[21–23]	[7,8]	[11]	[18–20]	[10]	[14]	[3]	[12,13]	[2]
Objective	Recommending queries	*	*	*	*	*	*	*	*	*
	Personalization query		*	*						
Data warehouse		*	*	*	*	*	*	*	*	
Spatial data warehouse									*	*
Cube	OLAP	*	*	*	*	*	*	*		
	SOLAP								*	*
Inputs of the algorithm	Log				*	*	*	*	*	*
	Profile	*	*	*						
	Schema	*		*	*	*	*		*	
	Instance	*	*	*	*	*	*		*	
	Current query	*	*	*	*	*	*	*	*	*
	Current session		*	*	*	*	*	*	*	*
	Visualization constraints		*							
	Previous sessions				*	*		*	*	*
Output of the algorithm	Query			*	*		*	*	*	
	Set of queries		*				*		*	*
	Set of tuples	*								
Intension approach			*	*	*		*	*	*	*
Extension approach		*				*	*			
Approach	Content-based method	*	*	*				*	*	
	Collaborative method				*	*	*		*	*
Filter	Select candidates recommendations	*	*	*			*	*	*	*
	Compute candidates recommendations	The maximum entropy theory	A graphic model	A graphic model	A Markov model	A graphic model	The maximum entropy theory and a distance model	A Markov model	A distance model	A distance model
Guide				*	*				*	
Manipulation language	SQL	*		*	*	*	*	*		
	MDX		*			*	*	*	*	*
	Spatial MDX								*	*

Manipulation language: We find that the methods proposed by [3,10,11,14, 18–23] used the SQL language, the methods proposed by [3,7,8,10–14] used the MDX language and the methods proposed by [2,12,13] used the MDX language with Spatial functions. Table 1 reports a comparison of the above approaches according to the presented criteria.

5   Conclusion

In this paper, we have done an overview of the developed and suggested query recommendation approaches. Each approach is presented and discussed, then, a comparative study between the different proposed works is presented in order to compare and evaluate them in terms of some criteria.The relative novelty of the domain leaves many challenges, opportunities and extended studies open for future work, which we addressed most of them in our deduced research gaps. The proposed work allows us to have a overall vision on the different proposals and takes advantage of the studied contributions in an optimized way in order to introduce our future work which is the proposal of a new approach on a trajectory query recommendation to describe the object movement in space over the time.

References

1. Adomavicius, G., Tuzhilin, A.: Toward the next generation of recommender systems: a survey of the state-of-the-art and possible extensions. IEEE Trans. Knowl. Data Eng. **17**(6), 734–749 (2005)
2. Aissi, S., Gouider, M.S., Sboui, T., Said, L.B.: Enhancing spatial data warehouse exploitation: a solap recommendation approach. In: Computer and Information Science, pp. 131–147. Springer (2016)
3. Aufaure, M., Kuchmann-Beauger, N., Marcel, P., Rizzi, S., Vanrompay, Y.: Predicting your next OLAP query based on recent analytical sessions. In: Proceedings Data Ware-housing and Knowledge Discovery - 15th International Conference, DaWaK 2013, Prague, Czech Republic, 26–29 August, pp. 134–145 (2013)
4. Badard, T.: L'open source au service du géospatial et de l'intelligence d'affaires. Geomatics Sciences Department (avril 2011)
5. Badard, T., Dubé, E.: Enabling geospatial business intelligence. Geomatics Sciences Department, Semptember 2009
6. Bédard, Y., Han, J.: Geographic Data Mining and Knowledge Discovery, 2e edn. Taylor & Francis, Boca Raton (2009)
7. Bellatreche, L., Giacometti, A., Marcel, P., Mouloudi, H., Laurent, D.: A personalization framework for OLAP queries. In: Proceedings DOLAP 2005, ACM 8th International Workshop on Data Warehousing and OLAP, Bremen, Germany, 4–5 November, pp. 9–18 (2005)
8. Bellatreche, L., Mouloudi, H., Giacometti, A., Marcel, P.: Personalization of MDX queries. In: 22èmes Journées Bases de Données Avancées, BDA 2006, Lille, 17–20 octobre 2006, Actes (Informal Proceedings) (2006)
9. Burke, R.: Hybrid recommender systems: survey and experiments. User Model. User-Adap. Inter. **12**(4), 331–370 (2002)
10. Giacometti, A., Marcel, P., Negre, E., Soulet, A.: Query recommendations for OLAP discovery-driven analysis. IJDWM **7**(2), 1–25 (2011)
11. Jerbi, H.: Personnalisation d'analyses décisionnelles sur des données multidimensionnelles. Ph.D. thesis, Institut de Recherche en Informatique de Toulouse - UMR 5505, France (2012)
12. Layouni, O., Akaichi, J.: A novel approach for a collaborative exploration of a spatial data cube. IJCCE Int. J. Comput. Commun. Eng. **3**(1), 63–68 (2014)

13. Layouni, O., Alahmari, F., Akaichi, J.: Recommending multidimensional spatial olap queries. In: Intelligent Interactive Multimedia Systems and Services 2016, pp. 405–415. Springer (2016)

14. Marcel, P., Missaoui, R., Rizzi, S.: Towards intensional answers to OLAP queries for analytical sessions. In: Proceedings DOLAP 2012, ACM 15th International Workshop on Data Warehousing and OLAP, Maui, HI, USA, November 2, pp. 49–56 (2012)

15. Marketos, G.: Data Warehousing & Mining Techniques for Moving Object Databases. Ph.D. thesis, Department of Informatics, University of Piraeus (2009)

16. Melville, P., Sindhwani, P.: Recommender systems. In: Encyclopedia of Machine Learning, pp. 829–838 (2010)

17. Negre, E.: Exploration collaborative de cubes de données. Ph.D. thesis, Université François Rabelais of Tours, France (2009)

18. Sapia, C.: On modeling and predicting query behavior in olap systems. In: Proceedings INT'L Workshop on Design and Management of Data Warehouses (DMDW 99), SWISS LIFE, pp. 1–10 (1999)

19. Sapia, C.: PROMISE: Predicting query behavior to enable predictive caching strategies for OLAP systems. In: Kambayashi, Y., Mohania, M., Tjoa, A (eds.) Data Warehousing and Knowledge Discovery, Lecture Notes in Computer Science, vol. 1874, pp. 224–233. Springer, Heidelberg (2000)

20. Sapia, C., Alexander, F., Erlangen-nürnberg, U.: Promise: modeling and predicting user behavior for online analytical processing applications. Ph.D. thesis submitted, Technische Universität München (2001)

21. Sarawagi, S.: Explaining differences in multidimensional aggregates. In: Proceedings of the 25th International Conference on Very Large Data Bases, VLDB 1999, pp. 42–53. Morgan Kaufmann Publishers Inc., San Francisco (1999)

22. Sarawagi, S.: User-adaptive exploration of multidimensional data. In: VLDB, pp. 307–316. Morgan Kaufmann (2000)

23. Sathe, G., Sarawagi, S.: Intelligent rollups in multidimensional olap data. In: Proceedings of the 27th International Conference on Very Large Data Bases, VLDB 2001, pp. 531–540. Morgan Kaufmann Publishers Inc., San Francisco (2001)

Regions Trajectories Data: Evolution of Modeling and Construction Methods

Marwa Massaâbi[1(✉)], Olfa Layouni[1], Assawer Zekri[1],
Mohammad Aljeaid[2], and Jalel Akaichi[2]

[1] BESTMOD Laboratory, Institut Supérieur de Gestion de Tunis, Tunis, Tunisia
massaabi.marwa@gmail.com, layouni.olfa89@gmail.com,
assawer.zekri@gmail.com
[2] College of Computer Science, King Khalid University, Abha, Saudi Arabia
moh.k.alotaibi@gmail.com, jalel.akaichi@kku.edu.sa

Abstract. Tracking movement and trajectory data analysis are very important in the era of sensor devices and technological evolution. The movement can be produced by an object represented by a point, a line or a region. The region can be in movement, but its movement is special in some way because it changes its position, shape and extent unpredictably when moving (such as tumors, massive rainfalls, etc.). However, representing moving regions trajectories without interfering or modifying their unstable aspect is more or less ignored by the most recent literature. Therefore, this paper investigates trajectories evolutions, construction and modeling techniques, in order to highlight the gap concerning regions' trajectory. Subsequently, we focus on regions types and their trajectories modeling techniques.

Keywords: Moving region · Modeling · Trajectory · Construction

1 Introduction

Moving objects are becoming equipped with GPS chipsets to track their movements and collect data. These tracking devices such as GPS, smartphones and RFID produce enormous amounts of movement data. This data represents in fact a series of spatio-temporal positions. The sequence of positions form the path that the moving object has taken, which is its *trajectory*. Trajectories constitute a very interesting subject to studies by research community since it reports the object's positioning history. Trajectories are used in several domains: wildlife tracking, traffic management, human behavior analysis, etc. They track all kinds of moving objects in their three possible forms: points, lines or regions. Moving points are considered in multiple works in the literature. However, regions are not getting enough attention. Natural catastrophes are the best example for moving regions. Hurricanes, forest fires, floods, storms, they all start from one place then move unpredictably while changing their shape. Therefore, studying their trajectories leads to understanding their behavior which helps preventing

© Springer International Publishing AG 2018
G. De Pietro et al. (eds.), *Intelligent Interactive Multimedia Systems and Services 2017*,
Smart Innovation, Systems and Technologies 76, DOI 10.1007/978-3-319-59480-4_34

catastrophes. For this reason, we will be interested in surveying the literature dealing with modeling and analyzing trajectories in general for a better understanding of the mobility concept. Then, we will define spatio-temporal regions and introduce real life examples to understand the importance of studying such phenomena. Also, we present their movements modeling techniques to highlight the fact that they are nearly neglected by research community. The rest of the paper is organized as follows: Sect. 2 presents the trajectory's basic definitions and evolution. Sections 3 and 4 survey the modeling and construction techniques. Section 5 focuses on presenting moving spatio-temporal regions, together with their application scenarios and modeling techniques. Section 6 concentrates on regions' trajectory modeling techniques. Finally, Sect. 7 concludes our conducted study.

2   Trajectories Data: From Raw to Semantic

A trajectory can be defined as a series of points that trace or describe a moving object's path, i.e. changing position, through time. These points can be obtained from tracking devices such as a GPS. The concept of trajectories has known several evolutions. It began by raw then structured followed by semantic trajectories as shown in Fig. 1.

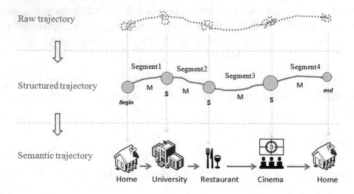

Fig. 1. Trajectory evolution.

2.1   Raw Trajectories

A raw trajectory is a series of points describing the movement of an object over time [5]. In fact, raw data is the basis of every trajectory data since it is the one directly generated from the sensors capturing the position and the time stamp of the moving object. The points are represented as $\{x_i, y_i, t_i\}$ where x_i and y_i are the spatial coordinates of the point in the 2D plane while t_i represents the temporal coordinate [13]. However, this type of trajectories is meaningless, it needs more structure and signification.

2.2   Structured Trajectories

Despite the improvement from raw to structured trajectories, there is still a lack in the semantic aspect. Therefore, structured trajectories are considered as a transition from raw to semantic trajectories, since they clarify and structure raw trajectories [4, 26]. Actually, they are considered as the partitioning of raw data into episodes or segments starting by *begin* and ending with *end* with some *stops* and *moves* in-between.

2.3   Semantic Trajectories

Raw trajectories lack semantics. Their analysis does not lead to satisfactory information. For instance, the observation of the trajectory of a tourist recorded using a tracking device without any tags is meaningless. It does not give information about his destination, the places he visited, the means of transport he used, etc. So, it is somehow useless. However, when adding geographic annotations and sense to the trajectory several observations and conclusions can be made. Here came the need for semantic trajectories. In fact, they are the evolution of structured trajectories by semantically enriching the episodes. It consists of a segmented raw trajectory tagged with meaningful annotations.

3   Trajectory Data Modeling Techniques

Trajectory data modeling has gained lots of interest and is attracting researchers as a thriving field due to the massive availability of tracking devices. Therefore, we present the main modeling techniques cited in the literature divided into two categories, *abstract* and *ontology/conceptual* modeling.

3.1   Abstract Modeling of Trajectory Data

The trajectory is seen as an operation called *trajectory* that reflects the projection of moving points on the plane [10], expressed with the Kernel algebra. In fact, if the moving object is a single point, then the operation *trajectory* is expressed as $moving(point) \rightarrow line$ and its semantics are expressed as following $rng(\mu) \backslash f_{locations}(\mu)$. The latter expression signifies that the range of the operation *trajectory* is the sequence of the point's locations. Whereas, if the moving object is a set of points, then the operation *trajectory* is expressed as $moving(points) \rightarrow line$ and its semantics is $\cup rng(\mu) \backslash f_{locations}(\mu)$. The last expression signifies that the range of the operation *trajectory* is the union of the sequences of the points sets' locations.

3.2   Conceptual/Ontological View of Trajectory Data

The first work to conceptually model trajectories [10] tried to find an equilibrium between the specifications of the moving objects and the needed associated semantics for the applications. It aims to conceptually represent the trajectory as a sequence of stops and moves. Other works [1, 20] were proposed as an evolution of this model. While, in [32], the author proposed a conceptual multilayer trajectory data model. This approach takes into consideration spatio-temporal features together with semantic ones in order to meet the trajectory data modeling requirements. Another conceptual model named *CONSTAnT* is proposed in [2] to represent semantic trajectories. This model considers the trajectory as a set of different semantic sub-trajectories generated according to the object's movements. From an ontological point of view, [31] proposed an ontological framework that semantically models trajectory data. It relies on three main ontological modules: Geometric trajectory ontology, Geography ontology and Application domain ontology.

4   Semantic Trajectory Construction

Trajectory construction or computation is building the trajectory from GPS data. The construction begins by data preprocessing. Once data is ready for the processing, the following step is its segmentation into episodes. The final step is the semantic enrichment.

4.1   Data Cleaning

Data cleaning is an essential task that initializes every data management system. It prepares data for the processing task and ameliorates the system's efficiency. Therefore, several works took interest in studying data cleaning steps and algorithms either based on unsupervised machine learning approaches [3] or using the maximum likelihood in order to obtain the best possible estimation of the erroneous values [34].

4.2   Data Segmentation

Data segmentation consists of splitting cleaned GPS feeds into segments, subtrajectories or episodes according to the trajectory stops identified by segmentation approaches or algorithms [8]. Several methods have been proposed. They can be categorized into three main categories: space, time and speed based methods [8, 29]. A brief review of these methods is given below.

- **Space based segmentation.** It consists of segmenting the trajectory according to its spatial properties. The division may be based on geographical locations or on a fixed distance.
- **Time based segmentation.** It permits the division of the trajectory according to the periods of time. The predefined time periods can for instance be hours, days, weeks or months. An other way of time based segmentation is to use a minimum stop duration which will be the basis of the stop identification. The determination of the stop duration threshold depends on the application's context and is fixed in advance.

– **Speed based segmentation.** During its movement, the speed of the moving object varies depending on its activity (running, resting, etc.). Therefore, it is one of the indicators of stops and moves. A speed threshold is fixed according to the application context. If the speed exceeds the threshold, the point is considered as moving, if not, it is considering as a stop.

4.3 Semantic Enrichment

Semantic enrichment is the phase during which the trajectory segments are enriched with semantic annotations and additional information. For instance, trajectory segments can represent an activity (walking, jogging, reading, etc.) or a place (at home, at work, etc.).

Semantic moves. Few researches such as [30, 33] payed attention to annotating moves since moving is not always meaningless, on the contrary, it enriches and completes the trajectory's significance.

Semantic stops. Stop annotation allows knowing the geographic location of the stop. This location has three possible forms. It could be shaped as a point (points of interest), a line (lines of interest) or a region (region of interest).

– *Annotating with points:* This type of annotation concerns geographic places represented as Points Of Interest (POIs). The most popular way to infer POIs is the Hidden Markov Model (HMM).
– *Annotating with lines:* This type of annotation concerns geographic networks represented as Lines Of Interest (LOIs). The lines describe the path taken by the moving object, such as roads or walkways. In order to infer LOIs from raw GPS data, researchers mainly use map matching algorithms [29, 35].
– *Annotating with regions:* This annotation concerns geographic places represented as Regions Of Interest (ROIs). In this case, the geographic place has to be a free form region, such as a campus.

5 Spatio-Temporal Regions

Some regular moving objects such as individuals, means of transport, etc., are represented by points and are moving in lines. And some are objects that move as regions (or in groups) such as hurricanes or diseases. This particular type of objects does not have an exactly delimited shape, thus it cannot be represented with exact coordinates (dimensions). So, they require a particular representation that takes into consideration their specificities.

5.1 Definition

Regions are shapes that have interiors and boundaries. It may be either motionless or unstable. The instability lies in the possibility of growing or shrinking. Modeling regions is a delicate task because every object is unique and needs a modeling that represents it. This representation has to be general to enclose all similar region objects but specific enough to preserve the object's specifications (size, speed, etc.). Several definitions were proposed to describe the region [6,10,27]. A region can be defined as *an abstraction of an object that has boundaries and interior (a forest, an oil spill, etc.). Its shape and extent are not defined with precision. A region may also be composed of faces which in turn may have holes.* In fact, the regions can be classified into three categories as shown in Figure 2: simple region, simple region with holes and complex region. A *simple region* is composed of a single component which has a connected interior and a connected boundary with no holes. A *simple region with holes* is a simple region that has holes inside. The holes do not belong to the region but they are enclosed by it. A *complex region* is a set of faces. It is composed of simple regions together with regions with holes. These faces meet at some point. The region can also be in movement. For instance, a storm is a moving region since it moves while changing its shape. It also starts from one place then moves unpredictably. A moving region changes its position discretely or continuously while changing its shape. In the following section, we will present some real life examples of moving regions to intensify their existence around us and to highlight the importance of dealing with them.

Simple region   Simple region with holes   Complex Region

Fig. 2. The three types of regions.

5.2   Regions Modeling Techniques

Region's abstractions represent lots of important aspects of our life. Therefore, researchers among the years tend to propose appropriate modeling techniques that could be classified into three main categories:

- *Conceptual modeling.* The conceptual modeling offers a clearer vision and understanding of the treated issues. Conceptual diagrams and data models indeed depict and describe complex objects and their relations [17,18,25].
- *Mathematical modeling.* In [6,11,12], the authors considered moving objects as geometries changing over time, and they defined corresponding algebraic operations. Then, in [10], the authors proposed a data model for moving objects in the two-dimensional space. While in [7], the authors completed the previous work [10] by presenting the discrete model. The work elaborated in [16] combined the three previously cited works [6,7,10] and enriched them. In fact, the authors developed powerful algorithms able to deal with a large set of operations and created a basis for the development of a reliable DBMS extension package for moving objects. The fuzzy aspect was also integrated. In [21],

the researcher offered a fuzzy spatial data type for fuzzy points, fuzzy lines and regions. He also proposed novel fuzzy spatial operations. He continued his work [22] by proposing metric operations on fuzzy spatial objects. Then, he defined a conceptual model of fuzzy spatial objects with its implementation [23]. The particularity of this work is the use of the discrete geometric domain (grid partition) instead of the Euclidean space. Afterwards, he proposed a fuzzy algebra to represent fuzzy data types while focusing on treating spatial vagueness [24].

– *Shape-based modeling.* The polygon representation was used in [27] to model every object moving both discretely and continuously. In [15], the authors proposed an approach that models regions and their trajectories, and predicts their future movements as well. In this work, regions are delimited by a rectangle named *enclosing box* to which was assigned a center point. While in [19], the authors presented their big data based system *Storm DB* used to answer rain fall amount queries in the continental United States. And in [28], the authors presented a data model for route planning in the case of forest fire. They used a fire simulation model that considers the wind properties since it influences the fire state and direction.

5.3 Application Scenarios

The application scenarios are classified into four main categories: natural catastrophes, environmental phenomena, diseases and animal swarming.

– *Natural catastrophes.* Natural catastrophes are the most common examples of moving regions since they move with constantly changing shapes. As examples, we can cite hurricanes, forest fires and massive rainfall.
– *Environmental phenomena.* Environmental phenomena such as air, sea and oil pollution can be represented by moving regions. Exposure to such kinds of pollution is dangerous and threatens all kinds of living species. Tracking and studying their mobility patterns helps preventing them and reducing their risks.
– *Diseases.* Cancer can be considered as a moving region since it is composed of multiple cells and moves inside the human body. Accordingly, a serious study of their behavior is crucial to increase the chance of curing the patients. Similar to cancer, alzheimer can be represented as a region. It attacks the brain and kills its cells in an unpredictable way.
– *Animal swarming.* Animals usually move in groups with coherent motion in order to achieve a common goal such as migration. Every species has its own way and reason of collective movement. It is important in this case to study their movement which is characterized by the group's position, orientation, speed, density, etc.

6 Regions' Trajectories Modeling Techniques

Modeling regions is important, however, modeling moving regions is more useful. Somewhat surprisingly, this aspect is largely ignored by the most recent literature, in other words, the representation of moving region trajectory without interfering or modifying their real unstable aspect. Therefore, moving regions trajectories attracted our interest. In fact, only few researchers appreciated this particular type of spatio-temporal data by proposing corresponding modeling techniques. In [9], a flock is a set of multiple moving point objects. It is defined by three main attributes: m which represents the minimum number of entities in a snapshot during an interval of time k, and r that represents the radius containing the entities. While in [14], the authors were interested in studying herds movements and behaviors. In fact, they tried to overcome the limitations of the work proposed in [9]. They identified three limitations. The first expresses that the radius doesn't always accurately represent the entities exact locations at an exact time. The second concerns flocks overlaps. And the third takes interest in the continuously changing flocks positions and shapes. Therefore, the authors proposed a new concept called *Herd* for which they associated four main activities: expand, join, shrink and leave. In order to describe the herds interactions, the authors used the graph representation. The vertices represent the herds, i.e., the clusters of entities. The edges represent the herds interactions. The interaction could be one of the four activities mentioned before. The graph representation is clearly descriptive, and the four chosen interactions are suited to describe the herds evolvements. However, the mobility aspect (location, direction, etc.) is not considered enough in this work. The authors took more interest in representing herds evolvements than their mobility. An another method is to describe the region's trajectory as a sequence of regions tracked over time [15]. The trajectory in this case is denoted by: $Traj(R) = R_t, R_{t+1}, R_{t+2}, ..., R_{t_{now}}$, where t_{now} is the most recent time in which the last snapshot was taken. Every region R is represented by an enclosing box. The trajectory is constructed by connecting the boxes' center points, which reduces regions to points at the end.

7   Conclusion

In this paper, we described the evolution of trajectories from raw to semantic data. Then, we discussed the modeling techniques by analyzing the past research contributions followed by a description of the trajectory construction steps. The tracked moving object has three possible forms: point, line or region. In the latter case, we were interested in region's trajectory to highlight the lack of researches in this specific field. In fact, moving regions, in particular, are characterized by their continuously changing path and shape. Therefore, they need specific modeling and representations that respect their specifications. Accordingly, we defined spatio-temporal regions together with their application scenarios and modeling techniques. Finally, we concluded our survey by focusing on regions' trajectories modeling techniques, which highlighted the lack of works in this particular type of mobility. As a future work, we intend to overcome the limitations of the discussed works by proposing a framework that deals with moving regions while respecting their specificities.

References

1. Bogorny, V., Heuser, C.A., Alvares, L.O.: A conceptual data model for trajectory data mining. In: Geographic Information Science, pp. 1–15. Springer (2010)
2. Bogorny, V., Renso, C., Aquino, A.R., Lucca Siqueira, F., Alvares, L.O.: Constant-a conceptual data model for semantic trajectories of moving objects. Trans. GIS **18**(1), 66–88 (2014)
3. Cetateanu, A., Luca, B.A., Popescu, A.A., Page, A., Cooper, A., Jones, A.: A novel methodology for identifying environmental exposures using GPS data. Int. J. Geogr. Inf. Sci. **30**(10), 1–17 (2016)
4. Chakri, S., Raghay, S., et al.: Enriching trajectories with semantic data for a deeper analysis of patterns extracted. In: International Conference on Hybrid Intelligent Systems, pp. 209–218. Springer (2016)
5. Damiani, M.L., Valdes, F., Issa, H.: Moving objects beyond raw and semantic trajectories. In: Proceedings of the 3rd International workshop on Information Management for Mobile Applications (IMMoA 2013), Riva del Garda, Italy. Citeseer (2013) http://ceur-ws.org/Vol-1075/00.pdf
6. Erwig, M., Güting, R.H., Schneider, M., Vazirgiannis, M.: Spatio-temporal data types: An approach to modeling and querying moving objects in databases. Geoinformatica **3**(3), 269–296 (1999)
7. Forlizzi, L., Güting, R.H., Nardelli, E., Schneider, M.: A data model and data structures for moving objects databases, vol. 29. ACM (2000)
8. Gong, L., Sato, H., Yamamoto, T., Miwa, T., Morikawa, T.: Identification of activity stop locations in gps trajectories by density-based clustering method combined with support vector machines. J. Mod. Transp. **23**(3), 202–213 (2015)
9. Gudmundsson, J., van Kreveld, M.: Computing longest duration flocks in trajectory data. In: Proceedings of the 14th Annual ACM International Symposium on Advances in Geographic Information Systems, pp. 35–42. ACM (2006)
10. Güting, R.H., Böhlen, M.H., Erwig, M., Jensen, C.S., Lorentzos, N.A., Schneider, M., Vazirgiannis, M.: A foundation for representing and querying moving objects. ACM Trans. Database Syst. (TODS) **25**(1), 1–42 (2000)
11. Güting, R.H., De Ridder, T., Schneider, M.: Implementation of the rose algebra: Efficient algorithms for realm-based spatial data types. In: Advances in Spatial Databases, pp. 216–239. Springer (1995)
12. Güting, R.H., Schneider, M.: Realm-based spatial data types: The rose algebra. VLDB J.-Int. J. Very Large Data Bases **4**(2), 243–286 (1995)
13. Hu, Y., Janowicz, K., Carral, D., Scheider, S., Kuhn, W., Berg-Cross, G., Hitzler, P., Dean, M., Kolas, D.: A geo-ontology design pattern for semantic trajectories. In: Spatial Information Theory, pp. 438–456. Springer (2013)
14. Huang, Y., Chen, C., Dong, P.: Modeling herds and their evolvements from trajectory data. In: Geographic Information Science, pp. 90–105. Springer (2008)
15. Junghans, C., Gertz, M.: Modeling and prediction of moving region trajectories. In: Proceedings of the ACM SIGSPATIAL International Workshop on GeoStreaming, pp. 23–30. ACM (2010)
16. Lema, J.A.C., Forlizzi, L., Güting, R.H., Nardelli, E., Schneider, M.: Algorithms for moving objects databases. Comput. J. **46**(6), 680–712 (2003)
17. Ma, Z., Zhang, F., Yan, L.: Fuzzy information modeling in uml class diagram and relational database models. Appl. Soft Comput. **11**(6), 4236–4245 (2011)
18. Massaâbi, M., Akaichi, J.: Modeling moving regions: Colorectal cancer case study. In: Intelligent Interactive Multimedia Systems and Services 2016, pp. 417–426. Springer (2016)

19. Olsen, B., McKenney, M.: Storm system database: A big data approach to moving object databases. In: 2013 Fourth International Conference on Computing for Geospatial Research and Application (COM. Geo), pp. 142–143. IEEE (2013)
20. Parent, C., Spaccapietra, S., Renso, C., Andrienko, G., Andrienko, N., Bogorny, V., Damiani, M.L., Gkoulalas-Divanis, A., Macedo, J., Pelekis, N., et al.: Semantic trajectories modeling and analysis. ACM Comput. Surv. (CSUR) 45(4), 42 (2013)
21. Schneider, M.: Uncertainty management for spatial datain databases: Fuzzy spatial data types. In: Advances in Spatial Databases, pp. 330–351. Springer (1999)
22. Schneider, M.: Metric operations on fuzzy spatial objects in databases. In: Proceedings of the 8th ACM International Symposium on Advances in Geographic Information Systems, pp. 21–26. ACM (2000)
23. Schneider, M.: Design and implementation of finite resolution crisp and fuzzy spatial objects. Data Knowl. Eng. 44(1), 81–108 (2003)
24. Schneider, M.: Fuzzy spatial data types for spatial uncertainty management in databases. In: Handbook of Research on Fuzzy Information Processing in Databases, vol. 2, pp. 490–515 (2008)
25. Singh, S., Agarwal, K., Ahmad, J.: Conceptual modeling in fuzzy object-oriented databases using unified modeling language. Int. J. Latest Res. Sci. Technol. 3, 174–178 (2014)
26. Spaccapietra, S., Parent, C., Damiani, M.L., de Macedo, J.A., Porto, F., Vangenot, C.: A conceptual view on trajectories. Data Knowl. Eng. 65(1), 126–146 (2008)
27. Tøssebro, E., Güting, R.H.: Creating representations for continuously moving regions from observations. In: Advances in Spatial and Temporal Databases, pp. 321–344. Springer (2001)
28. Wang, Z., Zlatanova, S., Moreno, A., van Oosterom, P., Toro, C.: A data model for route planning in the case of forest fires. Comput. Geosci. 68, 1–10 (2014)
29. Yan, Z., Chakraborty, D., Parent, C., Spaccapietra, S., Aberer, K.: Semantic trajectories: Mobility data computation and annotation. ACM Trans. Intell. Syst. Technol. (TIST) 4(3), 49 (2013)
30. Yan, Z., Giatrakos, N., Katsikaros, V., Pelekis, N., Theodoridis, Y.: Setrastream: Semantic-aware trajectory construction over streaming movement data. In: Advances in Spatial and Temporal Databases, pp. 367–385. Springer (2011)
31. Yan, Z., Macedo, J., Parent, C., Spaccapietra, S.: Trajectory ontologies and queries. Trans. GIS 12(s1), 75–91 (2008)
32. Yan, Z., Spaccapietra, S.: Towards semantic trajectory data analysis: A conceptual and computational approach. In: VLDB Ph.D. Workshop. Citeseer (2009)
33. Yu, F., Ip, H.H.: Semantic content analysis and annotation of histological images. Comput. Biol. Med. 38(6), 635–649 (2008)
34. Zhang, A., Song, S., Wang, J.: Sequential data cleaning: A statistical approach (2016)
35. Zheng, Y.: Trajectory data mining: An overview. ACM Trans. Intell. Syst. Technol. (TIST) 6(3), 29 (2015)

Integrating Trajectory Data in the Warehousing Chain: A New Way to Handle the Trajectory ELT Process

Noura Azaiez[1]([✉]) and Jalel Akaichi[2]

[1] Department of Computer Science, ISG Institute, Tunis University, Tunis, Tunisia
noura.azaiez@gmail.com
[2] Department of Computer Science, College of Computer Science, King Khalid University, Abha, Saudi Arabia
jalel.akaichi@kku.edu.sa

Abstract. Companies that build data warehouses and use business intelligence for decision-making ultimately save money and increase profit. The warehousing technology proceeds a consolidation of data from a variety of sources that is designed to support strategic and tactical decision making. Data gathered from different sources have to flow in the target area to be analyzed. ETL process is responsible to perform this task. It pulls data out of the source systems and placing it into a data warehouse. However, Geographical Information Systems, pervasive systems and the positioning systems impose moving beyond the traditional warehousing management towards what is called Trajectory Data Warehousing. This later supports trajectory data. Therefore, traditional ETL processes are unable to perform their tasks when the mobility aspect is integrated. Towards this inadequacy, Trajectory ETL process emerges. Few are the works that dealt with. In this paper, we present a taxonomy of works that deeply investigate the ETL process modeling whatever the data kind that the warehousing chain supports, then we express how the trajectory ELT process based on the Model Driven Architecture approach aims at enhancing decision making.

Keywords: Trajectory ELT process, Extraction, Loading, Transformation · Trajectory construction · Trajectory Data Source model · Trajectory Data Mart model · Model Driven Architecture, Hadoop clusters

1 Introduction

Business Intelligence (BI) is used to enhance decision-making capabilities for managerial processes. With a strong BI, companies can support decisions with more than just a gut feeling. Satisfying the BI needs depends on data warehouse designs and data integration workflows. Data warehousing emphasizes the capture of data from diverse sources for useful analysis and access. Performing the BI goal requires the transformation of these collected raw data into analytical data. Thanks to its features, the ETL process presents the core of the warehousing chain. This later pulls data out of the source systems and placing it into a data warehouse passing through the extraction, transformation and loading tasks; in general, classic data are targeted. However, thanks to

© Springer International Publishing AG 2018
G. De Pietro et al. (eds.), *Intelligent Interactive Multimedia Systems and Services 2017*,
Smart Innovation, Systems and Technologies 76, DOI 10.1007/978-3-319-59480-4_35

technology development in the Geographic Information Systems and pervasive ones realm, the trajectory data (TD) emerged. TD reflect the movements of mobile objects in the real world. Supporting this data kind by traditional Decision Support Systems impedes the coherence of the analysis results since these later features lack of the mobility aspects. Building a repository that supports this kind of data becomes a necessity. For that, the notion of a Trajectory Decision Support Systems (TDSS) arises [1]. Before their availability for decisional purposes, raw TD have to flow through a set of tasks under the general title Extract-Transform-Load trajectory ETL (T-ETL) process. T-ETL process is the art to integrate the trajectory aspect to handle the extraction, transformation and loading tasks. Few are the research works that investigate the problem of ETL process modeling, especially, when the trajectory data are targeted.

In this paper, our goal is to highlight a new vision to manage the warehousing chain when integrating mobility data. For that, we present taxonomy of works that investigate the ETL process modeling whatever the data kind that the warehousing chain supports. We expect the absence of a method that really serves the trajectory ETL modeling taking into account performance, coherence and maintenance. Thus, we describe a new way that handles the Trajectory ELT (T-ELT) process relying on the Model Driven Approach (MDA) [7] aiming to enhance the decision making systems and therefore achieve the BI goal in a better environment.

Our paper is described as follow. Section 2 presents works that deal with the ETL process modeling supporting classic and trajectory data. Section 3 discusses these works and presents a comparative study between them. In Sect. 4, we express our new vision towards ELT process modeling and its effectiveness on the decision making systems. Section 5 concludes the paper.

2   State of the Art

ETL process comes from Data Warehousing and stands for Extract-Transform-Load. It covers a process of how the data are loaded from the source system to the data warehouse. An ETL system provides a development environment, tools for managing operations of maintenance, discovering and analyzing. In particular, it allows extracting data from heterogeneous sources, cleaning and standardizing data according to business rules established by the company, loading them into a data warehouse and/or propagate them to the data marts. Contrary of the ETL process modeling that presents the focus of many research works, the Trajectory ETL process modeling doesn't receive the interest that deserves although its importance. In this section, we present a set of research works that propose solutions to deal with the ETL and T-ETL process modeling practical problems.

2.1   ETL Process Modeling in Traditional Way

The research community that interests on the ETL process modeling realm offers a set of solutions to deal with practical problems in the domain.

Let us start with the work of [6] where authors proposed a method that is based on RDF/OWL ontologies and design tools. The main steps of their work start from the mapping between the source data and its OLAP form which is done by converting the data first to RDF using ontology maps. Then, the data are extracted from its RDF form by queries that are generated using the ontology of the OLAP schema. As a final step, the extracted data are stored in the database tables and analyzed using an OLAP software.

Web services also play its whole role to support the ETL processes. Authors in [4] offered a new approach that is based on web services in order to perform the BI goal. The proposed architecture relies on the idea of splitting the classical wrapping module into a source specific and a target specific part and establishing the communication between these components based on Web Service technology. Relying on Web Service technology is explained by the description and the integration the participated data sources and the deployment within a specific database system.

Authors in [10] facilitated the modeling hard effort for ETL workflows relying on standard methods; they employed Unified Modeling Language (UML) for this purpose. The main advantage of their proposal is ensuring the seamless integration of the design of the ETL processes with the DW conceptual schema. Authors employed class diagrams and not activity diagrams for their modeling. The main reason for dealing with class diagrams is explained by the reason that the focus on modeling is done on the activities interconnection and data stores and not on the actual sequence of steps that each activity performs. Authors defined a set of UML stereotypes that represents the most common ETL tasks such as the integration of different data sources, the transformation between source and target attributes, the generation of surrogate keys, and so on. As illustration, authors implemented their approach in through the Rose Extensibility Interface (REI) [12].

Another manner to investigate the ETL modeling is offered in [2]. In fact, authors proposed a new approach for complex data integration, based on a Multi-Agent System (MAS) associated with the data warehousing approach. The proposed framework consists of physically integrating complex data into a relational database, named ODS (Operating Data Storage) considered here as a buffer ahead of the data warehouse. This system is composed of a set of intelligent agents offering the different services that are necessary to achieve the three major tasks for ETL modeling. As a first task, Data Extraction is performed by the agent in charge to extract data characteristics from complex data. The obtained characteristics are then transmitted to the agent responsible for Data Structuring that present the second task. To do so, the responsible agent deals with the organization of the data according to a well-defined data model. Then, the model is transmitted to the agent responsible for data storage which is the last task of the ETL process. Data Storage is performed by the agent that feeds the database with data by using the model supplied by the data structuring agent.

In order to create an active ETL tool, authors in [8] defined a model covering different types of mapping expressions. Authors specified a model to represent mapping guideline and expressions. Mapping guideline means the set of information defined by the developers in order to achieve the mapping between the attributes of two schemas. Mapping expression of an attribute is the information needed to recognize how a target attribute is created from the sources attributes. Authors' idea consists to define a model for the mapping guideline. Then, they proved this model by creating a warehousing tool (QELT). Authors are based on queries to represent the mapping between the source and the target data. Therefore, the proposed approach enables a complete interaction between mapping meta-data and the warehousing tool. In their work, authors described how it could be easier to acquire and transform data by using SQL queries which are automatically generated from mapping meta-data.

2.2 Trajectory ETL Process Modeling

Authors in [5] investigated the problem of TDW building, especially, the ETL process handling. Hence, they proposed a framework that emphasizes, at the first level, the trajectories reconstructing. This task requires to collect raw data generated from the mobile objects movements; they represent time-stamped geographical locations. In their work, authors assumed a filter to be part of a trajectory reconstruction manager, along with a simple method for determining different trajectories, which applies it on raw positions. Once trajectories have been constructed and stored in a MOD, the ETL phase is executed in order to feed the TDW. This is ensured through the CELL-ORIENTED-ETL algorithm. The proposed approach has been tested and validated using a large real dataset.

Author in [3] offers a new method that aims at designing a TDW model, having the ability to store and analyze trajectories' data. Hence, the author investigated deeply the ETL process modeling problem. The proposed framework collects streams of spatio-temporal observations related to the position of moving objects; this presents the first task to reconstruct TD model. The next step presents the ETL phase handling where such reconstructed trajectories will be used in order to load the aggregated data into the proposed TDW. To this end, the author relied on a Moving Object Database (MOD) [13] which is a special database able to handle trajectories data and to perform spatio-temporal queries on them. Given a MOD, in order to feed-up the Trajectory Data Warehouse, the measure associated to the base granule can be computed by finding out the trajectories portions that lie inside each base granule, using the MOD spatio-temporal capabilities, and then calculating the needed values starting from these partial data. Authors in [3] handled this task using two different ways of ad-hoc procedures such as cell-oriented (COA) and trajectory-oriented approach (TOA) for storing and loading the Trajectory Data Warehouse.

The research work [11] also served the Trajectory ETL modeling. Authors presented a Trajectory ETL model that is able to clean and manage Trajectory Data. They relied on the UML language in order to describe the ETL process as a flow of activities. This model is based on two levels. The highest level which describes the general process of data warehousing and the lowest level that details the Trajectory ETL process. Authors focused on the TDW conceptual modeling in order to facilitate the analysis of Trajectory Data in a multidimensional context. In their case, they need to analyze the activities of a mobile medical delegate. Authors proposed two algorithms. The first one implements trajectory ETL tasks and the second one aims to construct trajectories. They implemented those algorithms using the Geokettle Platform.

3   Discussion

In this section, we present a recapitulative table (Table 1) which offers a comparative study between some solutions that aim at solving the ETL and Trajectory ETL processes modeling problem. Following the study of this sample of the works described above, it becomes possible to rely on a set of principal criteria: ETL modeling, T-ETL modeling, ELT modeling, T-ELT modeling and mechanisms responsible for modeling performing.

Table 1. A taxonomy

	ETL modeling	T-ETL modeling	ELT modeling	T-ELT modeling	Mechanisms
Niinimäki et al. (2009) [6]	✓				RDF/OWL ontologies
Schlesinger et al. (2005) [4]	✓				Web service
Trujillo et al. (2003) [10]	✓				Unified Modeling Language (UML)
Boussaid et al. (2003) [2]	✓				Multi-Agent System (MAS)
Rifaieh et al. (2002) [8]			✓		Mapping guideline and expressions
Marketos et al. (2008) [5]		✓			CELL-ORIENTED-ETL algorithm
Leonardi (2014) [3]		✓			Cell-Oriented (COA) Approach + Trajectory-Oriented Approach (TOA)
Zekri et al. (2014) [11]		✓			Trajectory construction algorithms
Our proposition				✓	Hadoop clusters + Trajectory Construction algorithm + Model Driven Architecture

After a deep study in the ETL modeling realm, it becomes possible to extract the strengths and weaknesses of each one and to express our position.

Starting with ETL modeling, as it is mentioned above, a set of works investigated this problem and proposed a set of solutions to solve it. Many authors relied on classic techniques to perform the ETL process handling such as UML [10], MAS [2] or on mapping guideline and expressions [8]. Others challenged the existing solutions and relied on a set of powerful mechanisms that leads to relevant results. For example, relying on RDF/OWL ontologies as the work [6] or basing on the web services as the work [4] presents original approaches and therefore a real challenge in the domain of ETL modeling.

As it is described above, the majority of works treating the ETL modeling takes the Extraction, Transformation and Loading processes scheduling as the basis of their works. In fact, authors investigated the problem of the ETL modeling relying on the same strategy; Extraction, Transformation and then loading processes. Relying on the ETL processes is a gainful strategy at the time consuming level. In fact, only data relevant to the presentation are extracted and processed. Hence, the target area contains only data relevant to the presentation. This potentially reduce development and therefore time. Hence, reduced warehouse content simplifies the security regime implemented and thus the administration overhead. However, the ETL strategy complains from several weaknesses. Starting with the flexibility; targeting only relevant data for output means that any future requirements, that may need data that was not included in the original design, will need to be added to the ETL routines. As a result, this increases the time and costs involved.

We emphasized the works that integrate the trajectory notion and we noted that this later offers the possibility of reducing geographical disparities related to organization services. This has a direct impact on existing Information Systems to evolve towards a new type of repository that supports what is called trajectory data. Trajectory data reflect the movements and stops of mobile objects in the real world. Few are the authors that integrated the mobility aspect in their works. For example, both the work of [3, 5] relied almost on the same approach: a set of measures and equations that translate the use of cell-oriented (COA) and trajectory-oriented approach (TOA) for storing and loading the Trajectory Data Warehouse. Authors in [11] also introduced the trajectory aspect in their work and proposed two algorithms for implementing the trajectory ETL process tasks. Frameworks that are created to serve the Trajectory ETL process modeling are complex and need high level experts to handle all the process tasks.

A deep study of the research works shows that whatever the data kind supported (classic or trajectory), authors used classic techniques (transportable tables…) for loading data in the warehouse area.

As it is mentioned in the Table 1, we contribute to model a trajectory ELT process relying on a set of powerful mechanisms such as Hadoop clusters and the Model Driven Architecture approach. In the next section, we will approve how our proposition can enhance the organizations decision making.

4   A New Vision Towards the Trajectory ELT Modeling

The ETL process seems quite straight forward. As with every application, there is a possibility that the ETL process fails. This can be caused by missing extracts from one of the systems, missing values in one of the reference tables, or simply a connection or power outage. Therefore, it is necessary to model the ETL process keeping fail-recovery in mind.

4.1   Why Relying on ELT Process Is More Lucrative?

The ETL strategy ensures the transformation of the trajectory raw data towards decisional data. Unfortunately, it is not enough. In fact, for a knowledge worker, nothing is more important than better and faster decision making. Besides, the business analysis main purpose is to ensure the best knowledge extracted quality at minimal costs and time consuming. For that, we propose to move beyond the traditional ETL process strategy towards swapping its tasks scheduling by isolating the extract and load processes from the transformation process. In other word, instead of transforming (T) the data before they are written, the Extract, Load and Transform (ELT) process leverages the target system to perform the transformation task. Isolating the load process from the transformation process removes an inherent dependency between these stages. In addition of including the necessary data for the transformations, the extract and load process can include elements of data that may be required in the future. In fact, the load process could take the entire source and load it into the warehouse. Isolating the transformations from the load process fosters a more staged approach to the warehouse design and implementation; this embraces the ongoing changing nature of the warehouse build. Thanks to the ELT processes strategy, all data provided in the sources are extracted and loaded into the warehouse. This, combined with the isolation of the transformation process, means that future requirements can easily be incorporated into the warehouse structure. This proves the future effectiveness of the ELT processes. Another issue that ELT process can resolve is minimizing risks. In fact, removing the close interdependencies between each stage of the warehouse build process enables the development process to be isolated, and the individual process design can thus also be isolated. This provides an excellent platform for change, maintenance and management. The ELT process handling seems as a tremendous payoff strategy for decision making systems.

4.2   How Hadoop Clusters and the MDA Approach Can Contribute in the Trajectory ELT Modeling?

Handling the T-ELT processes is far from the traditional ETL processes management.

Integrating the concept of trajectory offers the possibility of reducing geographical disparities related to organization services. This has a direct impact on existing information systems to evolve towards a new type of repository named Trajectory Decision Support System (TDSS) that supports what is called trajectory data. These later reflect the moves and stops of mobile objects in the real world. Aiming at exploiting its advantages, we propose to integrate the trajectory aspect in our work in order to model the Trajectory ELT (T-ELT) process. This later is responsible to extract, load and transform data provided from mobile objects movements; those movements build mobile objects trajectories. Create the environment to analyze those trajectories is our main goal.

The integration of the mobility concept requires relying on powerful mechanisms that are able to support trajectory aspects. Besides, the main idea is not restricted to swap the rank of the processes tasks, but to express the power of each ELT task how to take part to achieve the main decision-makers goal from its position. Let us start with the trajectory extraction task. Raw trajectory data that describe mobile objects movements are collected from a generator tool related to the mobile device. Once data are extracted, the loading phase takes place. Performing this task can not be ensured by the traditional loading tools. Thus, ELT works with high-end data engines such as Hadoop clusters. In fact, Hadoop Distributed File System (HDFS) can store everything from structured data and unstructured data. Trajectory ELT process allows extracting and loading all gathered TD into HDFS to prepare them for the transformation task. Due to its complexity, we propose to divide this later in two sub-tasks: T1 and T2. Each sub-task is managed according to its own goal. Hence, T1 is performed through an algorithm that converts the raw extracted positions into a full mobile object trajectory. The resulting trajectories are stored into Trajectory Database (TDB) as Trajectory Data Source model (TDSrc).

The T2 goal is to identify how a target field could be generated from the source field.

To this end, we have already defined a set of textual rules in [14] that results the generation of a set of TDM elements directly from TDSrc at the aim to analyze mobile objects trajectories obtained in T1. However, these textual existing rules are inadequate to achieve this task since their application in their state leads to waste time and this is contradictory to our goal. Hence, their translation basing on a transformation language becomes a must. After taking a look on the works that takes the Model Driven Architecture a basis of their work, we note that the MDA approach expresses its powerful features in some domain such as the data warehouse evolution [9]. Therefore, we propose to rely on this mechanism to perform the transformation of the source models into target ones. The MDA approach is a specific proposition for implementing the Model Driven Engineering (MDE). MDE is heavily building on model transformations in converting models. It is a general methodology for software development, which recognizes the use of recurrent patterns to increase compatibility between systems. The MDA based development of a software system starts by building Platform Independent Models (PIM) of that system which are refined and transformed into one or more Platform Specific Models (PSMs) which can be used to generate source code. It is destined to application design and implementation. It encourages efficient use of system models in the software development process, and it supports reuse of best practices when creating families of systems. As defined by the Object Management Group (OMG), MDA is a way to organize and manage enterprise architectures supported by automated tools and services for both defining the models and facilitating transformations between different model types; in our case, transforming TDSrc that describes the obtained trajectories from T1 into Trajectory Data Mart (TDM) models.

5   Conclusions

In this paper, we presented a set of research work treating the problem of ETL process modeling when the warehousing chain supports both classic and trajectory data. Then, we presented a taxonomy to compare strategies adopted in each work. We discussed their strengths and weaknesses to define gaps that still remain open topics. This high-lights a new way to deal with the ETL process modeling realm. Therefore, we expressed our vision towards a powerful Decision Support System modeling. As a first trend, we proposed to integrate trajectory notion in the warehousing chain. Besides, thanks to its features, we presented the effectiveness of the ELT process to enhance decision making. We emphasized the power of Hadoop clusters and the MDA approach to handle the trajectory ELT tasks. This responds not only the current needs but also it serves the long-term vision since it offers an easy feasibility of incorporating future requirements into the DW structure; what proves the future efficiency of the ELT process. Our proposition described above is based, purely, on theoretical proofs. Therefore, we are working to validate our approach in order to create a framework which handles the ELT process modeling when integrating trajectory data.

References

1. Azaiez, N., Akaichi, J.: What is the impact of mobility data integration on decision support systems' modelling and evolution? Int. J. Inf. Syst. Serv. Sect. (IJISSS) **8**(1), 1–12 (2016)
2. Boussaid, O., Bentayeb, F., Darmont, J.: An MAS-Based ETL approach for complex data. In: 10th ISPE International Conference on Concurrent Engineering: Research and Applications, Madeira, Portugal, pp. 49–52 (2003)
3. Leonardi, L.: A framework for trajectory data warehousing and visual OLAP analysis. Doctoral thesis Ca'foscari university of Venice (2014)
4. Schlesinger, L., Irmert, F., Lehner, W.: Supporting the ETL-process by web service technologies. Int. J. Web Grid Serv. (IJWGS) **1**(1), 31–47 (2005)
5. Marketos, G., Frentzos, E., Ntoutsi, I., Pelekis, N., Raffaetà, A., Theodoridis, Y.: Building real-world trajectory warehouses. In: The International ACM Workshop on Data Engineering for Wireless and Mobile Access, Vancouver, BC, Canada, pp. 8–15 (2008)
6. Niinimäki, M., Niemi, T.: An ETL process for OLAP using RDF/OWL ontologies. In: Spaccapietra, S., Zimányi, E., Song, I.-Y. (eds.) Journal on Data Semantics XIII. LNCS, vol. 5530, pp. 97–119. Springer, Heidelberg (2009)
7. O.M.G.: Model Driven Architecture (MDA) (2004). http://www.omg.org/cgibin/doc?formal/03-06-01
8. Rifaieh, R., Benharkat, N.A.: Query-based data warehousing tool. In: 5th ACM International Workshop on Data Warehousing and OLAP, pp. 35–42 (2002)
9. Taktak, S., Alshomrani, S., Feki, J., Zurfluh, G.: An MDA approach for the evolution of data warehouses. IJDSST **7**(3), 65–89 (2015)
10. Trujillo, J., Luján-Mora, S.: A UML based approach for modeling ETL processes in data warehouses. In: 22nd International Conference on Conceptual Modeling, Chicago, IL, USA, pp. 307–320 (2003)
11. Zekri, A., Akaichi, J.: An ETL for integrating trajectory data a medical delegate activities use case study. In: International Conference on Automation, Control, Engineering and Computer Science, pp. 138–147 (2015)
12. Rational Software Corporation: Using the Rose Extensibility Interface (REI). Technical report (2001)
13. Pelekis, N., Frentzos, E., Giatrakos, N., Theodoridis, Y.: HERMES: a trajectory DB engine for mobility-centric applications. Int. J. Knowl.-Based Organ. (IJKBO) **4**(1), 19–41 (2014)
14. Azaiez, N., Akaichi, J.: How trajectory data modeling improves decision making? In: Proceedings of the 10th International Conference on Software Engineering and Applications, Colmar, Alsace, France, pp. 87–92 (2015)

Detection of Opinion Leaders in Social Networks: A Survey

Seifallah Arrami[1(✉)], Wided Oueslati[2], and Jalel Akaichi[3]

[1] High School of Business, Manouba University, Manouba, Tunisia
arrami.seifallah@gmail.com
[2] High Institute of Management, Tunis University, Tunis, Tunisia
widedoueslati@live.fr
[3] King Khalid University, Abha, Saudi Arabia
jalel.akaichi@isg.rnu.tn

Abstract. With the development of new media such as social networking sites, content sharing sites, blogs and micro blogs a profound transformation in terms of communication between consumers and companies has been created. In fact, this great revolution of the web has allowed different users to interact, express their opinion on a product or a service and post comments. Then, internet users have gone from passive to active actors who are able to produce information and make data available on the web as a rich opinions source. Therefore companies must deal with this reality, to know what others may say about their products of competing brands because the only and the best way to sell their products in good condition is to produce what consumers want. Along with this phenomenon, recent years have seen the birth of a generation of Internet users elected by the company to help it to manage its on-line reputation. Those users are called opinion leaders or influencers; they have a high capacity to influence those around them because they are considered to be more experienced, objective and able of provoking the emotions of someone else. Therefore the necessity of identifying opinion leaders has been proved more and more crucial. The goal of this paper is to present different research works that aimed to detect opinions leaders in social network.

Keywords: Social networks · Opinion leaders · Opinion leaders detection · Centrality technique · Maximization technique

1 Introduction

With the explosive growth of social networking websites, people have become more communicative by disseminating data about products, services, brands… These social media are new sources of important information for users and for companies. Moreover, social networks often serve as platforms for information dissemination and promotion through viral marketing or product placement. The success rate in this type of marketing could be increased by targeting specific individuals, called "opinion leaders or influential users". in this paper, we will present different approaches that solved the problem of opinion leaders detection in social networks such as Facebook, twitter, blogs… the paper is structured as follows: in Sect. 2, we will give the definition and the role of the opinion

© Springer International Publishing AG 2018
G. De Pietro et al. (eds.), *Intelligent Interactive Multimedia Systems and Services 2017*,
Smart Innovation, Systems and Technologies 76, DOI 10.1007/978-3-319-59480-4_36

leader. In Sect. 3, we will expose research works that used the centrality technique to detect opinion leaders. In Sect. 4, we focus on approaches that used the maximization technique for the same aim. In Sect. 5 we will elaborate a comparative study between those approaches. In Sect. 6, we conclude the work and present some perspectives to be done in the future.

2   Opinion Leader: Definition and Role

With the explosive growth of social media (blogs, forums, micro-blogs and social networks) on the Web, people have become more communicative by disseminating data about products, services, brands... Those social media are new sources of important information for users and for companies. Indeed, several analysis researches of social media were conducted for detection of opinion leaders. This latter was in [1] as a person who influence his environment and/or who exchange oral information about product and brands. Authors of [1] have replaced the term opinion leaders by the term influencer because in their point of view the first term refers to the exchange of information in one direction when we are often in the case of reciprocal exchange in both directions.

According to several authors this new term is not so convincing because influencer is not necessarily an opinion leaders. In fact, a TV star, a family member or even a friend can influence us but it is not necessarily an opinion leader. The opinion leader in marketing was defined by [2] as an attractive person, by his qualities of psychological, physical and social, whose knowledge in a given product category are considered credible. Believes and behavior may influent the believes and choice of brands around him in the product category.

In [3] an opinion leader is a person who influences the behavior of those around him informally for a desired objective. His opinions draw their attention and awake their curiosity about a product or service and sometime request them to buy it.

The marketing was dealt with opinion leaders because they are sought after by potential buyers when purchasing new products and, in a general manner, for product that purchase this character involving.

Authors in [4], consider that opinion leaders cause more consumer interest than advertising because they are considered to be more experienced on a product or service and their suggestion are more objective.

3   Detection of Opinion Leaders with the Centrality Technique

In this section, we will expose different approaches that were used to detect opinion leaders in social network based on the centrality technique.

The Internet world has witnessed a staggering growth of social networking websites. A social network can be defined as "a social structure made up of users called "nodes" and "links" (relationships/interactions) between them". These graphs based on studies gave birth to a notion of centrality, measuring the importance of each node in a given graph. The concept of centrality aims to highlight the important nodes that represent the opinion leaders or influential users. let's mention that there are different types of centrality as presented in Fig. 1. In fact, we find the Betweenness centrality (A), the Closeness centrality (B), the Eigenvector centrality (C), the Degree centrality (D), The Harmonic Centrality (E) and the Katz centrality of the same graph.

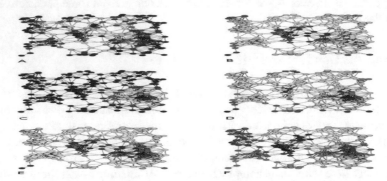

Fig. 1. Example of centrality types in the same graph

In the study of the notion of centrality, it is important to distinguish the case of oriented graph from the case of non-oriented graph. In oriented graphs, nodes have two types of links that are the inbound links and the outbound links. For a given definition of the notion of centrality, each node will then have two measures of importance: a measure relating to its outgoing links, called a measure of centrality, influence, or outgoing centrality, and another measure relating to its inward links, called measure of prestige, popularity, authority or incoming centrality [2].

In a non-oriented graph, each node has only one type of links or relationships with other nodes. Each node then has only one measure of importance called a measure of centrality [2]. Note that centrality is based on the choices made by a user whereas prestige depends on choices that a given user received from others.

Concerning the notion of centrality, there are several measures including: degree centrality, closeness centrality, betweenness centrality, degree of density, coefficient clustering, PageRank.

3.1 Degree Centrality

Authors in [3] presented the simplest and most intuitive form of the notion of centrality. It is based on the idea that the importance of an individual within a group depends on the total number of people he or she knows or interacts with. According to this measure, determining the importance of a node in a graph amounts to calculating the number of its neighboring vertices, or in an equivalent way, to calculate the number of links that are incident to it. In the theory of graphs, this number is called the degree of the node, hence the name of Degree centrality. $G = (V, E)$ is a graph of order N represented by its adjacency matrix A.

In the case where the graph G is not-oriented, the degree centrality of a node $v_l \in V$ is defined by:

$$C^{deg}(v_i) = \frac{1}{N-1} \sum_{j=1}^{N} a_{ij}$$

In the case where the graph G is oriented, each node $v_I \in V$ then has two measures of centrality of degree: One to the outbound links and one to the inbound links. They are defined respectively by:

$$C_{out}^{deg}(v_i) = \frac{1}{N-1} \sum_{j=1}^{N} a_{ij} \quad ; C_{in}^{deg}(v_i) = \frac{1}{N-1} \sum_{j=1}^{N} a_{ij}$$

Degree centrality is also called a measure of local centrality [1] because it does not take into account the global structure of the graph and is calculated only from the immediate vicinity of a vertex.

3.2 Closeness Centrality

The closeness centrality (CC) [4], in contrast to proximity prestige, pinpoints how close a member is to all the others within the social network. The main idea is that the member takes the central position if they can quickly contact other members in the network. This measure emphasizes quality (position in a network) rather than quantity (number of links, like in a centrality degree measure). The member with high CC is a good propagator of ideas and information.

A similar idea was studied for hypertext systems. The closeness centrality CC(x) of member x tightly depends on the geodesic distance, i.e. the shortest paths from member x to all other people in the social network and is calculated as follows:

$$CC(x) = \frac{m-1}{\sum_{y \neq x, y \in M} c(x,y)}$$

Where: $c(x,y)$ is a function describing the distance between nodes x and y (i.e. max, min, mean or median), m the number of nodes in a network.

3.3 Betweeness Centrality

Betweenness centrality (BC) [4], of member x pinpoints to what extent x is between other members. Members with high BC are very important to the network because others actors can connect with each other only through them. It can be calculated only for undirected relationships by dividing the number of shortest geodesic distances (paths) from y to z by the number of shortest geodesic distances from y to z that pass through member x. This calculation is repeated for all pairs of members y and z, excluding x:

$$BC(x) = \frac{\sum_{i \neq x \neq j, i,j \in M} b_{ij}(x)}{b_{ij}}$$

Where $b_{ij}(x)$ is the number of shortest paths from i to j that pass through x.

b_{ij} is the number of all shortest path between i and j. m is the number of nodes in a network.

If a member obtains high value of BC then it means that he/she is the node without which the network will split into subnetworks.

3.4 Eigenvector Centrality

The approach proposed by Bonacich [5] for calculating Centrality in a graph is very different from the other approaches that we have presented. It has indeed suggested the idea that the centrality of a node is determined by the centrality of the nodes to which it is connected. In a social network, this corresponds to the idea that an actor is all the more important as it is connected to actors who are themselves important. It is like recursive version of degree centrality:

$$x_v = \frac{1}{\lambda} \sum_{t \in M(v)} x_t = \frac{1}{\lambda} \sum_{t \in G} a_{v,t} x_t$$

For a given graph $G: = (V, E)$ with $|V|$ number of vertices. Let $A = (a_{v,t})$ be the adjacency matrix, i.e. $a_{v,t} = 1$ if vertex v is linked to vertex t, and $a_{v,t} = 0$ otherwise. The relative centrality score of vertex can be defined as: $\mathbf{Ax} = \lambda \mathbf{x}$.

PageRank and the Katz centrality are variants of the eigenvector centrality. Several researchers have considered using the links between documents is to improve the performance of search engines on the web. The relevance of a document to a query is then calculated by combining the document's similarity to the query with the centrality of the query document [1].

We present below the most known algorithms for the calculation of centrality in the graphs of documents. These algorithms are all based on the idea that the importance of a document is related to the importance of the documents to which it is connected.

The PageRank [6] is one of the link analysis algorithms that has most marked the field of information retrieval on the web. It was also ranked among the top 10 data mining algorithms [7]. It is due to this algorithm that the search engine of Google has known the success that it has today. In its simplified version, the PageRank algorithm considers that the importance (also known as popularity or PageRank) of a page depends on the popularity of the pages that point it or cite it. Specifically, the simplified PageRank of a page pi is given by [8]:

$$PR_S(p_i) = \sum_{p_j \in int(p_i)} \frac{PR_S(p_j)}{d^{out}(p_j)}$$

Where (p_i) is the set of pages that point to page p_i and $d^{out}(p_j)$ represents the outgoing degree of the page p_j.

The algorithm [8] indicates the different stages of calculation of the simplified PageRank. The Steps 1 to 2 of the algorithm transform the adjacency matrix A into a stochastic matrix A_I. This transformation is achieved by standardizing the lines of the adjacency matrix A (step 5) such that:

$$(a_i)_{ij} = \frac{a_{ij}}{\sum_{K=1}^{N} a_{iK}}$$

The goal of this normalization is to mitigate the effect of pages with a large number of outbound links. In other term, this amounts to considering that each page has only one "Vote" which will be distributed equally to the pages pointed.

The steps 3 to 7 of the algorithm correspond to the application of the method of the powers for the calculation of the dominant eigenvector of the matrix A_I^T.

In the last step, the vector of the PageRank **p** is normalized so that the sum of the PageRank of all the pages is equal to 1.

Several studies have tried to identify influential users; however, most have used page rank centrality [4] or adaptions of the Page Rank algorithm [9].

In [9], the TwitterRank algorithm, which is an extension of PageRank, was proposed to measure the user influence on Twitter taking both the topical similarity between users and the link structure into account.

TunkRank [10] is another adaptation of PageRank. It makes the assumption that if a user reads a tweet from his friend he will retweet it with a constant probability. The influence is calculated recursively considering the attention that a user can give to his friends and that their followers could attribute to their influencer as well. These methods do not consider users interaction in posts. Yet, it is interesting to judge their influences not by their friendship relations in static structure, but based on the dynamic interaction in online contents.

The search for information is also concerned with the analysis of messages based on ontology of opinions specific to the field of word-of-mouth marketing (e.g. product) that was used in [11] to identify Opinion leaders in social networks. This research takes into account the exploitation of existing social activities in social networks such as tweet, retweet, sharing, I love, comments... to estimate the relevance of a resource vis-à-vis of a given request.

4   Detection of Opinion Leaders with the Maximization Technique

Another method is influence maximization; the problem of finding a small set of most influential nodes in a social network so that their aggregated influence in the network is maximized.

In [12], authors study influence maximization in the linear threshold model, one of the important models formalizing the behavior of influence propagation in social networks.

Wei Chen et al. [12] show that computing exact influence in general networks in the linear threshold model is P-hard. As a contrast, they also show that computing influence in directed acyclic graphs (DAGs) can be done in time linear to the size of the graphs.

Based on the fast computation in DAGs, Wei Chen et al. [12] propose the first scalable influence maximization algorithm tailored for the linear threshold model.

Domingos et al. [13] are the first to study influence maximization as an algorithmic problem. Their methods are probabilistic, however. Kempe et al. [14] are the first to formulate the problem as a discrete optimization problem. they. proposed two basic stochastic influence cascade models, the independent cascade (IC) model and the linear threshold (LT) model, which are extracted from earlier work on social network analysis, interactive particle systems, and marketing. The two models characterize two different aspects of social interaction. The IC model focuses on individual (and independent) interaction and influence among friends in a social network. The LT model focuses on the threshold behavior in influence propagation, which we can frequently relate to when enough of our friends bought a smartphone, played a new computer game, or used a new online social networks, we may be converted to follow the same action, but one of the issues of their work is the scalability of their greedy algorithm.

In [15], authors proposed the CELF algorithm a "lazy-forward" optimization in selecting new seeds to significantly reduce the number of influence spread evaluations, but it is still slow and not scalable to large graphs with hundreds of thousands of nodes and edges.

Several recent works proposes different heuristic algorithms specifically designed for the IC model. In fact, in [16], Kimura and Saito propose shortest-path based influence cascade models and provide efficient algorithms to compute influence spread under these models. In [17], Chen et al. propose degree discount heuristics for the uniform IC model in which all edge probabilities are the same, the efficient influence maximization from two complementary directions. One is to improve the original greedy algorithm of [14] and its improvement [15] to further reduce its running time, and the second is to propose new degree discount heuristics that improves influence spread.

In [18], Chen et al. propose maximum influence arborescence (MIA) heuristic for the general IC model. The new heuristic algorithms proposed in [16, 17, 18] are orders of magnitude faster than the fastest implementation of the greedy algorithm while maintaining a competitive level influence spread, making it a very promising direction. However, all of these heuristic algorithms are designed using specific properties of the IC model. In contrast, for the equally important LT model, the above heuristics do not apply and there is no scalable heuristic designed by utilizing special features of the LT model.

5   Comparative Study

In the following table, we present the advantages of the centrality measures resulting from the analysis of social networks and their limits. Each type of centrality is treated such as the degree centrality, the closeness centrality, the betweens centrality and the Eigenvector centrality (Table 1).

Table 1. Comparative study between different types of centrality techniques.

Measures of centrality resulting from the analysis of social networks	Limits of the centrality measures resulting from the analysis of social networks	Advantages of the centrality measures resulting from the analysis of social networks
Degree centrality	• It is a local measure that does not take into account the overall structure of the graph.	• Represents the simplest and most intuitive form of the notion of centrality. • Is adapted to the graphs of documents because these arc oriented and generally unrelated (at best, they can be weakly connected).
Closeness Centrality	• It is only usable if the graph is strongly connected, i.e. it is therefore not possible to calculate the proximity centrality since the geodesic distances between certain nodes are indefinite (there is no path Between two nodes of a graph). • It is not suitable for the analysis of graphs of documents because the latter are oriented and generally not related.	• A measure of global centrality based on the intuition that a node occupies a strategic (or advantageous) position in a graph if it is globally close to the other nodes of this graph. • Takes into account the overall structure of the graph
Betweenness Centrality	• According to the definition, it is evident that any node that does not have at least one inbound link and one outbound link will have zero importance, that is, zero centrality. • It is not suitable for the analysis of graphs of documents because these are oriented and generally not related	• A measure of global centrality proposed by Freeman. The intuition of this measure is that in a graph, a node is all the more important as it is necessary to cross it to go from one node to another. • Takes into account the overall structure of the graph
Eigenvector centrality	• A zero centrality when the graph is oriented and some nodes possess only one type of links (i.e., either incoming or outgoing). These nodes cannot then contribute to the calculation of the centrality. • It is not suitable for the analysis of graphs of documents because these are oriented and generally not related	• Very different from the other approaches we have presented. It has indeed suggested the idea that the centrality of a node is determined by the centrality of the nodes to which it is connected a measure of global centrality • Takes into account the overall structure of the graph

6   Conclusion

In this paper, we presented the definition and the role of the opinion leader and we exposed research works that used the centrality technique and the maximization technique to detect opinion leaders in social networks. In addition, we elaborated a comparative study between different centrality technique types. In fact, we presented the advantages and the limits of each type. As future work we will present our new approach to detect opinions leaders in a facebook group.

References

1. Chikhi, N.F.: Calcul de centralité et identification de structures de communautés dans les graphes de documents. Ph.D. thesis, Université de Toulouse, Université Toulouse III-Paul Sabatier, pp. 17–40 (2010)
2. Wasserman, S., Faust, K.: Social Network Analysis: Methods and Applications. Cambridge University Press, Cambridge (1994). p. 825
3. Freeman, L.C.: Centrality in social networks conceptual clarification. Soc. Netw., 215–239 (1979)
4. Musiał, K., Kazienko, P., Bródka, P.: User position measures in social networks. In: Proceedings of the 3rd Workshop on Social Network Mining and Analysis, ACM, New York, NY, USA, article no. 6 (2009)
5. Bonacich, P.: Some unique properties of eigenvector centrality. Soc. Netw., 555–564 (2007)
6. Brin, S., Page, L.: The anatomy of a large-scale hypertextual web search engine. Comput. Netw. ISDN Syst. **30**, 107–117 (1998)
7. Wu, X., Kumar, V., Ross Quinlan, J., Ghosh, J., Yang, Q., Motoda, H., McLachlan, G., Ng, A., Liu, B., Yu, P., Zhou, Z.-H., Steinbach, M., Hand, D., Steinberg, D.: Top 10 algorithms in data mining. Knowl. Inf. Syst. **14**, 1–37 (2008)
8. Zhang, Y., Yu, J.X., Hou, J.: Web Communities: Analysis And Construction. Springer, p. 185 (2005)
9. Weng, J., Lim, E., Jiang, J., He, Q.: TwitterRank: finding topic-sensitive influential twitterers. In: Proceedings of the 3rd ACM International Conference on Web Search and Data Mining, WSDM 2010, pp. 261–270 (2010)
10. Avello, G., Brenes Martínez, D., José, D.: Overcoming spammers in twitter - a tale of five algorithms. In: I Spanish Congress of Information Retrieval, Madrid, pp. 41–52 (2010)
11. Sun, B., Ng, V.T.Y.: Identifying influential users by their postings in social networks. Springer, pp. 128–151 (2013)
12. Chen, W., Yuan, Y., Zhang, L.: Scalable influence maximization in social networks under the linear threshold model, pp. 88–97 (2010)
13. Domingos, P., Richardson, M.: Mining the network value of customers. In: KDD, pp. 57–663 (2001)
14. Kempe, D., Kleinberg, J.M., Tardos, É.: Maximizing the spread of influence through a social network. In: Proceedings of the 9th ACM SIGKDD Conference on Knowledge Discovery and Data Mining, pp. 137–146 (2003)
15. Leskovec, J., Krause, A., Guestrin, C., Faloutsos, C., Van Briesen, J., Glance, N.S.: Cost effective outbreak detection in networks. In: Proceedings of the 13th ACM SIGKDD Conference on Knowledge Discovery and Data Mining, pp. 420–429 (2007)
16. Kimura, M., Saito, K.: Tractable models for information diffusion in social networks. In: ECML PKDD, pp. 259–271 (2006)
17. Chen, W., Wang, Y., Yang, S.: Efficient influence maximization in social networks. In: KDD, pp. 199–208 (2009)
18. Chen, W., Wang, C., Wang, Y.: Scalable influence maximization for prevalent viral marketing in large scale social networks. In: KDD, pp. 1029–1038 (2010)

A Real Time Two-Level Method for Fingertips Tracking and Number Identification in a Video

Ouissem Ben Henia[✉]

Department of Computer Science, College of Computer Science, King Khaled University Abha, Abha, Kingdom of Saudi Arabia
o.benhenia@gmail.com

Abstract. This paper presents a real time method to estimate the number of fingers observed in a video. The method tracks the fingertips and exploits the shape of the hand contour to determine the number of fingers observed in a sequence of images. The first step of the proposed method is to detect the hand observed in the input image by segmentation into foreground and background areas using skin colour detection method. The foreground corresponds to the area representing the hand to be tracked. Due to the problem of the lighting variation, HSL colour space was used to represent the colour. The second step consists of computing the hand contour. Then a convex Hull and convexity defects are calculated to detect the fingertips. Principal components analysis (PCA) [13] method is applied on the convex hull to deal with the cases in which only one finger is observed in the image or when the hand is closed. The proposed method could be used to produce different Human Computer Interaction systems (HCI). Experimental results obtained from real images demonstrate the potential of the method.

Keywords: Hand tracking · PCA technique · Convex hull

1 Introduction and Related Work

Vision based hand motion tracking is a subject belonging to the computer vision domain which is a subfield of artificial intelligence. In fact the main purpose of the computer vision domain is to make computer intelligent by understanding what it can see through a video camera. In general, the computer uses features from the images to achieve a task [17]. The computer will be able then to: recognise and track objects, recognise faces, track vehicles and to estimate their speed, recognise the licence plate of cars, estimate the body poses and so on... Hand motion tracking using a video camera is one of the most challenging subject in computer vision domain [1,12]. The purpose is to extract different information about the hand motion observed in a sequence of images. The information extracted could be the hand position and/or orientation observed in the 2D(image)/3D space. The fingertips and the joint angles could also be estimated to achieve the tracking and to recognise the gestures of the hand [19].

© Springer International Publishing AG 2018
G. De Pietro et al. (eds.), *Intelligent Interactive Multimedia Systems and Services 2017*,
Smart Innovation, Systems and Technologies 76, DOI 10.1007/978-3-319-59480-4_37

Vision based hand motion tracking brought on a wage a large number of applications like human machine interaction (HMI) [12,23], new smart TV [2] and smart phones [3] where the device control is performed by a hand gesture system. Other applications belonging to sign language recognition were proposed in [21]. To achieve the tracking, different devices could be used. One of the most famous devices are the data gloves and the video cameras. Data gloves are commonly used as input devices to capture and estimate the human hand motion by means of sensors attached to the hand in order to measure the joint angles and the spatial position of the hand directly. Unfortunately, this kind of devices is very expensive, frail and not user friendly. An alternative solution proposed by vision computer researchers is to use video camera which is more affordable and user friendly. However, building a fast and effective vision-based hand motion tracking system is very challenging due to several issues like the high number of degrees of freedom of the hand (around 26) [16,29], the ambiguities due to self occlusion of the hand and the variation of the lighting which can definitively affect the tracking process. To produce a vision-based hand motion tracking system, the researchers proposed several methods which could be divided into different categories depending on how the hand tracking is defined and the application to produce.

The first category of methods uses a 3d hand parametric model that is compared to real hand poses. The idea consists of finding the parameters of the model assuring the matching between the hand poses observed in a video sequence and the model ones [1]. Different models were prosed in the literature based on planar patches [30], deformable polygon meshes [10] or generalized cylinders [7]. Stenger et al. [27] used quadric surfaces like cones and ellipsoids. Other authors proposed a fine mesh obtained by scanning the hand to track [5]. Sridhar et al. [26] proposed a model based on a sum of Sum of Anisotropic Gaussians(SAG). To achieve the tracking the model poses are compared with the ones observed through a cost function which is minimized by adapting the model parameters. The simpliest function are base exploit either silhouette or edge feature extracted from image using affordable 2D cameras [1,16,27]. Other functions are based on 3d features obtained by the use of 3d cameras [5,19,24,25]. The most important problem of this kind of methods to be highlighted is the initialisation part needed to start the tracking or after tracking lost. The second kind of approaches is based on the used of a database of gestures previously built. The methods of this category are also called view-base methods and are usually used to estimate the hand poses through classification or regression techniques [20,21]. A set of hand feature is labeled with a particular hand pose, and a classifier is trained from this data. Different kind of features were used as 2D ones with silhouette or edges [20,21] and colour ones [28] as well as 3d features with the use of 3d points [4]. The process consists of seeking the best match between the hand gesture observed in an input image and the gesture stored in the database. The problem with this kind of methods is the important number of possible hand poses which makes difficult or even impossible to perform dense sampling.

Another kind of approaches consists of tracking few information about the hand motion like the position and the fingertips [8]. This kind of methods are usually in real time [31] and don't need neither a 3d hand model nor a big database. It could use Active Contours [22], depth-based features [15], search window method [18]. This approach allows to produce real time applications like Human computer interaction systems (HCI) but is usually not able to achieve a complex tracking with estimating the joint angles of the fingers and the orientation of the hand in the 3D space.

In this work, a fast method is proposed to track the fingertips of the hand in real time and estimate their number in a sequence of images. The tracking process consists of segmenting images, then detecting the hand contour to compute of the convexity defects [9] to estimate the number of fingers observed in the video. Principal component analysis (PCA) [13] is used to deal with the cases where the hand is closed or when only one finger is observed in an image. The remainder of this paper explains the steps of the proposed process. The next section explains the first level of our method which is to detect the convex hull of the hand. The third section explains the second level which uses a PCA method. Experimental results obtained from real images are shown in Sect. 4. The Sect. 5 summarises the proposed process and gives an overview of improvement in a future work.

2   Fingertips Detection Through Convex Hull Calculation

The first step of the fingertips detection process is to compute the hand contour of the hand observed in each image. For that purpose, a segmentation technique based on skin colour detection is used to select the area of the image representation the hand. The hand contour is then detected using the canny method [6]. After that, the convex hull (Fig. 1) is computed using the Graham's algorithm [9]. In a case of 2D points representing the hand contour, the convex hull could be considered as the smallest polygon containing the 2D points. The convexity defects represent the points where the contour of the hand is concave, in other words the points between the fingers (Fig. 1).

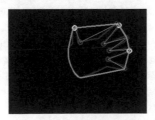

Fig. 1. Hand contour (in blue colour) with convex hull (in green), and the convexity defects (points between fingers)

Computing the convex hull of the hand shape is very important in our work, that's why we describe the method to compute a convex hull on a set of 2D points in (Algorithm 1) and the different steps of the used algorithm are summarized in the figure (Fig. 2)

Input   : a set S of N points (See example Fig. 2a)
select the point P with highest y coordinate.
Let $P_0 = P$.
Sort and label the remaing points by polar angle with the point P_0 (See
example Fig. 2b).
Push P_0 and P_1 onto a stack B
Let $i = 2$
while $i < N$ **do**
   | Let B_1 the top point on the stack B
   | **if** $B_1 == P_0$ **then**
   | | Push P_i onto B
   | | i++
   | **end**
   | Let $B_1 =$ the second top point on the stack B
   | **if** B_0, B_1 *and* P_i *are oriented clockwise* **then**
   | | (See example Fig. 2b)
   | | Push P_i onto B
   | | i++
   | **end**
   | **else**
   | | Pop the top point B_0 from B. (See example Fig. 2d)
   | **end**
end
Output: Convex Hull points contained in the stack B. (See example Fig. 2e)

Algorithm 1. Convex Hull Graham Algorithm

The convexity defect could be characterised by three points: start, end and defect points. The defect point has the farthest distance to the convex hull (Fig. 3).

It's important to notice that many convexity defects could be selected by the method even those which are not representing the fingertips. That's why, it's important to define thresholds to be able to select only the hand fingertips. Let's A, B, C three points (start, end, defect point) representing a convexity defect in an image as shown in (Fig. 3). The parameters that we define to be able to know if these points are belonging or not to the hand fingers are the length of the vectors AC and BC as well as the angle between these vectors. Thresholds are defined manually for these parameters, so if each parameter is between the thresholds, the candidate points A and B are considered as the fingertips of the hand. These thresholds are chosen carefully through empirical method to detect the fingertips of the hand. Convexity defects doesn't deal when only one finger is observed in an image. Fingertips are detected if at least two fingers are observed. To deal with the cases where hand is closed or when only one finger is observed, principal component analysis (PCA) [13] is used to know if the hand is closed or not. This method is important in our work, that's why it's described in the next section.

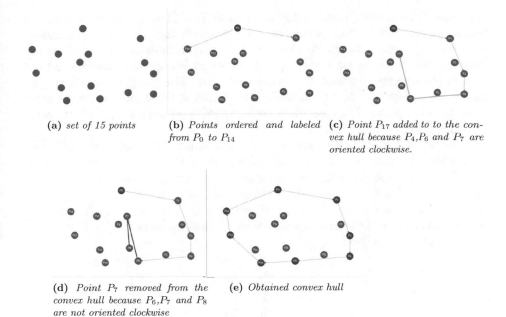

(a) *set of 15 points*

(b) *Points ordered and labeled from P_0 to P_{14}*

(c) *Point P_{17} added to to the convex hull because P_4, P_6 and P_7 are oriented clockwise.*

(d) *Point P_7 removed from the convex hull because P_6, P_7 and P_8 are not oriented clockwise*

(e) *Obtained convex hull*

Fig. 2. Steps of the convex hull algorithm

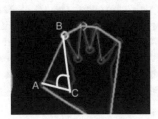

Fig. 3. Convexity defects represented by three points A, B and C: the start, the end, and defect point.

3   Principal Component Analysis to Detect Closed Hand

Principal component analysis is a mainstay of modern data analysis [13]. It is very known as a statistical technique to detect and emphasize variations in datasets. The PCA technique is very common used today in a host of applications such as computer vision and graphics, data and shape analysis [11]. In computer vision, it was used to simply to problem of hand motion tracking by reducing the high number of degrees of freedom of the articulated hand [14]. It was also used in [11] to obtain an interesting basis to describe the hand shape represented as a 3d point cloud. In this paper, the PCA technique is used to obtain information about the hand shape observed in an image in order to detect if the hand is opened or closed. The principal component analysis is applied on contour representing the hand. It produces an interesting basis which is used to calculate the smallest box containing the contour of the hand. If the length and width of the bounding box are very close it means that the hand is closed otherwise it means that the hand is opened and at least one finger is detected. We applied also PCA on the convex hull and the obtained results are very close. We highlight that the convex hull contains less that 10 points while the contour of the hand consists of more than hundred points. It's clear that applying PCA on the convex hull is better in terms of computing time.

The use of the PCA technique is very important in our work and that's why we decided to describe below the different steps of the PCA method. Let $P = \{P_1, P_2, ..., P_n\}$ a set of n 2D points representing the hand contour where $P_i = (x_i, y_i)_{i=1..n}$. The first step of the PCA method is to calculate the barycenter $G(g_x, g_y)$ of the set $\bar{P} = \{\bar{P}_1, .., \bar{P}_n\}$ where $\bar{P}_i = (x_i - x_g, y_i - y_g)$. The second step is to calculate the covariance matrix M defined as follows: $M = \bar{B}\bar{B}^t$ where B is a matrix defined as:

$$B = \begin{bmatrix} \bar{x}_1 & \bar{y}_1 \\ \bar{x}_2 & \bar{y}_2 \\ \vdots & \vdots \\ \bar{x}_n & \bar{y}_n \end{bmatrix}$$

The next step of the PCA method is to calculate the eigenvectors ve_1 and ve_2 of the matrix M which will represent the PCA basis vectors. This new basis is used to find the bouncing box of the point set P which represents the smallest box containing the point set P. The size of the bouncing box gives information about the shape. If the length of the bounding box is very bigger than the width, it means that the hand is opened. Otherwise, the bounding box is close to a square which means that the hand is closed.

4   Experimental Results

The performance of our two-level method is first evaluated for tracking the fingertips and estimating the number of fingers appearing in a real sequence of images. The size of the images is 640 × 480.

The obtained results are shown in (Table 1). The first column represents the input images containing a real hand. The second column shows the results of detecting the fingertips after calculating the convex hull and convexity defects on hand contour. The third column shows the result of applying PCA on the hand contour to detect if the hand is closed or not. If the hand is closed it means that there is no finger detected otherwise it means that at least one finger is detected. The final column contains similar results that the third column but after applying the PCA on the convex hull which gives similar results. The convex hull contains less points that the hand contour and that's why the use of PCA on the convex hull allows to improve the processing rate.

In the next section we make a conclusion of our work and discuss about future work and applications in which our method could be used.

Table 1. Results obtained with real images.

Input images	result convex hull technique	result after applying PCA on the hand contour	result after applying PCA on the Convex Hull 4

5   Conclusion and Future Work

In this work, a vision-based method for fingertips tracking was proposed. The process can be divided in three steps. In the first one, the input image is segmented to detect the hand area from which the contour is computed. After calculating the convex Hull and convexity defects the fingertips are detected. PCA method is used to deal with the cases where the hand is closed or when only one finger is observed in an image. Detecting the fingertips could be the first step to create different applications such as human computer interaction (HCI). One of these applications could be to produce a remote control like a tv remote to change the channel according to the number of detected fingers. In video games, the player could interact with characters using his hand motion. In future work, we plan to use a 3D video camera to improve the accuracy of the tracking and acquire more information such as the depth information which will allow to track the 3D position of the hand in the space and create more interaction with the computer.

References

1. Ben Henia, O., Hariti, M., Bouakaz, S.: A two-step minimization algorithm for model-based hand tracking. In: WSCG (2010)
2. http://www.samsung.com/ph/smarttv/common/guide_book_3p_si/waving.html
3. http://www.zdnet.com/article/control-your-mobile-without-your-hands-gesture-tech-coming-to-a-mobile-near-you-soon/

4. Ben Henia, O., Bouakaz, S.: A new depth-based function for 3D hand motion tracking. In: Proceedings of the International Conference on Computer Vision Theory and Applications (VISIGRAPP 2011), pp. 653–658 (2011)

5. Bray, M., Koller-Meier, E., Mueller, P., Van Gool, L., Schraudolph, N.N.: 3D hand tracking by rapid stochastic gradient descent using a skinning model. In: Chambers, A., Hilton, A. (eds.) 1st European Conference on Visual Media Production (CVMP), pp. 59–68. IEEE, March 2004

6. Canny, J.: A computational approach to edge detection. IEEE Trans. Pattern Anal. Mach. Intell. 8(6), 679–698 (1986)

7. Delamarre, Q., Faugeras, O.: Finding pose of hand in video images: A stereo-based approach. In: IEEE Proceedings of the third International Conference on Automatic Face and Gesture Recognition, pp. 585–590. IEEE Computer Society (1998)

8. Dhawan, A., Honrao, V.: Implementation of hand detection based techniques for human computer interaction. Int. J. Comput. Appl. 72(17), 6–13 (2013)

9. Graham, R., Francesyao, F.: Finding the convex hull of a simple polygon. J. Algorithms 4(4), 324–331 (1983)

10. Heap, T., Hogg, D.: Towards 3D hand tracking using a deformable model. In: Face and Gesture Recognition, pp. 140–145 (1996)

11. Ben Henia, O., Bouakaz, S.: 3D hand model animation with a new data-driven method. In: Proceedings of the Workshop on Digital Media and Digital Content Management, DMDCM 2011, pp. 72–76. IEEE Computer Society, Washington, DC (2011)

12. Ike, T., Kishikawa, N., Stenger, B.: A real-time hand gesture interface implemented on a multi-core processor. In: MVA, pp. 9–12 (2007)

13. Jolliffe, I.T.: Principal Component Analysis. Springer, New York (2002)

14. Kato, M., Chen, Y.-W., Gang, X.: Articulated hand motion tracking using ica-based motion analysis and particle filtering. J. Multimedia 1(3), 52–60 (2006)

15. Liang, H., Yuan, J., Thalmann, D.: 3D fingertip and palm tracking in depth image sequences. In: Proceedings of the 20th ACM International Conference on Multimedia, MM 2012, pp. 785–788. ACM, New York (2012)

16. Kato, M., Xu, G.: Occlusion-free hand motion tracking by multiple cameras and particle filtering with prediction. IJCSNS Int. J. Comput. Sci. Netw. Secur. 6(10), 58–65 (2006)

17. Montalvão, J., Molina, L., Canuto, J.: Robust hand image processing for biometric application. Pattern Anal. Appl. 13(4), 397–407 (2010)

18. Oka, K., Sato, Y., Koike, H.: Real-time fingertip tracking and gesture recognition. IEEE Comput. Graph. Appl. 22(6), 64–71 (2002)

19. Qian, C., Sun, X., Wei, Y., Tang, X., Sun, J.: Realtime and robust hand tracking from depth, June 2014

20. Rosales, R., Athitsos, V., Sigal, L., Sclaroff, S.: 3D hand pose reconstruction using specialized mappings. In: ICCV, pp. 378–385 (2001)

21. Shimada, N., Kimura, K., Shirai, Y.: Real-time 3-D hand posture estimation based on 2-D appearance retrieval using monocular camera. In: Proceedings of the IEEE ICCV Workshop on Recognition, Analysis, and Tracking of Faces and Gestures in Real-Time Systems (RATFG-RTS 2001), p. 23. IEEE Computer Society, Washington, DC (2001)

22. Silanon, K., Suvonvorn, N.: Fingertips tracking based active contour for general HCI application, pp. 309–316. Springer, Singapore (2014)

23. Sridhar, S., Feit, A.M., Theobalt, C., Oulasvirta, A.: Investigating the dexterity of multi-finger input for mid-air text entry. In: Proceedings of the 33rd Annual ACM Conference on Human Factors in Computing Systems, CHI 2015, pp. 3643–3652. ACM, New York (2015)
24. Sridhar, S., Mueller, F., Oulasvirta, A., Theobalt, C.: Fast and robust hand tracking using detection-guided optimization. In: Proceedings of Computer Vision and Pattern Recognition (CVPR) (2015)
25. Sridhar, S., Oulasvirta, A., Theobalt, C.: Interactive markerless articulated hand motion tracking using RGB and depth data. In: Proceedings of the IEEE International Conference on Computer Vision (ICCV), December 2013
26. Sridhar, S., Rhodin, H., Seidel, H.-P., Oulasvirta, A., Theobalt, C.: Real-time hand tracking using a sum of anisotropic gaussians model. In: Proceedings of the International Conference on 3D Vision (3DV), December 2014
27. Stenger, B., Mendonca, P.R.S., Cipolla, R.: Model-based 3D tracking of an articulated hand. In: Proceedings of the 2001 IEEE Computer Society Conference on Computer Vision and Pattern Recognition, CVPR 2001, vol. 2, pp. II-310–II-315 (2001)
28. Wang, R.Y., Popović, J.: Real-time hand-tracking with a color glove. In: ACM SIGGRAPH 2009 papers, pp. 1–8. ACM, New York (2009)
29. Wu, Y., Lin, J., Huang, T.S.: Analyzing and capturing articulated hand motion in image sequences. IEEE Trans. Pattern Anal. Mach. Intell. **27**(12), 1910–1922 (2005)
30. Wu, Y., Lin, J.Y., Huang, T.S.: Capturing natural hand articulation. In: ICCV, pp. 426–432 (2001)
31. Yeo, H.-S., Lee, B.-G., Lim, H.: Hand tracking and gesture recognition system for human-computer interaction using low-cost hardware. Multimedia Tools Appl. **74**(8), 2687–2715 (2015)

Trajectory ETL Modeling

Assawer Zekri[1]([✉]), Marwa Massaâbi[1], Olfa Layouni[1], and Jalel Akaichi[2]

[1] Computer Science Department, BESTMOD, High Institute of Management,
University of Tunis, Tunis, Tunisia
assawer.zekri@gmail.com, massaabi.marwa@gmail.com,
layouni.olfa89@gmail.com
[2] College of Computer Science, King Khalid University, Abha, Saudi Arabia
jalel.akaichi@kku.edu.sa

Abstract. Extraction-Transformation-Loading (ETL) tools are pieces of software responsible for the extraction of data from heterogeneous sources, their cleansing, customization and insertion into data warehouse. In fact, ETL are the key component of data warehousing process because incorrect or misleading data will produce wrong business decisions. Therefore, a correct design of the ETL process at early stages of data warehouse project is absolutely necessary to improve data quality. So, it is essential to overcome the ETL modeling phase with elegance in order to produce simple and understandable models. Many researches are dealt with the modeling of ETL process but today with the advent of trajectory data we need interactive modeling that manages the mobility aspect. In this paper, we extend existing model by incorporating a trajectory as a first class concept. So, we propose to model the ETL workflows as directed acyclic graphs that comprise three main components which are: the data sources, all the activities and the data stores.

Keywords: Trajectory data · ETL · Modeling · Graph · Activity · Moving object

1 Introduction

Extraction Transformation Load (ETL) plays an important role in data warehouse process. In fact, extraction is the retrieval of data from various sources. The second important phase in the ETL is Transformation. It is the change that must be made on the data source before it's loading on the data warehouse. The last phase is loading which aims to load the cleaned and transformed data to the warehouse. To summarize, the overall process consists of three major steps which are the identification of the source, the cleaning and transformation of data and finally moving resultant to the target location. In the last few years there has been a growing interest in trajectory data generated by sensors, mobile computing techniques and positioning technologies, there becoming available in massive and huge quantities. They describe the mobility of diversity moving objects, such as people, vehicles, animals, ships and airplanes. They are characterized by their complex structure including spatiotemporal attributes that values represent a data stream generated in a continuous way and at high speed. So, the vastness of data

© Springer International Publishing AG 2018
G. De Pietro et al. (eds.), *Intelligent Interactive Multimedia Systems and Services 2017*,
Smart Innovation, Systems and Technologies 76, DOI 10.1007/978-3-319-59480-4_38

volumes, the quality problems and the evolution of sources and data warehouse are the key factors that characterize the trajectory data.

Nevertheless, the difference on data structures imposes new requirements on the ETL modeling and implementation. What makes these tasks even more challenging is the fact that data continue to grow rapidly and business requirements change over time. So, to feed the data warehouse, data must be identified and extracted from their original locations. Consequently, data must be transformed and verified before being loaded into the data warehouse. Also, the large amount of data from multiple sources causes a high probability of errors and anomalies. This increases the need of a new ETL which is able to be adapted with the constant changes.

Based on the above, traditional modeling approaches need to be reconsidered to support the mobility aspect and incorporate the trajectory as a first class concept. However, little attention has been given to the trajectory modeling of ETL processes. So, we need interactive modeling that decreases the complexity of the ETL. Based on these reasons, this paper suggests modeling the trajectory ETL as a directed acyclic graph (DAG).

The remainder of the paper is organized as follows: Sect. 2 provides an overview of ETL modeling in the literature, Sect. 3 discusses in detail our modeling approach, Sect. 4 summarizes the work and draws conclusions.

2   Related Works

The conceptual model proposed by Panos Vassiliadis is the pioneering and first one designed for ETL process [13]. So, the authors describe a conceptual model which is customized for the tracing of the interrelationships of attributes, concepts, the transformations during the loading of the warehouse and the ETL activities. Another paper proposed by [5] which aims to the modeling of ETL based on an extension of UML notations by decomposing a complex ETL process into a set of mechanisms. In [6] the authors develop an approach based on the data mapping UML profile to trace the flow of data. The work of [11] shows a static conceptual model for identifying the transformation in the ETL process. It includes the transformation entity, constraints and notes. In these approaches most of the semantic transformations are given through notes, in a natural language. According to [7] an object oriented approach is proposed to accomplish the data extraction modeling of ETL process using UML 2.0. In [2] the authors propose a conceptual language for modeling ETL processes based on the Business Process Modeling Notation (BPMN). The model provides a large set of primitives that cover the design requirements of ETL processes. This work was pursued by [3] proposing the first modeling approach for the specification of ETL processes in a vendor-independent way and the automatic generation of its corresponding code in commercial platforms. In [4] the authors propose a web based framework model for representing extraction of data from one or more data sources, use transformation business logic and loading the data within the data warehouse. Another work proposed by [12] in which they suggest a conceptual model for ETL based on an enriched business process model with Qox annotations. The paper of [14] describes a generated weaving metamodel

(GWMN) that gives the complete mapping semantics through specific link types and appropriate Object Constraint Language (OCL). The work of [1] aims to adopt the MR model to improve the ETL against big data [9]. In [10] the authors propose a method that contains Two-ETL phases, one treats the pre-treatment phase and another deals with the actual ETL. Their method consists on determining the correspondence table, modeling new operations using the Business Process Modeling Notation (BPMN) and implementing these operations with Talend Open Source (TOS). Based on the literature review, most of previous studies do not take account the trajectory ETL design. So, studies in this context are still lacking. The first work in this context is discussed by [8]. In his thesis, the author focuses on trajectory ETL which is used to feed the data cube with trajectory data. So, once trajectories have been constructed and stored in a Moving Object Database, the ETL phase is executed in order to feed the trajectory data warehouse. So, the ETL procedure for feeding the fact table of the trajectory data warehouse is described by two alternatives: a cell-oriented approach and a trajectory-oriented approach. The second approach can results the distinct counting problem in which a trajectory can be found sum in both base cells. Another work is proposed by [16] in which the authors propose a conceptual model for trajectory ETL based in UML langage. The authors propose a generic model that describes the flow of activities of a trajectory ETL process. Also, reference [15] discusses briefly the steps performed to populate a Mob-Warehouse. The ETL process includes a semantic enrichment step, whose goal is to associate semantic information from the application domain with the trajectory data.

Based on the scarcity of work concerning the modeling of trajectory ETL and with the importance of this area especially with technological evolutions, designing an efficient, robust and evolvable trajectory ETL workflow became very relevant. Thus, we believe that the research on trajectory ETL processes is a valid research goal.

3 Contribution

Several studies in the literature have concentrated on traditional ETL modeling. Today, with the advent and the current developments in the pervasive and ubiquitous systems, we need to manage huge quantities of data called trajectory data. So, it is necessary to take an interest in the trajectory ETL modeling. For this reason, we design a new modeling for trajectory ETL. In general, an ETL may be represented as flows of activities. So, in our case, we choose to model the trajectory ETL as directed acyclic graph (DAG) with constraints where the vertexes are the activities and the edges represent the relationships between them. In fact, the full layout of a trajectory ETL scenario, such as activities and relationships between them can be modeled by a directed acyclic graph. Formally, let G (V, E) be the directed acyclic graph of a trajectory ETL scenario where V = Vertex and E = Edge. In our case, the graph comprises three main components which are: the data sources, all the activities and finally the data stores. So, the flow of data from the sources towards the data warehouse is performed through the composition of activities in a large scenario. The data sources may be database, OLTP, legacy, mainframe, flat files such as XML, CSV, Excel, etc. The activities in our case are transformation or task such as: extraction, alimentation, trajectory Flow Control, lookup or search trajectories, file management, aggregation, add and duplicate sequences, Cloning trajectory data, execute a process, replacing null values, calculation, trajectory analysis, cloning trajectory data, deduplication, sorting; load etc.

Usually, the graph modeling can involve both for the modeling of the internal structure of activities and for the modeling of the ETL scenario enable the treatment of the ETL environment from different points of view.

The graph of a trajectory ETL scenario comprises nodes and edges. In fact, the nodes may be: activities and/or attribute. The interactions between the entities are called relationships. So, $V = A \cup At$ and $E = Rs$ where A is activity, At is attribute and Rs is relationship.

The terminologies that we will use are listed as follows:

- An entity: is an activity which is characterized by a name and parameters etc.
- An activity may be filtering, fusion, cleaning redundancy data, aggregation, extracting, loading trajectory data in other words any activity of a trajectory ETL etc.
- Attributes are characterized by name and data type.
- Parameters are data which act as regulators for the functionality of the activity.
- Data type is the type of the attribute such as integer, double etc.

In this context, the graph covers the information on the typing of the involved entities and the regulation of the execution of scenario through specific parameters. Firstly, we describe a formal model of trajectory ETL process: the data store, activities of trajectory ETL and their constituent parts are formally defined. Secondly, we model all the entities as nodes and different kinds of relationships such as topological relationship, type checking information and provider relationship as edges. Finally, we construct the directed acyclic graph.

To summarize these descriptions, we will propose graphical notations of trajectory ETL. So, Table 1 shows the different entities of our model.

Table 1. Entities of the model

Type	Meaning	Symbol
Activity type	The type of the main activity of trajectory ETL which may be: extraction, transformation or loading	E: ● T: △ L: ▭
Parameters	Set of data which is necessary to execute an activity	A (P)

The Table 2 illustrates the different relationships between the entities of our model.

Table 2. Relationships of the model

Type	Meaning	Symbol
Topological relationship	Refers to the relationship between things. In our case, It is used to describe the relationships between activities of trajectory ETL. We have two kinds of topological relationships: • Inside topological relationship such as Is In • Adjacency topological relationship such as From, To etc.	Is In From To
Type checking information	Shows the type of the activity. In this context, the activity may be: a task or a transformation	T: Task Tr: Transformation
Provider relationship	Connects a data store to an activity or an activity to either another activity	⟶

Most restrictions at the target warehouse can be expressed by a set of constraints, which are also referred as data quality rules. These are summarized in Table 3 as follows:

Table 3. Constraints of the model

Type	Meaning	Symbol
Structure validation	• The type and the format of a data field should be same in the source and the target system	SV
Validate constraints	• Ensuring that constraints are applied on the tables	VC
Data completeness validation	• Checking if all the data is loaded to the target system • Validating the unique values of primary keys	DCV
Data consistency check	• Checking the integrity constraint	DCC
Data correctness validation	• Inaccurate data does not found in table	DCRV
Data transform validation	• Validating if the data types in the warehouse are the same in the data model	DTV
Data quality validation	• Precision check, data check, null check, etc.	DQV
Null validation	• Checking the null values	NV
Data cleaning	• Missing data should be removed before their loading in the target	DCN

Note:
The note aims to add textual information in a model to explain for example a constraint.

To summarize, our model is composed for four main components which are: the entities, the relationships, the note and the constraints which referred as data quality rules.

• The scenario of insertion trajectory data (feeding):

1. Extraction and moving the source data from staging to work database
2. Check the lookup tables for new items
3. If Is In (exist) table then remove the row
4. Else update existing rows
5. Insert new ones

6. Apply the rules and update target table
7. Insertion into target table

   So, the graph is represented as follows (Fig. 1):

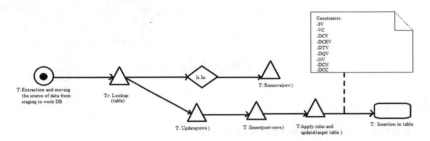

Fig. 1. Insertion of trajectory data

- Aggregation of trajectories (summarize, group by)

1. Extraction of trajectory data from table
2. Check the lookup tables
3. Apply the rules and cleaning data (redundant and missing data)
4. Construction of trajectories
5. Aggregation of trajectories
6. Loading trajectories in the trajectory data warehouse (Fig. 2).

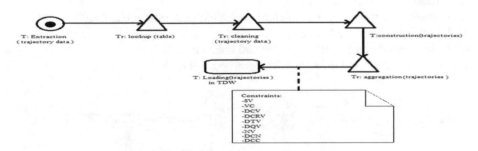

Fig. 2. Aggregation of trajectories

- Filtering trajectory data

1. Extraction of trajectory data from table
2. Cleaning the trajectory data to eliminate missing and redundant data
3. Treat them
4. Loading them in the trajectory data warehouse (Fig. 3).

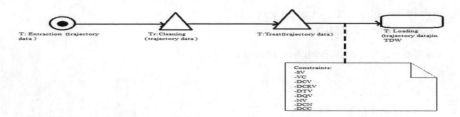

Fig. 3. Filtering trajectory data

- Migrate trajectory data

1. Extraction of trajectory data from DB
2. Cleaning the trajectory data to eliminate missing and redundant data
3. Migrate them from trajectory database to another DB [From…To] (Fig. 4).

Fig. 4. Migrate trajectory data

- Calculating trajectories

1. Extraction of trajectory data
2. Cleaning the trajectory data to eliminate missing and redundant data
3. Construction of the trajectories
4. Calculating distances between start and finish trajectories
5. Loading them in the trajectory data warehouse (Fig. 5).

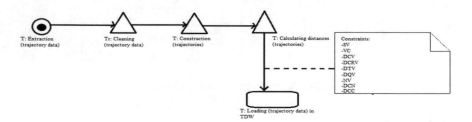

Fig. 5. Calculating trajectories

- Merge and sort geographic points

1. Lookup in the table
2. Cleaning trajectory data to eliminate missing data
3. Merge the geographic points (trajectory data)

4. Sort them
5. Construct trajectories
6. Loading them in the trajectory data warehouse (Fig. 6).

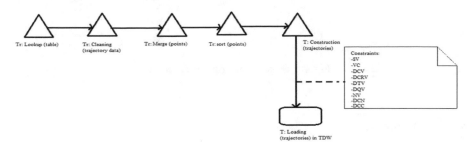

Fig. 6. Merge and sort geographic points

- Extraction of trajectory data

1. Extraction of trajectory data from trajectory database
2. Cleaning them to eliminate missing and redundant data
3. Treat them
4. Loading them in the trajectory data warehouse (Fig. 7).

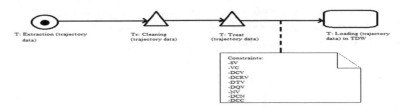

Fig. 7. Extraction of trajectory data

- Empty trajectory flow detection

1. Extraction of trajectory data from trajectory database
2. Search the empty flows
3. Detection of empty flows
4. Delete them
5. Loading them in the trajectory data warehouse (Fig. 8).

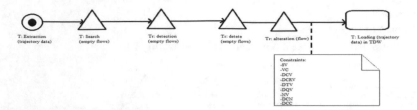

Fig. 8. Empty trajectory flow detection

4   Conclusion

In conclusion, it is evident that this study has shown the design of the trajectory ETL scenario of a data warehouse. This model includes the data stores, activities, their constituent parts and data warehouse. So, we model the trajectory ETL scenario as a directed acyclic graph. It is composed for four main components which are: the entities that represent the nodes of the graph, the relationships between them that are the edges, the note and finally the constraints which referred as data quality rules. Then, we have proposed some possible scenario which are: Insertion of trajectory data (feeding), Aggregation of trajectories (summarize, group by), filtering trajectory data, migrate trajectory data, calculating trajectories, merge and sort geographic points, Extraction of trajectory data, Empty trajectory flow detection. The next stage of our research will cover a trajectory data warehouse modeling then we will implement the proposed scenario by extending the Geokettle platform to support the trajectory data.

References

1. Bala, M., Alimazighi, Z.: Modélisation de processus ETL dans un modèle MapReduce. In: Conférence Maghrébine sur les Avancées des Systèmes Décisionnels (ASD 2013), Marrakech, Maroc, pp. 1–12 (2013)
2. El Akkaoui, Z., Zemányi, E.: Defining ETLWorfklows using BPMN and BPEL. In: DOLAP 2009, Hong Kong, China, pp. 41–48 (2009)
3. El Akkaoui, Z., Zimányi, E., Mazón, J.N., Trujillo, J.: A model-driven framework for ETL process development. In: DOLAP 2011, Glasgow, Scotland, UK, pp. 45–52 (2011)
4. Krishna, R., Sreekanth: An object oriented modeling and implementation of web based ETL process. IJCSNS Int. J. Comput. Sci. Netw. Secur. **10**(2), 192–196 (2010)
5. Lujan-Mora, S., Trujillo, J.: A UML based approach for modeling ETL processes in data warehouses. In: Song, I.-Y., Liddle, S.W., Ling, T.-W., Scheuermann, P. (eds.) Conceptual Modeling - ER 2003, vol. 2813, pp. 307–320. Springer, Heidelberg (2003)
6. Lujan-Mora, S., Vassiliadis, P., Trujillo, J.: Data mapping diagrams for data warehouse design with UML. In: Atzeni, P., Chu, W., Lu, H., Zhou, S., Ling, T.-W. (eds.) Conceptual Modeling – ER 2004, vol. 3288, pp. 191–204. Springer, Heidelberg (2004)
7. Mrunalini, M., Suresh Kumar, T.V., Evangelin Geetha, D., Rajanikanth, K.: Modelling of data extraction in ETL processes using UML 2.0. DESIDOC Bull. Inf. Technol. **26**, 3–9 (2006). DESIDOC

8. Marketos, G.: Datawarehousing and mining techniques for moving object databases. Thesis, Greece (2009)

9. Misra, S., Saha, S.K., Mazumdar, C.: Performance comparison of hadoop based tools with commercial ETL tools – a case study. In: Bhatnagar, V., Srinivasa, S. (eds.) Big Data Analytics (BDA 2013). LNCS, vol. 8302, pp. 176–184. Springer, Cham (2013)

10. Nabli, A., Bouaziz, S., Yangui, R., Gargouri, F.: Two-ETL phases for data warehouse creation: design and implementation. In: Morzy, T., Valduriez, P., Bellatreche, L. (eds.) Advances in Databases and Information Systems, vol. 9282, pp. 138–150. Springer, Cham (2015)

11. Simitsis, A.: Mapping conceptual to logical models for ETL processes. In: Proceedings of the 8th ACM International Workshop on Data Warehousing and OLAP, Germany, pp. 67–76 (2005)

12. Simitsis, A., Wilkinson, K., Dayal, U., Castellanos, M.: Leveraging business process models for ETL design. In: Proceedings of the 29th International Conference on Conceptual Modeling, Canada, pp. 15–30 (2010)

13. Vassiliadis, P., Simitsis, A., Skiadopoulos, S.: Conceptual modeling for ETL processes. In: Proceedings of the 5th ACM International Workshop on Data Warehousing and OLAP, USA, pp. 14–21 (2002)

14. Vučković, M., Petrović, M., Turajlić, N., Stanojević, M.: The specification of ETL transformation operations based on weaving models. Int. J. Comput. Commun. Control 7(5), 968–975 (2012)

15. Wagner, R., Macedo, J., Raffaeta, A., Renso, C., Roncato, A., Trasarti, R.: Mob-Warehouse: a semantic approach for mobility analysis with a trajectory data warehouse. In: Parsons, J., Chiu, D. (eds.) Advances in Conceptual Modeling. LNCS, vol. 8697, pp. 127–136. Springer, Cham (2014)

16. Zekri, A., Akaichi, J.: An ETL for integrating trajectory data: a medical delegate activities use case study. In: The International Conference on Automation, Control, Engineering and Computer Science (ACECS 2014), Monastir, Tunisia, pp. 138–147 (2014)

Computing Semantic Trajectories: Methods and Used Techniques

Thouraya Sakouhi[1]([✉]), Jalel Akaichi[2], and Usman Ahmed[2]

[1] Institut Supérieur de Gestion, Université de Tunis, Tunis, Tunisia
thouraya.sakouhi@isg.rnu.tn
[2] College of Computer Science, King Khalid University, Abha, Saudi Arabia
{jalel.akaichi,lusman}@kku.edu.sa

Abstract. The widespread use of mobile devices generates huge amount of location data. The generated data is useful for many applications, including location-based services such as outdoor sports forums, routine prediction, location-based activity recognition and location-based social networking. Sharing individuals' trajectories and annotating them with activities, for example a tourist transportation mode during his trip, helps bringing more semantics to the GPS data. Indeed, this provides a better understanding of the user trajectories, and then more interesting location-based services. To address this issue, diverse range of novel techniques in the literature are explored to enrich this data with semantic information, notably, machine learning and statistical algorithms. In this work, we focused, at a first level, on exploring and classifying the literature works related to semantic trajectory computation. Secondly, we capitalized and discussed the benefits and limitations of each approach.

Keywords: Mobility data · Trajectory · Semantic modeling · Ontology · Machine learning · Data mining · Activity recognition

1 Introduction

Mathematically, a trajectory is the path described by a point representing the moving object, usually his center of gravity. In biological sciences, it is the same applied for the living beings as moving objects. In human and social sciences, trajectories may represent the change of the state or the social position of the human being during his life. Modeling and analysis of these trajectories representing the mobility data are important issues to understand the behavior of the moving objects. Accordingly, advantages of mobility-oriented systems are manifold, including: sharing sensors derived status information in on-line social networks, capturing the characteristics and dynamics of everyday activities and enabling queries associated with physical space.

A straightforward idea to deal with this subject is extending the classical DBMS technology to handle such special type of data. Consequently, new technologies taking into account the specificities of such data have emerged. Initially,

© Springer International Publishing AG 2018
G. De Pietro et al. (eds.), *Intelligent Interactive Multimedia Systems and Services 2017*,
Smart Innovation, Systems and Technologies 76, DOI 10.1007/978-3-319-59480-4_39

the DBMSs were extended to consider the specificities of mobility data, which gave rise to the Moving Object Database (MOD). However, current DBMS ability, even with its extensions to support spatio-temporal data, is limited only to storing raw mobility data, represented as a set of spatio-temporal coordinates, which does not include any semantic information. To help researchers make efficient exploitation of this data, many attempts to integrate geographic and application domain knowledge, also referred to as semantic information, have been made to make it more comprehensible, manageable and understandable.

The semantic information added to mobility data could be geographic knowledge, such as application-oriented *Points of Interest* (PoIs) through which the moving object passes, or its activity during the travel. Activity is actually what the moving objects do at those interesting places, shopping in a mall for example. This natural link between traveling and activities motivates large number of research works to investigate novel approaches for enriching raw GPS data with moving objects' activities during travel. This could be useful in different applications, such as: social network applications, targeted advertising, public security, town planning and other related applications and services. In the following, we review the main works related to raw trajectory annotation with semantics. Contrary to the previous research works surveying this issue, our paper focuses on methods for computing semantic trajectories, explicitly transforming raw mobility data into meaningful knowledge about semantic trajectories, and introduces a new taxonomy of related work according to used techniques and typology of added semantics.

Accordingly, this paper is structured as follows. The first section, introduces the main concepts related to semantic trajectory modeling issues and presents a taxonomy of the different works tackling semantic enrichment of mobility data. After that, according to the previous taxonomy, we review and discuss related works, classified into three categories: *Raw Mobility Data Based Annotation*, *Static Trajectory Annotation* and *Dynamic Trajectory Annotation*. Finally, we conclude our discussion and present some interesting open issues.

2   Semantic Trajectory Computation: Taxonomy of Related Work

Answering sophisticated queries about trajectory data to understand moving objects' behavior requires higher level of analysis techniques. In fact, querying raw trajectory data is time consuming and increases the computational and formulation complexity from a user's perspective seeing that it requires spatial joins. Accordingly, it's important to start by adding semantic information to raw data, in a preprocessing step. Actually, enriching trajectories with semantic information leads to the discovery of richer and more interpretable semantic trajectory patterns that many data mining techniques, considering trajectories as sample points, may not be able to discover, and facilitates then the understanding of moving object behavior to domain experts.

Various techniques were implemented to enrich raw movement data with semantics. In fact, the addition of semantic information to raw data is referred to as *annotation*. Typically, the annotation process receives as input mainly raw mobility datasets, and could have in addition any contextual data repository as geographic information. And as output, it produces a set of *semantic trajectories*, referring to trajectories enriched with semantic information.

The main idea was, at first, introduced by [8]. In this work authors proposed a conceptual modeling approach to associate a layer of semantic information to trajectories. They introduced the trajectory concept as a set of segments: *moves* [8] (called trips in [7]) and *stops* [8] (called activities in [7] and stay points in [3]). The latter are essential components of any trajectory of a moving object. A *move* is defined as a time subinterval where the moving object's position changes. *Stops* are, in contrast, time subintervals where the moving object's position is fixed. Consequently, the process of trajectory annotation requires at first the identification of trajectory components then their annotation with semantic information. In the literature, in overall, two types of semantic information are used to annotate its so called *stops* and *moves*:

- *Activities*, what mobile objects do during their travel, while moving (*moves* activities: transportation mode, walking, running, etc.), and when they stop (*stops* activities: working, shopping, having lunch, etc.). Actually, the inference of an activity depends on the location of the moving object (associated PoI), the time in which it is carried out, its duration, the previous activity performed and other common truth constraints.
- *Geographic information*, gathering landmarks, hotspots, building places (museum, mall, restaurant, home, etc.), topography of the land (mountain, lake, etc.) for annotating the *stops*, and networks (highways, railways, etc.) and streets corresponding to the moving object's path for annotating the *moves*. They represent locations of interest or PoIs related to the application domain, whereby the moving object passes or where it stops for a period of time. Moreover, knowing the PoIs associated to the mobile object's trajectory, one could infer the possible activities practiced on each PoI (working in office, shopping or working in mall, etc.).

The exploitation of trajectories with additional semantic information is a recent but active field. Classically, in order to know what people do during their travels it is possible to ask them to record their activities, as well, it is possible for experts to manually associate semantic information to raw data points using their own words and expressions. This, however, is a tedious and boring task. This is evener more cumbersome when dealing with a large volume of mobility, complemented with the semantic heterogeneity and inconsistency derived from different agents annotating the data with different expressions having same meaning. Therefore, many attempts were made to automate the semantic trajectory computation process using techniques from different fields including machine learning and statistics. In the literature, related work for semantic trajectory annotation is mainly divided into three methods:

– *Raw mobility data based annotation*: machine learning and statistics are used to infer moving objects' activities from raw mobility data without additional semantic information associated. Supervised learning techniques, as Decision Tree (DT) for example, are mostly used to learn how to classify unlabeled data, in our case raw spatio-temporal coordinates, based on a set of labeled data, which are datasets already annotated with moving object's activities.

– *Static trajectory annotation*: geographic information, stored in spatial databases, is explicitly associated to raw mobility data. The joint between geographic information and trajectory spatial coordinates is computed using different techniques in the literature. Then, activities are inferred from the joint geographic information according to domain-related rules. In this case, raw mobility data is annotated with a set of fixed and predefined attributes, this is the reason why we referred to it as a static method.

– *Dynamic trajectory annotation*: The use of web resources is becoming more and more widespread all around the world, turning it into an unavoidable source of spatiotemporal information describing the movement and activities of human beings. Consequently, these big amounts of generated mobility-related information are of paramount importance for the understanding of human movement. However, their processing is not a straightforward task. For that, many works in the literature used text mining algorithms to extract space-time semantic information related to human movement from web documents and use it to annotate raw mobility data. We referred to this method as dynamic because added semantics change according to the information provided by social media, which is continuously updated by users.

Once raw mobility data is enriched with semantic information, knowledge discovery and data mining techniques are used to identify behaviors (such as tourist behavior) either from an individual or a group of moving objects' *semantic trajectories*. In the following, we discuss related works according to the aforementioned taxonomy of semantic trajectory computation techniques.

3   Raw Mobility Data Based Annotation

In these methods, extracted semantic information is mainly moving object activities such as transportation modes and low-level activities. Activities are recognized based on a set of raw sensor data and without external additional information such as PoIs. This falls within the context of activity recognition research field. In fact, in activity recognition, rules are used to recognize common human or animal's activities in real life settings. Those rules were beforehand provided by domain experts. Then, researchers started to apply machine learning and statistical algorithms to automate the rules generation task and avoid the expert's intervention. In the following, we review several related works, addressing this issue, categorized according to the recognized activities.

3.1   Annotation with Transportation Modes

As discussed above, a *move* is an essential part of a moving object's trajectory, and as well as a *stop*, it could be annotated with semantic information in order to describe the activities performed while moving. For recording the movement of human beings, this part is commonly annotated by the transportation modes the moving person uses to travel. Examples of target applications for this contribution are physical activity monitoring, transportation and mobility-based recruitment.

For instance, in [7], a post-processing procedure that takes as input only basic GPS raw data to infer a mobile object's transportation mode is proposed. The inferred transportation modes are: walking, cycling, in a car, in public transportation and in rail. As a first step, the trips (referring to moves) and activities (referring to stops) are determined. Then, the trips are segmented into single-mode stages. Finally, the transportation mode for each of the stages is identified using a fuzzy logic approach based on speed and acceleration features of the stages.

Differently, in [12] the proposed approach is based on supervised learning to infer users' transportation modes that could be driving, walking, taking a bus and riding a bike. A change point-based segmentation method is proposed to partition each GPS trajectory into separate segments of different transportation modes. Then, features of each segment are fed to a generative inference model to classify them into one of the transportation modes. The inference model is based on a DT and a bootstrap aggregating employed as a meta-algorithm to improve the accuracy of the model.

Also, in [6] authors introduced a classification system for inferring transportation modes when outside. However, in this work the inferred transportation modes include only: stationary, walking, running, biking and in a motorized transport, and did not distinguish the types of the motorized transportations such as: car, bus, rail, etc., as in the previous works. Moreover, added to the GPS data, authors used an accelerometer to measure acceleration. The used classification system consists of a DT and a first order Discrete Hidden Markov Model (DHMM). Actually, the DT/DHMM combination is considered to be the best accurate since DT is tuned to differentiate between boundaries of transportation modes while DHMM eliminates noise based on temporal knowledge of the transportation mode and the likelihood of transitioning into the next mode.

3.2   Annotation with Low-Level Activities

Many works in the literature used low-level activities to annotate raw mobility data such as: eating, cooking, working on a computer, talking, watching TV, etc. They are generally performed indoor and appeal physical movement with noticeable shift in position and direction of the body members. To be recognized, low-level activities require often, more than the raw time-location data provided by a GPS device, accelerometers tied to different parts of the body, microphones to differentiate between the sitting and watching TV activities for instance, light sensors, etc.

For example, in the work of [4], authors designed and built a multimodal sensor board that simultaneously captured data from seven different sensors: accelerometers, microphone, barometer, etc., gathering real-world activity traces to train activity classifiers. The recognized activities are date-to-date activities such as: walking, climbing stairs, cooking, working on a computer and so on.

Similarly, a user's activity recognition system is proposed in [2]. The system is based on an algorithm using a classification method to detect 20 physical activities including: brushing teeth, running, walking, watching TV, sitting and relaxing, etc., from data of biaxial accelerometers worn in different parts of the body. The system was tested on different classifiers including base-level and meta-level classifiers from WEKA as: decision table, IBL or nearest neighbor, naive Bayes, etc. The classifiers results are then compared and DT performed the best. In fact, the DT classifier is slow to train but quick to run. Consequently, a pre-trained DT can classify activities in real time quickly.

As in the previous work, the idea in the work of [5] is to recognize activities from triaxial accelerometer formulated as a classification problem. The objective is to analyze the performance of base level classifiers: decision tables, DT, KNN, SVM, naive Bayes, that are integrated in the Weka toolkit, and study the effectiveness of meta model classifiers: boosting, bagging, plurality voting and stacking in improving activity recognition accuracy by combining base-level classifiers. The activities to be recognized are: standing, walking, running, climbing upstairs, climbing downstairs, getups, vacuuming, brushing teeth. Experiments has shown that meta-level classifiers performed better than base-level classifiers, and plurality voting has the best performance.

In the aforementioned works, proposed approaches mainly used classifiers that took as input merely the raw mobility data and does not rely on external spatial/temporal information to infer moving objects' activities. Most often, DT classifiers showed the best performance. The annotation process starts with the splitting of trajectories into segments of different activities, the features selection and extraction and finally classifier's training. Generally, in low-level activities recognition, experiments are performed in laboratory settings. Then, it is already known where each activity starts and ends. Therefore, trajectories do not need segmentation as for transportation modes recognition.

In fact, it is non-trivial to accurately extract semantics from only raw mobility data since data requires too much processing before getting to the intended semantics, activities in this case. This also requires a very large amount of it to be processed to avoid noise and unmeaningful results, which might be impossible in some cases where big amounts of data may be unavailable. Therefore, domain prior knowledge as PoIs, social media, rules is required to improve annotation performance and compensate data sparsity. We noticed also the heterogeneity of the inferred activities. Then, a consensus of the research community about location-dependent/user specific models (ontology) would be required to unify the inferred activities.

4   Static Trajectory Annotation

Domain knowledge is represented in many cases as location information: landmarks, PoIs, etc. Such prior knowledge is proved to be helpful in the activity recognition process and improves then its results. In fact, using contextual knowledge about location is sometimes enough to infer coarse level activities. For example, being in restaurant outweighs activities such as eating, taking advantage of the user's spatial/temporal context to improve inference. In many works in the literature, statistical algorithms are used firstly to identify stops within a trajectory, then each stop is associated to an application-oriented PoI, finally activities are inferred. We refer to such annotation methods here as *static*.

Authors in [1] proposed a data preprocessing model aiming to integrate trajectories with geographic information. They used firstly the SMoT algorithm to compute *stops*. In fact, trajectories and geographic data overlap in space. Therefore, exploring trajectory *stops* is about finding the intersection between trajectory points and application candidate stops, referring to application-related PoIs, where the intersection duration is $\leqq \Delta_c$.

In the work of [10] an approach was proposed to automatically extract sequences of activities from large set of trajectory data and PoIs. They used for that two algorithms with the duplication reuse technique. At first, the Voronoi diagram is used to find subtrajectory parts influenced by a POI. After that, the duration of the influence of a PoI in a subtrajectory is calculated. Then, PAMS (PoI-Activity Mapping Set) provides the rules to map between user activities and PoIs using the calculated influence duration.

Likewise, a general framework for identifying top-k semantic locations significant to users from raw GPS data is introduced in [3]. The first step is to extract *stay points*, referring to *stops*, from GPS records based on a threshold duration between two consecutive records. Then resulting stay points are clustered to represent real word semantic locations using a semantic enhanced clustering algorithm SEM-CLS and additional semantic information from geocoded street addresses and yellow pages directory. For ranking the locations, a two layered graph is used for capturing location-location, location-user and location-user relationships using their location histories.

Compared to the first two works, the stops (referred to as stay points) identification method in the last work is more sophisticated appealing clustering, and not just considering the simple intersection between PoIs and trajectories. Clustering, as an unsupervised learning technique, is used in this context for automatically discovering interesting places, associated with stops in trajectories, computed based on density of positions in a spatial area for a certain period of time.

In the aforementioned works, raw mobility data is annotated with semantics (activities or PoIs) coming from predefined databases. Then, semantics used for annotation inhere represent a set of fixed and predefined semantics. However, the semantics of the same location can change according to time.

5   Dynamic Trajectory Annotation

In addition to spatial information as PoIs, semantic information could be represented as temporal information such as information about the event attended at a special location: concert, football game, etc., duration and time of activities: day, hour, weekend, holiday, etc. Added to the large amount of mobility data provided by mobile sensors, nowadays, there is a large-scale of crowd-generated social media data, as location-based social networks, weblogs, e-newspapers among other web resources, that could be used as a source of semantic information about location and time. For that, merging raw mobility data with social media information could help researchers to understand the spatial and temporal semantics of people's movement. Previous works on semantic trajectory computation used only mobility data and static geographic information to do so. It is only recently that some research works started to focus on merging the aforementioned resources together. We refer to this trajectory annotation method as *dynamic* since it considers, not only the spatial, but also the temporal semantics of a location.

For example, in the work of [9], researchers proposed a method to annotate the location history of a mobile user using spatiotemporal documents collected from social media e.g. tweets. The proposed system derives for each user's location history entry a list of words that best describe the purpose of the user's visit to that location at that time. In fact, researchers in this work annotated user location history with dynamic semantics using techniques from text processing research field. The proposed method for the annotation is time sensitive since that for each location r only documents with time window T are considered. Authors found, after experimentations on a large geo-tagged tweets dataset, that KDE is the more suitable model for annotating mobility data. Actually, the latter captures well the locality and relevance of the words w.r.t a record.

Actually, services could be provided to the user in two ways: location recommendation knowing the activity and activity recommendation knowing the location, to satisfy his information needs for daily routines as trip planning. To do so, authors in [11] modeled a location-activity matrix where each entry is a rating showing how often an activity is performed in a location. The matrix construction is based on users' GPS history data added to their available comments on the web to identify the activity of the moving object in each visited location according to the following process. The first step is to extract *stay points* from raw GPS data. Then, stay regions, which are the locations to be recommended, are calculated using grid-based clustering of stay points. Finally, comments attached to GPS points assigned to each of the resulting stay regions are parsed to extract activities. To compensate the sparsity of user-commented locations, authors exploit additional information sources: location features and activity-activity correlations. In this work, activities are extracted manually from comments. This task is too much time consuming, then text mining techniques could be used to automate the annotation process.

Dynamic annotation using social media is considered as a novel method for computing semantic trajectories, since the recent availability of required semantic information on the web. However, there are some improvements which we believe could be beneficial. Firstly, mining the web for extracting annotation words/activities could result in semantic inconsistencies (eating, food, restaurant for the same activity). Then, such proposals require integration of application-related concepts used for annotation (an ontology of activities). Statistical algorithms are used to mine texts in social media. However, social media data is particularly noisy. So treating this type of data requires using the right model to retrieve relevant words with regard to a mobility record. Also, likely proposals, require differentiating between landmark words and event related words using temporal characteristics. Finally, different types of contextual data sources: social media, geographic database, etc. could be used to compensate data sparsity.

6   Conclusion

A wide range of research works have been proposed in the literature to enrich trajectories with semantic information. In this work, we have been motivated by the need to enhance the foregoing attempts to compute semantic trajectories to respond to their limitations. To meet this need, we presented, in this paper, a survey and a taxonomy of the different works on semantic enrichment of trajectory data. The panoply of works proposed in the literature in this issue is consistent and diverse. However, shortcomings were deduced from the different methods presented inhere. Research works, classically, focused on adding spatial semantics to mobility data. Few of the proposed works, mainly dynamic annotation methods using social media, used temporal semantics. Recently, there have been some research efforts for incorporating domain knowledge, represented especially by formal semantics as ontology models, formalized, consensual and generally validated by experts to avoid annotation inconsistencies. In this case, the formality of semantic concepts provided by ontology models permits the automation of the whole process of trajectory semantic annotation going from raw GPS data to semantic trajectories. Also, the use of ontology induces a number of straightforward benefits, explicit semantics, well-defined reasoning mechanisms, great analysis by using warehousing techniques supporting the extraction of valuable information from raw trajectories. However, to the best of our knowledge, little work has been oriented to using ontologies for semantic trajectory computation. Consequently, as future work, we intend to experiment the effectiveness of ontology in helping to encode domain knowledge for computing semantic trajectories. Accordingly, we project to enhance the aforementioned works so that we could provide a complete framework for the construction of semantic trajectories going from raw mobility data to semantic trajectory analysis.

References

1. Alvares, L.O., Bogorny, V., Kuijpers, B., de Macedo, J.A.F., Moelans, B., Vaisman, A.: A model for enriching trajectories with semantic geographical information. In: Proceedings of the 15th Annual ACM International Symposium on Advances in Geographic Information Systems, GIS 2007, pp. 22:1–22:8. ACM, New York (2007)
2. Bao, L., Intille, S.S.: Activity recognition from user-annotated acceleration data. In: Ferscha, A., Mattern, F. (eds.) Pervasive Computing, pp. 1–17. Springer, Heidelberg (2004)
3. Cao, X., Cong, G., Jensen, C.S.: Mining significant semantic locations from GPS data. Proc. VLDB Endow. **3**(1–2), 1009–1020 (2010)
4. Choudhury, T., Borriello, G., Consolvo, S., Haehnel, D., Harrison, B., Hemingway, B., Hightower, J., Klasnja, P.P., Koscher, K., LaMarca, A., Landay, J.A., LeGrand, L., Lester, J., Rahimi, A., Rea, A., Wyatt, D.: The mobile sensing platform: an embedded activity recognition system. IEEE Pervasive Comput. **7**(2), 32–41 (2008)
5. Ravi, N., Dandekar, N., Mysore, P., Littman, M.L.: Activity recognition from accelerometer data. In: Proceedings of the 17th Conference on Innovative Applications of Artificial Intelligence, IAAI 2005, vol. 3. pp. 1541–1546. AAAI Press (2005)
6. Reddy, S., Mun, M., Burke, J., Estrin, D., Hansen, M., Srivastava, M.: Using mobile phones to determine transportation modes. ACM Trans. Sen. Netw. **6**(2), 13:1–13:27 (2010)
7. Schüssler, N., Axhausen, K.: Processing GPS raw data without additional information. Eidgenössische Technische Hochschule, Institut für Verkehrsplanung und Transportsysteme (2008)
8. Spaccapietra, S., Parent, C., Damiani, M., Demacedo, J., Porto, F., Vangenot, C.: A conceptual view on trajectories. Data Knowl. Eng. **65**(1), 126–146 (2008)
9. Wu, F., Li, Z., Lee, W., Wang, H., Huang, Z.: Semantic annotation of mobility data using social media. In: Proceedings of the 24th International Conference on World Wide Web, pp. 1253–1263. International World Wide Web Conferences Steering Committee (2015)
10. Xie, K., Deng, K., Zhou, X.: From trajectories to activities: a spatio-temporal join approach. In: Proceedings of the 2009 International Workshop on Location Based Social Networks, LBSN 2009, pp. 25–32. ACM, New York (2009)
11. Zheng, V.W., Zheng, Y., Xie, X., Yang, Q.: Collaborative location and activity recommendations with GPS history data. In: Proceedings of the 19th International Conference on World Wide Web, WWW 2010, pp. 1029–1038. ACM, New York (2010)
12. Zheng, Y., Chen, Y., Li, Q., Xie, X., Ma, W.Y.: Understanding transportation modes based on GPS data for web applications. ACM Trans. Web (TWEB) **4**(1), 1 (2010)

Ambulance Fastest Path Using Ant Colony Optimization Algorithm

Hazar Hamdi[1(✉)], Nouha Arfaoui[1], Yasser Al Mashhour[2], and Jalel Akaichi[2]

[1] BESTMOD, High Institute of Management, Tunis, Tunisia
hazarhamdi@yahoo.com, arfaoui.nouha@yahoo.fr
[2] College of Computer Science, King Khaled University, Abha, Saudi Arabia
almshhour@kku.edu.sa, j.akaichi@gmail.com

Abstract. The number of the accidents in Tunisia is terrifying and it is considered as the highest in the world while basing on accurate statistics. Such situation requires a very fast intervention. Therefore, it is necessary to promote the study in ambulance management, so as to optimize the strategy of response to a given accident. Based on the defect of the ambulance root choosing, this paper puts forward a new algorithm based on ant colony optimization algorithm to find the best way that minimizes the time while taking into consideration the cases of problems that can appear each time such as traffics, catastrophes natural, etc.

Keywords: ACO algorithm · Ambulance root choosing · Faster path · Shortest path

1 Introduction

According to the recent statistic in Tunisia during 2015, the number of accident is 7225, the number of injured is 10882 and the number of deaths is 1407. The causes of accidents are: crossing the road without paying attention 26,63%, speeding 16,08%, non-concentration 14,03%, and non-respect of priority 9,22% [24]. The numbers are terrifying and they are considered as the highest in the world. Such situation requires a very fast intervention.

In Tunisia, many regions suffer from the catastrophic situation of their infrastructures which makes the access to the locality of accident a very hard task. There is also the problem of the distance to the nearest hospital. We can give as example Jandouba where the patients have to move to Kairouan (the separated distance is 202 km) [25]. Besides, the problem of the lack of ambulances and means of transport [25] where sometimes, it takes six hours for an ambulance to arrive and to leave [26].

An ambulance has a start point that corresponds to the current position when the health staff gets information about the accident. It has, also, an end point that corresponds to the location of the accident.

In the literature, different works focus on defining the shortest path that corresponds the shortest distance between two points such as [17, 22], etc. Such path may not be always the best for an ambulance since many factors can influence the speed. For example, if there is traffic jam in the shortest path, or if there is any natural disaster or

© Springer International Publishing AG 2018
G. De Pietro et al. (eds.), *Intelligent Interactive Multimedia Systems and Services 2017*,
Smart Innovation, Systems and Technologies 76, DOI 10.1007/978-3-319-59480-4_40

even there are some works, the speed must decrease which influences on the time of arrival. For those reasons, sometime the shortest path is not the best path to be followed by an ambulance.

The idea in this work is to choose the fastest path. In order to achieve this task, we propose a new algorithm based on ACO (Ant Colony Optimization) algorithm to get the fastest path every time. Compared to the existing works ([5, 7, 14]), our solution is about partitioning the trajectory to be traveled to sub-trajectories taking into account the constraints that can occur during the displacement such the state of the road, the traffic jam, etc. So that, in each possibility of change of direction, the ACO is applied to determine the best path to be followed. Our solution looks for the fastest path that may not be the shortest one. Concerning the ACO algorithm, it has been used in various works such as [11, 12, 13, 16], etc. It was applied to ensure the choice of the optimal solution.

This work is organized as following: in Sect. 2 we present some works using the ACO algorithm to solve various type of problems such as the traveling salesman. In Sect. 3, we define the overview of ambulance vehicle management. Then, we give a motivated example where we explain the importance of choosing the fastest path rather than the shortest path for an ambulance to reach the accident zone. In Sect. 5, we propose our solution that determines the fastest path using the ACO algorithm. Section 6 contains the case study related to our solution. We finish this work with a conclusion.

2 State of the Art

In section, we present some existing works that use the ACO algorithm to solve different kind of problems.

2.1 Traveling Salesman Problem

Many computer scientists and mathematicians are interested by the Travelling Salesman Problem (TSP) which is considered as a typical NP-hard problem that consists on calculating the shortest round-trip with minimal cost for the salesman. Indeed, given a set of cities and the distance between them; the salesman must visit each city only once and after that he should return to the starting city in the end. In the following, we will focus on the application of the ACO to solve this problem, since ACO is considered among the most important computing methods for TSP [12].

In [16], authors presented a new model of ant system. They integrated ant with memory into ACO to get ants that can retain the best-so-far solutions.

In [12], the authors proposed a new version of ACO which is about putting initially ants on different cities to adjust the algorithm's parameters, also to improve solution for the local search algorithm. So the purpose of this study is to update dynamic heuristic parameter in order to avoid stagnation behavior and premature convergence.

The authors, in [21], combined ACO with culture algorithm (CA) framework using the dual inheritance mechanism in order to improve ACO performances.

In [15], junji and al treated the multiple TSP (MTSP) with ability constraint. These salesmen must visit "n" cities and each one must start and finish in the same city, also, each cities must visited once by only one salesman and the objective is to browse "the minimum of total distances travelled by all the salesmen.

In [4], the authors used the partitioning technique to solve TSP. This approach consists of partitioning the original problem into the group of sub problem. They divided small number of cities in each group of ants, then they considered each group as a separated problem that will be treated with ACO. At the end, they joined sub problems for obtain the overall solution.

In [20], authors introduced ACO approach with dynamic constraints for a special case of routing problem with a common topology to the TSP. In fact, they modeled the routing problem of cash machine "which clients of any bank can with draw money" such as a TSP and then they applied ACO to the problem.

In this work [6], Brezina et al., showed that ACO is useable for solving problem with practical applications such as TSP. They concluded that the number of ants can affect the quality of solution, since the higher accumulation of pheromone depends on the higher number of ants, and the lower accumulation of pheromone vaporize quickly implying the lower number of ants. The chosen path is one that possesses an elevated concentration of pheromone.

2.2   Application of ACO Algorithm in Others Domains

In [10], authors treated the case of query optimization for distributed data base systems especially optimization join queries. They proposed Max-Min Ant System Distributed Join Query Optimization (MMAS-DJQO) based on multi colony ant algorithm. The solution treats two cost models. The first one is about optimizing the total time and the second affects the response time. This algorithm introduces, in each iteration, four types of ants' hence four ant colonies to find the best decision made by each ant to achieve optimal execution plan.

In water resources and environmental management, we find many problems which needed to be resolved. For that, numerous versions of ant algorithms are proposed and applied to solve this problem like [13] who improved an ACO to find a solution to the reservoir operation problem by deriving the optimal period of records releases.

In [18] the authors introduced electrical cost problem for the water distribution networks. They applied an ACO framework for optimal scheduling of pumps. The scheduling of operation of pumps consists to control time during status (ON/OFF) for each pump. In the same context and in order to make more flexibility pumping application [11] used variable-speed pump and combined the ant system iteration best algorithm and EPNET2.0 for "minimize the size of search space".

This research [3] treated the continuous optimization problem. The authors applied the continuous ACO algorithm to get the optimal design of storm sewer network. They proposed two alternative approaches: constrained and unconstrained, that they added as a free parameter for each approach; a Gaussian probability density function for the unconstrained approach and a known value of the elevation at downstream node of pipe.

In [1] the authors solve the transient flow that is considered as a very serious problem causing by valve and pump shut-off in water pipeline system. For that they coupled the ACO algorithm with a hydraulic simulation module to design a water supply pipeline system.

The work presented in [2], also used ACO to find the ideal location of monitoring station in large scale water distribution networks; presenting a multi-objective versions of ACO consists to minimize the number of monitoring station with increasing the total coverage.

In the area of intrusion detection system, we quote the work of [9] that combined a clustering based on self-organized Ant Colony Network with Support Victor Machine (SVM) for realize real-time detection in high-speed network.

In [8], the authors chose one of different variants of the ACO to be the basis of the proposed approach (fuzzy approach for diversity control in ACO). They tried to avoid or slow-down full convergence and their objective is to "maintain diversity at some level to avoid premature convergence" is realized. Thereafter, they applied this method in the design of fuzzy controllers, primarily in the optimization of member ship function for a unicycle mobile robot trajectory. Also the algorithm ACO was tested with Tomera and al in [19], based on PID (Proportional-Integral-Derivative) control algorithm. They applied the ACO to optimize parameters of the ship course controller to achieve effective and rapid handling of a ship and fuel consumption reduction.

3   Overview of Ambulance Vehicle Management

In Tunisia, ambulances belong essentially to five different structures. Ambulances of hospital structure which deal with transport of patients in intra or extra hospitals to provide care in inter-service. Civil protection ambulances and SAMU ambulances that provide care in the event of disaster or a public road accident in close collaboration with two separate regulatory centers but which may be in direct collaboration. The SAMU ambulances can provide more care due to the compulsory presence of a physician and more supplied materials. There are also the private ambulances which intervene essentially at the request of an individual. Finally, the ambulances of structure belonging to particular structures such as the ministry of the interior or the airport that respond to unlimited ad hoc requests.

The figure below (Fig. 1) shows the process of the ambulance management since receiving an emergency call until the arrival of the ambulance.

First, a collaborating agent who can be any person in the scene of the accident such as a police officer, another car driver, a pedestrian, etc., makes an emergency call to a regulation center that can be the rescue police (in Tunisia the call number is 197), civil protection (the call number is 198), or else he calls the SAMU using 190. This center specifies the type of the ambulance to send according to the information provided (ambulance model). After having specified the type of the ambulance (ambulance model), the latter has as aim to reach its destination as rapidly as possible; therefore an ambulance management is necessary.

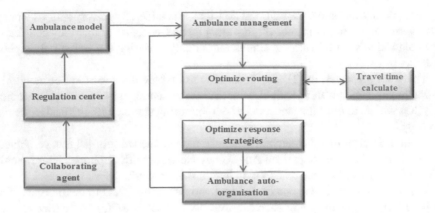

Fig. 1. Ambulance management process

The objective of the ambulance management is to carry out an auto-organization of the ambulance which consists in optimizing the strategy of response to a given accident. This optimization can only be carried out after optimization of choice of the route to be covered and which necessitates minimizing the travel time of the ambulance.

4   Motivated Example

In this section, we give a motivated example where there are two different trajectories to take (Trajectory 1 and Trajectory 2). We follow the behaviors of the drivers between two points: A and F. A corresponds to the begin of the trajectory and F corresponds to its end.

Figure 2 presents two different roads that exist between the two points A and F. Each road represents a specific trajectory.

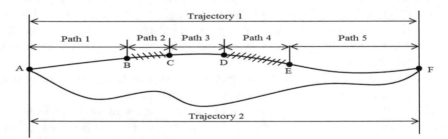

Fig. 2. Example of trajectories between two points

Under normal road conditions, Trajectory 1 corresponds the shortest road because it has the shortest distance between A and F that corresponds to 120 km. With a speed about 100 km/h, this trajectory takes 1 h 12 min.

Concerning Trajectory 2, it connects A and F with a distance equals to 180 km. It takes 1 h 24 min to move between the two extremities with the same speed (100 km/h). We assume that the state decides to make improvements on parts of Trajectory 1 since it is the most used. So, there are works between B and C, and between D and E. In those parts, the speed is 30 km/h. The distance between B and C is 10 km, it takes 20 min. And between D and E, it is 20 km, so it takes 40 min. Because of the works, Trajectory 1 is subdivided into portions according to the value of the speed. Each portion corresponds to a specific path. The duration (D) of Trajectory 1 is the sum of the duration of its corresponding paths as presented in (1).

$$D_{\text{Trajectory 1}} = \sum\nolimits_{i=1}^{n} D_{\text{Path } i} \text{ for } n > 0 \tag{1}$$

$$\text{So;} D_{\text{Trajectory 1}} = D_{\text{Path 1}} + D_{\text{Path 2}} + D_{\text{Path 3}} + D_{\text{Path 4}} + D_{\text{Path 5}} \text{ with } D = \frac{\text{Speed}}{\text{Distance}}$$

$$\text{Distance}_{\text{Path 1+Path 3+ Path 5}} = 90 \text{ km; Speed} = 100 \text{ km/h} \rightarrow D = \frac{90}{100} = 54 \text{ min}$$

$$\text{Distance}_{\text{Path 2}} = 10 \text{ km; Speed} = 30 \text{ km/h} \rightarrow D = \frac{10}{30} = 20 \text{ min}$$

$$\text{Distance}_{\text{Path 4}} = 20 \text{ km ; Speed} = 30 \text{ km/h} \rightarrow D = \frac{20}{30} = 40 \text{ min}$$

$$D_{\text{Trajectory 1}} = 54 + 20 + 40 = 1\text{h } 54 \text{ min}$$

$$\text{So } D_{\text{Trajectory 1}} > D_{\text{Trajectory 2}}$$

For an ambulance, it is much better to follow Trajectory 2 instead of Trajectory 1 because it takes less time.

5   Adaptation of the Ant Colony Optimization Algorithm

In this section, we propose our algorithm "Ambulance_Fastest_Path" that generates the fastest path for an ambulance. It starts by detecting the current position of the ambulance. It gives the hand to the user to specify the destination to extract the shortest path using the ACO algorithm. The ambulance' driver follows the specified path. For each possibility of change of direction such the roundabouts, there is a verification of the remaining portion of path. If it is fine, the ambulance continues its path otherwise, there is an extraction of the possibilities of paths to be followed. The choice of one of them depends on the shortest path and the lack of obstacles. The steps are repeated for each change of direction until the ambulance reaches the accident location. Concerning the ACO algorithm, it consists on randomly choice of a walk starting from the nest looking for the food. Pheromones are left if an ant finds food, consequently other ant can follow it. More the quantity of this substance is higher ants follow this path. In this section, we describe the algorithm basing on [23].

The ACO algorithm refers essentially to the TSP problem that is about looking for the shortest path by browsing each city once and return to the starting city for a given n's city set. Mathematically talking, it is about minimizing the following formula (Formula 2).

$$minD = \sum_{t=1}^{n-1} d(i, i + 1) + d(n, 1) \qquad (2)$$

With:

- n indicates n's cities for passing through
- $d(i, i + 1)$ indicates the distance between city i and city i + 1
- $d(n, 1)$ indicates the distance from n's city to return to the starting city.

Algorithm1. Ambulance_Fastest_Path
Detect the current position
Define the destination
Determine the shortest path using ACO Algorithm
For each change of path direction
Verify the situation of the remaining portion of the path
If there is no problem
The ambulance continues its path
Else
Determine the other possible paths
Choose one of them by applying the ACO algorithm
End If
End For
End

6 Case Study

In the previous section, we described our algorithm; in this section we present a case study. We take as example an accident that took place in front of the high school (Ezzahrouni high school). The shortest path takes 12 min going through the bleu path as presented in Fig. 3.

The problem is this application does not take into account the case of traffic jam when calculating the duration of the path. Indeed, in the roundabout of Ezzahrouni there are a lot of traffic jam because many vehicles come from different places (North West of Tunisia, Sidi Hcine, Ezzahrouni, …) also there is the project of the Fast Rail Network (FRN)[1] that disrupts the traffic of cars and creates endless plugs. Such situation cannot be taken into account by google maps and the drivers car are stuck with a loss of time considerable, hence the idea of our work which actualizes in real time the road situation and gives to the users an idea of the traffic allowing thus a saving in time which represents a major factor

---

[1] https://www.tustex.com/economie-actualites-economiques/le-reseau-ferroviaire-rapide-de-tunis-rfr-verra-le-jour-debut-2018-et-permettra-le-transport-de-2500.

Fig. 3. The shortest path giving by Google maps

in the management of the critical medical situations. So, the ambulance follows another path that corresponds to Ezzahrouni High school → B → C → D → E → F → G → Rabta Hospital. Thus, it goes through street 13 August as presented in Fig. 4.

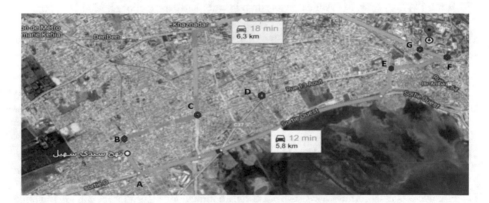

Fig. 4. The best path giving by our solution

7   Conclusion

The number of accident in Tunisia is terrifying which requires a very near real time intervention of ambulances to save the maximum number of persons.

For this reason, we focus in this work on the determination of the fastest path to be followed to reach the accident place as soon as possible. The fastest path is not always the shortest path since many factors can influence the duration of the trajectory such as the traffic jam, the weather, etc. So, we propose as solution a new algorithm based on ACO algorithm to improve the route choosing. It is about determining for each possibility of direction change during the path from the start point of the ambulance to the location of the accident the best path to follow taking into account the new information that can occur each time.

References

1. Abbasi, H., Afshar, A., Jalali, M.R.: Ant-colony-based simulation-optimization modeling for the design of a forced water pipeline system considering the effects of dynamic pressures. J. Hydroinf. **12**(2), 212–224 (2010)
2. Afshar, A., Mariño, M.A.: Multi-objective coverage-based ACO model for quality monitoring in large water networks. Water Resour. Manage **26**(8), 2159–2176 (2012)
3. Afshar, M.H.: A parameter free continuous ant colony optimization algorithm for the optimal design of storm sewer networks: constrained and unconstrained approach. Adv. Eng. Softw. **41**(2), 188–195 (2010)
4. Bajpai, A., Yadav, R.: Ant colony optimization (ACO) for the traveling salesman problem (TSP) using partitioning. Int. J. Sci. Technol. Res. **4**(09), 376–381 (2015)
5. Ben Abdouallah, M., Bojji, C., El Yaakoubi, O.: Deployment and redeployment of ambulances using a heuristic method and an ant colony optimization—case study. In: International Conference on Systems of Collaboration (SysCo), pp. 1–4. IEEE (2016)
6. Brezina Jr., I., Čičková, Z.: Solving the travelling salesman problem using the ant colony optimization. Manage. Inf. Syst. **6**(4), 010–014 (2011)
7. Bura, W., Boryczka, M.: Ant colony system in ambulance navigation. J. Med. Inf. Technol. **15**, 115–124 (2010)
8. Castillo, O., Neyoy, H., Soria, J., Patricia, P., Valdez, F.: A new approach for dynamic fuzzy logic parameter tuning in ant colony optimization and its application in fuzzy control of a mobile robot. Appl. Soft Comput. **28**, 150–159 (2015)
9. Feng, W., Zhang, Q., Hu, G., Huang, J.X.: Mining network data for intrusion detection through combining SVMs with ant colony networks. Future Gener. Comput. Syst. **37**, 127–140 (2014)
10. Golshanara, L., Rankoohi, S.M.T.R., Shah-Hosseini, H.: A multi-colony ant algorithm for optimizing join queries in distributed database systems. Knowl. Inf. Syst. **39**(1), 175–206 (2014)
11. Hashemi, S.S., Tabesh, M., Ataeekia, B.: Ant-colony optimization of pumping schedule to minimize the energy cost using variable-speed pumps in water distribution networks. Urban Water J. **11**(5), 335–347 (2014)
12. Hlaing, Z. C. S. S., Khine, M.A.: An ant colony optimization algorithm for solving traveling salesman problem. In: International Conference On Information Communication And Management, Singapore, IPCSIT (2011)
13. Jalali, M.R., Afshar, A., Marino, M.A.: Improved ant colony optimization algorithm for reservoir operation. Scientia Iranica **13**(3), 295–302 (2006)
14. Javidenah, A., Ataee, M., Alesheikh, A.A.: Ambulance routing with ant colony optimization. In: Proceedings of GIS 89 Conference, Iran (2010)
15. Junjie, P., Dingwei, W.: An ant colony optimization algorithm for multiple travelling salesman problem. In: First International Conference on Innovative Computing, Information And Control-Volume I (ICICIC 2006), pp. 210–213. IEEE (2006)
16. Li, B., Wang, L., Song, W.: Ant colony optimization or the traveling salesman problem based on ants with memory. In: 2008 Fourth International Conference on Natural Computation, pp. 496–501. IEEE (2008)

17. Lin, K.C., Chern, M.S.: The fuzzy shortest path problem and its most vital arcs. Fuzzy Sets Syst. **58**(3), 343–353 (1993)
18. López-Ibáñez, M., Prasad, T.D., Paechter, B.: Ant colony optimization for optimal control of pumps in water distribution networks. J. Water Resour. Plan. Manage. **134**(4), 337–346 (2008)
19. Silva, C.A., Runkler, T.A.: Ant colony optimization for dynamic traveling salesman problems. In: Arcs Workshops, pp. 259–266 (2004)
20. Tomera, M.: Ant colony optimization algorithm applied to ship steering control. Procedia Comput. Sci. **35**, 83–92 (2014)
21. Wei, X., Han, L., Hong, L.: A modified ant colony algorithm for traveling salesman problem. Int. J. Comput. Commun. Control **9**(5), 633–643 (2014)
22. Zhan, F.B., Noon, C.E.: Shortest path algorithms: an evaluation using real road networks. Trans. Sci. **32**(1), 65–73 (1998)
23. Zhang, P., Feng, L.U.: Application in emergency vehicle routing choosing of particle swarm optimization based ant colony algorithm. J. Comput. Inf. Syst. **9**, 8571–8579 (2013)
24. Accidents De La Route En Tunisie: Les Statistiques De 2014 et 2015. http://Efigure.Net/Accidents-De-La-Route-En-Tunisie-Les-Statistiques-De-2014-Et-2015/
25. Rapport De Diagnostic Des Gouvernorats De Kairouan, Siliana, Kef Et Jendouba. https://tunisia.iom.int/sites/default/files/activities/documents/Rapport%20de%20Diagnostique%20des%20gouvernorats%20de%20Kairouan,%20Siliana,%20Kef%20et%20Jendouba%20-%20START.pdf
26. Une Ambulance Pour La Tunisie. https://www.assen-asso.fr/projets/aide-durgence/132-une-ambulance-pour-ma-tunisie

Educational Assessment: Pupils' Experience in Primary School (Arabic Grammar in 7th Year in Tunisia)

Wiem Ben Khalifa[1,2(✉)], Sameh Baccari[2], Dalila Souilem[3], and Mahmoud Neji[2]

[1] University of Dammam, Dammam, Saudi Arabia
wiembenkhlifa@gmail.com
[2] MIRACL Laboratory, Sfax, Tunisia
sameh_baccari@yahoo.fr, mahmoud.neji@fsegs.rnu.tn
[3] Umm al-Qura University, Mecca, Saudi Arabia
dalila.souilem@yahoo.fr

Abstract. Regardless of the considered instruction level, the evaluation practices applied in classrooms and educational institutions raise a large number of questions. Do these practices reflect the learners' cognitive metamorphoses? Do formative and summative evaluations aim at accessing knowledge acquisition at the expense of understanding? Is the content of the evaluation approaches and instruments part of the learners' previous knowledge? What is the relation between this content and the learners' specific knowledge base? Does the structure of evaluation instruments give access to knowledge construction? Is this structure more oriented towards the evaluation of low-level cognitive skills than to the assessment of high-level cognitive skills? Does the context of the assessment introduce affective variables that may negatively influence learners' results? Do the used instruments allow systematically determining the learners' strengths and weaknesses? Do the correction criteria respect the value of each of the learner's responses to construct a particular knowledge? Are these criteria more influenced by objectivity requirements than by the cognitive value of the responses? All these questions show the complexity and the importance of the evaluation problem in the school environment and in the cognitive progress of the learners. The questions mentioned are rather related to the content of the evaluation instruments, their structure, their objectives and the context in which the evaluation takes place than to the correction criteria. In this paper, we introduce a personalized evaluation environment for mobile learning. We first introduce m-evaluation scenario. Then, we describe this personalization as well as the architecture of the evaluation environment composed of Web services and based on communication between these services carried out via Semantic web technologies. The proposed environment allows implementing an assessment of the learners' knowledge about a specific teaching content by researching, selecting and generating a test adapted to their level of knowledge. Once the evaluation is performed, a new course will be proposed. We also conducted an experiment based on determining the degree of student's satisfaction.

Keywords: Mobile technology · Mobile evaluation · Semantic web · Context

© Springer International Publishing AG 2018
G. De Pietro et al. (eds.), *Intelligent Interactive Multimedia Systems and Services 2017*,
Smart Innovation, Systems and Technologies 76, DOI 10.1007/978-3-319-59480-4_41

1   Introduction

The development of Information and communication technologies has greatly changed the organizational structures and processes learning over the last decade essentially because of the exponential growth of using network technologies and Internet in particular. These technologies resulted in the generalization of using mobile communication tools, such as laptops and mobile phones connected to the internet through the currently available wireless networks (Wi-Fi, 3G, etc.). The use of these mobile tools greatly transformed the traditional relationships of how pupils view time and space in the context of learning. Mobile technologies provide pupils with high capacities in the sense that they can carry out their academic activities potentially at anytime and anywhere. Thus, mobile technologies are beneficial for pupils. In fact, they allow them to improve their communication, coordination and collaboration skills. They facilitate their interactions with teachers and provide better responsiveness and greater autonomy.

Maximizing the effectiveness of m-evaluation experience can be achieved by considering many parameters in such learning approach including learner mobility, learner profile, organization of knowledge, learning time and gain, etc. Indeed, mobility in m-evaluation is observed especially in the concept of context which reflects the situation in which the learner finds himself/herself. It is therefore necessary to determine, depending on the context, what resources to send, in what way, at what time, on which interface, etc. Thus, the whole evaluation process must be adapted to these changing contexts. With the emergence of Web 3.0, the semantic web seems a promising technology to implement m-evaluation system. Indeed, the semantic Web is a Web comprehensible to both men and machines. Thus, it represents a vast field of applying works emanating from the formalisms of knowledge and reasoning representation [6].

In this manuscript, we propose a flexible environment for mobile evaluation that uses semantic Web technologies for description and reasoning. This paper is structured as follows: First, we present related works. Then, we introduce our work based on a mobile evaluation (m-evaluation) scenario in an open learning network. Afterwards, we analyze the proposed scenario and we describe the m-evaluation process. The architecture set up for the evaluation system is depicted in Sect. 5. In Sect. 6, we discuss the first uses of this system and we end the paper with a conclusion and our future works.

2   Related Work

Numerous studies on the use of mobile technologies have been carried out for the evaluation in mobile environments. For instance, Cheniti et al. developed a Mobile and Personalized Evaluation System (MPES) based on Web services and the Semantic Web. In fact, MPES associates and implements a set of web services that interact with each other to satisfy the needs of the system. This architecture is an efficient approach to exploit also the technology of the semantic web. Indeed, web services generate evaluation resources in accordance with a set of ontologies, the learners' needs and their interactions with the mobile environment [1]. In the same context [4] Coulombe and Phan suggested the MobileQuizz project which represents a multiplatform mobile application carried out in Ajax with the Google Web Toolkit. In the same way the Context aware adaptive and personalized mobile learning systems is presented by Sampson et al. in [9].

The approach of Chen [2] is based on the research results of self-assessment and peer review where he developed a Mobile Assessment Participation System (MAPS) used by PDA (Personal Digital Assistant). In addition, Chen proposed a model of MAPS implementation that facilitates and enhances the efficiency of self and peer assessments in classrooms. On the other hand, in [3], authors examined the effect of providing a skill-based assessment via PDAs to a group of students in their last year of medicine. This assessment showed the good experience of using mobile technologies in evaluating the students' level.

Similarly, Zualkernan et al. designed an architecture that takes a QTIv2.1 evaluation test based on an XML file and automatically generates an executable Flash Lite to be presented and executed on a mobile device [7].

3   A Mobile Evaluation Scenario (M-Evaluation) of a Learner

In this section, we describe the context of our research using the following evaluation scenario: Idriss, a 7th year Basic student, would like to prepare for his exam on the Arabic grammar course. This student wants, for example, to review the lesson on the components of the nominal sentence in the Arabic grammar course and evaluate his knowledge. He starts by selecting this lesson. The necessary prerequisites for the chosen lesson will first be examined by researching and presenting the corresponding questions to the student. These questions form a pre-assessment on the chosen lesson. The student will then have the opportunity to revise the lesson parts that are not known and for which the corresponding questions were not answered correctly.

Finally, to ensure that Idriss understands the important concepts of the previously-mentioned lesson, questions will be searched and introduced to constitute a personalized post-assessment test. The learning environment must show Idriss's progress in revising the selected lesson. The tests presented to Idriss should provide an accurate and precise assessment of his knowledge and degree of understanding. Thus, all the important parts of the course must be well understood. Whenever Idriss chooses a lesson to revise, the learning environment should look for appropriate assessment resources, especially the questions that should be put, after the examination of the final tests, the learner's preferences in terms of language of instruction and the parameters of the learning plan. Idriss wants to access the resources of this evaluation from a mobile interface. Served by a personal PDA, the student can, for instance, be in the school library for an hour. Idriss will, therefore, access a user interface where he can choose the course to be evaluated as illustrated in Fig. 1. Then, he enters a set of keywords showing the theme of the evaluation. Idriss should also specify the type of his location. Moreover, he must indicate the maximum time reserved for the self-evaluation test: 1 h. Finally, he gives his preferences in terms of the questions types. For example, he chooses the SOAQs (Short and Open Answer Questions).

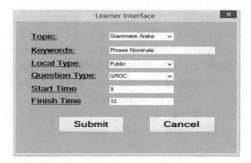

Fig. 1. Learner interface

All the information specified by Idriss must be taken into account when generating the assessment test. The evaluation environment presents itself as a context-sensitive system for the generation of personalized and mobile evaluation tests. It must also score and safeguard its progress and evaluation hints in order to use them in future evaluation activities. Assessment tests, presented to the students, should be adapted to their profiles, level and prerequisites.

The environment should, therefore, personalize the learning and evaluation content to the learner's level of knowledge based on two main concepts:

- Reuse and Interoperability through the design and implementation of the content in the form of learning objects: allows a dynamic generation of the content by identifying and searching teaching or evaluation resources to be reassembled. Through this design, the course will not be conceived in a monolithic way, but as a set of independent parts. These objects represent also the basis of new approaches and reflections on the possibilities of electronic learning systems standardization.
- Estimation of the learner's level of knowledge at each learning stage through a personalized evaluation.

Four functionalities must then be defined in such evaluation system:

- Presentation of the content to the learner.
- Assessment of knowledge acquired by the learner.
- The generation of a content adapted to the assessment of the learner's knowledge.
- Consideration of the context of the learner and the personalization of the evaluation according to the parameters provided by the evaluation context.

4   An Analysis of the Proposed Scenario

Mobile learning environments are characterized by unpredictable learning situations and circumstances. In such environments, context modeling is needed to better understand the learners' activities and to personalize the learning resources. Context-sensitive environments generally refer to a class of environments that can capture environmental parameters and adapt accordingly the set of decisions and behaviors. They are characterized essentially by their heterogeneous nature involving continuous change of context according to many circumstances depending mainly on the learner, location, time, place, etc. The ultimate goal of learning is always to provide the learner with context-appropriate learning and/or evaluation resources. Thus, the learning or assessment process must be changed and adapted to the context. Contextual reasoning is an important factor in studying the ambient intelligence. Its purpose is to deduce new knowledge based on the available data. It makes context-sensitive applications more intelligent and personalized according to the users' needs. Thus, context sensitivity results from the dynamic and heterogeneous nature of the ambient environments.

In order to provide contextual reasoning and supply the learner with a Mobile Assessment Object (MAO), First Order Logic (FOL) should be used. Indeed, FOL is a very powerful and effective means of reasoning about the context in a mobile environment. The set of Framework information is represented as first-order predicates. In fact, this representation is very expressive and can be used to show the different types of information.

We present personalized evaluation formalism with FOL developed from the formalism of hyper-media adaptive educational systems [8]. The choice of the predicates logic for this formal description is motivated by the fact that this type of FOL provides a precise design of the data representations. The evaluation personalization, in a mobile environment, is influenced by a set of parameters constituting our framework. Indeed, the Mobile Assessment Framework (MAF) must be described according to a set of information included in a group of ontological models, and the learner's interactions with the Framework must be considered in order to update the learner model and apply it in any evaluation activity. Similarly, the Framework must include a personalization component allowing personalizing the assessment activity according to a set of information.

5   Architecture of Context-Sensitive Mobile Evaluation System [5]

The Mobile learning environments are characterized by unpredictable learning situations and circumstances. In such environments, context modeling is needed to better understand the learners' activities and to personalize the learning resources. However, context-sensitive environments generally refer to a class of environments that can capture environmental parameters and adapt the set of decisions and behaviors accordingly. They are characterized essentially by their heterogeneous nature involving continuous change of context according to many circumstances depending mainly on the learner, location, time, place, etc. The ultimate goal is always to provide the learner with context-appropriate evaluation resources. The evaluation process must, therefore, be changed and adapted to the context.

Thus, we propose, in this paper, the Context-Aware Mobile Assessment System based on semantic web. This architecture represents an efficient approach to exploit the technology of the semantic web. Web services generate evaluation resources in accordance with a set of ontologies, the learners' needs and their interactions with the mobile environment.

The various exchanged documents are of RDF type. Therefore, in order to allow a personalized evaluation support, metadata about the field of application (courses), assessment resources (questions), learners (prerequisites and skills), context and evaluation history are necessary. The system uses the DC standard, a group of ontologies and specifications, to foster interoperability. It is also designed to comply with the IMS/QTI specification in order to ensure the exchange of evaluation resources. The architecture of the system is illustrated in Fig. 2.

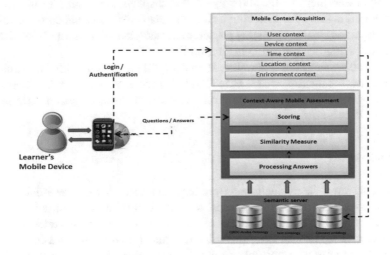

Fig. 2. A architecture of personalized m-evaluation system

Our system is concretely based on five modules:

1. Authentication module: consists in identifying the learner in his/her mobile environment. It is ensured by parameters which will be later communicated to the local database.
2. Context management module: This process is divided into three main levels:
   - Acquisition: The acquisition of the environment contextual information is made by modules that directly capture information from the environment through probes (à verifier), devices, means of interaction, etc.
   - Aggregation and storage: The captured data are meaningful and comprehensibly aggregated and interpreted for a specific use. This contextual information will be stored for later use and available for processing.
   - Processing: Context information processing consists in selecting resources from the user's request and applying an adaptation method to elaborate the selected resources. Contextual information and relevant resources are sent at the application level. The use of context information at this level consists in presenting this information to the user to interact, to notify him/her about the changes or to trigger the system event or action, etc.
3. Ontologies database: Contains the ontologies that can be used to model the context of the learner by indicating the user context, device context, location context and acquisition context). It also includes the ontologies of the learner response and those of the teacher.

4. Evaluation module: It corrects the answers and calculates the result of the learner according to these answers which will be segmented into words after performing the partial decomposition to compare them with the model answers already available for the score computing according to the similarity measure of Wu & Palmer in order to obtain the final results. Then, this module sends these results to the Interface agent which passes them to the tutor agent that displays the results to the learner. It performs this task after receiving it from the adapted test list and corrects this list by the adaptation Engine agent that prepares tests adapted to the learner's level. Finally, it saves the learner's result in his/her profile.
5. Test database and SOAQs-Arabic: It contains all the tests provided for each course and the test results for each learner. The tests are of SOAQs (Short and Open Answer Questions) type, which allows a semi-automatic correction. Indeed, each test has a defined deadline.

6 Discussion

We proposed a mobile evaluation system that requires taking into account the context and its components. Indeed, the concepts of context and context sensitivity are two key terms widely used in the exploitation of mobile environments. In order to provide a mobile evaluation, it is necessary to model the context. In our work, we chose to use the ontology-based approach not only to model the context, but also to think about the described data. This approach is combined with the logic-based one allowing the deduction of a set of facts and new inferences. Besides, we introduced a mobile assessment situation as a set of evaluation context information defined over a specific time period which can change the behavior of the evaluation system. As far as the evaluation objects are concerned, it is necessary to define a mobile evaluation object which considers the use of these resources in mobile environments.

To check the system efficiency and test the aspect of personalization and adaptation, mobile evaluation system should be designed. In this work, we performed our experiments on 7th year Tunisian pupils at Bouficha High School. The aim of our study is to assess the satisfaction and results of pupils by testing the system on different devices, such as mobile phone, touch pad, PC, employing the system at various time in order to evaluate pupils' prerequisites, and changing their locations.

The evaluation process was carried out on 30 pupils who were classified into two groups based on their Arabic grammar prerequisite: beginners and intermediate-level pupils who were given a system performance evaluation form containing 16 questions relying on 4 assumptions:

A: Assumption 1: personalization aspect assessment contains 4 questions to test both the portfolio and the level of the pupils.
B: Assumption 2: adaptation aspect assessment contains the same number of questions as the personalization aspect assessment.
C: Assumption 3: usability assessment includes also 4 questions. Students are invited to improve their opinions in terms of: navigation, generic use, etc.
D: Assumption 4: system guidance and improvement evaluation.

Figure 3 represents the histogram of the answered questions based the pupils' level.

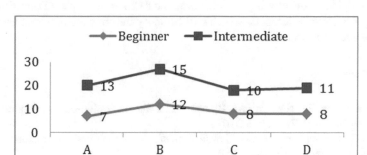

Fig. 3. Histogram representing answered questions according to pupils' levels.

From this figure, we notice that the majority of intermediate-level pupils recognized the criteria of personalization. However, beginners did not appropriately perceive it. In fact, to improve the personalization aspect, pupils had to utilize the system several times and correctly answer some given questions. B analysis shows that the aspect of adaptation was appreciated by the two afore-mentioned groups of pupils. Indeed, changing the utilized device or the type of location resulted in the generation of different tests.

As shown in D and C, we observe that it was not easy for beginners to use the system essentially because of the Arabic Grammar courses content complexity. Moreover, the majority of them liked to utilize large-screen device system because they were generally accustomed with employing small-screen devices for entertainment.

Besides, pupils found that using such system is efficient in mobile learning and self-assessment. They proposed that applying this system will be extended in other courses. The obtained findings are promising. They prove the possibility of employing this system by a wide range of pupils if the tool is introduced in various knowledge domains.

The choices of the parameters used in the discussion are marked by an open mind rather than partisan. They are guided not by a methodological blindness but by an intention to derive a rich and original understanding of the object of study.

7   Conclusion and Perspectives

We introduced, in this paper, a personalized evaluation system using the semantic web services allowing the selection and presentation of learning and evaluation resources adapted to the learner's level of knowledge. We presented also a formal description of the m-evaluation and discussed its main components to be considered when dealing with the mobile version. This description is based on FOL. Moreover, we detailed the framework and we suggested Logical rules based on first order logic. All these rules and other ones will be used later to personalize the assessment in a mobile learning environment and to adapt it to the assessment activity context. The architecture of the system was also developed.

The evaluation system, in its mobile version, was tested for a group of learners. The next step consists in developing mobile components for other types of questions and to test them on a group of learners, which will certainly improve the personalization functionalities offered to learners.

References

1. Cheniti-Belcadhi, L., Harchay, A., Braham, R.: Personnalisation de l'évaluation : de l'E-Evaluation vers la M-Evaluation (2013)
2. Chen, C.H.: The implementation and evaluation of a mobile self-and peer-assessment system. Comput. Educ. **55**(1), 229–236 (2010)
3. Coulby, C., Hennessey, S., Davies, N., et al.: The use of mobile technology for work-based assessment: the student experience. Br. J. Edu. Technol. **42**(2), 251–265 (2011)
4. Claude, C., Nobry, P.O.: MobileQUIZ : une application mobile multiplateforme réalisée en Ajax avec le Google Web Toolkit. Journée MATI Montréal 2011 (2011)
5. Boumiza, D.S., Ben Khalifa, W.: Design of an evaluation system in e-learning Arabic language. In: 2012 International Conference on Education and e-Learning Innovations (ICEELI). IEEE (2012)
6. Berners-Lee, T., Hendler, J., Lassila, O.: The semantic web. Scientific American, Mai 2001, pp. 29–37 (2001)
7. Zualkernan, I.A., Ghanam, Y.A., Shoshaa, M.F., et al.: An architecture for dynamic generation of QTI 2.1 assessments for mobile devices using Flash Lite. In: Proceedings of the Seventh IEEE International Conference on Advanced Learning Technologies, ICALT 2007, pp. 194–195. IEEE (2007)
8. Baccari, S., Mendes, F., Nicolle, C., Soualah-Alila, F., Neji, M.: A comparative study of the mobile learning approaches. In: International Conference on Mobile, Secure and Programmable Networking, pp. 76–85. Springer International Publishing, June 2016
9. Sampson, D.G., Zervas, P.: Context-aware adaptive and personalized mobile learning systems. In: Ubiquitous and Mobile Learning in the Digital Age, pp. 3–17. Springer, New York (2013)

Clustering Social Network Profiles Using Possibilistic C-means Algorithm

Mohamed Moussaoui[1]([⊠]), Montaceur Zaghdoud[2], and Jalel Akaichi[3]

[1] BESTMOD, Institut Supérieur de Gestion, University of Tunis, Tunis, Tunisia
mohamed.moussaoui.com@gmail.com
[2] Information System Department,
Prince Sattam Bin Abdulaziz University, Al-Kharj, Saudi Arabia
zaghdoud@psau.edu.sa
[3] Information System Department, King Khalid University, Abha, Saudi Arabia
jalel.akaichi@kku.edu.sa

Abstract. Social networking has become one of the most useful tools in modern society. Unfortunately, terrorists are taking advantage of the easiness of accessing social networks and they have set up profiles to recruit, radicalize and raise funds. Most of these profiles have pages that existing as well as new recruits to join the terrorist groups see and share information. Therefore, there is a potential need of detecting terrorist communities in social network in order to search for key hints in posts that appear to promote the militants cause. Community detection has recently drawn intense research interest in diverse ways. However, it represents a big challenge of practical interest that has received a great deal of attention. Social network clustering allows the labeling of social network profiles that is considered as an important step in community detection process. In this paper, we used possibilistic c-means algorithm for clustering a set of profiles that share some criteria. The use of possibility theory version of k-means algorithm allows more flexibility when assigning a social network profile to clusters. We experimentally showed the efficiency of the use of possibilistic c-means algorithm through a detailed tweet extract, semantic processing and classification of the community detection process.

Keywords: Social network analysis · Community detection · Possibility theory · Clustering

1 Introduction

During the last decade, social networking has become one of the largest and most influential components of the web. Therefore, the analysis of networks has become a challenge due to the growing demand for analyzing large amounts of data. The social network analysis is not limited to sociology or even the social sciences: the relationships between users can be studied in political science [5], economics [9], and engineering [18]. But until very recently, very thin

© Springer International Publishing AG 2018
G. De Pietro et al. (eds.), *Intelligent Interactive Multimedia Systems and Services 2017*,
Smart Innovation, Systems and Technologies 76, DOI 10.1007/978-3-319-59480-4_42

research tradition does not fit into dominant paradigms. After the terrorist attacks of Paris, Brussels, Orlando and Istanbul, it clearly shows that terrorists have increasingly used social networks to further their goals and spread their message. So, there is a potential need of analyzing social networks and detecting terrorists communities. Indeed, the community detection is an important task in the area of social networking, which aims to find all communities in the entire network [22]. Graph analytics have proven to be valuable tools in solving these challenges. The core study of the graphs is related to their famous characteristics to model complex structures such as social networks, where nodes represent people and edges represent interactions between them. So, analyzing these data and extracting results is a big challenge. Early work in graph partitions could be adopted as algorithms for community detection. One of the ways used to present a large or multi-graphs is to cluster a set of nodes that share some criteria. Clustering is a common technique for statistical data analysis used in many fields such as machine learning, pattern recognition, image analysis, information retrieval, etc. In the literature, several algorithms have been proposed to allow the clustering of nodes in an efficient way. Yet, these techniques are usually based on the structural aspect and do not integrate the semantic aspect of the graph. In other words, these techniques can cluster a large set of nodes efficiently, and provide basic statistical information. Besides, the lack of contextual information have made the interpretation of clusters very difficult. Moreover, graph data are generally imperfect and imprecise due to noise, incompleteness and inaccuracies. In case of community detection, we are not certain that a member of social network belongs or no to a specific community due the uncertainty that can be inherent in graph. This rises the need for a different way to consider the structural similarity such that both close and distant structural similarities would be detected with respect to the soundness of results. In order to overcome these drawbacks, we aim to detect four communities (leader terrorist, active terrorist, empathized with terrorist, non-terrorist) as well as the relation between communities considering the uncertainty that can be inherent in real-world networks. To achieve this objective, we propose the possibilistic c-means clustering (PCM) model [23] as well as a uncertainty clustering algorithm. The novelty of this work is in proposing a formal algorithmic study of organizational socialization. This method would help in the development of new and more effective models for personalization and recommendation of con-tent in social environments. For the sake of clarity, this paper will focus on the following contributions:

- Identify terrorists using possibilistic c-means clustering algorithm of their social network profiles (leader terrorist, active terrorist, empathized with terrorist, non-terrorist).
- Proving the effectiveness of our approach with an experimental evaluation of the models using a real life social network dataset.

The remainder of this paper is organized as follows. In Sect. 2, we formulate the problem of community detection. Section 3 presents related works in this field. In Sect. 4, we present general framework of our possibilistic approach and its modules. For each module a detailed description of functions and design choices will be provided.Then, we evaluate these approaches with a real data set and report the results and findings in Sect. 5, and conclude this study in Sect. 6.

2   Problem Statement

Even though the existing approaches for detecting communities only focus on the topological structure of the graph. Data clustering has been studied for a long time but most algorithms (e.g., k-means, EM) do not deal with relational data. The novelty of this work is in proposing a formal, algorithmic study of organizational socialization. More specifically we investigate the following network building problems. The general framework of our possibilistic approach is as follows. Given a graph G = (V, E) with vertex set V and edge set E, the graph clustering problem is to partition the graph into k disjoint clusters. Accordingly, we adopt a two-step clustering framework, where in the first step we encode each user profile into a description-vector of key terms containing a special weight. In the second step, we perform a clustering using the description-vectors in order to detect some clusters that represent terrorist communities. Throughout the paper, the terms set, cluster, and community are used interchangeably.

3   Related Works

Generally, community allows to extract a set of groups of interacting objects and the relations between nodes and the relations between them. Many works have been proposed for finding communities in large graph. They can be divided into optimization and heuristic methods. Both of the algorithms require a criterion to evaluate the community partition. In [6], the authors proposed an approach that used an indicator of a node's centrality in a network, betweenness centrality, to extract community limits. In [15], Newman and Girvan proposed a posteriori measure of the overall quality of a graph partition as a modularity. The latter has been very powerful in recent communities detection works. It can use spectral techniques to approximate it [14]. Modularity measures internal (and not external) connectivity, but it does so with reference to a randomized null model. In [12], the authors introduced an iterative borough strategy to identify communities that is coupled with a fast constrained power method that sequentially achieves tighter spectral relaxations. Some fuzzy community detection algorithms calculate the possibility of each node belonging to every community, such as [7, 8, 20, 24]. The authors of these approaches provide a fresh idea for the understanding of community detection. Therefore, coupling relationship and content information in social network for community discovery is an emerging research area because current methods do not focus on social graphs or they are not efficient for large-scale datasets.

4   Inferring Communities Using Possibilistic Clustering

In this section, we present the segments forming our framework and we describe how it contributes to the entire system. We demonstrate our choices out from predefined requirements and characteristics. We aim to detect four communities (leader terrorist, active terrorist, empathized with terrorist, non-terrorist) as well as the relation between communities. At a glance, our approach involves two steps. In the first step, each profile is encoded into a description-vector containing the corresponding values for a set of semantic attributes. In the second step, profiles with similar descriptions are clustered together. Basically, we observe the top terms of a given cluster and we specify to which group they belong. Our approach take a simplistic assumption that connections are approximately correlated with attributes to construct a vector-space model of user profiles. The key features used in the model are statistics on fields of user-generated text. These include both frequencies of patterns within the cluster and comparison of text frequencies across the entire user base. A more detailed description is given in the following sections.

4.1   TF-IDF Weighting

Generally, community detection in networks aims to find clusters of nodes that are connected between them. Clustering this social network to find out the terrorism trends, the attribute terrorist views of a person is more important than its name or gender. The clustering criteria can be as simple as grouping all accounts that share a common characteristic (post status, link, image, video, etc.). Thus, we preview the frequent terms on each cluster and we define manually it sentiment from sentiments we defined earlier. Before being able to run clustering method on a set of profile, the profiles have to be represented as mutually comparable vectors. To achieve this task, we employ to create these vector space models is the TF-IDF (Term Frequency Inverse Document Frequency) [17] to take into account term's importance to a tweet. This technique provides a numerical value for each term in a document, indicating the relative importance of a term in its contextual document corpus. The TF-IDF value increases proportionally to the number of times a word appears in the document, but is offset by the frequency of the word in the corpus, which helps to adjust for the fact that some words appear more frequently in general. Variations of the TF-IDF weighting scheme are often used by search engines as a central tool in scoring and ranking a document's relevance given a user query. TF-IDF can be successfully used for stop-words filtering in various subject fields including text summarization and classification. Stop words are the most common words they generally doesn't have a meaning and they have no effect on the sentence meaning. Here are some examples of English stop words: a, the, an, then, are, is, you, each, to, those, they, of, it, once, only, etc. Basically Stop words are useless. The analysis accuracy and speed will be improved by removing them. For that we loop through the tweet word-list and we exclude stop words from it. One of the simplest ranking functions is computed by summing the TF-IDF for each query term, many more sophisticated ranking functions are variants of this simple model. The result of TF-IDF method is a list of pairs representing the word and it proportional weight in document. Now that we're equipped with a numerical model with which to compare our data, we can represent each profile as a vector of terms using a global ordering of each unique term. After we have our data model, we have to compute distances between documents.

4.2   Possibilistic C-means Clustering (PCM)

Here, we discuss the second part of our approach which is the clustering step. The clustering method accesses the collected vectors and performs unsupervised clustering resulting in four clusters representing the typical topics viewed by terrorist users. We use PCM [23] which is a well known clustering algorithm that is widely used in unsupervised learning [10]. PCM determines a possibilistic partition, in which a possibilistic membership measures the absolute degree of typicality of a point in a cluster. The possibilistic membership is rational and it has a meaningful interpretation. Formally, given a set $X = \{x_1, ..., x_n\}$ of n patterns $x_i \in \mathbb{R}$, the set of centroids $V = \{v_1, ..., v_c\}$ and the membership matrix U are defined. The set V contains the prototypes/representatives of the c clusters. U is a $c \times n$ matrix where each element u_{ih} represents the membership of the pattern h to the cluster i. In the PCM, $u_{ih} \in [0, 1]$ and memberships of a pattern to all the c clusters are not constraint to sum up to one. In this possibilistic clustering means (PCM), the membership value of each data point can be interpreted as degree of compatibility (possibility). In [21], the authors indicated that if all the cluster centers are identical, then only the objective function to inverse distance functions between cluster centers. The objective function for PCM is given by [4]:

$$P_m(T, V, X, Y) = \sum_{i=1}^{N} \sum_{k=1}^{C} t_{ik}^m d_{ki}^2 + \sum_{i=1}^{C} Y_i \sum_{k=1}^{N} (1 - T_{ki})^m$$

where T_{ki} is the typicality of X_k to the cluster i, T is the typicality matrix, defined as $T = [T_{ki}]$, d_{ki} is a distance measure between X_K and C_i and Y_i denotes a user-defined constant: $Y_i > 0$ and i $\in \{1,...,C\}$. The PCM is more robust in the presence of noise, in finding valid clusters, and in giving a robust estimate of the centers.

5   Experiments and Results

In this section, we provide an example of a large real-world social network and present an experimental evaluation of the weighting methods discussed above. But first we introduce a validation metric, which allows us to quantify the quality of community structure in networks and enables a comparison between the different weightings. To evaluate our approach, we start by collecting tweets about terrorism. During the data collection, we excluded truncated tweets to reduce data duplication and SPAM. We used our previously defined keywords in order to target our topics. Then, we transformed the selected tweets into a graph of retweets with 85k nodes and 115k edges using an open graph viz platform named Gephi [1].

5.1 Evaluation Measures

We extract the communities from the Twitter dataset, using 4 different methods:

- **Attribute-based clustering:** K-means method [11] is used to group nodes based on the similarity in attributes (link information is ignored).
- **Random walks:** method proposed by Steinhaeuser et al. [19], based on random walks and hierarchical clustering. The walk length is set to the number of nodes.
- **Fast greedy:** method proposed by Clauset et al. [3] based on the greedy optimization of modularity. The graph is weighted by node attribute similarities.
- Our proposed algorithm based on possibility theory [2].

To evaluate the quality of these methods, we compare the number of communities, size of communities, modularity structure, modularity attribute and additional two measurements: density D and entropy E.

$$D = \sum_{c=1}^{K} \frac{m_c}{m}$$

where m_c is number of edges in community c, m is the number of edges in G, K is the number of communities. D reflects the proportion of community intra-links over total number of links. High density denotes good separation of communities.

$$E = \sum_{c=1}^{K} \frac{n_c}{n} \; Entropy(c)$$

$$Entropy(c) = \sum p_{ic} \; log(p_{ic})$$

where n_c is the number of nodes in community c, n is the number of nodes in G, pic is the percentage of nodes in c with attribute i. Communities with low entropy means they are more homogeneous with respect to the attribute a_i. Table 1 reports the compare of the average clustering entropy and density of our proposed approach with three approaches proposed in the literature. Average entropy of our possibilistic approach is lower than other approaches. This shows that possibilistic's communities are more homogeneous. In terms of density, we notice that possibilistic's communities are more dense.

5.2 Accuracy

For a stable classification model, we decided to build a bagging classifiers. It consist of using multiple classifiers that work in parallel to produce more accurate

Table 1. Comparison of entropy and density of our possibilistic approach with approaches proposed in the literature.

Method	Avg. entropy	Avg. density
Attribute-based	1.28	0.42
Random walks	0.86	0.51
Fast greedy	0.47	0.61
Possibilistic approach	0.23	0.62

predictions by combining all the classifiers results into one via a custom voting method. Using Sklearn library [16], we built 7 classifiers based on 7 different classification algorithms:

- **Classic Naive Bayes classifier:** A simple probabilistic classifier based on Bayes theorem with a strong independence assumption between features.
- **Multinomial Naive Bayes classifier:** A probabilistic classifier with a multinomial event model which enables the representation of the frequencies with which certain events have been generated.
- **Bernoulli Naive Bayes classifier:** A probabilistic classifier with a multi-variate event model. Features are processed as separate binary variables.
- **Logistic regression classifier:** A regression algorithm where the dependent variables (DV) is categorical. It is used to estimate the probability of a binary response based on one or more predictor feature.
- **Stochastic gradient descent classifier:** A linear classifier. Simple yet very efficient approach to discriminative learning of linear classifiers under convex loss functions.
- **Linear support vector machine classifier:** A non-probabilistic binary linear classifier. An SVM training algorithm builds a model that assigns new examples into one category or the other.
- **Nu support vector machine classifier:** Another SVM classifier with a parameter "nu" for controlling the number of support vectors.

We trained the classifiers using the prepared Word2Vec model [13]. Then, we stored our trained classifiers as a persistent models for future use. After that, we tested data against the trained classifiers. This will produce a class prediction which is the sentiment of the tested data and an accuracy rate for each classifier. Next, we built a voting algorithm that could be implemented in two different ways: the first approach consists of selecting the prediction of the classifier with the biggest accuracy rate. The second approach consists of defining the most common prediction as a final prediction and calculating the mean of classifiers with the defined common prediction. After training all our classifiers against it, we performed multiple tests on our classifiers using the same testing dataset to verify the accuracy rate given by each classifier and how stable are they are.

According to the obtained results(See Fig. 1) after performing 4 tests, it looks like the naive Bayesian classifiers are not doing well in terms of accuracy on

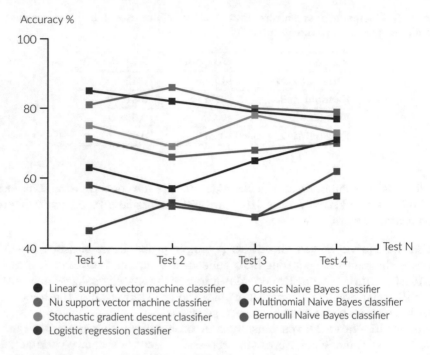

Fig. 1. Classifiers accuracy testing

average, we notice also that most of the classifiers are not steady during the tests: Most of the classifier accuracies showed great, during some tests they give a good accuracy rate and in some others they made a sharp drop. In addition, the sentiment prediction was mostly equivalent in the 4 tests. Based on these results, we can concluded that using a unique classifier is not stable or accurate enough and we decided to find a solution to fix the stability and accuracy issues. We opted the bagging classifiers technique which consist of unifying all the classifiers results using a voting algorithm which define the most common predictions given by the classifiers and confidence rate as we mentioned in the classification section. This technique enabled us to optimize the accuracy by unifying all the results and making a decision based on all classifiers' predictions which means that a prediction is only considered if the majority of the classifiers agreed to it. Also, it helped us to stabilize our model by adding a confidence rate to limit the variations by simply defining a minimum confidence rate.

6   Conclusion and Future Work

In this paper, we proposed an innovative knowledge-based methodology for terrorist activity detection on social network. More precisely, we studied the issue of community detection in attributed graphs. Besides, we present how to make diagnoses for our framework, and report significant improvements in the experimental study. The results indicate that the method is more accurate than traditional clustering methods, while does not presents weaknesses of the previous method.

Our approach is generic and could be used to detecting other types of criminals surfing the Web such as pedophiles accessing child pornography sites. The success of this theory is due to its ability to handle uncertainty and imprecision in simple way and to offer a classified semantics to natural language statements. The possibility theory provides some principal justifications over the usual mathematical modeling of probabilities in handling uncertainty which occur in many real-life problems. This concept takes advantage of the bipolar representation setting which distinguishes between positive information and negative information expressing impossibility. Therefore, our approach can well avoid the traditional method using the probabilistic approach. In fact, there is an extensive formal correspondence between probability and possibility theories, where the addition operator corresponds to the maximum operator. Hence, the modeling of positive and negative information in possibility theory [2] is studied and the observation is made precise in a well-defined way using the possibility theory. A promising future direction to understand the roles of links and content information in the formation of online communities in order to devise adapted discovery strategies and to model the dynamic of the networks. The future direction could be to remove the k constraint. This can be simply done using a parameter free clustering algorithm.

References

1. Bastian, M., Heymann, S., Jacomy, M.: Gephi: An open source software for exploring and manipulating networks (2009)
2. Benferhat, S., Dubois, D., Kaci, S., Prade, H.: Modeling positive and negative information in possibility theory. Int. J. Inf. Syst. (IJIS) **23**(10), 1094–1118 (2008)
3. Clauset, A., Newman, M.E.J., Moore, C.: Finding community structure in very large networks. Phys. Rev. E **70**, 066111 (2004)
4. Correa Farias, C., Valero Ubierna, C., Barreiro Elorza, P., Diago Santamaria, M.P., Tardaguila Laso, J.: A comparison of fuzzy clustering algorithms applied to feature extraction on vineyard, October 2011
5. Cranmer, S.J., Desmarais, B.A.: Inferential network analysis with exponential random graph models. Polit. Anal. **19**(1), 66–86 (2011)
6. Girvan, M., Newman, M.E.J.: Community structure in social and biological networks. Proc. Natl. Acad. Sci. **99**(12), 7821–7826 (2002)
7. Golsefid, S.M.M., Zarandi, M.H.F., Bastani, S.: Fuzzy duocentric community detection model in social networks. Soc. Netw. **43**, 177–189 (2015)
8. Gomez, D., Rodriguez, J.T., Yanez, J., Montero, J.: A new modularity measure for fuzzy community detection problems based on overlap and grouping functions. Int. J. Approximate Reasoning **74**, 88–107 (2016)
9. Jackson, M.O., Lpez-Pintado, D.: Diffusion and contagion in networks with heterogeneous agents and homophily. Netw. Sci. **1**, 49–67 (2013). 4
10. Jain, A.K.: Data clustering: 50 years beyond k-means. Pattern Recogn. Lett. **31**(8), 651–666 (2010). Award winning papers from the 19th International Conference on Pattern Recognition (ICPR)

11. Kanungo, T., Mount, D.M., Netanyahu, N.S., Piatko, C.D., Silverman, R., Wu, A.Y.: An efficient k-means clustering algorithm: Analysis and implementation. IEEE Trans. Pattern Anal. Mach. Intell. **24**(7), 881–892 (2002)

12. Li, W., Schuurmans, D.: Modular community detection in networks. In: Proceedings of the Twenty-Second International Joint Conference on Artificial Intelligence, IJCAI 2011, vol. 2, pp. 1366–1371. AAAI Press (2011)

13. Maas, A.L., Daly, R.E., Pham, P.T., Huang, D., Ng, A.Y., Potts, C.: Learning word vectors for sentiment analysis. In: Proceedings of the 49th Annual Meeting of the Association for Computational Linguistics: Human Language Technologies, HLT 2011, vol. 1, pp. 142–150. Association for Computational Linguistics, Stroudsburg (2011)

14. Newman, M.E.J.: Modularity and community structure in networks. Proc. Natl. Acad. Sci. **103**(23), 8577–8582 (2006)

15. Newman, M.E.J., Girvan, M.: Finding and evaluating community structure in networks. Phys. Rev. E **69**(2), 026113 (2004)

16. Pedregosa, F., Varoquaux, G., Gramfort, A., Michel, V., Thirion, B., Grisel, O., Blondel, M., Prettenhofer, P., Weiss, R., Dubourg, V., Vanderplas, J., Passos, A., Cournapeau, D., Brucher, M., Perrot, M., Duchesnay, E.: Scikit-learn: machine learning in python. J. Mach. Learn. Res. **12**, 2825–2830 (2011)

17. Rajaraman, A., Ullman, J.D.: Mining of Massive Datasets. Cambridge University Press, Cambridge (2011)

18. Schlauch, W.E., Zweig, K.A.: Influence of the null-model on motif detection. In: Proceedings of the 2015 IEEE/ACM International Conference on Advances in Social Networks Analysis and Mining 2015, ASONAM 2015, pp. 514–519. ACM, New York (2015)

19. Steinhaeuser, K., Chawla, N.V.: Identifying and evaluating community structure in complex networks. Pattern Recogn. Lett. **31**(5), 413–421 (2010)

20. Sun, P.G.: Community detection by fuzzy clustering. Phys. A **419**, 408–416 (2015)

21. Timm, H., Borgelt, C., Döring, C., Kruse, R.: Fuzzy cluster analysis with cluster repulsion. In: Proceedings of the European Symposium on Intelligent Technologies, Hybrid Systems and Their Implementation on Smart Adaptive Systems, eunite 2001, Puerto de la Cruz, Tenerife, Spain. Verlag Mainz, Aachen (2001)

22. Wang, M., Wang, C., Yu, J.X., Zhang, J.: Community detection in social networks: an in-depth benchmarking study with a procedure-oriented framework. Proc. VLDB Endow. **8**(10), 998–1009 (2015)

23. Yang, M.-S., Wu, K.-L.: Unsupervised possibilistic clustering. Pattern Recogn. **39**(1), 5–21 (2006)

24. Zhang, S., Wang, R.-S., Zhang, X.-S.: Identification of overlapping community structure in complex networks using fuzzy-means clustering. Phys. A **374**(1), 483–490 (2007)

Big Data Classification: A Combined Approach Based on Parallel and Approx SVM

Walid Ksiaâ$^{(\boxtimes)}$, Fahmi Ben Rejab, and Kaouther Nouira

BESTMOD, Institut Supérieur de Gestion de Tunis,
Université de Tunis, 41 Avenue de la Liberté, 2000 Le Bardo, Tunisie
ksiaawalid@gmail.com, fahmi.benrejab@gmail.com, kaouther.nouira@planet.tn

Abstract. This paper presents a combined solution for Big Data classification, by using one of the extended versions of the Support Vector Machines (SVM), known as the Parallel Support Vector Machines (PSVM). The main problem assumes that, once a PSVM model is obtained, a feature can be removed overtime, resulting in a decrease of the accuracy with the existing model. While Big Data is one of the interesting contexts, then training a new PSVM with the new data structure is time-consuming. The solution is to use an approach that approximates any SVM model based on the Radial Basis Function (RBF) kernel, and called the Approx SVM. In order to avert a new training step, this paper proposes to apply the Approx SVM in a parallel architecture. Despite that the Approx SVM was not purposely used to deal with large-scaled data set, the experimental results, which will be presented at the end of the article, are proofs that this approach is an appropriate choice for PSVM models.

Keywords: Support vector machines · Parallel computing · Big data · Approx SVM · Parallel SVM

1 Introduction

Machine learning algorithms tend to be used under Big Data context, whereas some of the classifiers are not recommended in this framework, due to some disadvantages. In our case, we are interested in using the Support Vector Machines (SVM) classifier known to have a good generalization performance [1]. However, it has a high computational complexity and excessive memory requirements for processing large-scaled data set [1]. In order to benefit from its performance, several modified versions of SVM have been proposed namely the LASVM [2], the ISVM [3] and RTSVM [4], and the interesting one called Parallel SVM (PSVM). PSVM splits data into many chunks, to train many SVM simultaneously. Such a task is done for the sake of reducing the computation time as well as improving the accuracy. This can be problematic if one of the features is deleted, leading to a need of adapting PSVM models. To avoid training again our models, the approx SVM is used to keep qualifying the PSVM as a solution for a dynamic change of our input in terms of features.

© Springer International Publishing AG 2018
G. De Pietro et al. (eds.), *Intelligent Interactive Multimedia Systems and Services 2017*,
Smart Innovation, Systems and Technologies 76, DOI 10.1007/978-3-319-59480-4_43

This paper will be disposed as follows: Sect. 2 focuses on the theoretical approach of SVM, parallel SVM, and approx SVM, considered as a starting point for the experimentation. Section 3 details our contribution through the parallel approx SVM. Section 4 shows experimental results. Section 5 concludes this paper.

2 Support Vector Machines and Its Modified Versions

In this section, we will present the main techniques used in this work namely the support vector machines (SVM) and its modified versions the parallel SVM and approx SVM.

2.1 Support Vector Machines

SVM is a popular machine learning method for classification, regression, and other learning tasks [5]. Its principle is to set a hyperplane, separating both group of samples, and whose theoretical approach is based on two main concepts, known as the maximum margin and the kernel trick.

Maximum Margin
The optimal hyperplane maximizes the minimal distance to the training examples, that are the support vectors [6].

The hypothesis noted $h(x)$ of the optimal hyperplane is defined as follows:

$$h(x) = Argmax_{\omega, \omega_0} min\{\|x - x_i\| : x \in \mathbb{R}, (\omega^T x + \omega_0) = 0, \ i = 1, ..., m\} \quad (1)$$

with:
ω: The weighting vector (also called the normal to the hyperplane).
x: An input.
ω_0: also noted b, is the bias [1]. In maximum margin, there is an important formula in geometry: $\|\frac{h(x)}{\omega}\|$.

It is the length of the shortest line segment, joining a training example (e.g. (x_i, u_i)) to a hyperplane $(h(x_i))$ [7].

In other words, for each hyperplane, the distance between an element (x_i, ui) and a hypothesis $h(x_i)$ is: $\|\frac{1}{\omega}\|$ [8].

In that case the margin is equal to $\|\frac{2}{\omega}\|$, and if we need to maximize the margin, we need to minimize $\|\frac{2}{\omega}\|$ [8]. The optimization is the formal approach used within the SVM framework for minimization.

When there is not a linear separation of data, we should use kernel trick which is presented as follows.

Kernel Trick

The nonlinear separation requires the transformation of the input space into a feature space using a transformation function ϕ, based on dot products:

$$\begin{cases} Max_\alpha\{\sum_{i,j=1}^m \alpha_i - \frac{1}{2}\sum_{i,j=1}^m \sum_{i,j=1}^m \alpha_i\alpha_j u_i u_j \langle \Phi(x_i).\Phi(x_j)\rangle\} \\ \alpha_i \geq 0, \quad i = 1,...,m \\ \sum_{i=1}^m \alpha_i u_i = 0 \end{cases} \tag{2}$$

with: $\Phi_i(x)$: The transformed input x of the ith element.

It may be impossible to compute the scalar product [7], so the kernel trick will replace the dot products. A proof of the equality is the Hilbert Schmidt Theory:

$$\langle \Phi(x).\Phi(x_i)\rangle = \sum_{i=1}^{\infty} \lambda_i \phi_i(x)\phi_i(y) = K(x,y) \tag{3}$$

with: λ_i: A weighting coefficient of the ith element.

In order to handle data that are not linearly separable, standard SVM needs a high execution time in training phase. Moreover, the SVM is unable to provide results in case of using big data. As a result, a modified version of SVM called the parallel support vector machines has been proposed. The following section detailed this method.

2.2 Parallel Support Vector Machines

The PSVM is an SVM extension, and it works on large scale data sets, which are split into smaller data sets, to use SVM upon them, and to output a set of support vectors (SV) [10]. Compared to the SVM, the PSVM's contribution is it reduces the training time and usually improve the accuracy [11,12].

PSVM's Distributed Optimization

The Quadratic Programming (QP) solution is also used in PSVM, but in a distributed approach. Indeed, the output will be two sets of support vectors, thus two optimization problems will be involved. The new problem is the combination of the previous optimization problems, from the merged inputs [13].

In [13] Graf et al. explained the principle of merge, by assuming that (given) W_i and G_i are, respectively, the objective function, the gradient of the objective function of the ith SVM.

Formally they are defined as:

$$W_i = -\frac{1}{2}\ \overrightarrow{\alpha}_i^T * Q_i\ \overrightarrow{\alpha}_i + \overrightarrow{e}_i^T\ \overrightarrow{\alpha}_i \tag{4}$$

with: Q_i: The kernel matrix of the ith SVM.
and

$$\overrightarrow{G}_i = -\overrightarrow{\alpha}_i^T * Q_i + \overrightarrow{e}_i^T \tag{5}$$

Merging the two gradients of SVM_1 and SVM_2, will initialize the optimization problem of SVM_3. Graf et al. affirmed that initializing the merged results is initiating a combination of previous set of optimization, through a matrix:

$$W_3 = -\frac{1}{2} \begin{bmatrix} \overrightarrow{\alpha}_1 \\ \overrightarrow{\alpha}_2 \end{bmatrix}^T \begin{bmatrix} Q_1 & Q_{12} \\ Q_{21} & Q_2 \end{bmatrix} \begin{bmatrix} \overrightarrow{\alpha}_1 \\ \overrightarrow{\alpha}_2 \end{bmatrix} + \begin{bmatrix} \overrightarrow{e}_1 \\ \overrightarrow{e}_2 \end{bmatrix}^T \begin{bmatrix} \overrightarrow{\alpha}_1 \\ \overrightarrow{\alpha}_2 \end{bmatrix} \tag{6}$$

$$\overrightarrow{G}_3 = - \begin{bmatrix} \overrightarrow{\alpha}_1 \\ \overrightarrow{\alpha}_2 \end{bmatrix}^T \begin{bmatrix} Q_1 & Q_{12} \\ Q_{21} & Q_2 \end{bmatrix} + \begin{bmatrix} \overrightarrow{e}_1 \\ \overrightarrow{e}_2 \end{bmatrix} \tag{7}$$

The next section will present the theoretical approach of Approx SVM, which will be applied within the PSVM experimentation, based on mapreduce framework using Hadoop. In this issue, the conditions of using the Approx SVM are SVM models based on the RBF kernel, and, at least, one feature deletion.

2.3   Approx Support Vector Machines (ASVM)

The Approx SVM focuses on the nonlinear separation. The kernel is seen as a relevant method that, permits us to use a linear classifier (like SVM) in a feature space, and generates models, which are nonlinear in an input space [14].

The RBF kernel has a simpler definition, when introducing a parameter γ: $\gamma = \frac{1}{2\sigma^2}$ [15].

Thus, the RBF it is equal to: $K(x,y) = \exp^{-\gamma \|x-y\|^2}$ [14].

The hypothesis based on the RBF kernel, yields: $h(y) = \sum_{i=1}^{m} \alpha_i^* u_i \cdot \exp^{-\gamma\|x-y\|^2} + \omega_0^*$

The temporal complexity of the kernel method with the RBF is $O(m * d)$, with m as the number of SV, and d as the number of dimensions [14].

If the computation time is too high, as well as the number of support vectors, then, the solution is to use a linear method that uses only $O(d)$, at the cost of reduced accuracy [16].

The Approx SVM keeps relying on the use of RBF kernel for lowering the run-time complexity, by introducing the second-order Maclaurin series approximation of exponential functions [14].

2.3.1   Second-Order Maclaurin Series

The origin of the Maclaurin series is the Taylor series.

A Taylor series is a representation of a function as an infinite terms of values that are calculated from the gradient of the function at a single point [14].

$$\sum_{n=0}^{\infty} \frac{f^{(n)}(a)}{n!} (x - a)^n \tag{8}$$

In addition to this assumption, if $a = 0$, then the Taylor series is called a Maclaurin series [14].

A Maclaurin series is a Taylor series expansion of a function about 0 [14].

Cleasen et al. has given, for exponential functions, like the RBF, the formal definition of Maclaurin series:

$$\exp^x = \sum_{k=0}^{\infty} \frac{1}{k!} x^k \tag{9}$$

In the approximation, the contributors proceed to the derivation in a matrix using the gradient and the Hessian matrix on the Maclaurin series of the RBF kernel.

As for the hypothesis (in Eq. 14), the RBF kernel was expanded:

$$h(y) = \sum_{i=1}^{m} \alpha_i^* u_i . \exp^{-\gamma \|x\|^2} . \exp^{-\gamma \|y\|^2} . \underbrace{\exp^{2\gamma x_i^T y}} + \omega_0^* \tag{10}$$

The following steps given by Claesen et al., describes the approximation process, using the second-order Maclaurin series:

1. First, we reorder the expanded Eq. 11, by moving the factor $\exp -\gamma \|x\|^2$ in front of summation (because this factor can be computed in $O(d)$):

$$h(y) = \exp^{-\gamma \|x\|^2} . (\sum_{i=1}^{m} \alpha_i^* u_i . \exp^{-\gamma \|y\|^2} . \exp^{2\gamma x_i^T y}) + \omega_0^* \tag{11}$$

2. Suppose that: $g(y) = \sum_{i=1}^{m} \alpha_i^* u_i . \exp^{-\gamma \|y\|^2} . \exp^{2\gamma x_i^T y}$.

   The second-order Maclaurin series is: $\exp^x \approx 1 + x + \frac{1}{2} x^2$ [14].

   If we apply the Eq. 10 in $g(y)$, then the inner product exponent $\exp^{2\gamma x_i^T y}$ is replaced by the following approximation:

$$\exp^{2\gamma x_i^T y} \approx 1 + 2\gamma x_i^T z + 2\gamma^2 (x_i^T y)^2 \tag{12}$$

3. The approximation 11 yields:

$$\hat{g}(y) = \sum_{i=1}^{m} \alpha_i^* u_i . \exp^{-\gamma \|y\|^2} . (1 + 2\gamma x_i^T y + 2\gamma^2 (x_i^T y)^2) \tag{13}$$

4. The Eq. 14, is a simpler formal definition of $\hat{g}(y)$ after being developed:

$$\hat{g}(y) = c + v^T . y + y.^T M.y \tag{14}$$

The three variables represents the developed dot products with the second part of the exponent, their formal definitions are highlighted in [14].

5. As a consequence, we obtain the new hypothesis $\hat{h}(y)$, using the approximated function, based on the second order Maclaurin series, $\hat{g}(y)$, in Eq. 15:

$$\hat{h}(y) = \exp^{-\gamma \|x\|^2} . \hat{g}(y) + \omega_0^* = \exp^{-\gamma \|x\|^2} . (c + v^T . y + y.^T M.y) + \omega_0^* \tag{15}$$

2.3.2   Approximations Accuracy

The authors stated that the relative error of the second-order Maclaurin series approximation for exponential function is less than $\pm 3.05\%$, for exponents x in the interval of $[-0.5, 0.5]$.

In order to ensure this approximation with exponents in this interval, a new constraint will be added in our optimization problem, to be able to generate the approximated hypothesis: $|2\gamma x_i^T y| < \dfrac{1}{2} \ \forall i$ [14].

The absolute inner product can be expressed otherwise, to avoid it, using the Cauchy-Schwarz inequality: $|x_i^T y| \leq \|x_i\|.\|y\|$, $\forall i$[14], which is a way to evaluate the validity of the approximation in terms of the support vector $\|x_M\|$ with maximal norm $(\forall i : \|x_M\| \geq \|x_i\|)$: $\|x_M\|^2 \|y\|^2 < \dfrac{1}{16\gamma^2}$.

3   Parallel Approx SVM for Big Data

This section assumes a combination between PSVM models and the Approx SVM. This fusion yields applying the approximation upon PSVM models, depending on the inclusion of the constraint defined in Eq. 16:

$$W = -\frac{1}{2} \begin{bmatrix} \overrightarrow{\alpha}_1 \\ \overrightarrow{\alpha}_2 \end{bmatrix}^T \begin{bmatrix} Q_1 & Q_{12} \\ Q_{21} & Q_2 \end{bmatrix} \begin{bmatrix} \overrightarrow{\alpha}_1 \\ \overrightarrow{\alpha}_2 \end{bmatrix} + \begin{bmatrix} \overrightarrow{e}_1 \\ \overrightarrow{e}_2 \end{bmatrix}^T \begin{bmatrix} \overrightarrow{\alpha}_1 \\ \overrightarrow{\alpha}_2 \end{bmatrix} \quad \begin{cases} 0 \leq \alpha_i \leq C \\ \sum_{i=1}^{m} \alpha_i u_i = 0 \\ \|x_M\|^2 \|y\|^2 < \dfrac{1}{16\gamma^2} \end{cases} \quad (16)$$

The Eq. 16 is a distributed optimization, assuming that we have divided our input data into two chunks. To apply an approximation, we removed a relevant attribute, using a feature selection (FS) method. The experiments' section displays the practical use of this solution based on an PSVM output model, from

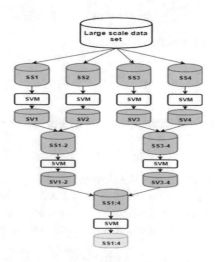

Fig. 1. Parallel Approx SVM architecture

the first data set. Within the parallel approx SVM, when we decrease the number of attributes, each two intermediate support vectors (SV) are merged and used as input in the next layer [13]. This process, described in Fig. 1, is repeated until only one set of SV is remaining.

4 Experiments

The next experimental results were generated using the Hadoop framework, due to the fact that one its main component mapreduce has the same characteristics as the PSVM approach. This software has been setup using the following hardware configuration:

4 machines were used; A master node, and the first slave node having an Intel Core i7 processor, an 8 GB RAM.

The second and the third slave nodes used Intel Pentium inside processors and 4 GB RAM. The Ubuntu Linux OS is installed in each machine.

SVMs suffer from a scalability problem in both memory (number of support vector SV) used and computational time. We compare the PSVM vs. SVMs to prove the performance of PSVM when splitting data sets. This capability, positively, affects the training time, the accuracy and the support vectors, thus, those three terms are chosen as the comparison criteria.

The training time is used to see which SVM extension is faster, the accuracy to see which SVM version is giving the best classification rate, the support vectors to have a look at which SVM extension is providing the most precise model.

For each data set, the maximum number of splits in Tables 1, 2, and 3, was set according to the accuracy. For example, in Table 1, the maximum number of splits is 8 because the accuracy started to decrease at 16 splits.

Three data sets are presented in Table 1, from the website of LIBSVM[1], were used for the experiments, and one feature has been deleted within each data set.

Table 1. Used data sets

Data sets	ijcnn1	W18A	Covtype.binary
Number of instances	191 681	412 928	581 012
Number of features	22	300	54
Training set	115 681	262 928	350 000
Training set in %	60%	64%	60%
Test set	76 672	150 000	231 012
Training set in %	40%	36%	40%
Sampling	Stratified	Random	Random
Label of the deleted feature	"20"	"200"	"10"

---

[1] https://www.csie.ntu.edu.tw/~cjlin/libsvm/.

Table 2 highlights the comparison between SVM and PSVM classifiers, from the data set **ijcnn1**.

Table 2. The SVM and PSVM results in "ijcnn1" data set

SVM version		Training time	Accuracy	Support vectors	
SVM		6 min, 14 s, 44 ms	93.7278%	19 095	1 : 9 533 −1 : 9 562
PSVM	4 splits	3 min, 51 s, 208 ms	94.5978%	14 601	1 : 7 300 −1 : 7 301
	8 splits	3 min, 12 s, 178 ms	94.3317%	14 585	1 : 7 292 −1 : 7 293

Training a PSVM is faster than the SVM. In this case, the partition giving the best results in terms of the 3 criteria cited above is kept. Thus, making a parallel execution of SVM with 4 subsets is the optimal partition. Furthermore, we stopped at 8 splits, due to the fact that the accuracy started to decrease at 16 splits.

The second experiment, highlighted in Table 4, deals with the "W18A" data set.

Table 3. The SVM and PSVM results in "W18A" data set

SVM version		Training time	Accuracy	Support vectors	
SVM		8 min, 45 s, 943 ms	98.2990%	11 750	1 : 5 726 −1 : 5 844
PSVM	4 splits	5 min, 26 s, 777 ms	98.3570%	11 474	1 : 5 720 −1 : 5 754
	8 splits	5 min, 6 s, 279 ms	98.4010%	11 399	1 : 5 679 −1 : 5 720
	16 splits	3 min, 45 s, 558 ms	98.462%	11 311	1 : 5 634 −1 : 5 677
	32 splits	3 min, 1 s, 701 ms	98.462%	11 224	1 : 5 595 −1 : 5 629

PSVM is faster and more accurate than the standard SVM. If we divide the data set into more subsets then we obtain a higher accuracy. Thus, training a PSVM with 16 splits is the best choice.

The "Covtype.binary" data set, is a bit different from the two first examples. In fact it has more precise attribute values, requiring higher computation time, and making prediction a more difficult job. However, a preliminary task, which scales the data set to get lower training time, but do not always guarantee a better precision in the test phase. All of this is described in the Table 4 below.

Table 4. The SVM and PSVM results in "Covtype.binary" data set

SVM version		Training time	Accuracy	Support vectors	
SVM		2 h, 51 min, 27 s, 723 ms	60.3813 %	186 021	1 : 93 003 −1 : 93 018
PSVM	4 splits	2 h, 15 min, 32 s, 483 ms	62.0072%	186 742	1 : 93 373 −1 : 93 369
	8 splits	2 h, 14 min, 2 s, 028 ms	62.0500%	186 739	1 : 93 372 −1 : 93 367
	16 splits	2 h, 19 min, 43 s, 780 ms	62.0916%	186 529	1 : 93 265 −1 : 93 264
	32 splits	2 h, 15 min, 37 s, 340 ms	62.2007%	186 139	1 : 93 065 −1 : 93 074
	64 splits	2 h, 14 min, 28 s, 105 ms	61.1301%	186 885	1 : 93 411 −1 : 93 474

Same as the previous data set, the "Covtype.binary" data set obtained better results from the PSVM than SVM, both in training time and accuracy. As a consequence, training a PSVM model with 32 splits is the best choice. Besides, in this third example we stopped at 32 splits because the accuracy started to decrease at 64 splits.

In the next step of this work, we focus on the choice of deleting one attribute. A Python script, which executes a feature selection method named the "Principle Component Analysis" (PCA), is used to output the list of attributes sorted by a degree of influence. For each data set, the removed feature is classified in the middle among the five first attributes (they are between relevant and non relevant features).

Table 5. Feature deletion

Data set	Removed feature	Accuracy		Rate difference
		Before deletion	After deletion	
ijcnn1	20	94.5978%	93.5791%	↘ 1.0187%
w18a	200	98.4620%	96.2040%	↘ 2.258%
Covtype	54	62.2007%	62.2007%	0%

Table 4 displays the comparison between both of the solutions in terms of accuracy: For each data set, there are 4 possible results:

Table 6. Retraining and approximating PSVM models after feature deletion

PSVM model/Data set	ijcnn1		W18A		covtype.binary	
Accuracy with the old PSVM model	93.5791%		96.2040%		62.2007%	
Retrain or approximate PSVM model	Retrain	Approx	Retrain	Approx	Retrain	Approx
Accuracy with the new PSVM model	93.8204 %	94.2743 %	98.4780 %	97.0710 %	62.4825%	66.0546%

1. An approximation giving a PSVM model with a higher computation time and a lower accuracy than the retraining solution ⇒ Worst case.
2. An approximation giving a PSVM model with a higher computation time and a higher accuracy than the retraining solution.
3. An approximation giving a PSVM model with a lower computation time and a lower accuracy than the retraining solution.
4. An approximation giving a PSVM model with a lower computation time and a higher accuracy than the retraining solution ⇒ Best case.

Based on Table 4, in "ijcnn1", the approximated model is a solution privileged to the retrained one, while it generates an increase of around 0.6952%, to the detriment of an increase with only 0.2413%. As a reminder the comparison is made not only through the accuracy, but also the execution time. Despite that retraining a PSVM is more accurate than the approximation in "W18A", the approximation is always faster than the retraining solution, thus, it would be suitable to get a higher accuracy (by winning only about 2.2740%), but at the cost of a higher training time. Although it provides an increase of the accuracy by only 0.8670%. In "covtype.binary", the approximation increased the accuracy in at a higher level (with 3.8539%) than the retraining alternative (only by 0.4818%). Based on Table 4 we remark that the PSVM reduces memory and increase accuracy based on approximate matrix.

5   Conclusion

In this paper, we proposed a new combined approach based on parallel and approx SVM for big data classification. This new approach overcomes two main issues of the SVM classification method which are the inability to handle large-scaled datasets and the change of attributes' numbers over time.

We take advantages of parallel SVM to decrease the execution time in order to build the classification model. As we deal with big data, the number of attributes is changing over time thus, we have to update the classification model. To obtain the new model which takes account of the dynamic changing of the number of attributes, we combined the approx SVM with the parallel SVM. We have tested this new approach using large datasets. We have obtained interesting results in terms of accuracy compared to the standard SVM. Besides, the parallel approx SVM considerably decreased the time needed to build the new model when we have new data over time. As future work, we aim to improve this new approach by considering the changing of instances over time.

References

1. Burges, C.J.C.: A tutorial on support vector machines for pattern recognition. Data Min. Knowl. Discov. **2**(2), 121–167 (1998)
2. Bordes, A., Ertekin, S., Weston, J., Bottou, J.: Fast kernel classifiers with online and active learning. J. Mach. Learn. Res. **6**, 1579–1619 (2005)
3. Cauwenberghs, G., Poggio, T.: Incremental and decremental support vector machine learning. Adv. Neural Inf. Process. Syst. (NIPS*2000) **13**, 409–415 (2000)
4. Ben Rejab, F., Nouira, K., Trabelsi, A.: Real time support vector machines. Sci. Inf. SAI **2014**, 496–501 (2014)
5. Hsu, C.-W., Chang, C.-C., Lin, C.-J.: A practical guide to supportvector classification, pp. 1–16 (2016)
6. Nalavade, K., Meshram, B.B.: Data Classification Using Support Vector Machines. In: National Conference on Emerging Trends in Engineering and Technology (VNCET), vol. 2 (2012)
7. Cornuéjols, A., Miclet, L.: Apprentissage artificiel: concepts et algorithmes. Editions Eyrolles (2011)
8. Fletcher, T.: Support vector machines explained. Tutorial paper (2009)
9. Sewell, M.: SVMdark: A Windows Implementation of a Support Vector Machine. UCL, London (2005)
10. Priyadarshini, A.: A map reduce based support vector machine for big data classification. Int. J. Database Theor. Appl. **8**(5), 77–98 (2015)
11. Joachims, T.: Making large-scale SVM learning practical. Technical report, SFB 475: Komplexittsreduktion in Multivariaten Datenstrukturen, No. 1998, 28. Universitt Dortmund (1998)
12. Collobert, R., Bengio, S., Marithoz, J.: Torch: a modular machine learning software library. No. EPFL-REPORT-82802. Idiap (2002)
13. Graf, H.P., et al.: Parallel support vector machines: the cascade SVM. In: NIPS, Vol. 17 (2004)
14. Claesen, M., et al.: Fast prediction with SVM models containing RBF kernels. arXiv preprint (2014). arXiv:1403.0736
15. Vert, J.-P., Tsuda, K., Schlkopf, B.: A primer on kernel methods. In: Kernel Methods in Computational Biology, pp. 35–70 (2004)
16. Maji, S., Berg, A.C., Malik, J.: Efficient classification for additive kernel SVMs. IEEE Trans. Pattern Anal. Mach. Intell. **35**(1), 66–77 (2013)
17. Lemaire, V., Salperwyck, C., Bondu, A.: A survey on supervised classification on data streams. In: Proceedings of Business Intelligence: 4th European Summer School, EBISS, pp. 88–125 (2015)

The Four Types of Self-adaptive Systems: A Metamodel

Luca Sabatucci[1]([⊠]), Valeria Seidita[1,2], and Massimo Cossentino[1]

[1] ICAR-CNR, Palermo, Italy
{luca.sabatucci,massimo.cossentino}@icar.cnr.it
[2] DIID, Università degli Studi di Palermo, Palermo, Italy
valeria.seidita@unipa.it

Abstract. The basic ideas of self-adaptive systems are not a novelty in computer science. There are plenty of systems that are able of monitoring their operative context to take run-time decisions. However, more recently a new research discipline is trying to provide a common framework for collecting theory, methods, middlewares, algorithms and tools for engineering such software systems. The aim is to collect and classify existing approaches, coming from many different research areas. The objective of this work is providing a unified metamodel for describing the various categories of adaptation.

1 Introduction

Today's society always more depends on complex distributed software systems available 24 h and with minimal human supervision and maintenance effort during the operating phase. The more software systems grow in complexity and size, the more management automation, robustness, and reliability become central: it becomes essential to design and implement them in a more versatile, flexible, resilient and robust way.

In [17] authors discuss how an ambient intelligent system (AmI) plunges into the real world. The authors underline that, in such complex systems, the boundary between software and society blends and often disappears. The social environment is enriched with artificial intelligence to support humans in their everyday life. The IBM manifesto of autonomic computing [11] suggests a promising direction for facing software complexity through self-adaptation.

A self-adaptive system is a system with the ability to autonomously modify its behavior at run-time in response to changes in the environment [5,7,21]. The vision of a computing system that can manage itself is fascinating [5,7]: to modify the behavior at run-time for maintaining or enhancing its functions [5]. This vision has deep roots in several research fields, as for instance, artificial intelligence, biologically inspired computing, robotics, requirements/knowledge engineering, control theory, fault-tolerant computing, and so on. In the last decade, the vast and heterogeneous number of works concerning self-adaptation investigated several aspects of the problem, for instance, specific architectures for

© Springer International Publishing AG 2018
G. De Pietro et al. (eds.), *Intelligent Interactive Multimedia Systems and Services 2017*,
Smart Innovation, Systems and Technologies 76, DOI 10.1007/978-3-319-59480-4_44

implementing adaptive control loops [14], self-organizing paradigms [1], adaptive requirements [6] and so on. However, to date, many of these problems remain significant intellectual challenges [5,7]. For instance, general purpose software engineering approaches are still missing for the provision of self-adaptation [5]. The long-term objective is to establish the foundations for the systematic development of future generations of self-adaptive systems.

This work resembles existing approaches for systematically engineering self-adaptive system and proposes a unified metamodel of the four types of adaptation. The objective is to provide a framework for identifying and classifying smart systems according to their self-adaptive properties. A metamodel supports this framework for aiding the designer to choose the most appropriate category of systems, depending on the problem statements. The description reports the main components and some illustrative case studies for each type of adaptation.

The paper is structured as follows: Sect. 2 analyzes definitions of self-adaptive systems in state of the art. Section 3 presents the unified metamodel and describes the four categories in details. Finally, Sect. 4 reports the conclusions.

2   Related Work

In last decade, the definition of adaptation has been deliberately generic to gather many sub-fields under a common umbrella and produce interesting synergy. However, this trend has generated some sub-definitions with sometimes significant differences. For instance, in [5,7] a self-adaptive system can modify its behavior in response to changes in the environment. For the models@run.time community [12], a dynamically adaptive system (DAS) can be conceptualized as a dynamic software product line in which variabilities are bound at runtime for improving the quality of service (QoS).

Unifying different definitions is required. Salehie and Tahvildari [20] identify two categories of self-adaptation based on impact (the scope of system effects) and cost factors (in terms of time, resources and complexity). The weak adaptation mainly involves modifying parameters using pre-defined static mechanisms (limited impact/low cost). Conversely, the strong adaptation deals with high-cost/extensive impact actions such as adding, replacing, removing components.

In [16], the authors provide a classification scheme for four categories of self-adaptive systems: *Type 1* consists in anticipating both changes and the possible reactions at design-time: the system follows a behavioral model that contains decision-points. For each decision point, the solution is immediately obvious given the current perceptions and the acquired knowledge about the environment.

Type 2 consists of systems that own many alternative strategies for reacting to changes. Each strategy can satisfy the goal, but it has a different impact on some non-functional requirement. Selecting the best strategy is a run-time operation based on the awareness of the different impact towards these external aspects. Typically the decision is taken by balancing trade-offs between alternatives, based on the acquired knowledge about the environment.

Type 3 consists of systems aware of its objectives and operating with uncertain knowledge about the environment. It does not own pre-defined strategies but it rather assemblies ad-hoc functionalities according to the execution context.

Type 4 is inspired by biological systems that are able of self-modifying their specification when no other possible additions or simple refinements are possible.

In [2], authors face the same point, but from a requirement engineering perspective. The premise is that requirement engineering task is to determine the kinds of input of a system and the possible responses to these inputs. Therefore they identify four types of requirement engineering activities concerning a self-adaptive system.

Level 1 activities are done by humans and resemble the traditional RE activities. The analysts determine all the possible domains to be considered by the system (inputs) and all the possibility system functionalities (reactions).

The system executes *Level 2*'s activities. Whereas the analysts have determined a set of possible behaviors (i.e. reactions), the system can identify the functionality to execute next, when the environment does not match any of the input domains.

Conversely, *Level 3* activities are done by humans for implementing the decision-making procedure that allows the system to apply level 2. Level 3 often includes a meta-level reasoning, that exploits determined program-testable correspondences to environmental changes that trigger adaptation.

Implementing the adaptation mechanism (i.e. the feedback loop) is a *Level 4* activity, for which humans are responsible.

3   The Proposed Metamodel

A smart system is often immersed in a pre-existing world (or environment) populated by objects and persons it interacts with, it influences and is influenced. Boundaries between the software realizing the smart system and the environment are becoming lighter, more and more, and they are almost disappearing. All this significantly affects the design of smart systems. In this section, we present a framework for identifying and illustrating which may be the minimal set of elements (Fig. 1) a system as to own for being classified self-adaptive of the four cited types.

The metamodel in Fig. 1 is composed of two parts: the first part includes all the generic elements of a smart system, whereas the second part embraces all the elements implementing the different types of self-adaptation. The first part comes from a previous study of some authors of this paper [17], where a set of abstractions for representing smart systems in an Ambient Intelligent context has been explored and experimented. All the elements in the second part have been identified reviewing the literature and definitions about managing and designing self-adaptive systems.

We argue that, in general, abstractions representing a smart system are mainly the environment and all the entities it is composed of. There may be, passive, dumb and smart entities, all of them present a state that may be perceived and changed during the running phase of the system. Smart entities are cognitive, intentional and rational entities, able to perform actions according to the principle of rationality. Dumb and passive entities are elements of the environment, the former are resources (software applications, physical devices, ...) and the latter are simply objects in the environment such as physical or digital objects; both of them are part of the environment and influence the actions of the smart system.

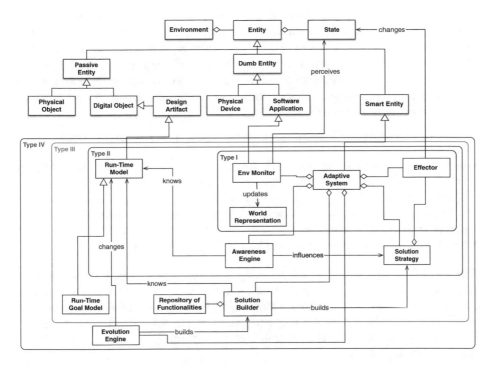

Fig. 1. Portion of metamodel that describes the four types of self-adaptive systems.

The second part of the metamodel shows all the elements of a self-adaptive system. As it may be deduced by the definitions, the set of elements for each type is contained in the higher ones. A smart system belongs to the first type of self-adaptation if it owns a kind of smart entity, the *Adaptive System* composed of the elements necessary to know the environment and to act on it. The second type principally requires an *Awareness Engine* qualified to know the model at run-time of the system and to change or influence the solution strategy built at design time. The third type involves a more complex element that is the *Solution Builder*, this is able to build a completely new solution strategy by using a repository of established functionalities. Finally, the fourth type requires the *Evolution Engine* that is able to set up a *Solution Builder*.

In the following, each type is broadly described. It is important to note that *Design Artifact, Run-Time Model, Awareness Engine, Solution Builder, Evolution Engine* are not classic elements of a problem domain you may find in a metamodel but are strictly related to the software solution. This is to stress the fact that self-adaptive systems cannot be conceived or designed if not as an alive part of the domain.

3.1   Type I

Adaptation of *Type I* is the simplest implementation of smart systems. It arises from a deep analysis of the domain for analyzing all the possible changes the system will observe and react to. The design activity includes the study of all the possible reactions the system will enact. The design leads to the definition of a behavioral model that contains decision-points: a decision point is analogous to a 'if...then...else' statement, whereas conditions typically depend on perceptions of the execution environment.

Problem Statement. The Adaptation of *Type I* is a good choice when:

- it is necessary a smart system able to operate in different ways, according to the state of the environment, acquired by perceptions;
- changes mainly affect observable attributes of the environment;
- it is possible to anticipate (at design-time) all the environment changes that are of interest for the system;
- dealing with uncertainty is not central in the implementation.

System Description. An Adaptive System of Type I is a specific class of artificial system, i.e. a smart entity that operates in an environment. An Adaptive System of Type I owns effectors for changing the state of the environment and perceptors (Env Monitor) for acquiring the current state of the environment. Decisions points are rules that connect knowledge and perception (cause) to an effector (consequence).

Known Usage. An illustrative scenario for Type I is the Robotic Navigation System presented in [5]. The scenario illustrates autonomous control software system of unmanned vehicles (UVs). The adaptation system must consider the regular traffic environment, including the infrastructures and other vehicles. A cause of adaptation can be an unexpected obstacle (people crossing the road). To this aim, it is necessary to monitor the environment (perceptors) and to detect possible obstacles in front of the vehicle. An example of a cause-consequence rule is: IF an obstacle is in front of the vehicle THEN maneuver around the obstacle for avoiding a collision.

Another example is a smart information system for airports [15]. A cause of adaptation occurs when a flight delays. According to the importance of the delay, the system may select different actions: inform the traveler, rebook the flight as soon as possible, book a hotel and re-plan the flight the next day.

Limits. Do not use Type I:

– if the system decision must deal with non-functional aspects; for instance, to rebook a service as soon as possible, or to change the itinerary for maintaining the traveler's satisfaction.
– if a flexible strategy for changing user's goals is required.

3.2 Type II

Adaptation of *Type II* consists of complex systems that offer many alternative strategies for addressing the same goal. Different strategies encompass different operative situations and provide a different impact on the non-functional requirements. This situation requires revising the traditional analysis of requirements phase for including the detailed exploration of a large problem space. Selecting the contextual strategy to apply is a run-time decision that considers complex trade-offs among the state of the environment and the desired quality of service. This decision requires a higher level of knowledge than Type I. It must include the awareness of functional/non-functional requirements, and the different way each operation impacts on them.

Problem Statement. The Adaptation of *Type II* is a good choice when:

– it is necessary a smart system that can detect and recover run-time deviations between behavior and requirements;
– it is not possible to anticipate all the environment changes at design-time, and therefore it is preferable to discover them at run-time;
– changes can also affect requirements;
– it is necessary to incorporate a degree of uncertainty in requirements.

System Description. An Adaptive System of Type II contains a set of effectors and perceptors as well as the Type I and extends it with a set of solution strategies and an awareness engine. A solution strategy is a plan which actions are the monitors and effectors. A solution strategy (or its components) is linked to some functional requirement that motivates its execution. Moreover, the solution strategy (or its components) may also be related to one or more non-functional requirements, describing the kind of expected impact of its execution. This complex model is often a run-time artifact that the Awareness Engine acquires and manages.

Known Usage. An example of Type II adaptation is contained in the scenario of London Ambulance Service (LAS) [10]. According to official recommendations, such a system must involve a full process of consultation between management, staff, trade union representatives and the Service's information technology advisers. Authors have identified many solutions for the main goal (to respond emergency calls within a specific time). A solution is composed of many tasks, each one addressing a specific subgoal. The system follows an adaptation of Type II because it may enact alternative functions. The system reasons on which task to select (and therefore which is the best solution) according to quality aspects, for instance, the time, the efficiency and the reliability. Authors used a goal model [9] for depicting the problem requirements together with tasks and quality aspects. This goal model is implemented as a run-time model the system receives as an input for taking contextual decisions.

Another example comes from dynamic workflow execution engines. In [4] the authors provide a workflow execution engine where tasks are related to abstract services. Each abstract service is the high-level description of many actual services provided by different providers. A real service owns some quality of services. The execution of the workflow requires selecting among alternative concrete services. The choice is to optimize one of the global quality of service assets [8]: different solutions occur if the user decides to optimize the total cost or the total time to complete.

Limits. Do not use Type II:

- if user's functional/non-functional requirements are dynamics;
- if you desire a flexible way for extending system functionalities.

3.3   Type III

Adaptation of *Type III* represents an advanced implementation of a smart system that is instructed with a set of basic functionalities. The system can use for assembling ad-hoc behaviors not contained in any of the predefined solution strategies. This kind of system is particularly suitable for working with uncertain knowledge about the environment and the requirements.

Problem Statement. The Adaptation of *Type III* is a good choice when:

- the problem domain is not entirely anticipated at design-time,
- requirements evolve frequently due to business rules or societal norms;
- it is necessary to assemble ad-hoc functionalities on the fly;
- system functionalities are open to third-party providers and dynamics (depending, for instance, from network conditions).

System Description. An Adaptive System of Type III owns the core characteristics of Type II, but it supports a new component, called Solution Builder. This element can access to a (dynamic) repository of functionalities and build one or more new solution strategies for addressing an unanticipated problem. Selecting among many solutions frequently implies an optimization phase that also considers non-functional aspects.

Known Usage. An example of Type III adaptation is a smart travel assistance system [18]. Such a system can compose a trip itinerary according to user's preferences and create travel service as the composition of multiple atomic services (flight booking, hotel reservation, local transportation tickets and so on). The system is also able to monitor either possible problems during the journey, or changes in user's preferences, and to optimize the travel itinerary consequently.

The new frontier of service composition over the cloud is called cloud mashup. The adaptive scenario –described in [19]– includes on-demand, on-the-fly application mashup for addressing a set of designer's goals. The customer does not provide goals, so the designer can not incorporate them into the orchestration model. An adaptation system of type III allows injecting goals at run-time. Consequently, the system builds a new mashup by aggregating existing cloud services, according to availability, cost and reliability [13].

Limits. Do not use Type III:

- for real-time systems;
- if the system should be able of inspecting itself and autonomously evolving its functionality.

3.4   Type IV

Adaptation of *Type IV* is the higher level of a smart system that is able of inspecting itself, learning from experience and self-modifying its specification. They are designed to afford the worst cases of adaptation: when the system does not own suitable actions/strategy to be used, and it is not capable of generating any one. In this case, the system is able of revising its run-time model, thus to produce a new version of the software. In this category of adaptation, it is more appropriate to refer to evolution. Indeed these systems are inspired by biological systems that own the ability to cope with environment variance by genetic changes.

Problem Statement. The Adaptation of Type IV is a good choice:

- when developers deal with incomplete information about the highly complex and dynamic environment, and, consequently, incomplete information about the respective behavior that the system should expose;
- the system must be able of interpreting incomplete run-time models, and applying –when necessary– changes to those, in order to regulate its behavior;
- the system must be able of generating, at the best of its possibilities, a suitable strategy even when some basic functionality lacks.

System Description. Whereas in classical conception of a self-adaptive system (Type I, II, and III) the system can modify its behavior according to the specifications and to the environment changes, the self-adaptive systems of Type IV are also able of changing their specification. It may be considered as a strong form of learning. This could include some run-time technique for validating the new specification, performing, when necessary, possible trade-off analysis between several potentially conflicting goals.

Known Usage. To the best of our knowledge, there are not popular examples of self-adaptive systems of Type IV.

A case study that could benefit from this kind of adaptation could be a smart firewall [5]. It is a system able to respond to cyber-attacks, but it can not possibly know all attacks in advance since malicious actors develop new attack types all the time. So far, this kind of systems is yet a challenge for artificial intelligence and computer science. It implies a high level of self-awareness and the ability to learn, to reason with uncertainty and to generate new code for coping with unexpected scenarios.

4 Conclusions

All the types of self-adaptive systems share a shift of some design decisions towards run-time in order to improve the control over the behavior.

In practice, the reader has to focus on run-time activities and in particular on the decision-making process. This latter is the algorithm/technique used for directing the system behavior.

This paper proposes a framework for classifying systems and their self-adaptive attributes by means of a set of abstractions grouped into a metamodel. The metamodel offers the designer the possibility to select the right elements related to the chosen self-adaptive type. From a designer point of view, the power of this metamodel is that it considers the smart system and the environment it operates in strictly tied each other: the environment is part of the software solution and in the same way the system at run-time, with its design artifacts, algorithms and so on, is part of the environment thus realizing the feedback loop regulating all the activities of a self-adaptive system [3]. The monitor senses the environment and collects relevant data and events for future reference. The analyzer compares data in order to evaluate differences between the actual and the expected behavior. The planner uses these data for taking decisions about the behavior to be executed. Finally, the execute (or act) module applies the planned decisions through its effectors.

Moreover, in order to classify a smart system according to the type of adaptation, the following guideline focuses on the kind of perception ability and decision-making process. If the run-time activity is the enactment of a set of hard-coded actions (selected and/or configured according to the operative context), then the adaptation is of Type I. If the system owns a set of pre-defined strategies (each strategy is an aggregation of actions) and if the strategy is selected and/or configured at run-time, according to quality aspects, then the adaptation is of Type II. If the system is able of assembling a new strategy at run-time, then the adaptation is of Type III. If the system can modify its run-time models for generating new functions, then the system is of Type IV.

References

1. Baresi, L. Guinea, S.: A3: self-adaptation capabilities through groups and coordination. In: Proceedings of the 4th India Software Engineering Conference, pp. 11–20. ACM (2011)
2. Berry, D.M., Cheng, B.H. Zhang, J.: The four levels of requirements engineering for and in dynamic adaptive systems. In: 11th International Workshop on Requirements Engineering Foundation for Software Quality (REFSQ), p. 5. (2005)

3. Brun, Y., Marzo Serugendo, G., Gacek, C., Giese, H., Kienle, H., Litoiu, M., Müller, H., Pezzè, M., Shaw, M.: Engineering self-adaptive systems through feedback loops. In: Cheng, B.H.C., Lemos, R., Giese, H., Inverardi, P., Magee, J. (eds.) Software Engineering for Self-adaptive Systems, pp. 48–70. Springer, Heidelberg (2009)
4. Casati, F., Ilnicki, S., Jin, L.-J., Krishnamoorthy, V., Shan, M.-C.: eFlow: a platform for developing and managing composite e-services. In: Proceedings of the Academia/Industry Working Conference on Research Challenges, pp. 341–348. IEEE (2000)
5. Cheng, B.H.C., Lemos, R., Giese, H., Inverardi, P., Magee, J., Andersson, J., Becker, B., Bencomo, N., Brun, Y., Cukic, B., et al.: Software engineering for self-adaptive systems: a research roadmap. In: Cheng, B.H.C., Lemos, R., Giese, H., Inverardi, P., Magee, J. (eds.) Software Engineering for Self-adaptive Systems. Springer, Heidelberg (2009)
6. Dalpiaz, F., Giorgini, P., Mylopoulos, J.: Adaptive socio-technical systems: a requirements-based approach. Requir. Eng. **18**(1), 1–24 (2013)
7. Lemos, R., Giese, H., Müller, H.A., Shaw, M., Andersson, J., Litoiu, M., Schmerl, B., Tamura, G., Villegas, N.M., Voge, T., et al.: Software engineering for self-adaptive systems: a second research roadmap. In: Lemos, R., Giese, H., Müller, H.A., Shaw, M. (eds.) Software Engineering for Self-adaptive Systems II, pp. 1–32. Springer, Heidelberg (2013)
8. C. Di Napoli, D. Di Nocera, and S. Rossi. Computing pareto optimal agreements in multi-issue negotiation for service composition. In: Proceedings of the International Conference on Autonomous Agents and Multiagent Systems, pp. 1779–1780. International Foundation for Autonomous Agents and Multiagent Systems (2015)
9. I. Jureta, A. Borgida, N. A. Ernst, and J. Mylopoulos. Techne: Towards a new generation of requirements modeling languages with goals, preferences, and inconsistency handling. In: RE, pp. 115–124 (2010)
10. Jureta, I.J., Borgida, A., Ernst, N.A., Mylopoulos, J.: The requirements problem for adaptive systems. ACM Trans. Manag. Inf. Syst. (TMIS) **5**(3), 17 (2015)
11. Kephart, J.O., Chess, D.M.: The vision of autonomic computing. Computer **36**(1), 41–50 (2003)
12. Morin, B., Barais, O., Jezequel, J.-M., Fleurey, F., Solberg, A.: Models@ run.time to support dynamic adaptation. Computer **42**(10), 44–51 (2009)
13. Napoli, C.D., Sabatucci, L., Cossentino, M., Rossi, S.: Generating and instantiating abstract workflows with QOS user requirements. In: Proceedings of the 9th International Conference on Agents and Artificial Intelligence (2017)
14. Patikirikorala, T., Colman, A., Han, J., Wang, L.: A systematic survey on the design of self-adaptive software systems using control engineering approaches. In: ICSE Workshop on Software Engineering for Adaptive and Self-Managing Systems (SEAMS), pp. 33–42 (2012)
15. Qureshi, N.A., Jureta, I.J., Perini, A.: Requirements engineering for selfadaptive systems: core ontology and problem statement. In: International Conference on Advanced Information Systems Engineering, pp. 33–47. Springer (2011)
16. Qureshi, N.A., Perini, A., Ernst, N.A., Mylopoulos, J.: Towards a continuous requirements engineering framework for self-adaptive systems. In: First International Workshop on Requirements@ Run.Time (RE@ RunTime), pp. 9–16. IEEE (2010)
17. Ribino, P., Cossentino, M., Lodato, C., Lopes, S., Seidita, V.: Requirement analysis abstractions for AmI system design. J. Intell. Fuzzy Syst. **28**(1), 55–70 (2015)

18. Sabatucci, L., Cavaleri, A., Cossentino, M.: Adopting a middleware for self-adaptation in the development of a smart travel system. In: Pietro, G., Gallo, L., Howlett, R.J., Jain, L.C. (eds.) Intelligent Interactive Multimedia Systems and Services 2016, pp. 671–681. Springer, Cham (2016)
19. Sabatucci, L., Lopes, S., Cossentino, M.: A goal-oriented approach for self-configuring mashup of cloud applications. In: International Conference on Cloud and Autonomic Computing (ICCAC), pp. 84–94. IEEE (2016)
20. Salehie, M., Tahvildari, L.: Self-adaptive software: landscape and research challenges. ACM Trans. Auton. Adapt. Syst. (TAAS) 4(2), 14 (2009)
21. Whittle, J., Sawyer, P., Bencomo, N., Cheng, B.H., Bruel, J.-M.L Relax: incorporating uncertainty into the specification of self-adaptive systems. In: 2009 17th IEEE International Requirements Engineering Conference, pp. 79 88. IEEE (2009)

Context Reasoning and Prediction in Smart Environments: The Home Manager Case

Roberta Calegari$^{(\boxtimes)}$ and Enrico Denti

Dipartimento di Informatica-Scienza e Ingegneria (DISI), Alma Mater Studiorum–Università di Bologna, Viale Risorgimento 2, 40136 Bologna, Italy
{roberta.calegari,enrico.denti}@unibo.it

Abstract. In Smart Environments computing systems are ubiquitous, intelligence pervades the space, and people's situatedness in time and space is exploited to provide a contextualised, adaptive user experience.

Their socio-technical nature calls for skills, concepts, methodologies, technologies from diverse fields – AI, coordination, distributed systems, organisational sciences, etc. –, promoting a multi-paradigm perspective.

This work explores the Smart Environments context moving from two basic bricks: *Butlers for Smart Spaces*, a technology-neutral reference framework focused on users' situatedness and interaction aspects; and *Home Manager*, a multi-paradigm, agent-based implementation platform for Smart Living contexts, particularly focused on the reasoning aspects.

For concreteness, we take a Smart Kitchen as our running example, discussing how it can be devised in Butlers for Smart Spaces and deployed on Home Manager, focusing on context reasoning and prediction aspects.

1 Introduction

Smart Environments (also called Smart Spaces) [7,31] aim at augmenting apartments, offices, museums, hospitals, schools, malls, outdoor areas with smart objects and systems: both people and smart objects are immersed in time and space, and computer systems seamlessly integrate into people's everyday lives "anywhere, anytime" [24]. Intelligence pervades the environment, and space and time awareness sets the base for a contextualised, adaptive user experience.

A key aspect of these scenarios is that the technology complexity is amplified by the organisational complexity of the application domain: such systems need to be conceived, designed and developed taking into account both the technological and the human/organisational aspects from the earliest stage, combining different dimensions and behaviour from *pervasive, distributed, situated* and *intelligent* computing—altogether [23]. Because of their characteristics, Smart Environments assume the co-existence of heterogeneous entities, differing in terms of execution platform, development language, enabling technologies, support infrastructures, models, roles, objectives, etc.—which is why they inherently call for a *multi-paradigm* approach: in fact, their development calls for skills, concepts, methodologies, technologies from a variety of fields (such as, for instance, AI, coordination, distributed systems, organisational sciences, etc.).

© Springer International Publishing AG 2018
G. De Pietro et al. (eds.), *Intelligent Interactive Multimedia Systems and Services 2017*, Smart Innovation, Systems and Technologies 76, DOI 10.1007/978-3-319-59480-4_45

The Internet of Things [18,23,26] is providing the fundamental bricks to make such scenarios concrete: appliances and devices of any sort are networked together and can possibly embed some form of (limited) intelligence to provide suggestions from the user's context and habits (examples include Apple [2], Amazon [1], Google [13], Microsoft [18], Samsung [25], to name just a few).

Yet, most applications today merely provide some nice app, very much like a novel form of remote-control–which is probably the main reason for the gadget-like feeling that still affects most of the available prototypes [1,2,13,18,25]. To make one step further, three key preconditions seem to be *(i)* the availability of an effective coordination middleware, going beyond the basic support to interoperability; *(ii)* an effective support to *situatedness*, intended as awareness of the environment and chance to react to changes—the basic brick to support contextualised reasoning and possibly prediction; and *(iii)* guidelines and enabling techniques exploiting skills, concepts, methodologies, technologies from the most diverse (socio-technical) fields, in a multi-paradigm perspective.

In this paper we explore the construction of Smart Environments (Sect. 2), moving from *(i)* the *Butlers for Smart Spaces* framework [5,6], i.e. the specialisation to this context of the *Butlers* vision [8], providing the conceptual reference for the design and development of advanced services to users immersed and situated in time and space; and *(ii) Home Manager* [17], a multi-paradigm, agent-based platform for the implementation of Smart Living contexts—particularly focused on the reasoning aspects, mainly to anticipate the users' needs.

For concreteness, we take a Smart Kitchen (Sect. 3) as our running example: we first discuss how it can be devised via Butlers for Smart Spaces, and then deployed on the Home Manager platform—highlighting in particular the context reasoning, prediction and adaptation aspects.

2   Smart Environments in the Butlers Perspective

2.1   The Butlers Vision in the Smart Spaces Context

The Butlers architecture [8] defines a reference framework made of seven layers, relating technologies with the features and value-added for users (Fig. 1, left). Its abstract, technology-neutral nature makes it inherently well suited to a multi-paradigm approach, accounting for a multiplicity of different programming paradigms, languages, models and platforms.

Leaving the full discussion of the Butlers approach to [8], the bottom layers concern the enabling technologies – communication-enabled sensors, meters, actuators, etc. – while the infrastructural/middleware layers provide coordination and geographical information services, and the top layers focus on intelligence, sociality, gamification aspects (not necessarily to be taken in the sequence).

Butlers for Smart Spaces [6] is the specialisation of Butlers to the Smart Spaces context: Fig. 1 (center) shows its relationship with the general Butlers vision. In this scenario, lower-level functionalities are typically provided by the underlying infrastructure, while some envisioned upper functionalities are either

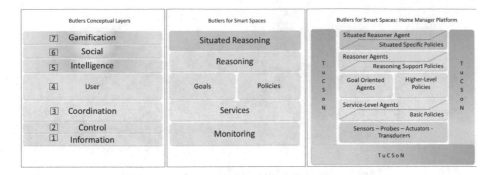

Fig. 1. Butlers, Butlers for Smart Spaces, and the Home Manager specialisation

too far from the foreseeable future or from the current state of the art: so, their layers can be conveniently collapsed. Accordingly:

– the information (1) and control (2) Butlers layers are grouped together in a single *Monitoring* layer;
– the new *Services* layer, in-between information (Butlers layer 1) and coordination (layer 3), pre-processes raw information into exploitable knowledge;
– coordination (3) and user-aware (4) Butlers layers are on the one hand grouped into a single layer, because coordination in a Smart Space must necessarily take users – the main actors – into account; on the other, the foreseeable complexity of such coordination leads to split such a layer into *Goals* and *Policies* side-by-side, for both practical and conceptual reasons;
– the intelligence (5) Butlers layer is also split into two *Reasoning* and *Situated reasoning* on top of each other, to separate the reasoning which exploit only the local/user knowledge from the ones which exploit also the surrounding environment—which features the very nature of a Smart Space.

As detailed below, Home Manager further concretises the *Butlers for Smart Spaces* layers onto a TuCSoN-based MAS: so, the framework is re-shaped as in Fig. 1 (right). The TuCSoN infrastructure surrounds all layers, enabling the seamless integration of heterogeneous entities, bridging among technologies and agents' perceptions, and supporting situated intelligence.

Leaving details to [5,6], the key point to be highlighted here is that all the *Butlers for Smart Spaces* layers, except for the bottom one, concretely split into agents and policies when deployed as a TuCSoN MAS, thus accounting for the full separation of agents and policies also at the implementation stage.

2.2   The Home Manager Platform

Home Manager [17] is an open source platform for Smart Spaces, inspired to the above architecture and explicitly conceived to be open, and deployable on a wide variety of devices (PCs, smartphones, tablets, up to Raspberry PI 2). The basic idea is to go beyond the mere monitoring and remote control of house appliances via app, enabling the development of advanced services to users immersed

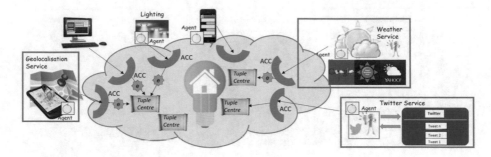

Fig. 2. The Home Manager intuitive architecture

in/interacting with the surrounding environment, providing the native ability to reason on potentially any kind of possibly situated data (*context reasoning*)—mainly, to anticipate the users' needs (a limited form of *context prediction*).

More concretely, Home Manager refers to a smart house immersed in a Smart Living context, with devices (air conditioners, lights, etc.) and users of different categories and (RBAC-based) roles. At the basic operation level, the goal is to satisfy the users' desires (e.g. room light, temperature) while respecting some global constraints (e.g. energy saving, temperature range, etc.); at a higher level, however, the goal is more ambitious—that is, to perform contextual reasoning to anticipate the user's needs, taking into account the user's situation in time and space, his habits, and more generally any other user-related information, including the environment where he lives, travels, purchases goods, etc.

The platform is built on top of TuCSoN [21,29], a multi-agent coordination infrastructure enabling intelligence to be spread both on agents and coordination artifacts, thanks to the underlying tuProlog [11,30] technology. Being light-weight, Java-based, as well as inherently multi-paradigm, tuProlog makes it possible to develop any kind of declarative (Prolog), imperative (Java,C#) and hybrid (Java+Prolog) intelligent agents [10], providing designers with the full range of degrees of freedom as concerns spreading intelligence anywhere needed.

The intuitive architecture is shown in Fig. 2: each device is assumed to be equipped with an agent, which is connected to a TuCSoN boundary artifact (an *Agent Coordination Context*—ACC) that defines its admissible operations and roles in the society. These agents embed the individual intelligence. The "social" intelligence is embedded in TuCSoN tuple centres—programmable coordination media that embed the ability to react to internal, external and situated events [20]. Overall, the infrastructure embeds and enforces the coordination laws to mediate among agents, governing the agent-agent and the agent-environment interaction. Thus, it provides the "glue" to cope with heterogeneity, bridging among programming paradigms and languages, interaction protocols, and ontologies—as a *multi-paradigm enabler*. For instance, heterogeneous entities, such as legacy agents, can be integrated by charging the infrastructure of bridging the gap between the common ontology and the specific agents' representations and ontologies [20]: agents can thus be heterogeneous in nature, implementation language, etc.—the only requirement being that they coordinate via the TuCSoN APIs.

As shown in Fig. 2 (details in [5]), Home Manager provides services to:

- exploit the user's location, tracked by the smartphone GPS, to enable an intelligent reasoner agent to take autonomous "situated" decisions;
- explore the environment around the user's location, extracting information on shops, services, etc., to be taken as a further reasoning knowledge base;
- get information about the surrounding environment (e.g. weather) so as to tailor decisions to the user's habits and needs;
- interact with selected social networks (e.g. Twitter) to grab information that could later be exploited fur further reasoning;
- track the human presence for intrusion detection or elderly applications (e.g. to detect falls, stand or walk status, etc.).

2.3   The Home Manager Technology

The underlying presence of a TuCSoN-based MAS is intentionally invisible for Home Manager users, who can control the configuration and interact with the system with no need to know anything, nor operate directly, on the underlying machinery—i.e., the inner tuple-based representation of data and policies.

Notably, the declarative, tuple-based approach is what makes it easy to evolve the system incrementally from a purely-simulated environment on a personal computer (with simulated house, inhabitants, and sensors) to an "increasingly-real" system, interfaced e.g. to a smartphone [5,9] to exploit its geo-localisation feature to support advanced services based on the user's situatedness.

The system can also run "out of the box" on low-cost platforms like the Raspberry PI 2, thus dropping the requirement of a personal computer for the hosting environment, and exploiting the many sensors and devices available for that platform. For our intended Smart Kitchen application (Sect. 3), we coupled the Raspberry with the GrovePi [14] board, which provides displays, switches, temperature sensors, in an all-in-one pack with a customized Raspbian version; we also added an RFID reader, to simulate the presence of RFID-tagged items.

Orthogonally, the Raspberry can be exploited as an implementation platform for smart devices—e.g., Smart Fridge and Smart Oven [5,6]. Although Java and a Raspbian-based Raspberry are the most obvious choices, we also experimented with Microsoft Windows 10-IoT Core [18], since going beyond a single-platform, single-language environment is a key issue in the IoT context.

In order to enable cross-platform communication and interaction between Home Manager, rooted on the (Java-based) TuCSoN infrastructure, and Windows 10 clients, written e.g. in C#, an ad-hoc bridge was developed. To be TuCSoN-compliant, the bridge, written in Java, communicates via UTF-8 strings: an XML configuration file specifies the data required by the client (bridge IP and port numbers, target tuple centre name, etc.) and maps its primitives onto the TuCSoN ones. While only simple experiments driving LEDs have been performed so far [5], their success seems promising, showing that Home Manager can work in the perspective as a Multi-Paradigm Programming platform.

3 The Smart Kitchen Case Study

3.1 Implementing Smart Environments in Home Manager

Following the intuitive architecture in Fig. 2, defining a Smart Environment in Home Manager means *(i)* to identify the relevant device and service categories; *(ii)* to define a tuple-based representation of the relevant knowledge; *(iii)* to define the agent interaction protocols—which need not match the knowledge representation 1–1, since tuple centres can be programmed to bridge the gap; and *(iv)* to develop an agent for each device category and for each service to interact with. The resulting architecture is then implemented as a TuCSoN MAS.

This process makes it possible to keep the social/individual intelligence and the mechanisms/policies clearly separate, as well as to allow independent testing and debugging of agents and policies, effectively exploiting the data-driven, multi-paradigm software development approach.

3.2 Application Scenario

In this Section, we discuss the case of the Smart Kitchen, made of a *Smart Fridge*, a *Smart Pantry*, a *Smart Oven*, a *Smart Mixer*, integrated with a *Smart Shopper* butler aimed at managing the food supply.

Figure 3 shows the envisioned scenario, highlighting the layers described in the previous section. Bottom-up, the Smart Fridge and the Smart Pantry are capable of monitoring the quantity of food, and of collecting historical data on user's habits, e.g. the most commonly eaten food and preferred meals. All data are reified on selected tuple centres to create a knowledge base, usable by upper-level reasoning agents (like the Smart Shopper) to predict the user's needs or

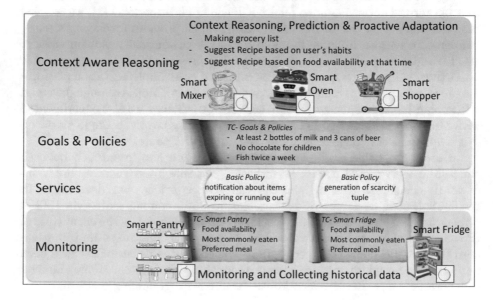

Fig. 3. The Smart Kitchen Scenario

make contextualised suggestions. In the Services level, whenever a product is taken from the fridge, suitable Basic Policies check its quantity against user-defined, per-product thresholds, and generate the corresponding buy tuple if necessary (details in [5]). No Service-Level agents are devised at this layer.

Goals and Policies refer to user preferences and constraints, like making sure that there are at least 2 bottles of milk, that fish is cooked twice a week, etc.

At the upper layer, the Smart Shopper butler (Fig. 4) compiles the shopping list based on the above tuples, contacting the "proper" vendor via the "appropriate" means: what "proper" and "appropriate" stand for depends on context-aware policies. Vendor selection could be based e.g. on fidelity cards, promotional campaigns, distance, consumer habits, etc., while contact means could be email, online shopping, up to "ask your neighbour", etc. Policies could also require that the shopping cart total reaches a minimum amount to get free home delivery, that multiple markets are compared to find the most convenient—possibly taking into account fidelity cards and special offers; and so on.

In its turn, the Smart Oven aims to support the user's food cooking—in principle, exploiting any available technology to identify and cook the food; the user profile is supposed to include information about his/her dietary requirements. The Smart Mixer manages the recipe instructions, interacting with both the Smart Fridge – to check that the ingredients for the selected recipe are actually available – and with the Smart Oven – to check its ability to cook that food and potentially synthesise the proper control instructions. Since recipes are prepared based on the current content of Smart Pantry and Smart Fridge, the Smart Mixer behaviour aims to exhibit a (primitive) form of context adaptation.

In the current experimental prototype (Fig. 4), the Smart Oven, Mixer and Pantry are simulated in software, while the Smart Fridge integrates software-only and software+Raspberry+GrovePI hardware (for display, sensors and LEDs), plus an RFID tag reader for tracking the content [5]. Given the proof-of-concept nature of this prototype, policies are intentionally kept simple. The Smart Fridge monitors the fridge content, reified as fridge_content/4 tuples: a product is *scarce* when its quantity falls below a pre-defined threshold, expressed as another suitable tuple. So, whenever an item is taken out of the fridge, the policy performs the above check and generates a scarcity/5 tuple if this is the case. Based on this information, other policies generate the purchase orders, in the form of buy/3 tuples: the current (trivial) policy is to produce an order when "enough" (user-definable) scarcity tuples have been accumulated. The Smart Shopper finally consumes these tuples and, based on its own policies, decides the quantities to be purchased, and sends the order to the "proper" (currently: the pre-defined) vendor via the "appropriate" means (currently: by email).

The middleware infrastructure encapsulates the coordination laws, enabling interoperability and integration with third-part services, like the mailing service that sends shopping orders. Apart from the knowledge base, the declarative approach provides a *lingua franca* to bridge among the different forms of heterogeneity, supports the agent uncoupling and the separation between policies and mechanisms, and supports context reasoning.

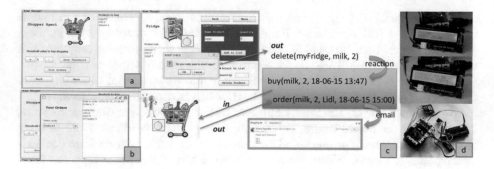

Fig. 4. Policies setup (a), order monitoring (b), Shopper Agent (c), hardware (d)

Of course, it is up to the designer to balance between the intelligence to be embedded in higher-level policies and to be put onto reasoning agents: generally speaking, agents can be expected to focus on opportunistic behaviour, while policies on synthesising the information for higher-level reasoning—especially to take into account environmental time- and space-situatedness. For instance, opportunistic behaviour could exploit the user's location to alert of a nearby market, minimising the time spent in traffic (and therefore also fuel consumption and cost), or suggest an alternative market to avoid traffic jams; and so on.

4   Related Work and Conclusions

Smart Environments are one of today's most challenging fields: IoT applications for smart grids and smart homes are hot research areas. Key issues range from the complexity of architectures to the lack of unifying frameworks and affordable infrastructures, configuration complexity, security aspects, interoperability among hardware and software components from different manufacturers [3].

Several commercial solutions and advanced Demand Side Management systems exist, but are mainly focused on large industrial consumers [12,22]: residential consumers are mostly unconsidered. In [19], novel architectures with state-of-the-art software technologies are proposed, that focus on domestic environments and habitat monitoring: design guidelines are presented for collecting and integrating household data, thus enabling data interoperability. In [28], the problem of temperature regulation inside commercial and administrative buildings is faced by focussing on the design and implementation of specific network topologies and node localisation within the system. Holistic IoT-based frameworks are also proposed [15,16,27,32], both in smart home and other contexts, such as the monitoring of production processes in industry [27].

The Butlers for Smart Spaces approach takes a different starting point: instead of moving from a specific need or application area, it defines a technology-neutral reference framework for Smart Environments in pervasive IoT contexts, which takes situated reasoning capabilities as a fundamental requirement from the very beginning. The Home Manager platform concretises this view by enabling the independent design and development of the Butlers layers, delegating the infrastructure to bridge among the diverse agents' ontologies, APIs, knowledge representations, interaction protocols, etc. Moreover, by suitably reifying in tuple centres both the relevant knowledge and the system policies, Home Manager lays the foundations to support context reasoning and prediction [4].

Of course, we are well aware that our prototype suffers from many limitations [5] and that the above case study is quite basic: actually, the prediction of future context based on recorded past context is just embryonic at this stage. Yet, early results are encouraging, both as concerns Butlers for Smart Spaces as the reference framework and Home Manager as the support technology—thus indirectly strengthening the effectiveness of our multi-paradigm approach.

Future work will focus *(i)* on a deeper exploration of the context reasoning aspect, considering technologies like machine learning as third-party components; *(ii)* on cross-platform interoperability, both as concerns Java/Windows 10 on the Raspberry and in the perspective with emerging standards; and *(iii)* on the further development of the Smart Kitchen application, developing more complex policies and implementing other advanced situated services.

Acknowledgments. Authors would like to thank Dr. Ambra Molesini for her suggestions and feedback on the manuscript of this paper.

References

1. Amazon: Smart Home (2017). https://www.amazon.com/smart-home/b?node= 6563140011
2. Apple: Apple Home Kit (2014). https://developer.apple.com/homekit/
3. Borgia, E.: The Internet of Things vision: key features, applications and open issues. Comput. Commun. **54**, 1–31 (2014)
4. Boytsov, A.: Context reasoning, context prediction and proactive adaptation in pervasive computing systems (2011)
5. Calegari, R., Denti, E.: Building smart spaces on the Home Manager platform. ALP Newsletter, December 2016. https://www.cs.nmsu.edu/ALP/2016/12/ building-smart-spaces-on-the-home-manager-platform/
6. Calegari, R., Denti, E.: The Butlers framework for socio-technical smart spaces. In: INSCI 2016. LNCS, vol. 9934, pp. 306–317. Springer (2016)
7. Chen, H., Finin, T., Joshi, A., Kagal, L., Perich, F., Chakraborty, D.: Intelligent agents meet the semantic web in smart spaces. IEEE Internet Comput. **8**(6), 69–79 (2004)
8. Denti, E.: Novel pervasive scenarios for home management: the Butlers architecture. SpringerPlus **3**(52), 1–30 (2014)
9. Denti, E., Calegari, R.: Butler-ising Home Manager: a pervasive multi-agent system for home intelligence. In: 7th International Conference on Agents and Artificial Intelligence (ICAART 2015), pp. 249–256. SCITEPRESS, Lisbon, 10–12 January 2015
10. Denti, E., Omicini, A., Calegari, R.: tuProlog: making Prolog ubiquitous. ALP Newsletter, October 2013. http://www.cs.nmsu.edu/ALP/2013/10/tuprolog-making-prolog-ubiquitous/

11. Denti, E., Omicini, A., Ricci, A.: tuProlog: a light-weight Prolog for Internet applications and infrastructures. In: Practical Aspects of Declarative Languages. LNCS, vol. 1990, pp. 184–198. Springer (2001)
12. Finn, P., Fitzpatrick, C.: Demand side management of industrial electricity consumption: promoting the use of renewable energy through real-time pricing. Appl. Energy **113**, 11–21 (2014)
13. Google: Works with Nest (2014). http://techcrunch.com/2014/06/23/google-makes-its-nest-at-the-center-of-the-smart-home/
14. GrovePi Home (2017). http://www.dexterindustries.com/grovepi/
15. Gubbi, J., Buyya, R., Marusic, S., Palaniswami, M.: Internet of Things (IoT): a vision, architectural elements, and future directions. Future Gener. Comput. Syst. **29**(7), 1645–1660 (2013)
16. He, W., Yan, G., Xu, L.D.: Developing vehicular data cloud services in the IoT environment. IEEE Trans. Ind. Inf. **10**(2), 1587–1595 (2014)
17. Home Manager (2017). apice.unibo.it/xwiki/bin/view/Products/HomeManager
18. Microsoft: the Internet of your things (2015). https://dev.windows.com/en-us/iot/
19. Monacchi, A., Egarter, D., Elmenreich, W.: Integrating households into the smart grid. In: 2013 Workshop on Modeling and Simulation of Cyber-Physical Energy Systems (MSCPES), pp. 1–6, May 2013
20. Omicini, A., Denti, E.: From tuple spaces to tuple centres. Sci. Comput. Program. **41**(3), 277–294 (2001)
21. Omicini, A., Zambonelli, F.: Coordination for Internet application development. Auton. Agents Multi-Agent Syst. **2**(3), 251–269 (1999)
22. Palensky, P., Dietrich, D.: Demand side management: demand response, intelligent energy systems, and smart loads. IEEE Trans. Ind. Inf. **7**(3), 381–388 (2011)
23. Ricci, A., Piunti, M., Tummolini, L., Castelfranchi, C.: The mirror world: preparing for mixed-reality living. IEEE Pervasive Comput. **14**(2), 60–63 (2015)
24. Saha, D., Mukherjee, A.: Pervasive computing: a paradigm for the 21st century. Computer **36**(3), 25–31 (2003)
25. Samsung: Samsung Smart Things (2015). https://www.smartthings.com
26. Schaffers, H., Komninos, N., Pallot, M., Trousse, B., Nilsson, M., Oliveira, A.: Smart cities and the future Internet: towards cooperation frameworks for open innovation. In: The Future Internet, pp. 431–446. Springer (2011)
27. Shrouf, F., Miragliotta, G.: Energy management based on Internet of Things: practices and framework for adoption in production management. J. Cleaner Prod. **100**, 235–246 (2015)
28. Stojkoska, B.R., Avramova, A.P., Chatzimisios, P.: Application of wireless sensor networks for indoor temperature regulation. Int. J. Distrib. Sens. Netw. **10**(5) (2014). Article No. 502419
29. TuCSoN: Home page (2017). http://tucson.apice.unibo.it/
30. tuProlog: Home page (2017). http://tuprolog.apice.unibo.it
31. Wang, X., Dong, J.S., Chin, C., Hettiarachchi, S., Zhang, D.: Semantic space: an infrastructure for smart spaces. IEEE Pervasive Comput. **3**(3), 32–39 (2004)
32. Xu, B., Xu, L.D., Cai, H., Xie, C., Hu, J., Bu, F.: Ubiquitous data accessing method in IoT-based information system for emergency medical services. IEEE Trans. Ind. Inf. **10**(2), 1578–1586 (2014)

Social Activities Recommendation System for Students in Smart Campus

Sabrine Ben Abdrabbah$^{(\boxtimes)}$, Raouia Ayachi, and Nahla Ben Amor

LARODEC, Université de Tunis, ISG Tunis, 2000 Bardo, Tunisia
abidrabbah.sabrine@gmail.com, raouia.ayachi@gmail.com,
nahla.benamor@gmx.fr

Abstract. Smart campus is generally defined as an academic institution that uses the Information and Communication Technologies (ICT) to enhance the quality of students' life through buildings infrastructure, the way of learning and social activities planning. A large aspect of smart campus solution requires an intelligent information system which should be able to identify students' requirements and make accordingly specific recommendations. This paper addresses the question of how smart campus can organize and plan events and activities for students by exploiting ICT. The proposed solution is based on a group recommender system of students' social activities allowing them to express their preferences via complete orders. A new aggregation method (so-called Avg-Pos) is proposed to produce the global preference ordering relative to the whole group members. Experimental study carried out on a real-world data obtained from students of Lille1 University (France) points out the promising results on the effectiveness of the proposed recommender system.

Keywords: Smart campus · Students' social life · Group recommendation · Ranking aggregation methods

1   Introduction

A smart city is a set of multidimensional components of technology (e.g. wireless, sensors, etc.), people (e.g. tourists, citizens, etc.) and institutions (hospitals, campuses, etc.). An increasing number of researchers investigated in developing smart campuses to help students to comply in ease with the ongoing faculty support and consequently, make the students' campus life easier, comfortable and attractive. A smart campus provides students with reliable services and activities from different components including campus infrastructure (e.g. smart energy, smart water distribution system, etc.), students' academic life (e.g. smart courses, smart timetable, etc.) and students' social life (e.g. sporting, associative and cultural activities).

We are in particular interested in those which are part of the students' social life since the campuses afford each year an important investment fund for the setting up of these activities (e.g. the campus of Lille1[1] affords around 160000

---

[1] www.univ-lille1.fr.

© Springer International Publishing AG 2018
G. De Pietro et al. (eds.), *Intelligent Interactive Multimedia Systems and Services 2017*,
Smart Innovation, Systems and Technologies 76, DOI 10.1007/978-3-319-59480-4_46

euros each year). The agent responsible for the organization of social activities is not able to capture all students' interests and consequently, students' expectations in term of the presented activities can not be intended. In this context, group recommender systems can be an efficient tool able to provide students with recommendations of items that can better meet their needs and tastes. These items correspond to social activities selected to be consumed by a group of students.

This paper presents a group recommender system which provides an intelligent planning of social activities in smart campus-wide environments. The proposed system is designed to serve as a supportive decision-making system to help/assist the decision maker seeking to invest in a social activity. This system considers the case where students express their preferences qualitatively by defining an order of interest over activities. Our idea is to use ranking aggregation methods to combine individual orderings into a single one that captures the collective preference of students. However, due to the nature of the available data, existing rank aggregation methods may suffer from one or many weaknesses such as the cycle of pairwise comparison problem, the pairwise match-ups problem, etc. As a solution, we propose an average of positions method (denoted by Avg-Pos) which consists in generating the group ordering based on items positions in the individual orderings. The experimental results obtained by analyzing preferences data of students of Lille1 University (France) show that Avg-Pos method performs better when the orderings length is important.

This paper is structured as follows. Section 2 presents the review of the literature related to recommendation methods in the context of smart campus. Section 3 defines the proposed social activities recommender system. Finally, Sect. 4 presents the experimental study.

2   Recommender Systems in Smart Campus Environment

Recommender systems have been proposed to provide users with relevant items that they might be interested in order to address the information overload problem. It can be used in different fields including the campus environment. Such systems are mainly focused on providing intelligent services to improve the students' life. For instance in [5], the authors proposed a course enrollment recommender system to recommend personalized courses with respect to each student's skills, interests and free time slots in his timetable. This system aims to help students to make decisions about courses to be selected during the enrollment process at the beginning of each semester. In [7], authors presented a location-based recommender system using mobile device applications. They generated recommendations of seminars to attend and events around the campus based on users' interest or research area and users' current position. In [8], authors developed and integrated a situation-aware recommender system to assist mobile users in the campus environment. The proposed system finds relevant and available resources (buildings, terminals and people) based on the user's current activity, position and profile. These recommendations are updated according to the dynamic evolution of the user situation over time. In [9], the time series information of borrowing books and book circulation times and the students' learning trajectories have been considered to generate book recommendations for students in the campus academic library. This system helps students to find relevant books quickly.

The major part of these researchers has focused on improving the academic life of students through enhancing the process of teaching and learning, helping students with searching for suitable courses, etc. But, they have overlooked an important component in the smart campus which is "the students' social life". In fact, a campus is a place of student life, work, trade and culture. They carry all the services dedicated not only to training, guidance, employability but also those which are part of the students' social life such as sporting, associative and cultural activities. It is necessary that the campus will be able to propose relevant social activities and new options that can fit the students' preferences and interests. Moreover, the above recommender systems support the individual recommendations which consist in recommending items to individual users. However, in many circumstances, these recommendations are intended to be consumed collectively (e.g. students want to attend a course, students want to participate in an activity, etc.). Hence, group recommender systems will be emphasized on.

Group recommender systems look for a consensus among users to support their different tastes and preferences. In literature, there are two main group recommendation strategies [3] including (i) preferences aggregation which consists in combining the group members' prior ratings into virtual user's profile and then generating recommendations and (ii) recommendations aggregation which consists in generating the members' individual recommendations using an individual recommendation method, then combining them to return a single recommendation list for the group. Nevertheless, these group recommendation strategies may only support the quantitative preferences of users which are generally expressed on a numerical scale. In fact, the quantitative votes given by users may be influenced by many factors, namely: user's mood when one user may react differently with the same item according to his mood, the limited scale when the user may give the same rating to two items that he appreciates differently since the scale of possible values is generally reduced, etc. Thus, numerical ratings given by users are not reliable and cannot represent the precise degree of users' liking [4]. Indeed, providing a preference order over items is easier than expressing numerical ratings over items. Few works [2,4] have studied the qualitative expression of users' preferences in the context of recommender systems using ranking aggregation techniques.

In what follows, we will propose a group recommender system which studies the students' preferences in order to capture their needs in term of activities and help the decision maker to choose the best alternative.

3 A New Group Recommender System for Social Activities in Smart Campus

By exploiting students interests provided as a set of preference orders over activities, we propose a group recommender system of social activities that can be used to provide the decision maker with the top relevant activities that the campus students might be interested in. Our idea is to generate an order of preference that reflects the group preference based on the students' ordinal rankings, then aggregate the group members' individual orderings to capture the relevance group ranking over all the activities and finally recommend activities appearing in the top positions.

Formally, let us consider a finite set of n activities $I = \{i_1, i_2, ..., i_n\}$, a group G composed of m students $U = \{u_1, u_2, ..., u_m\}$ such as each student u gives an order that expresses his preferences over the activities I. Then, the group ordering over the n activities, denoted by $>_G$, is obtained by mapping/combining the m individual orderings of all members of the group of students as follows:

$$>_G = RAP_{k=1}^{k=m} >_k \tag{1}$$

where RAP is the aggregation operator used to combine the m individual orderings into a unique group ordering. It is demonstrated that combining the individual orderings using rank aggregation methods addresses the problem of finding consensus ranking between items for the group [14]. In fact, different ranking aggregation methods can be used including Plurality-with-elimination, Borda rule and kemeny rule since they are commonly used in the recommendation context.

- *Plurality-with-elimination rule* [10] eliminates, in each iteration, the first item ranking by the fewest number of users, and continues with preferences of the remaining items. This is repeated until one item has a simple majority. The Plurality with Elimination Method satisfies the *Majority Criterion* and it uses the information contained in the preference schedule of all the candidates.
- *Borda rule* [11] works by assigning a score of $m - 1$ to the item in the first-ranked position, $m-2$ to the item in the second-ranked position, until reaching zero points for the last item. The item with the highest borda score is selected as a winner. Borda rule is monotone and it satisfies the *consistency criterion* (i.e. when the item which is the winner within the first group of voters and also within the second group of voters, is also the winner when all the votes are combined) and the *Plurality criterion* (i.e. if the number of voters ranking A as the first preference is greater than the number of voters ranking B, then candidate A should be the winner).
- *The kemeny rule* [12] works by minimizing disagreements and also maximizing agreements among the resulting order and the users' ranked lists. The order of items which corresponds to the higher kemeny score is selected as the best order for the whole group. Kemeny score is computed as follows:

$$score_k(>_G) = \sum_{j,k \in pairs} (n_{agg}(j, k, >_g) - n_{dis}(j, k, >_g)) \tag{2}$$

where $>_G$ is the group ordering, $n_{agg}(j, k, >_G)$ (resp. $n_{dis}(j, k, >_G)$) denotes the number of individual orderings that agree with j $>_G$ k (resp. disagree with j $>_G$ k). *The kemeny rule satisfies neutrality* (i.e. when the ranking aggregation rule is invariant to changing the identity of the items), *consistency*, and the *Condorcet criterion* (i.e. when one item is preferred in every one-to-one comparison with the other items). Nevertheless, the fact of computing kemeny score for the $m!$ possible orderings (where "m" is the number of items) is NP−hard.

Example 1. *Let us consider the following preference ordering given by the members of a group of 9 users for 3 items A, B and C: such that 3 say A > C > B, 3 say B > C > A, 2 say C > B > A and 1 says A > B > C.*

- *4 users have placed item A in the top of their preferences while three users have considered item B as their first choice and just 2 users considered item C as the better. So, Plurality-with-elimination drops C in the first iteration since it has the least number of the first-ranked position. By only considering the remaining items, 5 users consider item B in the first-ranked position, so item B has the majority of votes. The group ordering is then: $B >_g A >_g C$ and the top one recommendation for the group G is item B.*
- *Borda rule assigns scores of 8 (i.e. $(2*3)+(0*3)+(0*2)+(2*1)$) to A, 9 (i.e. $(0*3)+(2*3)+(1*2)+(1*1)$) to B and 10 (i.e. $(1*3)+(1*3)+(2*2)+(0*1)$) to C. Thus, the resulting preference order of the group is $C >_g B >_g A$. Based on the obtained group ordering, the top one recommendation for the group G is item C.*
- *The Kemeny rule computes the score for each possible group ranking of A, B and C. Kemeny rule selects group ranking $C >_G B >_G A$ (i.e. $score_k(C >_G B >_G A) = (5 − 4) + (5 − 4) + (5 − 4) = 3$) as the best order choice over the items for all the group members since it corresponds to the highest score. The top 1 recommendation for group G is then item C.*

Ranking aggregation methods work usually with full lists of items (i.e. the individual orderings contain the same items) and they may suffer from one or more weaknesses. For instance, *plurality-with-elimination* rule may suffer from the pairwise match-ups problem (i.e. when the rule fails to select an alternative C as a winner even though it defeats other alternatives in pairwise match-ups), *borda count* treats ordinal preferences as cardinal preferences and it often fails to elect the Condorcet winner and the problem is NP−hard since it needs to compute the kemeny score for the $m!$ possible orderings. To overcome these limits, we propose an average of positions method (denoted by Avg-Pos) to generate the consensus group ordering based on the ordinal preferences of students.

Typically, the basic preferences aggregation strategy computes the group preference of an item as the average sum of all the users' individual preferences.

Intuitively, the average of positions method consists in computing the position of each alternative 'k' in the group ordering as the average of its positions in all the individual orderings as follows:

$$Pos_{k,>_G} = \frac{\sum_{i=1}^{m} Pos_{k,>_i}}{|m|} \tag{3}$$

where m is the number of group members.

Clearly, the alternative having the minimum average of positions is declared as the winner in the group ordering. If there exist two alternatives (i.e. activities) that have the same average of positions, then the alternative which is ranked first by the majority of students (i.e. appearing in the first position of the most orderings) is selected as the winner in the whole group ordering. The basic steps of the proposed method are presented in Fig. 1.

Example 2. *Let us consider the following preference ordering given by the members of a group of 2 users for 4 items A, B, C and D: $A >_1 B >_1 C >_1 D$ and $D >_2 B >_2 A >_2 C$. After computing the average of positions of items A (i.e. $\frac{1+3}{2} = 2$), B (i.e. $\frac{2+2}{2} = 2$), C (i.e. $\frac{3+4}{2} = 3.5$) and D (i.e. $\frac{1+4}{2} = 2.5$), the average of positions method selects $A >_g B >_g D >_g C$ as the ordering corresponding to the group preference. 'A' is selected as the winner since it has the minimum average of positions value and it has the maximum number of the first positions.*

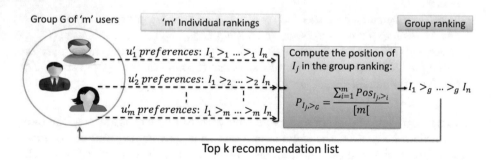

Fig. 1. Group recommendation using the average of positions

Proposition 1. *Avg-Pos method satisfies the Condorcet criterion (i.e. if the majority of the voters prefer alternative 'A' over every other alternative, then 'A' is selected as the winner) and the majority criterion (i.e. if a candidate receives the majority of the first places, then it should be declared as the winner).*

Proof. Suppose, to derive a contradiction, that K has a strict majority over j and j appears before k in the group ordering (j $>_G$ K) and a variable '*maj*' in $[1, m]$

- j $>_G$ K

$Pos_{j,>_G} < Pos_{k,>_G}$

$\frac{\sum_{i=1}^{m} Pos_{j,>_i}}{|m|} < \frac{\sum_{i=1}^{m} Pos_{k,>_i}}{|m|}$

$\sum_{i=1}^{m} Pos_{j,>_i} < \sum_{i=1}^{m} Pos_{k,>_i}$

$\sum_{i=1}^{maj} Pos_{j,>_i} + \sum_{i=maj}^{m} Pos_{j,>_i} < \sum_{i=1}^{maj} Pos_{k,>_i} + \sum_{i=maj}^{m} Pos_{k,>_i}$

$\sum_{i=1}^{maj} Pos_{j,>_i} - \sum_{i=1}^{maj} Pos_{k,>_i} < \sum_{i=maj}^{m} Pos_{k,>_i} - \sum_{i=maj}^{m} Pos_{j,>_i}$

$\Rightarrow \sum_{i=1}^{maj} Pos_{j,>_i} - Pos_{k,>_i} < \sum_{i=maj}^{m} Pos_{k,>_i} - Pos_{j,>_i}$

- K has a strict majority over j

$\sum_{i=1}^{maj} Pos_{k,>_i} < \sum_{i=1}^{maj} Pos_{j,>_i}$

$and \sum_{i=maj}^{m} Pos_{j,>_i} < \sum_{i=maj}^{m} Pos_{k,>_i}$

$\sum_{i=1}^{maj} Pos_{k,>_i} - \sum_{i=1}^{maj} Pos_{j,>_i} < 0$

$and \sum_{i=maj}^{m} Pos_{k,>_i} - \sum_{i=maj}^{m} Pos_{j,>_i} > 0$

$\Rightarrow \sum_{i=1}^{maj} -(Pos_{j,>_i} - Pos_{k,>_i}) < \sum_{i=maj}^{m} Pos_{k,>_i} - Pos_{j,>_i}$

So, it is contradictory and then, the average of positions method satisfies the *condorcet criterion* and consequently the *majority criterion.*   □

4   Experimental Study

In order to evaluate the performance of the generated recommendations, we propose to perform experiments on real-world data extracted from campus students. All the experiments are implemented in Java JDK 1.8 on windows 7 based PC with Intel Core $i3$ processor and compiled in eclipse framework.

4.1   Experimental Protocol

To the best of our knowledge, there is no available database that contains users' preferences under the form of orderings. For this reason, we prepared a questionnaire to obtain data from real students. The questionnaire was filled in by 100 students of the university of Lille1. Each student was asked to express an order of preference over 10 activities (e.g. organize group visits to museums or industrial sites, organize international representative matches against other campuses, organize dance party...etc).

We focus on performing an *offline evaluation* which consists in randomly selecting groups of students of different sizes (i.e. 40, 60, 80 and 100). We divide the group into two sets: the training set which will be used to learn the group ordering and the test set (i.e. 20 students) which will be used to evaluate the generated group recommendations. We run 5-fold cross-validation on data. We adapt the precision evaluation metric to be able to measure the performance of the generated group ordering based on the individual orderings of the members in the test set. The precision is defined in such a way that it may detect the average of the true positioned alternatives relative to the total number of the alternative in the group ordering.

$$Precision = \frac{TP}{TP + FP} \tag{4}$$

where TP is the true positioned items and FP is the false positioned items.

4.2   Experimental Results

The precision values of the group ordering generating using kemeny, borda, plurality-with-elimination and Avg-Pos method for different group sizes are shown in Fig. 2.

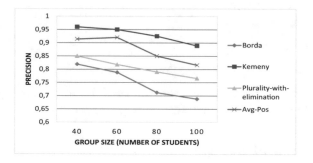

Fig. 2. The precision of the group recommendation for different group sizes (n = 10)

Figure 2 shows that kemeny is unsurprisingly more efficient than Borda and plurality-with-elimination for different group sizes because it satisfies neutrality and consistency as stated in [13]. Moreover, Kemeny provides a better ranking quality over alternatives compared to the Avg-Pos method. This can be explained by the fact that kemeny may take into account the pairwise disagreements and agreements between the group ordering and the individual orderings of the training set. We note that Avg-pos overcomes both Borda and plurality-with-elimination since it also satisfies the *Condorcet* criterion.

As a second experiment, we simulate students' data to generate 100 individual orderings of 20 activities. We propose to increase the number of activities 'n' which will be ordered by each student in order to study the sensitivity of the aggregation methods. Figure 3 shows the corresponding experimental results.

As expected, Fig. 3 shows that the accuracy of the group recommendation list generated by all methods decreases when the group size increases. In fact, finding a consensus order for a large size of a random group is more difficult

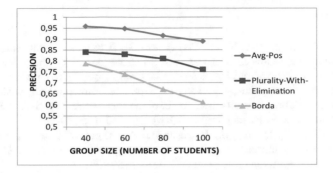

Fig. 3. The precision of the group recommendation for different group sizes (n = 20)

since the dispersion of users' preferences is more significant. Moreover, we can observe from Fig. 3 that Avg-Pos method has a significant improvement (i.e. a better precision) compared to ranking aggregation methods (i.e. Borda, kemeny and plurality-with-elimination) for different group sizes. In other words, Avg-Pos method provides the most accurate group recommendation list in term of ranking quality (i.e. it provides a better classification accuracy). This improvement is justified by the fact that Avg-Pos method satisfies both *Condorcet* criterion and *majority* criterion and it is able to support large number of activities (i.e. we can consider even the case where $n = 50$) contrarily to Kemeny rule which is not able to produce any group ordering when $n = 20$ because it is of O(n! mn log(n)) (i.e. it needs to compute the kemeny score for 20! possible orderings). The proposed Avg-Pos method is able to generate more interesting recommendations when increasing the number of activities presented in the individual orderings (i.e. precision when $n = 20$ is better than precision when $n = 10$).

5   Conclusion

This paper proposes a group recommender system that combines the individual preferences of students into a consensus ordering representing the group preference. The top k interesting activities are then selected for students to help the campus decision maker to satisfy the needs and the expectations of students. The group ordering is generated using the ranking aggregation methods and the average of positions method. Our experiment shows that kemeny rule produces the best group recommendation list when the database is limited to a small number of alternatives (i.e. not scalable). While Avg-Pos method performs better when there are many alternatives since it satisfies *Condorcet* criterion and *majority* criterion and it is not sensitive to the orderings length. As a future work, we propose to consider the individual ordering over items as partial to take into account some circumstances such as the incomparable items and the items which could not be ordered by the students.

References

1. Ben Abdrabbah, S., Ayachi, R., Ben Amor, N.: A dynamic community-based personalization for e-Government services. In: Proceedings of the 9th International Conference on Theory and Practice of Electronic Governance, pp. 258–265 (2016)
2. Baltrunas, L., Makcinskas, T., Ricci, F.: Group recommendations with rank aggregation and collaborative filtering. In: Proceedings of the Fourth ACM Conference on Recommender Systems (RecSys 2010), pp. 119–126 (2010)
3. De Pessemier, T., Dooms, S., Martens, L.: An improved data aggregation strategy for group recommendations. In: Proceedings of the 7th ACM Conference on Recommender Systems (RecSys 2013), pp. 36–39 (2013)
4. Brun, A., Hamad, A., Buet, O., Boyer, A.: Towards preference relations in recommender systems. In: Proceedings of the ECML/PKDD Workshop on Preference Learning (2010)
5. Bydzovská, H.: Course enrollment recommender system. In: Proceedings of the 9th International Conference on Educational DatanMining, EDM 2016, pp. 312–317 (2016)
6. Ray, S., Sharma, A.: A collaborative filtering based approach for recommending elective courses. In: Information Intelligence, Systems, Technology and Management, vol. 141, no. 1, pp. 330–339. Springer, Heidelberg (2011)
7. Hlaing, H.H., Ko, K.T.: Location-based recommender system for mobile devices on University Campus. In: Proceedings of 2015 International Conference on Future Computational Technologies (ICFCT 2015), Singapore, 29–30 March, pp. 204–210 (2015)
8. Bouzeghoub, A., Do, N.K., Krug Wives, L.: Situation-aware adaptive recommendation to assist mobile users in a campus environment. In: AINA 2009: The IEEE 23rd International Conference on Advanced Information Networking and Applications, pp. 503–509 (2009)
9. Zhang, F.: A personalized time-sequence-based book recommendation algorithm for digital libraries. IEEE Access. **4**, 2714–2720 (2016)
10. Cary, D.: Estimating the margin of victory for instant-runoff voting. In: Proceedings of the Conference on Electronic Voting Technology/Workshop on Trustworthy Elections (2011)
11. Saari, D.G.: The Borda dictionary. Soc. Choice Welfare **7**, 279–319 (1990)
12. Davenport, A., Kalagnanam, J.: A computational study of the Kemeny rule for preference aggregation. In: Proceedings of the 19th National Conference on Artificial Intelligence, pp. 697–702 (2004)
13. Young, H.P., Levenglick, A.: A consistent extension of condorcet's election principle. SIAM J. Appl. Math. **35**(2), 285–300 (1978)
14. Dwork, C., Kumar, R., Naor, M., Sivakumar, D.: Rank aggregation methods for the web. In: Proceedings of the 10th International Conference on World Wide Web, New York, NY, USA, pp. 613–622 (2001)

A Deep Learning Approach for Scientific Paper Semantic Ranking

Francesco Gargiulo, Stefano Silvestri[⊠], Mariarosaria Fontanella,
Mario Ciampi, and Giuseppe De Pietro

Institute for High Performance Computing and Networking, ICAR-CNR,
Via Pietro Castellino 111, 80131 Naples, Italy
{francesco.gargiulo,stefano.silvestri,mariarosaria.fontanella,
mario.ciampi,giuseppe.depietro}@icar.cnr.it

Abstract. In this paper we proposed a novel Deep Learning approach to realize a Word Embeddings (WEs) similarity based search tool, considering the medical literature as case study. Using the compositional properties of the WEs we defined a methodology to aggregate the information coming from each word to obtain a vector corresponding to the abstracts of each PubMed article. Through this paradigm it is possible to capture the semantic content of the papers and, consequently, to evaluate and rank the similarity among them. The preliminary results with the proposed approach are obtained analysing a subset of the whole the PubMed collection. The results correctness has been verified by human domain experts, showing that the methodology is promising.

Keywords: Deep Learning · Word Embeddings · Natural Language Processing · Information retrieval · Document similarity

1 Introduction

Plenty of digital scientific papers have been made available since the introduction of electronic publishing systems. Nowadays the scientific research community is supported by specific dedicated tools and, among them, the academic search engines earned a paramount importance for all researchers. An Academic Search Engine (ASE) is a document retrieval system used for finding and assessing articles in journals, repositories, databases, archives, or other collections of scientific papers. Actually many ASE are available, ranging from the general domain ones, like Google Scholar or Scopus, to the domain specific ones, like PubMed, which is devoted to Biomedical and Life Sciences scientific digital library, or EconLit, specialized in Economics science literature.

The main function of an ASE is to provide a list of papers as result of a query. As like every common search engine, the user writes a search query specifying some keywords, or a whole sentence related to the papers he is searching for. Finding the desired results is of crucial important for a researcher, for both developing and assessing his own work. However, with the growth in number

© Springer International Publishing AG 2018
G. De Pietro et al. (eds.), *Intelligent Interactive Multimedia Systems and Services 2017*,
Smart Innovation, Systems and Technologies 76, DOI 10.1007/978-3-319-59480-4_47

of digitalized articles, researchers might find it hard to find appropriate and necessary papers. A set of instruments for refining the search results are available in all ASE, but actually there is a lack of semantic-based option, a need for a human user to better describe the meaning of his requests.

Another problem related to search results is their ranking, considering that they actually are based on the occurrence of a search term in the title, in the keywords and in the full text of each article and even on the number of its citations [3]. It means that papers containing an high frequency of the same words used in the query have an high ranking in results. On the other hand, papers that have an high frequency of words semantically similar to the ones used in the search query will not appear in the highest position of the results, or could even never be found. It is very hard to express a concept through a sentence and to search for papers that have sentences semantically similar to the one used as query [2]. The automatic semantic analysis of a text is an Artificial Intelligence (AI) problem whose main challenge is the bridging of the semantic gap among contexts of papers.

To build an AI-based automatic scientific paper similarity assessment and ranking methodology, able to capture their semantic content and to search for semantically similar items, we propose a Deep Learning (DL) methodology based on Word Embeddings (WEs) [17] to represent each word involved in the titles and in the abstracts of whole PubMed archive with a real valued array. WEs are a neural net model that produces word vectors with deep learning via skip-gram and Continuous Bag Of Words (CBOW) models implemented in word2vec, using either hierarchical softmax or negative sampling methodologies [17,18]. WEs most interesting characteristic consists in the fact that mutual position of words in a metric space strongly depends on their meaning, so that words having similar semantics have high similarity. In addition to that, word vectors have the property of compositionality [17,21]: the sum of word vectors preserves the original properties of its composing words. Given the geometrical characteristics of WEs, it makes sense to compute for each paper the average of the corresponding WEs [12], or a normalized sum [11] of them. This new vector summarize the meaning of the paper, because it captures the semantic content of the whole group of words. So the distance between these vectors can be even used to assign a rank to paper similarity, based on cosine similarity.

Due to the dimension of PubMed, used as experimental assessment in this paper, or, more in general, of the whole digitally available scientific library, it is necessary to use Big Data Analytics (BDA) techniques, making possible to apply the proposed methodology to a huge amount of data.

The paper is structured as follows: Sect. 2 will be devoted to the discussion of related works; in Sect. 3 all the details about the approach implementation will be explained; Sect. 4 will describe the architecture of the system realized for the experimental assessment; Sect. 5 will show the preliminary qualitative results obtained and in Sect. 6 will be summarized the conclusions and some future works.

2   Related Works

The task of assessing and ranking the semantic similarity between research papers is of crucial interest in a variety of application and so it has attracted the attention of the scientific community. Many different approaches has been recently proposed to address this problem. A Deep Learning approach has been proposed by [23], where two distributional sentence models based on Convolutional Neural Networks (CNNs) work in parallel mapping documents formed by pieces of text to their distributional vectors, which are used to learn the semantic similarity between them. In [16] they described and verified several context-based models to compare research paper abstracts, based on Vector Space Models (VSMs) and language modelling. In the first case they built unigram features for each document, enriching the information using Explicit Semantic Analysis (ESA). In the latter case, they estimated unigram language models for each document and they calculated their divergence, identifying topic information by using Latent Dirichlet Allocation (LDA).

Focusing on WEs based techniques, in [8] the authors proposed a neural probabilistic model that jointly learns the semantic representations of citation contexts and cited papers, training a multi-layer neural network. They used WEs negative sampling method proposed in Skip-gram model [17] for word representation learning and Noise-contrastive estimation [7] for document representation learning. In [25] the authors proposed to model distributions of word vectors in documents or document classes. Their experiments confirmed the possibility of modelling documents in the semantic space starting from original WEs. The study of [12] aimed to overcome the problems of computing a vector assigned to a piece of text from the corresponding WEs vectors. They proposed Paragraph Vector method, an unsupervised framework that learns continuous distributed vector representations for whole pieces of texts of variable length, ranging from sentences to documents. This representation is trained to be useful for predicting words in a paragraph. In [19] they proposed a *Dual Embedding Space Model* to implement a document ranking methodology. Starting from a WEs, they mapped the query words into the input space and the document words into the output space, and then they computed a relevance score by aggregating the cosine similarities across all the query-document word pairs. This methodology proved that a document is about a query term, in addition to and complementing the traditional term frequency based approach. In [24] is described a methodology to combine ESA representations and WEs as a way to generate a denser representations for short texts, obtaining a better similarity measure between them. In [11] they represented text documents as a weighted point cloud of embedded words. Then they defined a new metric, called *Word Mover's Distance* (WMD), as the minimum distance between two text documents A and B as the distance that words from document A need to travel to match exactly the point cloud of document B. They used WMD to assess the semantic distance among texts. The authors of [13] proposed to obtain a text document vector by weighting the WEs corresponding to each word of a document with its Term Frequency-Inverse Document Frequency (TF-IDF). They obtained better performances in text classification using a linear Support Vector Machines if compared to a summation of vectors. In [5] they showed that Paragraph Vectors proposed in [12] are superior to LDA for measuring semantic similarity on Wikipedia articles. In [14] they presented a new feature set and a mechanism to calculate the scores related to short text similarity, proposing a slight variant of the WEs model to represent words. In [9] is described a methodology for determining short text similarity. A document is represented through the average of WEs related to the corresponding words, but the similarity, obtained by means of cosine similarity of two document vectors, is computed after discarding noisy information with the binning of some dimensions of the two vectors.

3 Methodology

In this work we present an innovative approach to realize a Deep Learning WEs similarity based search and ranking tool, using the PubMed articles as information source. The semantic content of a word or a sentence can be well captured by the neural net of WEs, in particular when the input training text is sufficiently large and it belongs to a specific domain [10]. It is possible to freely download all the abstract from the PubMed collection and use them to create our training data set for WEs. Through the deep learning neural network of word2vec is it possible to obtain word vectors. In the proposed methodology the training process can be divided into two different stages: (i) Word Embeddings Dictionary Creation (WEDC) and (ii) Document Vector Creation (DVC), explained in details in the following of this Section.

In Fig. 1 is given a graphical overview of the proposed methodology. Both WEDC and DVC stages share the *Text Extraction* and *Pre-processing* modules, in which the documents, formed by titles and abstracts text of each paper, are preliminary extracted using BDA techniques; then the text of documents is normalized, applying Tokenization, Stop Word Filter and Lemmatization processes, exploited with Natural Language Processing (NLP) and BDA tools. More in details, after extracting the title and the abstract of each paper indexed by PubMed, we normalize the text using NLP tools. These operations are necessary to reduce data sparsity, obtaining a less noisy training set for WEs.

In WEDC stage the normalized text is used to train the WEs model, mapping each word in an N-dimensional space. The main characteristic of this vector space is that words *semantically* closed each other are represented with two near vectors in terms of distance [17]. The WEs obtained are even compositional, so all vector properties remain true also considering the vector sum related to a multi-word expression or to a sentence, as showed in [1,12]. In other words, the vector obtained as the sum of vectors corresponding to the words of a multi-word expression, of a sentence, or even of a whole piece of text still preserves the property of semantic similarity between elements close in the space.

Using these properties, in DVC stage we build for each document a vector, averaging the WEs vector of its corresponding normalized words. More formally, in WEDC stage each word is mapped in an N-dimensional space

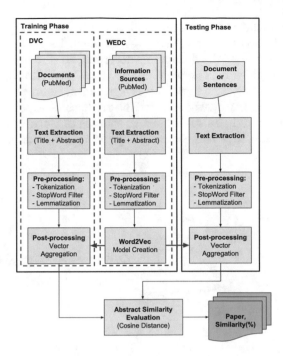

Fig. 1. The figure represents the conceptual schema of the proposed approach.

$wordVEC_i = d_{i,1}, d_{i,2}, \ldots, d_{i,N}$ where N is the number of components selected and $d_{i,j}$ is a real number referred to the jth component of the $wordVEC_i$. Then defining each document as a collection of words, $doc_j = word_1, word_2, \ldots, word_{M_j}$ where M_j is the number of words within doc_j, in DVC stage we evaluate the vectors associated to each document of the *Documents Base* as:

$$docVEC_j = \frac{1}{M_j} \cdot \sum_{i=1}^{M_j} d_{i,1}, \frac{1}{M_j} \cdot \sum_{i=1}^{M_j} d_{i,2}, \ldots, \frac{1}{M_j} \cdot \sum_{i=1}^{M_j} d_{i,N} \qquad (1)$$

In this way we obtain a mapping between the documents and the WEs space that preserves all semantic properties of original word vectors. So it possible to verify the semantic similarity between two $docVECs$ using the cosine similarity (Eq. 2). We can build the corresponding $docVEC$ for an input document, or even a manually inserted sentence, using the described methodology. Then we compute the cosine similarity with all other $docVECS$ obtained during the training phase, evaluating the semantic similarity between them. The cosine similarity ($Csim$) between two vectors $docVEC_i$ and $docVEC_j$ is evaluated according to the following equation:

$$Csim = cos(\theta) = \frac{docVEC_i \cdot docVEC_j}{\|docVEC_i\|\|docVEC_j\|} = \frac{\sum_{k=1}^{N} d_{i,k}d_{j,k}}{\sqrt{\sum_{k=1}^{N} d_{i,k}^2}\sqrt{\sum_{k=1}^{N} d_{j,k}^2}} \qquad (2)$$

Cosine similarity is defined in the range $[0, 1]$ and if $Csim$ is equal to 1 the maximum similarity between the two vectors is reached. Using this metric we rank the semantic similarity of the documents according to $Csim$. For seek of clarity in the Fig. 2 we graphically represent the $wordVEC$ and the $docVEC$, considering only two dimensions. The Figure gives a graphical interpretation of the main idea behind the proposed methodology. The analysis of experimental results in Sect. 5 show that building a $docVEC$ as an average of $wordVECs$ trained using the proposed methodology can correctly capture its semantic content.

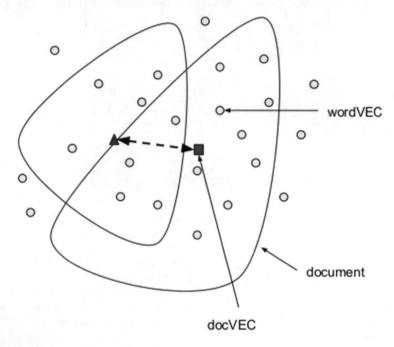

Fig. 2. The circles represents wordVECs and the blue and red sets collect the wordVEC of two generic documents. The red triangle and blue square represents respectively the docVECs associated to the red and blue sets.

4 Architecture

The architecture implemented for the experimental assessment of the above described methodology follows the conceptual schema depicted in Fig. 1. At the higher layer of the workflow there are the PubMed documents, needed for both DVC and WEDC training phases. To obtain them we used Entrez utility tools [22], a PubMed official free suite that provides an integrated search and retrieval system to access to all databases simultaneously with a single query string and user interface. Entrez can efficiently retrieve related sequences, structures, and references. Some textbooks are also available online through the Entrez system. In this way, we were able to download a structured xml file containing the whole collection of papers indexed in PubMed with the text of their titles and abstracts, besides various additional related information. At the moment of the download (the beginning of January 2017) more than $16,000,000$ of entries were available, obtaining a file bigger than $160\,$GB.

To select and process the information buried in this huge file, we need to apply BDA methods. A flexible and powerful solution to exploit BDA techniques is Hadoop [6], an open source framework that allows for the distributed processing of large data sets across clusters of computers using simple programming models. We mainly take advantage of its two major functionalities: (i) the distributed file system named *Hadoop Distributed File System* (HDFS) and (ii) the *MapReduce* framework, used to implements our algorithms in a distributed environment. In details, we considered an abstraction of the MapReduce framework known as *Apache Pig*, an open source technology that offers a high-level mechanism for the parallel programming of MapReduce jobs to be executed on Hadoop clusters. We used these frameworks to implement the Text Extraction module, selecting only English languages texts of the titles and the abstracts of all the collected papers, aggregating them with their corresponding IDs, obtaining automatically a structured NoSQL set. Even the subsequent Pre-processing module has been implemented using Pig, allowing to run as MapReduce tasks the tokenization, stop word filtering and the lemmatization Java modules of the Stanford NLP software tool [15]. The stop word list used is the one from Onix Text Retrieval Toolkit[1], reducing in this way the noise without deleting too much information.

The obtained normalized text is used to train the neural net of WEs after WEDC phase. We used word2vec deep learning algorithm [17] implemented in *Gensim* [20] package, a Python library for topic modelling, document indexing and similarity retrieval. In our preliminary experiments, following the results of [1,10], we use skip-gram model, which is more efficient when a large training data set is used [17], setting the window context window equals to 10 and the vector size equals to 500; we discarded from the vector vocabulary the words whose frequency is less than 3 and we set the sub-sampling parameter equals to 10^{-5}, improving accuracy and speed of the training. We performed 6 iterations. As shown in the next Sect. 5, these settings produced promising results for the WEs training. After the WEDC we perform the DVC phase, associating to each document a vector obtained averaging its corresponding word vectors. This task has been performed using a MapReduce task implemented through a Pig script. The same modules for Pre-processing and DVC are used even to obtain a vector associated to an input text (left part of Fig. 1), which can be either a document corresponding to a paper, or one or more sentences manually written by user. Finally, the computation of cosine similarity between input vector and the whole document vector space has been performed using KNIME environment [4]. KNIME provides a graphical user interface that allows to assembly nodes for data preprocessing (ETL: Extraction, Transformation, Loading) for modelling and data analysis and visualization. Giving in input the obtained vector spaces, we realized an interactive system that makes it possible the visu-

---

[1] http://www.lextek.com/manuals/onix/stopwords1.html.

alization of the documents and their similarity scores, used for the qualitative assessment described in the next Sect. 5.

5   Preliminary Experimental Results

The preliminary experiments, used to evaluate the effectiveness of the proposed approach, have been performed on two subsets of the whole PubMed Dataset. Due to the huge size of the whole PubMed dump (more than 160 GB), we decided to perform a preliminary evaluation and assessment phase on these subsets. In details, we trained the WEs model using a dataset composed by 364.589 abstract papers (36.590.774 words), sufficiently large to obtain a correct VSM with a dictionary composed by 255.671 vectors. We selected a subset of this reduced dataset, formed by 10.939 papers, to create the corresponding document vectors. To assess the correctness of our methodology we performed two different kinds of test, considering a qualitative evaluation due to the lack of gold standard for this task. In the first case we search for the semantic similar papers, giving in input a target paper. The result example depicted in Fig. 3 shows the titles and the ranking of the first six papers found in one of these tests. It is worth noting that the cosine similarity is equal to 1 for the paper itself and the titles (the abstract text has been omitted in the Figure) show a set of keyword semantically related to the input, suggesting a good result. Two medical domain experts supported us for testing phase: after the analysis of the abstract obtained with our system they confirmed the correctness of the results and they found in many cases different papers from the ones obtained with the official PubMed search engine.

The other test (an example of results is in Fig. 4) has been performed giving as input a generic sentence written by domain expert users, instead of an abstract of a paper. Also in this case the papers found by our system show clearly semantic similarity with the input sentences in all tests, as confirmed by domain experts. In this case it is important to underline that even in this preliminary implementation the system helped the experts to find useful results for their own work, not found with PubMed standard tools. Definitely all the experiments confirm that the proposed approach is already really promising.

S title	D ▼ simil
Venous thromboembolism in outpatients with cancer receiving chemotherapy for solid tumors.	1
Venous thromboembolism (VTE) in patients with advanced gastric cancer: An Asian experience.	0.946
Role of hypoalbuminemia in the development of venous thromboembolism in cancer chemotherapy patients.	0.945
Role of personal history in the development of recurrent venous thromboembolism during cancer chemotherapy.	0.943
Incidence and predictors of venous thromboembolism (VTE) among ambulatory patients with lung cancer.	0.938
Incidence and risk factors predictive of recurrent venous thromboembolism in patients with cancer at a comprehensive canc...	0.929

Fig. 3. Top six similar papers. The first row, with grey background, is the input. The first column is the title and the second contains the similarity.

S title	D ▼ simi
Muscle interleukin 6 and fasting-induced PDH regulation in mouse skeletal muscle.	0.77
Diminished force production and mitochondrial respiratory deficits are strain-dependent myopathies of subacute limb ischemia.	0.758
Sleep-Disordered Breathing Exacerbates Muscle Vasoconstriction and Sympathetic Neural Activation in Patients with Systolic...	0.71
Smooth Muscle Endothelin B Receptors Regulate Blood Pressure but Not Vascular Function or Neointimal Remodeling.	0.71
Characterizing the sympatholytic role of endothelial-dependent vasodilator signalling during handgrip exercise.	0.699
Inorganic Nitrate Mimics Exercise-Stimulated Muscular Fiber-type Switching and Myokine and GABA Release.	0.681

Fig. 4. Top six similar papers, considering as input the sentence: *heart vs. skeletal muscle differential expression.*

6   Conclusion and Future Works

In this paper we proposed an approach to realize a WEs semantic similarity based search tool for PubMed articles. We used the compositional properties of WEs to profitably aggregate the information coming from each word, obtaining a vector representation of each document. We decided to verify the main idea on a medical context using PubMed papers collection. At this aim, given the huge amount of data involved, we realized a framework based on the Hadoop infrastructure. The Deep Learning approach used to obtain WEs word vector is the word2vec model, inherited by the Python package *Gensim*. The Similarity among papers abstract from PubMed collection was computed using the cosine similarity and the final output and the visualization of the results, for a qualitative assessment, was made using KNIME environment.

The preliminary experiments have been performed on a subset of the whole PubMed collection. The obtained results have been qualitatively assessed by domain experts, that used the system for their own purposes. All the tests confirmed that the proposed methodology is promising, because the systems always returned semantic similar documents to sentences or to the document in input. More important, the experts confirmed that the results are useful for their work, obtaining papers different from the ones outputted by PubMed tools. As future works we are planning to extend the analysis on the whole PubMed archive (160 GB) and to use, when available, as information source even the full text of paper instead of the title and abstract. We plan to analyse and experiment other Deep-Learning techniques such as *Paragraph Vector*, *doc2vec* or other document vector extraction methodologies, integrating them in a complete pipeline.

References

1. Alicante, A., Corazza, A., Isgrò, F., Silvestri, S.: Semantic cluster labeling for medical relations. Innov. Med. Healthcare **2016**(60), 183–193 (2016)
2. Amato, F., Gargiulo, F., Mazzeo, A., Romano, S., Sansone, C.: Combining syntactic and semantic vector space models in the health domain by using a clustering ensemble. In: Proceedings of the International Conference on Health Informatics, pp. 382–385 (2013)
3. Beel, J., Gipp, B.: Google scholar's ranking algorithm: an introductory overview. In: Proceedings of the 12th International Conference on Scientometrics and Informetrics, vol. 1, pp. 230–241 (2009)

4. Berthold, M.R., Cebron, N., Dill, F., Gabriel, T.R., Kötter, T., Meinl, T., Ohl, P., Sieb, C., Thiel, K., Wiswedel, B.: KNIME: the Konstanz information miner. In: Studies in Classification, Data Analysis, and Knowledge Organization (GfKL 2007). Springer (2007)

5. Dai, A.M., Olah, C., Le, Q.V.: Document embedding with paragraph vectors. arXiv preprint arXiv:1507.07998 (2015)

6. Ghazi, M.R., Gangodkar, D.: Hadoop, MapReduce and HDFS: a developers perspective. Procedia Comput. Sci. **48**, 45–50 (2015)

7. Gutmann, M., Hyvärinen, A.: Noise-contrastive estimation: a new estimation principle for unnormalized statistical models. J. Mach. Learn. Res. **13**, 307–361 (2012)

8. Huang, W., Wu, Z., Chen, L., Mitra, P., Giles, C.L.: A neural probabilistic model for context based citation recommendation. In: AAAI, pp. 2404–2410 (2015)

9. Kenter, T., de Rijke, M.: Short text similarity with word embeddings. In: Proceedings of the 24th ACM International on Conference on Information and Knowledge Management, pp. 1411–1420. ACM (2015)

10. Krebs, A., Paperno, D.: When hyperparameters help: beneficial parameter combinations in distributional semantic models. In: Proceedings of the Fifth Joint Conference on Lexical and Computational Semantics (*SEM 2016), pp. 97–101 (2016)

11. Kusner, M.J., Sun, Y., Kolkin, N.I., Weinberger, K.Q., et al.: From word embeddings to document distances. ICML **15**, 957–966 (2015)

12. Le, Q.V., Mikolov, T.: Distributed representations of sentences and documents. ICML **14**, 1188–1196 (2014)

13. Lilleberg, J., Zhu, Y., Zhang, Y.: Support vector machines and word2vec for text classification with semantic features. In: 14th International Conference on Cognitive Informatics and Cognitive Computing, pp. 136–140. IEEE (2015)

14. Ma, W., Suel, T.: Structural sentence similarity estimation for short texts. In: FLAIRS Conference, pp. 232–237 (2016)

15. Manning, C.D., Surdeanu, M., Bauer, J., Finkel, J., Bethard, S.J., McClosky, D.: The Stanford CoreNLP natural language processing toolkit. In: Association for Computational Linguistics (ACL) System Demonstrations, pp. 55–60 (2014)

16. Martín, G.H., Schockaert, S., Cornelis, C., Naessens, H.: Using semi-structured data for assessing research paper similarity. Inf. Sci. **221**, 245–261 (2013)

17. Mikolov, T., Sutskever, I., Chen, K., Corrado, G.S., Dean, J.: Distributed representations of words and phrases and their compositionality. In: Advances in Neural Information Processing Systems, pp. 3111–3119 (2013)

18. Mikolov, T., Sutskever, I., Chen, K., Corrado, G.S., Dean, J.: Distributed representations of words and phrases and their compositionality. In: Proceedings of 27th Annual Conference on Neural Information Processing Systems 2013, pp. 3111–3119 (2013)

19. Nalisnick, E., Mitra, B., Craswell, N., Caruana, R.: Improving document ranking with dual word embeddings. In: Proceedings of the 25th International Conference Companion on World Wide Web, pp. 83–84 (2016)

20. Řehůřek, R., Sojka, P.: Software framework for topic modelling with large corpora. In: Proceedings of the LREC 2010 Workshop on New Challenges for NLP Frameworks, pp. 45–50. ELRA, May 2010

21. Salehi, B., Cook, P., Baldwin, T.: A word embedding approach to predicting the compositionality of multiword expressions. In: HLT-NAACL, pp. 977–983 (2015)

22. Sayers, E., Miller, V.: Entrez programming utilities help [internet]. The E-utilities in-depth: parameters, syntax and more (2014)

23. Severyn, A., Moschitti, A.: Learning to rank short text pairs with convolutional deep neural networks. In: Proceedings of the 38th International ACM SIGIR Conference on Research and Development in Information Retrieval, pp. 373–382. ACM (2015)
24. Song, Y., Roth, D.: Unsupervised sparse vector densification for short text similarity. In: HLT-NAACL, pp. 1275–1280 (2015)
25. Xing, C., Wang, D., Zhang, X., Liu, C.: Document classification with distributions of word vectors. In: Annual Summit and Conference on Asia-Pacific Signal and Information Processing Association, pp. 1–5. IEEE (2014)

neOCampus: A Demonstrator of Connected, Innovative, Intelligent and Sustainable Campus

Marie-Pierre Gleizes[1(✉)], Jérémy Boes[1], Bérangère Lartigue[2], and François Thiébolt[1]

[1] IRIT – University of Toulouse, Toulouse, France
`{gleizes,boes,thiebolt}@irit.fr`
[2] PHASE - University of Toulouse, Toulouse, France
`Berangere.Lartigue@univ-tlse3.fr`

Abstract. The progress in communication technologies and data storage capacity has brought tremendous changes in our daily life and also in the city where we live. The city becomes smart and has to integrate innovative applications, and technology to improve the quality of life of its citizens. To experiment and estimate these innovations, we show that a university campus is a ground "in vivo" experiments adequate. The infrastructure deployed in the Toulouse III Paul Sabatier University campus, enabling its transformation in a smart and sustainable campus, is detailed. The approach to answer the main challenges a smart campus has to deal with, is described. Some challenges are illustrated with a case study on the balance between the energy efficiency and the comfort in a class.

Keywords: IoT · Ambient systems · "In vivo" testing grounds · Self-adaptive · Self-organizing systems

1 Introduction

The progress in communication technologies and data storage capacity has brought tremendous changes in our daily life. Currently, the connections between people and machines have their speed considerably increased. The IoTs (Internet of Things) with all its connected objects and the mobile smartphones, provides a huge amount of heterogeneous data. The cloud and the farms of servers enable to store always more and more data; currently we speak about trillion gigabytes of data. All these advanced technologies, pervasive networks, are transforming our urban areas in **smart cities.** The possibility to get data from our environment thanks to communications and to analyse them enable better management of the resources and better services to citizens. A promising consequence of the access and the division of the data is to inform better, and thus, to involve the citizens as the **actors** of their city in a transparency way. Some examples are providing feedback on the quality of services, adopting a more sustainable and healthy lifestyle, or participating to different initiatives with other citizens.

Nevertheless, innovative technologies and services must be tested and evaluated before their use application in the city daily life. As it will be motivated in the next section, a university campus is the right scale and the right location to experiment and

© Springer International Publishing AG 2018
G. De Pietro et al. (eds.), *Intelligent Interactive Multimedia Systems and Services 2017*,
Smart Innovation, Systems and Technologies 76, DOI 10.1007/978-3-319-59480-4_48

evaluate them before a complete deployment of innovations in a city. Our aim is to transform our university campus in Toulouse as a large scale and in vivo testing ground. This implies the participation of the university community (students, faculty, and staffs) and a well as companies in original partnerships. It has been conceived and is developed with the strong willingness to share materials, software and results for research, education and projects with companies. The ongoing design of our new campus, called neOCampus, follows some guidance in order to answer the main challenges of smart cities. This new view of the campus as a smart campus can, easily, be replicated in other existing campuses.

In this paper, Sect. 2 provides definitions of smart cities from literature and explains why a campus can be adequate to evaluate experiments. Section 3 defines our vision of a smart campus and describes the smart campus neOCampus. Section 4 goes deeper with the main challenges that arises when dealing with smart cities and explains the approach followed in neOCampus to tackle them. Section 5 presents with a case study, before we conclude in Sect. 6.

2   From Smart City to Smart Campus

Being a multi-disciplinary subject, there is not an accurate and consensual definition of what a smart city is. In general, the definition is related to a specific dimension of the city (like "Intelligent City", "Creative City", etc.) [1]. Most of the time, the use of information and communication technologies is an important part of the design of a smart city. Harrison and all [1] consider that three characteristics are common to all smart cities: instrumented, interconnected and intelligent. Instrumented means that the city can use a large set of data-acquisition systems in order to produce data from the real world. Interconnected means that these data can be used across different services and users in the city, and finally, intelligent emphasizes the ability to analyses and use efficiently these data. The ITU-T FG-SSC (International Telecommunication Union Focus Group on Smart Sustainable Cities) [3] provides a definition based on the analysis of around one hundred definitions. They define a **smart sustainable city** as "*an innovative city that uses Information and Communication Technologies (ICTs) and other means to improve quality of life, efficiency of urban operation and services, and competitiveness, while ensuring that it meets the needs of present and future generations with respect to economic, social and environmental aspects.*" A city is an eminently complex system [1, 2] because it is characterized by multiple dynamics, non-linear relation, feedback, unpredictability, strong inertia… where the non-linearity leads to the impossibility to plan all the consequences of a change, even small.

The concept sends back to a type of urban development capable of answering the evolution or the emergence of the needs for institutions, companies and citizens, both on the plan economical, social, political, legal, cultural and environmental level. Therefore, the smart city **is not static**, it is alive, evolves with citizens, with policies, with its environment, and must be resilient to answer quickly to new challenges.

The services to be provided to institutions, companies and citizens are based on smart technology and infrastructure. They concern several domains [5]: health, citizens, governance and education, energy, buildings, mobility.

These domains must design solutions in a **different way** comparing with traditional ones, which have developed solutions in **a "siloed" fashion**. The services and the data must be shared to allow a mutual enrichment and to promote innovations. For example the data on the water consumption used for the management of an intelligent building can also inform about activity of an individual.

A university campus is generally spread on several hectares, with a lot of buildings, lawns, and a roads infrastructure for different kinds of vehicles (buses, cars, bicycles…). The buildings are dedicated to different activities: research, education, sports, administration, energy plants, and habitations… In general, a university is well equipped with communication networks, with servers and storage capacity. It is frequented by several types of end-users: students, faculty, staff, and visitors, who work, live and move on the campus. These activities consume energy, water and produce wastes.

Due to their size, users and mixed activities, university campuses can be considered as districts or small cities. The two main motivations to experiment innovative services and technics in the context of smart cities in a campus are based on its community. Firstly, the university community is usually aware about innovations. Consequently, it is easier to form evaluation groups, to experiment the use of new cyber-physical systems, or new services, and to involve people in experiments. Then, the university community is able to design these new services and technics and therefore to design the campus of the future.

3   neOCampus: A Smart Campus

The first question to ask about smart campus is: what is a smart campus, and the second question is how to create and develop it for impacting positively its users? This section gives our point of view on these two questions and then presents neOCampus.

3.1   Definition of a Smart Campus

A **smart campus** is an existing University campus with its end-users, having high quality infrastructures for communication and data storage and numerous software and material devices interconnected (IoT). The smart campus is a digital campus, sustainable and intelligent campus, allying innovative teaching equipment, sensors, and systems of communication, storage, location, simulation. Innovative materials used in the university buildings and outside can increase the quality of life of the users and reduce the consumptions of resources. The fact to be smart is not a static state, and it means that the campus must continuously evolve and self-adapt to its environment: the end-users, the technologies, the transportation, the city…. in being sustainable.

In a smart campus, the end-users must be involved, they are actors to reach the objectives. Researchers and students have a main role in the transformation of the campus. The smart campus equipment will be based on marketed materials and it will be enriched by the last researches stemming from the research laboratories, favouring a short circuit between the research and the potential applications. The operation of research done in the laboratories of the university, supported by the equipment has for objective to design products and services associated to the ambient socio-technical systems.

A smart campus must be qualified as connected, innovative, sustainable and self-adaptive.

The main constraints to create a smart campus are in one hand to deal with the existing buildings, existing infrastructure and functioning, and on the other hand to not be able to predict of what will be made the campus of the future in the 2030 s. Therefore, it must be incremental, open and deals with interoperability. **Incremental** in the sense where a device (material or software) can be added without questioning what already exists. It is considered as **open** because it will remain operational while integrating new products of the digital technology of the market and the products stemming from research. The campus will be permanently innovative because the innovations will be added as they will exist. The **interoperability** means that the different elements of the system can communicate and interact. The campus will favour open source software and open hardware.

The operation neOCampus (www.irit.fr/neocampus) started in June, 2013. The teachers-researchers and the researchers of 11 laboratories of the UPS: CESBIO, CIRIMAT, ECOLAB, IRIT, LA, LAAS, LAPLACE, LCC, LERASS, LMDC, PHASE intervene in neOCampus. These laboratories aim to cross their skills to improve the comfort to the everyday life for the university community while decreasing the ecological footprint of buildings and by reducing the costs of functioning (fluid, water, electricity…).

The Toulouse III Paul Sabatier University, within the framework of the operation neOCampus, which aims at building the campus of the future at the level services and quality of life is a platform for innovative experiments done at large scale and "in vivo" (with real end-users, in real situations) and welcome partnerships with companies. The heart of these collaborations is a research and innovation project. In this context, companies can benefit of the neOCampus platform, the novelty is that after the project the materials and software stay in the campus to upgrade it and will be used in other projects. Of course, juridical convention are signed to guarantee the intellectual properties.

The campus of Paul Sabatier University (Fig. 1) extends over around 150 ha and contains more than 407 000 m^2 of buildings, with 36 000 users who frequent it daily. Inside the campus, several solutions of mobility exist: pedestrian, bicycles, vehicles, buses and subway. The services provided in the different buildings are: administration, education, research, catering, housing, a market, a kinder garden. All the activities on the campus consume 140 GWh a year and produce 23 250 tons of CO_2 (diagnosis made in 2010).

Fig. 1. Toulouse III Paul Sabatier University Campus

3.2 Architecture

Toulouse III Paul Sabatier University has the advantage to have the appropriate management of a large part of its infrastructures. The university already has a good quality digital communication network on a large number of its buildings and main data storage capacity. Several classrooms are equipped with various sensors and actuators. Data from those sensors are geared toward the neOCampus main server through the neOCampus network, an IoT dedicated network that spreads over almost all buildings of the campus. In addition, neOCampus also benefits from the CloudMIP platform (http:// cloudmip.univ-tlse3.fr/), a powerful OpenStack platform that is part of the France-Grilles federated cloud initiative. On its own, CloudMIP features 280 physical cores, 1.6 TB RAM and enables up to 45 TB of persistent data.

From an infrastructure point of view, the neOCampus platform features three major equipments:

- a 16 cores, 64 GB RAM and 2.7 TB of high speed storage main server,
- a brand new 45 TB storage server shared with the CloudMIP platform,
- a both wired and wireless dedicated IoT network.

The main server holds the MQTT broker that enables all of the sensors/actuators/ ambient applications to get connected together. The sensors/ actuators get tied to the IoT network through devices like Raspberry Pi and ESP8266. These devices get authenticated through a powerful mechanism on the main server involving the sensOCampus application (http://sensocampus.univ-tlse3.fr) and a database.

In addition, a wireless LoRaWAN network is on way to get deployed at the scale of the campus. It will enables LoRa devices to send back data from the sensors in a range >2 km. Data will then get routed to the neOCampus main server.

It is to mention that a lot of time has been dedicated to the ability for the devices to get either fully reinstalled or updated through the network in an automatic way. This is part of the sensOCampus features.

The sensors deployed in the classrooms enable to collect ambient parameters (temperature, CO_2, hygrometry, luminosity, presence, sound), in addition to energy-meters (lighting system and power-plugs), thermal camera, cameras, and air flow meters. Actuators are able to control the lights, the blinds and some displays.

One of the requirements was the ability to either connect digital sensors (i2c bus) but also industrial and building sensors featuring an analog output (0–10 v): this aim was reached through a home-made open-source PLC, the neOCampus concentrators http://neocampus.univ-tlse3.fr/projets/concentrator.

4 Challenges for Smart Campus

In this section, the different challenges to design a smart campus are described. Designing a smart campus asks to tackle some challenges: social, economical, juridical and technological. We focus on challenges to reach the smart city/campus objectives with the digital and technical point of view. As previously said, the smart campus is a complex systems and by consequence, all applications/services/systems are plunged into a complex context. This complexity leads the designed systems to cope with the following challenges: the large scale and maintenance, the non-linearity, the openness, the heterogeneity, the cognitive overload, the privacy of data and the unpredictable dynamics.

4.1 Large-Scale and Maintenance

Smart campuses are deemed to generate huge masses of data. For instance, an energy management of all the buildings of the campus has to collect an amount of data from various and numerous sensors (temperature, electricity, heating, water, lights…) in each rooms/buildings. A centralized management system seems to be not relevant to take quickly a decision. In neOCampus, the solution proposed is based on the distribution of the control, a bottom up approach and has to rely on autonomous unit (agent) taking local decisions. The control is distributed inside the devices of the system. To maintain all these devices, it is also needed to provide self-healing property to the devices and to use collective behaviour between the devices for detecting failures.

4.2 Non-linearity

When a small change in the input of a system may result in a big change in its output, the system is said non-linear. Controlling such a system is a difficult task. Unfortunately, voltage in smart grids, heating in buildings and housing, traffic, and many smart cities/campuses systems we seek to control are non-linear. In a linear system, the distribution of data is such that it can be exactly abstracted using only simple mathematical functions. On the contrary, machine-learning algorithms have to be sophisticated and fine-tuned to be able to learn non-linear patterns and perform with contextual data from smart campus [6].

4.3 Openness

A system is said open when its components may dynamically enter and exit the system. With the IoTs, smart campuses are inherently open [8]. New devices, new sources of contextual data, are continually added, and some old ones are deleted, or suffer fatal failures. An intelligent algorithm designed to handle contextual data in a smart campus/ city should be able to easily incorporate new sources and delete old ones. For instance, sensors measuring the luminosity are added in different locations in an instrumented classroom to get more detailed information. They add their data to the management system, which had learnt the optimal behaviours to balance energy consumption and comfort. Now, it has to incorporate this new data source to its decision making process in an autonomous way. Otherwise, administrators would have to reset and restart the whole learning process. It would be very time costly. Such openness can lead difficulty for many types of machine-learning algorithms, particularly artificial neural networks and evolutionary algorithms.

4.4 Heterogeneity

Contextual data in smart campuses include a large variety of data types, whether they are numerical or not, continuous or discrete, multidimensional or not. Not all algorithms are able to deal with such heterogeneity. While numerical data are usually easily handled by current algorithms, modelling and treating ontologies and abstract concepts is a whole field of research. If some algorithms work very well with continuous numerical data, they are utterly unable to deal with data like the colour of an object, the smell of an animal, or more abstract concepts like the emotion of people. Sophisticated methods have to be employed to process together heterogeneous data in a single algorithm. The heterogeneity appears also when several devices designed by different companies or laboratories have to communicate or to interact. In neOCampus, we promote the use of open sources software and open sources hardware to facilitate the implementation of communications.

4.5 Cognitive Overload

The end-user on the campus can employ a lot of connected objects and can rapidly be overloaded. It takes a lot of time before obtaining satisfactory conditions. For example, before each class the teacher has to adjust a lot of devices: videos, tablet computers, lights, blinds, heatings…Firstly, he has to learn a lot of users' guides, and then he has to spend time to control all these elements before his class. It is the reason why in neOCampus, we want to experiment providing autonomy for all devices in the system. This means that each device has the capacity to perceive, take a decision and act as an agent in a multi-agent system. The price of Arduino or Rasberry Pi enables this today.

4.6 Privacy

Smart campuses are able to collect and gather large amounts of information, and this could harm the privacy of citizens [9]. Yet, whatever the services brought by new connected technologies, nobody should have to give up its privacy. Data should not be used outside of what, the considered services need. In neOCampus, a good alternative to enforce this would be for the data to be processed physically close to its source, and not being transferred and stored when it is not necessary, for example by leaving the execution of the processing of a scene on the video camera.

4.7 Unpredictable Dynamics

Due to the changes in the weather, attendance levels, traffic, users' preferences, and so on, the smart campus has to deal with dynamics. Current algorithms can easily find predictive relationships in smooth linear and complete data set. However, in the real world, the way data change is difficult to handle, and often unpredictable. The dynamics you have learnt at a given time may not be true sometimes later. This stems from several factors such as non-linearity, partial perception (there is not always a sensor for every relevant data), unreliable or missing data, failures, openness, and so on. In neOCampus, we develop new machine-learning algorithms to perpetually self-adapt to the ever-changing data. These algorithms are based on self-adaptive multi-agent systems [6, 10].

5   In Vivo Experimentation: The Case Study consOCampus

The aim of the consOCampus application is to balance energy consumption and comfort regarding the luminosity in a classroom. The classroom is equipped with several sensors: luminosity, temperature, presence detection (through video camera and local treatment) and several effectors acting on the lights and the blinds. Each sensor and effector is embedded on a Rasberry Pi and agentified, and are called agents (Fig. 2).

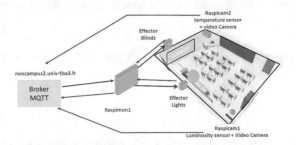

Fig. 2. consOCampus in a classroom

In this version, the agents do not communicate directly but use the MQTT broker. One agent subscribes to one or several agents and by consequence when an agent sends a message all subscriber agents receive it.

Each agent is autonomous and is able to make a decision in cooperating with the other. Therefore, in this ambient application, the devices are autonomous and have to decide alone what they have to do. This is very different from the pre-established scenarios provided by commercialized applications.

The blinds and the lights interact to be able to decide which device has to act. The blinds have to decide if they have to open or shut down and the lights have to decide if they must be on or off. In function of the temperature, the presence of people in the room and the luminosity, the blinds and the lights have not to act all together: open the blinds and switch off the lights. The behaviour of these agents is based on a cooperative behaviour developed in the Adaptive Multi-Agent Systems theory [11].

We have launched different experiments where a human acts to modify the luminosity in the room in hiding or lighting the luminosity sensor. We have observed that the blinds always open the first and then if it is not sufficient the lights switch on. The next version of this experiment is to enable the system to learn the right behaviour in observing the human activity [12].

6   Conclusion

Today, the cities are in a mutation to become smart in using ICTs and IoTs. In this paper, we motivated the fact that a University campus is a small city where innovative applications and technology can be experimented and evaluated "in vivo" and at large scale. The campus is a great ground of in vivo experiments and large scale. To be smart and sustainable, it has to be built to deal with the following main challenges: the large scale and the maintenance, the non-linearity, the openness, the heterogeneity, the cognitive overload, the privacy of data and the unpredictable dynamics. The campus of Toulouse III Paul Sabatier University is currently built to take into account these challenges. All devices are agentified and are able to perceive information in their environment, to autonomously decide and to act. neOCampus is incrementally built and includes the last results of research from the laboratories of the university. It is also accessible to companies to enable them to experiment and evaluate their innovation in cooperation with researchers.

Acknowledgements.   neOCampus is funded by Paul Sabatier Toulouse III University and by the INS2I CNRS organism.

References

1. Nam, T., Pardo, T.A.: Conceptualizing smart city with dimensions of technology, people, and institutions. In: Proceedings of the 12th Annual International Digital Government Research Conference: Digital Government Innovation in Challenging Times, pp. 282–291. ACM (2011)
2. Harrison, C., Eckman, B., Hamilton, R., Hartswick, P., Kalagnanam, J., Paraszczak, J., Williams, P.: Foundations for smarter cities. IBM J. Res. Dev. **54**(4), 1–16 (2010)

3. Bai, X., McAllister, R.R.J., Beaty, R.M., Taylor, B.: Urban policy and governance in a global environment: complex systems, scale mismatches and public participation. Curr. Opin. Environ. Sustain. **2**(3), 129–135 (2010)
4. ITU-T Focus Group on Smart Sustainable Cities, An overview of smart sustainable cities and the role of information and communication technologies, Focus Group Technical report 10/2014
5. http://www.slideshare.net/FrostandSullivan/smart-cities-from-concept-to-reality
6. Boes, J., Nigon, J., Verstaevel, N., Gleizes, M.-P., Migeon, F.: The self-adaptive context learning pattern: overview and proposal (regular paper). Dans: International and Interdisciplinary Conference on Modeling and Using Context (CONTEXT 2015), Larnaca, Cyprus, 02/11/15–06/11/15. LNAI, vol. 9405, pp. 91–104. Springer 2015
7. Vlahogianni, E.I., Kepaptsoglou, K., Tsetsos, V., Karlaftis, M.G.: A real-time parking prediction system for smart cities. J. Intell. Transp. Syst. **20**(2), 192–204 (2016)
8. Jin, Jiong, Gubbi, Jayavardhana, Marusic, Slaven, Palaniswami, Marimuthu: An information framework for creating a smart city through internet of things. IEEE Internet Things J. **1**(2), 112–121 (2014)
9. Martinez-Balleste, A., Perez-Martinez, P.A., Solanas, A.: The pursuit of citizens' privacy: a privacy-aware smart city is possible. IEEE Commun. Mag. **51**(6), 136–141 (2013)
10. Gleizes, M.-P.: Self-adaptive complex systems. In: Cossentino, M., Kaisers, M., Tuyls, K., Weiss, G. (eds.) European Workshop on Multi-Agent Systems (EUMAS 2011), Maastricht, The Netherlands, 13/11/2011–16/11/2011, vol. 7541, pp. 114–128. Springer 2012
11. Georgé, J.-P., Gleizes, M.-P., Camps, V.: Cooperation. In: Di Marzo Serugendo, G., Gleizes, M.-P., Karageogos, A., (eds.) Self-organising Software, Natural Computing Series, pp. 7–32. Springer, Heidelberg (2011)
12. Nigon, J., Glize, E., Dupas, D., Crasnier, F., Boes, J.: Use cases of pervasive artificial intelligence for smart cities challenges, In: IEEE Workshop on Smart and Sustainable City, Toulouse, July 2016

MUSA 2.0: A Distributed and Scalable Middleware for User-Driven Service Adaptation

Luca Sabatucci[✉], Salvatore Lopes, and Massimo Cossentino

ICAR-CNR, Palermo, Italy
{luca.sabatucci,salvatore.lopes,massimo.cossentino}@icar.cnr.it

Abstract. MUSA is an agent-based middleware for user-driven self-adaptation. It is based on the separation of concerns between user's goals and system's capabilities. This work analyses some architectural problems of the current implementation and illustrates a new architecture based on the agents and artifacts paradigm.

1 Introduction

The Middleware for User-driven Self-Adaptation (MUSA) has arisen from a couple of pressing objectives in the research agenda of dynamic workflow execution: managing run-time business process evolution and adaptivity [19]. Traditional languages as BPMN or BPEL provide a narrow space of solution that does not give enough freedom of action to overcome exceptional events. We adopted the solution of relaxing the rigid imperative-based workflow specification thus to increase the opportunity of exploring a wider space of solutions. The key concept is a clear separation between 'what the system has to address' and 'how it will operate for addressing it'. The result is a multi-agent system implemented in the Jason [3] agent-oriented programming language.

This paper analyses the current implementation drawbacks, and the known limits in the running in a distributed and highly dynamic environment such as the Cloud. The main problem is that the architecture has been thought to run in a centralized way. We designed the new version of the middleware for exploiting the distributed nature of multi-agent systems. The architecture is based on a distributed and open environment, where agents may be provided by third-party in an open-market paradigm. This architecture opens new scenarios in which external developers may add/evolve/remove their services according to a pay-per-use model.

The paper is structured as follows: Sect. 2 describes the first version of MUSA. Section 3 identifies the limits of the current approach and describes the new architecture for overcoming them. Some related works are described in Sect. 4 and finally conclusions are drawn in Sect. 5.

2 Background: The MUSA Vision

The enablers of MUSA core vision are: (i) representing *what* and *how* as run-time artifacts the system may reason on (respectively goals and capabilities);

© Springer International Publishing AG 2018
G. De Pietro et al. (eds.), *Intelligent Interactive Multimedia Systems and Services 2017*,
Smart Innovation, Systems and Technologies 76, DOI 10.1007/978-3-319-59480-4_49

(ii) a reasoning system for connecting capabilities to goals; (iii) finally a common grounding semantic, represented with some formalism.

2.1   Declarative Specification of Goals

The first aspect of MUSA is the ability to accept run-time requirements as a set of goals to be injected into the system [21]. The characteristics of being autonomous and proactive make the agents able to explore a solution space, even when this space dynamically changes or contains uncertainty.

To this aim, we proposed GoalSPEC, a language for describing user's goals to delegate to the agents of the system. *GoalSPEC* is based on the natural language, and it is specifically conceived to be attractive for a business audience.

GoalSPEC supports Adaptivity. The language is intended to be used for making explicit the business objectives. Indeed, a goal does not specify how to operate but it rather defines the expected results.

GoalSPEC supports Evolution. Traditionally, business process changes during time. It could happen because of new business goals, laws or functionalities. Any revision implies to check inter-dependencies among related (sub)processes with a consequent hard work of ensuring coherence. The model@run.time approach tries to automatize this activity.

Considerations on the Generality. GoalSPEC can be used in a range of different application domains. The task for instantiating the system for a specific domain includes the definition of an ontology/conceptual diagram.

2.2   Declarative Specification of Capabilities

The concept of capability comes from the AI (planning actions [7]) and service-oriented architecture (micro-services [14]). MUSA presents this dual nature by the separation of the abstract capability – a description of the effect of an action that can be performed – and the concrete capability – a small, independent, composable unit of computation that produces some concrete service.

Some examples of capabilities are provided in [20] (the smart travel domain). In this context, each capability encapsulates a web service for reserving some travel service (hotel, flight, local events).

This approach has a number of benefits: – each capability is relatively small, and therefore easier for a developer to implement, – it can be deployed independently of other capabilities, – easier to organize the overall development effort around multiple teams, – it supports self-adaptation because of improved fault isolation.

Moreover, we focus on the idea that capabilities make it easier to deploy new versions of the software frequently. Indeed, from experience acquired in three different applications of MUSA, we discovered the easiness of continue evolving a system. The conclusions are that providing capabilities as run-time entities constitute the basis for continuous data exchange between human and agents and therefore system evolution.

2.3 Self-configuring a Solution

Goals and capabilities are two fundamental elements for supporting self-adaptation through run-time models. A run-time model is inspired by the mechanism of reflection: (i) inspect the code, (ii) generate new code, or (iii) change the existing code [22]. However, MUSA also requires an automatic reasoner that selects and associates capabilities to goals.

Our approach is based on Goals and Capabilities as first-class entities to be used for agent deliberation. The premise for an agent to address an unanticipated goal is to decide which capability (or a combination of capabilities) to execute (and in which order) for achieving the desired goals. To this aim, it is required an automatic reasoner that selects and associates capabilities to goals [18]. The basic idea is that of exploring a space of solutions, where goals represent points of the space that must be reached, and capabilities provide evolution functions that allow moving through this space.

Therefore self-configuration is defined as a space search problem. The algorithm used in [18] is very close to a symbolic planning algorithm. The proposed strategy is that a designated agent incrementally builds a *computational graph* model by exploring different combinations of capabilities. This agent requests the list of capabilities at the beginning and then it works alone (reducing the necessary communications). The resulting structure hopefully contains many possible compositions and therefore different resulting behaviors, the whole multi-agent system may enact.

2.4 Holon Formation and Run-Time Orchestration

The solution produced by the proactive means-end reasoning is a description of the assignment of responsibilities required to address a set of goals (sometimes referred as *goal operationalization*). Responsibilities are instances of (simple or composed) capabilities. Agents of the system are up to enact these responsibilities. They constitute the core unit of the system behavior, such as executing a capability, sending or receiving a message, or waiting for a specified amount of time. More sophisticated capabilities are recursively composed of other capabilities. Holons provide an elegant and scalable method to manage this recursion, guaranteeing –at the same time– knowledge sharing, distributed coordination, and robustness. A holon is a system (or phenomenon) which is an evolving self-organizing structure, consisting of other holons [11]. A holon has its individuality, but at the same time, it is embedded in larger wholes (principle of duality or *Janus effect*).

MUSA adopts holonic multi-agent systems for generating compositions of services. The advantage is the possibility to manage multi-layer services as a single service, thus hiding the complexity of service composition. Moreover, holons are suitable for adaptation because a sub-holon may change without affecting the whole structure.

In [6] holon formation is described as an emergent phenomenon. The approach is that holarchy is formed as the recursive replication a basic schema, i.e. a template that defines roles a holon is made of service-broker, state-monitor, and goal-handler.

The **service broker** is the role in charge of establishing conversations with remote services (by calling end-points). The service broker must also be able to catch exceptions and failures and to raise the need for self-adaptation. The **env monitor** is the role responsible for controlling the user environment (both physical and simulated, including persons acting inside). The knowledge is necessary for invoking services by providing the right input/output parameters. It is also responsible for analyzing inconsistencies due to contrasting beliefs that could generate service failures. The **goal handler** is responsible for analyzing user's goals for checking the satisfaction level.

2.5 The Feedback Loop

The main property of a self-adaptation system is the feedback loop [4]. In MUSA the feedback loop is made of two nested loops. The service holon enacts the inner loop: *Monitor, Analyze, Plan,* and *Act* (MAPE [5]). The inner loop is responsible for handling with periods of stability of the system. During this time a configuration has been generated, and the MAPE sub-cycle is active for enacting the corresponding behavior by invoking and observing the services specified in a configuration. If everything goes as planned, the goal will eventually be addressed.

However, it is possible that some malfunctioning occurs or that a monitor reveals some violation of requirements (for example due to an unanticipated exogenous event). In both cases the system becomes unstable and going on without any fixing could be unfeasible or even too risky. These events raise a need for adaptation. The system blocks the current execution for executing the outer loop made of monitor goal injection, monitor environment/requirements, and proactive means-end reasoning. The objective is that of switching from a configuration to another one by exploiting the proactive means-end reasoning.

3   The Evolution of a Middleware: MUSA 2.0

The first version of MUSA –described in Sect. 2– is realized in Jason [3], an agent-oriented programming language based on the BDI paradigm [17]. The middleware has been adopted in several research projects and case studies (see Table 1).

Despite its flexibility, the current implementation has some drawbacks that limit its employment in a distributed and highly dynamic environment such as the Cloud. In particular, the proactive means-end reasoning algorithm has been developed in a centralized way: one agent is responsible for calculating the configurations that address the user's goals. This agent gathers all the available capabilities from all the participating agents and then executes the space exploration algorithm. This approach is suitable for a small number of goals and capabilities, but it is not scalable. The algorithm does not exploit the distributed

Table 1. Summary of research projects and case studies where the MUSA middleware has been employed between 2013 and 2016.

Acronym	Type	Description
IDS (Innovative Document Sharing)	Research project	The aim has been to realize a prototype of a new generation of a digital document solution that overcomes current operating limits of the common market solutions. MUSA has been adopted for managing and balancing human operations for enacting a digital document solution in a SME
OCCP (Open Cloud Computing Platform)	Research project	The aim was the study, design, construction and testing of a prototype of cloud infrastructure for delivering services on public and private cloud. MUSA has been employed, in the demonstrator, in order to implement an adaptive B2B back-end service for a fashion company
Smart Grid (Cloud Mashup)	Research project	MUSA has been adopted as mashup engine, i.e. for self-configuring ad-hoc orchestration of existing services in order to address run-time business
SIGMA (Multi-Risk Emergency)	Research project	MUSA has been adopted in order to improve environmental monitoring for the prevention and the management of risks. It performs the acquisition, integration and processing of heterogeneous data coming out from several sensor network (Meteo, Seismic, Volcanic, Hydrological, Rainfall, Car/Naval Traffic, Environmental, Video)
Smart Travel	Case study	MUSA provides the planning engine that creates a travel-pack as the composition of several heterogeneous travel services. The planning activity is driven by traveler's goals
Smart Home	Case study	MUSA is the core for implementing the behavior of smart devices deployed in a simulated environment
Exhibition Center (draft)	Case study	MUSA has been adopted for managing the big smart space of an exhibition center with thousand of visitors with different interests

nature of multi-agent systems. It assumes all the agents are collaborative and trusted, hence is not robust to a malicious behavior. Thus, it is enough for a closed environment, where agents come from well-known providers.

We designed the new version of the middleware for working in a distributed and open environment, where several providers dynamically provide agents (and capabilities) in an open-market paradigm. External developers may deploy/evolve/ remove their capabilities according to a pay-per-use model. This section discusses how the new architecture addresses these new requirements.

3.1   The Agent and Artifact Architecture

The new architecture is based on the JaCaMo approach to multi-agent programming. JaCaMo extends Jason with an infrastructure for environment programming: the environment is modeled by an independent computation layer that encapsulates functionalities the agents can access and manipulate. These entities are called artifacts and are programmed as traditional Java objects.

Artifacts represent resources and tools that agents can create, discover, perceive, and use at runtime. Each artifact provides some operations for the agents (interfaces) and may generate observable events. The new architecture exploits agents and the environment by mapping (i) agent actions into artifact operations and (ii) artifact properties and events into agents percepts.

Figure 1 shows a clear separation of concerns among the definition of the environment, the agent layer, and the organization layer.

The environment layer is organized in four workspaces: (i) core, (ii) configuration, (iii) broker and (iv) orchestration. The core workspace provides general facilities for the registration of agents (Domain Directory) and services (Marketplace). The configuration workspace is responsible for the distributed discovery of solutions. All the artifacts in this workspace are instantiated for each set of goals that is injected by the user. The Solution Graph artifact contains the computational graph that is incrementally built by all the agents of the system.

The Graph Access Manager regulates the write access to the graph thus avoiding concurrent accesses. The Configuration Selector browses the graph in order to detect suitable solutions, and the QoS Manager uses local QoS values to negotiate a valid global QoS [15].

The broker workspace is responsible for exploring the space search problem, according to the available services of the marketplace. An instance of the Problem Exploration artifact is created for any cluster of services[1]. Many instances of the Problem Exploration artifact are deployed on various computation node. The Capability Instance artifact is responsible for interacting with the real service, when necessary.

Finally, the orchestration workspace contains services for executing the selected composition of capabilities and applying run-time adaptation when necessary.

The agent layer describes roles that agents of the system may play:

- the specification manager is responsible for ensuring a set of goals is eventually addressed,
- the discovery agent is responsible for building the computation graph,
- the negotiator tries to optimize the identified solutions according to the QoS factors,
- the case manager coordinates the execution of the workflow,
- the workers are responsible for executing the capabilities of the selected solution.

In Fig. 1, the organization layer shows only the two missions corresponding to the self-configuration and adaptive orchestration phases. This layer does not reveal the organizational complexity that is embedded in the holonic structure [6].

3.2   Distributed Continuous Proactive Means-End Reasoning

The centralized version of the *proactive means-end reasoning* algorithm runs on a single node. Checking pre/post conditions is the most time-consuming part of

---

[1] Typically, services are clustered by service providers.

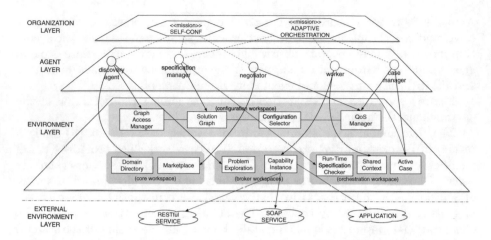

Fig. 1. The Agents and Artifacts view of the MUSA 2.0 architecture.

the algorithm. This approach revealed not scalable when the number of goals and capabilities increases. It was required a new version of the algorithm that distributes the resource-consuming elaboration among many nodes.

The solution was to implement the *computational tree graph* as a shared blackboard artifact (Solution Graph in Fig. 1). Each agent is allowed to read the graph, but it can not modify it: the agent generates possible expansions for the graph on its own (storing new nodes and arcs in a private Problem Exploration artifact). The synchronization of the shared graph with the many private ones is done via a periodic auction [12]. Winning the auction provides the write-access permission in order to regulate the grown of the graph.

The Solution Graph artifact represents a blackboard service, where all the agents can build the computation tree in a collaborative fashion. A new Solution Graph is created for each new goal-set. In the beginning, the graph only contains the initial state of the world node.

A *Graph Access Manager* artifact regulates read/write permissions, thus avoiding concurrent modification. The strategy is to adopt a cycle of auctions for deciding the priority of writing access. At each stage, the Graph Access Manager sends to all the agents a request to participate in a blind auction protocol [12].

Each discovery agent executes an *expand-and-evaluate* loop. It consists of working on a private clone of the shared graph. Each agent produces new states and transitions, thus expanding the cloned graph. Participating the auction allows pushing all the changes to the shared graph.

Each new state is evaluated with the set of goals to give priority to the many push actions. The objective is to predict how much the new state is promising for addressing the goal-set. Here it is possible to use some heuristics. Therefore, the more the state of the world is close to one of the 'final states' of the system, the higher is the score assigned to the node. It is important that all the agents use the same scoring function.

The expand-and-evaluate cycle is described in the following:

1. each discovery agent pulls changes from the shared Solution Graph artifact;
2. each discovery agent selects the most promising node, among those that satisfy one of its capability pre-conditions;
3. each discovery agent calculates the effects of the capability by generating new states and transitions;
4. each discovery agent generates the score of the new nodes.
5. nodes with score and transitions are stored in its own Problem Exploration artifact.

Periodically, when a call-for-bid arrives, each discovery agent selects the most promising from the personal Problem Exploration artifact. The state's score is used for placing the bid to the auction. This strategy rewards those capabilities that potentially improve the solution graph (according to the selected heuristic). The auction closes after a fixed time without counter-offer. The highest bid wins the auction and therefore the bidding discovery agent pushes the updates to the Solution Graph.

The whole procedure cycles permanently even after one solution is discovered, selected and executed. The motivation is to continue expanding the graph for future usage. Indeed, it is plausible alternative solutions may be selected for either re-configuration or optimization purposes. Clearly, after one solution is executed, both the auction and the expand-and-evaluate cycles become background processes (consuming minimal computation resources).

4   Related Work

The current state of the art in cloud computing delineates *mashup* as an innovative technology for the integration of cloud applications [1, 13, 16]. Compared to traditional 'developer-centric' composition technologies, a mashup is inspired by principles of flexibility and user-friendliness.

Among the system affine to Musa, we cite Colombo [2], a framework for automatic web service composition that exploits relational database schema, atomic processes, message passing and a finite-state transition system. As well as our approach, they introduce the goal service, i.e. they make explicit that a composite service is aggregated for addressing a goal. The main difference is that in Colombo a service goal is directly represented as a transition system, that demands a user to learn very technical skills.

Another execution model is provided by OSIRIS [23], an Open Service Infrastructure for Reliable and Integrated process Support that consists of a peer-to-peer decentralised service execution engine and organises services into a self-organizing ring topology. Conceptually close to our orchestration model, Hahn and Fischer, in [10] illustrate how a service choreography can easily be implemented through holons. Their approach is a design-to-code technique based on model-driven transformations which result is a holonic multi-agent system.

However, there are many works that study strategies for dynamic formation of holons for executing complex activities.

In the context of the SASSY research project [9], Gomaa and Hashimoto use software adaptation patterns for Service-Oriented Applications. The goal is not only to execute a mashup but to dynamically adapt distributed transactions at run-time. This aim is made possible by separating the concerns of individual components of the architecture from concerns of dynamic adaptation. In their approach, the solution has been using a central manager that works as connector adaptation state-machine.

On the same direction, Ghezzi et al. [8] propose ADAM (ADAptive Model-driven execution) a mixed approach between model transformation techniques and probability theory. The modelling part consists of creating an annotated UML Activity diagram whose branches can have a probability assigned, plus an annotated implementation. Then an activity diagram becomes an MDP (Markov Decision Process). It is possible to calculate the possible values for the different executions and thus navigate the model to execute it.

5   Conclusions

In this paper, we presented a new version of agent-based middleware MUSA that supports user-driven self-adaptation. We developed MUSA with the fundamental concept of separating users goals from systems capabilities. This work analyses some problems of the prior implementation, such as the difficulty to scale up the means-end reasoning centralized algorithm to the number of agents and capabilities. Thus, we developed a distributed algorithm where agents located on several nodes collaborates for defining a new solution. A periodic auction regulates the coordination and allows the write-access to the agent pushing the most promising sub-solution. We expect that the new version performs much better than the older in more complex application domain.

References

1. Aubonnet, T., Henrio, L., Kessal, S., Kulankhina, O., Lemoine, F., Madelaine, E., Ruz, C., Simoni, N.: Management of service compositionbased on self-controlled components. J. Internet Serv. Appl. **6**(1), 1–17 (2015)
2. Berardi, D., Calvanese, D., De Giacomo, G., Hull, R., Mecella, M.: Automatic composition of transition-based semantic web services with messaging. In: Proceedings of the 31st International Conference on Very Large Data Bases, pp. 613–624. VLDB Endowment (2005)
3. Bordini, R., Hübner, J., Wooldridge, M.: Programming Multi-agent Systems in AgentSpeak Using Jason, vol. 8. Wiley-Interscience (2007)
4. Brun, Y., Serugendo, G.D.M., Gacek, C., Giese, H., Kienle, H., Litoiu, M., Müller, H., Pezzè, M., Shaw, M.: Engineering self-adaptive systems through feedback loops. In: Software Engineering for Self-adaptive Systems, pp. 48–70. Springer (2009)

5. Cheng, B.H., De Lemos, R., Giese, H., Inverardi, P., Magee, J., Andersson, J., Becker, B., Bencomo, N., Brun, Y., Cukic, B. et al.: Software engineering for self-adaptive systems: a research roadmap. In: Software Engineering for Self-adaptive Systems, pp. 1–26. Springer (2009)
6. Cossentino, M., Lodato, C., Lopes, S., Sabatucci, L.: Musa: a middleware for user-driven service adaptation. In: Proceedings of the 16th Workshop "From Objects to Agents", Naples, Italy, 17–19 June 2015
7. Gelfond, M., Lifschitz, V.: Action languages. Comput. Inf. Sci. **3**(16), 1–16 (1998)
8. Ghezzi, C., Pinto, L.S., Spoletini, P., Tamburrelli, G.: Managing non-functional uncertainty via model-driven adaptivity. In: Proceedings of the 2013 International Conference on Software Engineering, pp. 33–42. IEEE Press (2013)
9. Gomaa, H., Hashimoto, K.: Dynamic self-adaptation for distributed service-oriented transactions. In: 2012 ICSE Workshop on Software Engineering for Adaptive and Self-Managing Systems (SEAMS), pp. 11–20 (2012)
10. Hahn, C., Fischer, K.: Service composition in holonic multiagent systems: model-driven choreography and orchestration. In: Holonic and Multi-Agent Systems for Manufacturing, pp. 47–58. Springer (2007)
11. Kay, J.J., Boyle, M.: Self-Organizing, Holarchic, Open Systems (SOHOs). Columbia University Press, New York (2008)
12. Krishna, V.: Auction Theory. Academic Press, San Diego (2009)
13. Marston, S., Li, Z., Bandyopadhyay, S., Zhang, J., Ghalsasi, A.: Cloud computing—the business perspective. Decis. Support Syst. **51**(1), 176–189 (2011)
14. Namiot, D., Sneps-Sneppe, M.: On micro-services architecture. Int. J. Open Inf. Technol. **2**(9), 24–27 (2014)
15. Napoli, C.D., Sabatucci, L., Cossentino, M., Rossi, S.: Generating and instantiating abstract workflows with QOS user requirements. In: Proceedings of the 9th International Conference on Agents and Artificial Intelligence (ICAART 2017), Porto, Portugal, 24–26 February 2017
16. Papazoglou, M.P., van den Heuvel, W.-J.: Blueprinting the cloud. IEEE Internet Comput. **6**, 74–79 (2011)
17. Rao, A.S.: Agentspeak (l): BDI agents speak out in a logical computable language. In: Agents Breaking Away, pp. 42–55. Springer (1996)
18. Sabatucci, L., Cossentino, M.: From means-end analysis to proactive means-end reasoning. In: Proceedings of 10th International Symposium on Software Engineering for Adaptive and Self-Managing Systems, Florence, Italy, 18–19 May 2015
19. Sabatucci, L., Lodato, C., Lopes, S., Cossentino, M.: Towards self-adaptation and evolution in business process. In: AIBP@ AI*IA, pp. 1–10. Citeseer (2013)
20. Sabatucci, L., Lodato, C., Lopes, S., Cossentino, M.: Highly customizable service composition and orchestration. In: Dustdar, S., Leymann, F., Villari, M. (eds.) Service Oriented and Cloud Computing. LNCS, vol. 9306, pp. 156–170. Springer International Publishing (2015)
21. Sabatucci, L., Ribino, P., Lodato, C., Lopes, S., Cossentino, M.: Goalspec: a goal specification language supporting adaptivity and evolution. In: Engineering Multi-Agent Systems, pp. 235–254. Springer (2013)
22. Sawyer, P., Bencomo, N., Whittle, J., Letier, E., Finkelstein, A.: Requirements-aware systems: a research agenda for re for self-adaptive systems. In: 2010 18th IEEE International Requirements Engineering Conference (RE), pp. 95–103. IEEE (2010)
23. Stojnic, N., Schuldt, H.: Osiris-sr: a safety ring for self-healing distributed composite service execution. In: 2012 ICSE Workshop on Software Engineering for Adaptive and Self-Managing Systems (SEAMS), pp. 21–26 (2012)

Approximate Algorithm for Multi-source Skyline Queries on Decentralized Remote Spatial Databases

Hideki Sato$^{(\boxtimes)}$, Shuichi Hirabayashi, and Masaya Takagi

School of Informatics, Daido University,
10-3 Takiharu-cho, Minami-ku, Nagoya 457-8530, Japan
hsato@daido-it.ac.jp

Abstract. A multi-source skyline query about spatial data considers several query reference points at the same time, while a single-source skyline query refers to only one query point. Nowadays, supporting multi-source skyline queries on decentralized remote spatial databases is much important, since (1) multi-source skyline queries about spatial data are effective for supporting spatial decision making in various tasks and (2) various kinds of spatial datasets are available on the Internet and are accessible via Web services provided. This paper proposes Approximate Algorithm for Multi-Source Skyline Queries on Decentralized Remote Spatial Databases (*AMUSE*). *AMUSE* finds the MIN-SUM data point regarding a set of query points first. Then, it makes a series of k-NN queries move on a half line connecting both a MIN-SUM data point and each query point. According to the experimental evaluation, *AMUSE* is excellent in *Precision*. Additionally, it is also excellent in *Recall*, except cases that the search area is large and/or k of k-NN queries to search for data points is small.

Keywords: Approximate algorithm · *AMUSE* · Multi-source skyline query · Decentralized remote spatial database · Web service · Aggregate query

1 Introduction

Skyline queries [1] have received considerable attention in the database and data mining fields. Given a set of tuples D, a skyline query finds a set of skyline tuples from D, such that every tuple in the skyline set is not dominated by any other tuple in D. That is to say, if tuple t is a member of the skyline set, there exists no other tuple t' in D such that t' is pair-wise smaller than t for values of all the skyline attributes. Without loss of generality, it is assumed that smaller values are preferable over larger ones. The skyline queries can be either *absolute* or *relative*, where '*absolute*' means that dominance is based on the static skyline attribute values of tuples in D and '*relative*' means that the difference between a tuple in D and a user-given query value needs to be computed for dominance. *Relative* skyline query is also known as *dynamic* skyline query [2].

© Springer International Publishing AG 2018
G. De Pietro et al. (eds.), *Intelligent Interactive Multimedia Systems and Services 2017*,
Smart Innovation, Systems and Technologies 76, DOI 10.1007/978-3-319-59480-4_50

An important category of skyline query applications is related to a spatial data which is associated with a location represented by address and/or latitude and longitude. Skyline queries about spatial data are useful for supporting spatial decision making in various tasks. One familiar and everyday example of such queries is to find restaurants which are cheap and close to oneself, where a distance between a restaurant and a person is used to make the answer. The query is called single-source skyline one, because it refers to only one query point, namely a relative query in relation to user's location. In contrast, a multi-source skyline query considers several query points at the same time (e.g., to find hotels which are cheap and close to the Station, the Museum, and the Restaurant).

To answer a spatial query, it is necessary to access both location data of query points and a set of data points. Regarding a set of data points, various kinds of spatial datasets are nowadays available on the Internet and are accessible via Web services provided. Actually, new Web services are created by mashing up existing Web services. Accordingly, it is practical and realistic to make use of such datasets managed in a decentralized fashion. However, accesses to datasets on the Internet are limited by simple and restrictive Web API interfaces.

This paper proposes Approximate Algorithm for Multi-Source Skyline Queries on Decentralized Remote Spatial Databases (*AMUSE*). Hereafter, it is assumed that the distance between two points a and b is approximated using the *Euclidean* distance. A k-Nearest Neighbor (NN) query with a single query point is used to access decentralized remote spatial databases (i.e., datasets managed by other sites on the Internet), since it is a typical interface of Web services. *AMUSE* finds the MIN-SUM data point regarding a set of query points first. Then, it makes a series of k-NN query searches move on a half line connecting both a MIN-SUM data point and each query point. The experimental evaluation of *AMUSE* shows that *Precision* is nearly equal to 1.0. Additionally, *Recall* is over 0.9, except cases that the search area is large and/or k of k-NN queries to search for data points is small.

The rest of this paper is organized as follows. Section 2 briefly reviews multi-source skyline queries and k-NN queries. Section 3 proposes a method for multi-source skyline query processing on decentralized remote spatial databases. Section 4 experimentally evaluates the approximate algorithm using both synthetic datasets. Section 5 mentions related work. Finally, Sect. 6 concludes the paper.

2 Preliminaries

In this section, multi-source skyline queries and k-NN queries are briefly reviewed.

2.1 Multi-source Skyline Queries

Given a set Q of query points, a data point p is said to spatially dominate another point p', denoted by $p \succ_Q p'$, if Eq. (1) is satisfied. Without loss of generality, it is assumed that smaller values are preferable over larger ones. Here, $d(p,q)$ is used to represent the distance between p and q. Given a set Q of query points, Eq. (2) defines a *skyline* query Ψ regarding Q. In other words, a data point p belongs to a skyline set if no other data points spatially dominate it.

$$(\exists q_i \in Q, d(p, q_i) < d(p', q_i)) \wedge (\forall q_j \in Q, d(p, q_j) \leq d(p', q_j)) \tag{1}$$

$$\Psi(D, Q) = \{p \in D| \nexists p' \in D, p' \succ_Q p\} \tag{2}$$

Consider the example of Fig. 1, where $D(= \{p_1, p_2, p_3, p_4, p_5, p_6\})$ is a set of data points and $Q(= \{q_1, q_2\})$ is a set of query points. The number on each edge connecting a data point and a query point represents any distance cost between them. Figure 2 shows a skyline query result $\{p_1, p_2, p_3\}$ from D regarding Q.

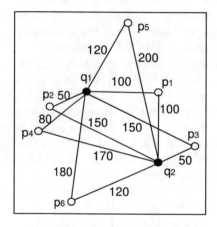

Fig. 1. Example of query points (solid circles) and data points (hollow circles)

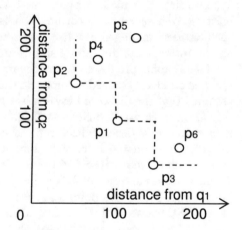

Fig. 2. Multi-source skyline query result

2.2 k-Nearest Neighbor Queries

A k-NN query is used to process multi-source skyline queries on decentralized remote spatial databases. Given a query point q and a set of data points D, a k-NN query answers the top k nearest neighbors to q from D. The circle of Fig. 3 is a searched region of a 5-NN query, where q is the query point and set $\{p_1, p_2, p_3, p_4, p_5\}$ is the query result. The radius of the circle is equal to the distance r between q and the 5th nearest neighbor p_5. Inside the circle, it is evident that there exists no other data point except $\{p_1, p_2, p_3, p_4, p_5\}$.

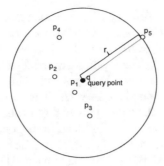

Fig. 3. Searched circle of 5-NN query (query point (solid circle) and data points (hollow circles))

3   Multi-source Skyline Query Processing Method

In this section, *AMUSE* is proposed. Before introducing the algorithm, search areas are discussed first. Then, *AMUSE* is presented.

3.1   Search Area

A search area is a region for finding data points to make an answer of a multi-source skyline query, by using k-NN queries. Given a set Q of query points and a set D of data points, a search area is determined based on a MIN-SUM data point. For a data point $p(\in D)$, SUM distance $d_{sum}(p, Q)$ regarding Q is defined as $\sum_{q_i \in Q} d(p, q_i)$, where $d(p, q)$ is a distance between p and q. Consider the example of Fig. 4, where $Q(= \{q_1, q_2\})$ is a set of query points and $D(= \{p_1, p_2, p_3, p_4, p_5, p_6, p_7\})$ is a set of data points. The MIN-SUM data point of the example is p_1 whose SUM distance is minimal among D.

Let p_1 be the origin on the plane of Fig. 4. It is apparent that p_1 spatially dominates any data points in the first quadrant (i.e., p_4, p_5). Also, there exists no dominance relationships between p_1 and any data point in either the second quadrant or the fourth quadrant (i.e., p_2, p_3, p_6, p_7). Furthermore, there exists no data point in the third quadrant. Accordingly, p_1 is a skyline point regarding Q, since no other data point dominates p_1.

Figure 5 shows a contour graph of distance function $d_{sum}(p, Q)$ regarding a set of 10 query points whose location are uniformly generated. If the number of query points is over 2, $d_{sum}(p, Q)$ is a convex function. In that case, there certainly exists a single point, namely a MIN-SUM point, at which its function value is minimal. However, there does not necessarily exist any MIN-SUM data point at a MIN-SUM point. A MIN-SUM point can be computed neither by an analytical method nor by an algorithm based on differentiation[1]. Instead, it can be obtained by employing Nelder-Mead method [3] for nonlinear programming problems, which does not rely on gradients of a function. If the number of query

---

[1] Distance function $d_{sum}(p, Q)$ is not differentiable at each query point of Q.

points is 2, any point on a line-segment between one query point and another can be MIN-SUM one. In that case, *AMUSE* uses the middle point on the line-segment as a representative of MIN-SUM points.

To obtain the MIN-SUM data point regarding Q, a k-NN query with the MIN-SUM point as a query point is requested. Then, the data point with the minimal SUM distance among the query result is chosen as the MIN-SUM data point. It is not guaranteed, however, it is highly probable for the most case that a query result contains an exact MIN-SUM data point, if k of k-NN queries is adequate. Even if that were not the case, it can be expected that a data point comparable to a MIN-SUM data point is chosen. Hereafter, note that a data point with the minimal SUM distance among a query result is also meant by a MIN-SUM data point.

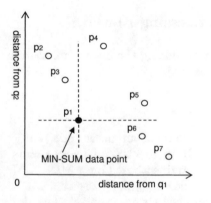

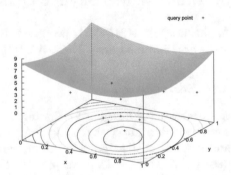

Fig. 4. MIN-SUM data point   **Fig. 5.** SUM distance function

Figure 6 shows a target search area, namely a union of a set of circles. A center of a circle corresponds to each query point and its radius is the length between the query point and the MIN-SUM data point. Consider a data point located outside a target search area. It is evident that it cannot be a skyline point, since it is dominated by a MIN-SUM data point. That is to say, any distance from each query point is longer than a corresponding distance between the query point and the MIN-SUM data point. Accordingly, it is sufficient to search a target area for computing a multi-source skyline query.

Figure 7 shows an actual search area (drawn with solid lines) to cover a target search area (drawn with a dotted line) corresponding to query point $q_i (\in Q)$. To search the target area, a series of k-NN queries is requested to cover the area. An actual search area is a union of a set of circles, each of which corresponds to a searched circle of a distinct k-NN query and moves on a half line connecting both a MIN-SUM data point and q_i. A series of k-NN queries starts with the first search circle which finds a MIN-SUM data point (as is mentioned in the above). According to a search direction, it repeats until an intersection between a half line and a target search circle is covered.

The query point of the $(i+1)$-th k-NN query is an intersection between a half line and C_i, if the searched circle of the i-th k-NN query result is C_i. A family of such a series of searches progresses in parallel for a set of query points. Even if an actual search area does not cover a corresponding target search area, it can be highly expected that the former approximates the latter effectively.

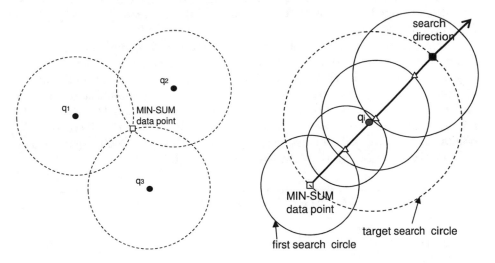

Fig. 6. Target search area   **Fig. 7.** Actual search area

Figure 8 shows searched circles C_i and C_{i+1} of two consecutive k-NN queries, where C_i is a searched one of the i-th k-NN query and C_{i+1} is a searched one of the $(i+1)$-th k-NN query. Since query point O_{i+1} of the $(i+1)$-th k-NN is on a circumference of C_i, C_i and C_{i+1} surely overlap. This is a careful setting of a query point for a next search, although it is possible to consider different settings. Let an average radius of a searched circle be r, although a radius of one searched circle is not necessarily equal to that of the other. The shaded region of C_{i+1} is a newly searched one, not overlapped with C_i. The area is $(\pi/3 + \sqrt{3}/2)r^2$ and the expected number of newly searched data points is $k \cdot (\sqrt{3}/2\pi + 1/3)$. For example, it is expected with the equation that a 50-NN query newly returns 30.45 data points.

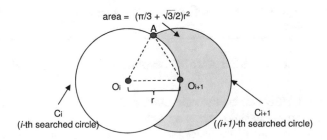

Fig. 8. Newly searched area

3.2   Approximate Algorithm *AMUSE*

Figure 9 shows pseudo codes of *AMUSE*. First, the MIN-SUM point regarding a set Q of query points is computed using Nelder-Mead method [3] (line 01). Using the point as a query point, the first k-NN query is requested (line 02). The searched circle is computed from the query result (line 03) and the MIN-SUM data point is set by using the query result (line 04). The initial set of half lines are arranged with the MIN-SUM data point and a set Q of query points (line 05–06).

Consecutive k-NN queries are repeatedly requested to find data points in the actual search area (line 07–14). In the loop, the half line with the longest unsearched segment is selected first (line 08) and next query point on the line is computed (line 09). With the query point, a k-NN query is requested (line 10) and the searched circle is computed for the query result (line 11). Completely retrieved half lines are removed from a list of half lines (line 12). Searched data points are collected without duplicates (line 13). Finally, a skyline query result is computed by using all the data points searched (line 15).

Approximate Algorithm for Multi-Source Skyline Queries *AMUSE*(Q, k)
Input:
   a set of query points   Q
   number of data to be returned for k-NN query   k
Output:
   skyline data points list   Slist

```
01    MIN_SUM_point := compute a MIN-SUM point regarding Q;
02    DPlist := NEAREST_NEIGHBOR_SEARCH(k, MIN_SUM_point);
03    circle := MAKE_CIRCLE(MIN_SUM_point, last element of DPlist);
04    MIN_SUM_data_point := find MIN_SUM_point from DPlist regarding Q ;
05    HLlist := make a list of half_lines between MIN-SUM_data_point and each of Q;
06    remove half_lines from HLlist , each of which is fully searched by circle;
07    while(LHlist is not empty){
08        half_line := select the half_line withthe longest unsearched segment from HLlist;
09        query_point := find next query_point on half_line;
10        DPlist' := NEAREST_NEIGHBOR_SEARCH(k, query_point);
11        circle := MAKE_CIRCLE(query_point, last element of DPlist');
12        remove half_lines from HLlist , each of which is fully searched by circle;
13        DPlist:= append DPlist' to DPlist and remove duplicates;
14    }
15    Slist = compute skyline data points from DPlist regarding Q;
16    return Slist;
```

Fig. 9. Approximate algorithm *AMUSE* for processing multi-source skyline queries

Note that *AMUSE* can be directly extended to efficiently process more general cases where non-spatial attributes are also considered (e.g., hotel prices). The non-spatial attributes are static values. Therefore, they can be treated as normal attributes which have pre-computed "distances" to all data objects.

4 Experimental Evaluation

In this section, *AMUSE* is experimentally evaluated. First, the number of a MIN-SUM data point in the line of nearest neighbors from a MIN-SUM point is examined. Then, *AMUSE* is evaluated regarding performance measures.

Each experiment is conducted 100 times and measured values are averaged. A synthetic dataset of 10,000 data points is used, each of whose location is uniformly generated. A set of query points is synthesized, each of whose location is also generated uniformly. To manage an area where a query point is generated, a query point area parameter n is given. Figure 10 shows a selected area for generating a coordinate (x, y) of a query point. Given n, x-axis and y-axis are divided into n segments of an equal length, respectively. As a result, the plane is divided into n^2 rectangles of an equal size. First, one rectangle (i.e., shaded region) is randomly selected among n^2 ones. Then, a set of query points are uniformly generated inside the selected rectangle.

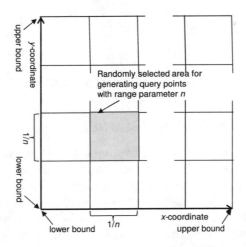

Fig. 10. Randomly selected area for generating query points

4.1 Number of MIN-SUM Data Point in Line from MIN-SUM Point

The number of a MIN-SUM data point in the line of nearest neighbors from a MIN-SUM point is experimentally measured. Experiments are conducted by varying the number of query points and query point area parameter. The result is allowable enough, since the average number is less than 3 in all the cases (See Fig. 11(a)). Furthermore, it is evident that an exact MIN-SUM data point can be obtained in most cases (See Fig. 11), if a 10-NN query is requested once.

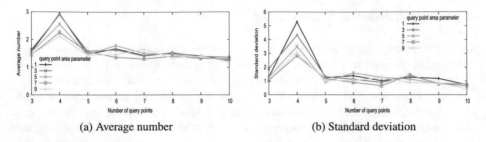

(a) Average number   (b) Standard deviation

Fig. 11. The number of MIN-SUM data point in line of nearest neighbors from MIN-SUM point

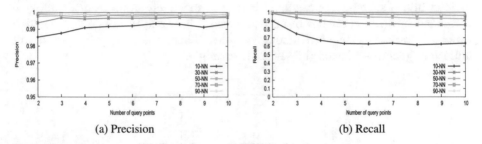

(a) Precision   (b) Recall

Fig. 12. Precision and Recall (query point area parameter = 5)

4.2   Performance Evaluation of *AMUSE*

Performance of *AMUSE* is experimentally evaluated by measuring *Precision*, *Recall*, and *NOR* (Number Of k-NN Requests). Both *Precision* and *Recall* are accuracy criteria. Let A be an exact answer set and B be an answer set with *AMUSE*. *Precision* is defined as $|A \cap B|/|B|$ and *Recall* is defined as $|A \cap B|/|A|$. On the hand, *NOR* is a requested number of k-NN queries. Since the processing time for a multi-source skyline query with *AMUSE* is approximately proportional to *NOR*, it can be used as another criteria for evaluating performance.

An experiment is conducted by setting query point area parameter at 5 and by varying the number of query points and k of k-NN queries. According to the result, *Precision* is over 0.98 (See Fig. 12(a)). Additionally, *Recall* is over 0.9, except cases that k is either 10 or 30 (See Fig. 12(b)). A series of k-NN queries moves on a half line connecting both a MIN-SUM data point and each query point. However, it might not be sufficient to search in the direction orthogonal to a half line in case that k is less and/or the number of query points is more.

Another experiment is conducted by varying the number of query points and query point area parameter. According to the result, *Precision* is over 0.99 (See Fig. 13(a)). Additionally, *Recall* is over 0.9, except cases that query point area parameter is either 1 or 3 (See Fig. 13(b)). The less query point area parameter is, the larger a set of query points is distributed. Accordingly, it might not be sufficient to search in the direction orthogonal to a half line in such a case.

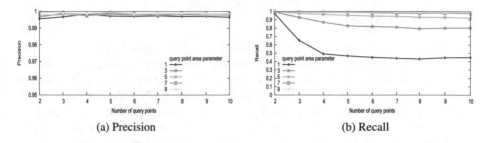

(a) Precision

(b) Recall

Fig. 13. Precision and Recall (50-NN queries)

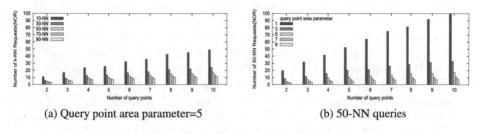

(a) Query point area parameter=5

(b) 50-NN queries

Fig. 14. Number of k-NN queries requested (NOR)

Both Fig. 14(a) and (b) show NOR. The former is the result of an experiment conducted by setting query point area parameter at 5 and by varying the number of query points and k of k-NN queries. The less k of k-NN queries is, the more NOR is. The latter is the result of an experiment conducted by varying the number of query points and query point area parameter. The less query point area parameter is, the more NOR is.

Figure 15 shows the number of exact skylines by varying the number of query points and query point area parameter. By using the measured values in Figs. 13(b) and 15, approximated number of exact recalled skylines can be calculated. Furthermore, by combining "30.45 data points are newly returned by a 50-NN query" mentioned in Sect. 3.1 and the measured values in Fig. 14(b), approximated number of total data points searched can be also calculated. Finally, Fig. 16 shows the approximated number of searched data points per exact recalled skyline. It ranges from 10 to 22 approximately.

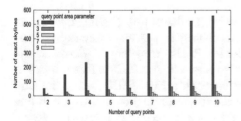

Fig. 15. Number of exact skylines

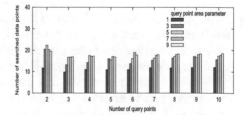

Fig. 16. Number of searched data points per skyline (50-NN queries)

5   Related Works

Papadias et al. introduced Aggregate Nearest Neighbor (ANN) queries and query processing methods [4–6]. They are relied on the aggregate distance between a data point and user-given multiple query points, by using either SUM function or MAX function. First, it was dedicated to the case of *Euclidean* distance [4,5]. Then, it was extended to the case of network distance [6]. However, their work differs from ours in the query type and the use of centralized local databases.

Multi-source skyline queries were studied by Deng et al. [7]. They used network distance and proposed 3 algorithms for processing such queries. However, they were relied on spatial datasets managed by centralized local databases.

Morimoto et al. also studied spatial skyline queries. First, they put ANN queries and skyline queries together for a group of users [8]. Given a set Q of query points, skyline computation regarding both SUM distance from Q and each distance between $q_i (\in Q)$ and a data point is carried on. Another work was dedicated to selecting not good data points but good grid areas, by making use of skyline queries regarding proximity relationships of desirable facilities and undesirable facilities [9] Both works are based on *Euclidean* distance. However, they are relied on spatial datasets managed by centralized local databases.

Both of the works in [10,11] were related to spatial queries on decentralized remote spatial databases, because they provided users with location-dependent query results by using Web API interfaces on the Internet. The former [10] proposes k-NN query processing algorithm that uses one or more range queries to search the nearest neighbors of a given query point. The latter [11] proposes two range query processing algorithms by using k-NN search. However, ours differs from theirs in dealing with multi-source skyline queries on decentralized remote spatial databases.

Sato et al. developed approximate algorithm *RPSA* first, to process ANN queries on decentralized remote spatial databases [12]. Then, they also applied *RPSA* to process Aggregate Range queries on decentralized remote spatial databases [13,14]. They are relied on the aggregate distance between a data point and user-given multiple query points, by using either SUM function or MAX function. *RPSA* is based on k-NN queries as a means of accessing decentralized remote spatial databases.

6   Conclusion

This paper proposed *AMUSE* for processing multi-source skyline queries on decentralized remote spatial databases. According to the experimental evaluation, it is excellent in *Precision*. Additionally, it also excellent in *Recall*, if the search area is large and/or k of k-NN queries is small. *AMUSE* makes a series of k-NN queries move on a half line connecting both a MIN-SUM data point and each query point. As a result, searches in the direction orthogonal to a half line are not sufficient in such a case. Our future work includes (1) improving *Recall* and (2) extending *AMUSE* to the version of network distance for Road network applications.

References

1. Borzsonyi, S., Kossmann, D., Stocker, K.: The skyline operator. In: Proceedings of International Conference on Data Engineering, pp. 421–430 (2001)
2. Papadias, D., Tao, Y., Fu, G., Seeger, B.: An optimal and progressive algorithm for skyline queries. In: Proceedings of ACM SIGMOD International Conference on Management of Data, pp. 467–478 (2003)
3. Nelder, J.A., Mead, R.: A simplex method for function minimization. Comput. J. **7**, 308–313 (1965)
4. Papadias, D., Shen, Q., Tao, Y., Mouratidis, K.: Group nearest neighbor queries. In: Proceedings of International Conference on Data Engineering, pp. 301–312 (2004)
5. Papadias, D., Tao, Y., Mouratidis, K., Hui, C.K.: Aggregate nearest neighbor queries in spatial databases. ACM Trans. Database Syst. **30**(2), 529–576 (2005)
6. Yiu, M.L., Mamoulis, M., Papadias, D.: Aggregate nearest neighbor queries in road networks. IEEE Trans. Knowl. Data Eng. **17**(6), 820–833 (2005)
7. Deng, K., Zhou, X., Shen, H.T.: Multi-source skyline query processing in road networks. In: Proceedings of International Conference on Data Engineering, pp. 796–805 (2007)
8. Arefin, M.S., Ma, G., Morimoto, Y.: A spatial skyline queries for a group of users. J. Softw. **9**(11), 2938–2947 (2014)
9. Annisa, A., Zaman, Y.: Morimoto: area skyline query for selecting good locations in a map. J. Inf. Process. **24**(6), 946–955 (2016)
10. Liu, D., Lim, E., Ng, W.: Efficient k-Nearest neighbor queries on remote spatial databases using range estimation. In: Proceedings of SSDBM, pp. 121–130 (2002)
11. Bae, W.D., Alkobaisi, S., Kim, S.H., Narayanappa, S., Shahabi, C.: Supporting range queries on web data using k-nearest neighbor search. In: Proceedings of W2GIS, pp. 61–75 (2007)
12. Sato, H., Narita, R.: Approximate search algorithm for aggregate k-nearest neighbor queries on remote spatial databases. Int. J. Knowl. Web Intell. **4**(1), 3–19 (2013)
13. Sato, H., Narita, R.: Approximate processing for aggregate range queries on remote spatial databases. Int. J. Knowl. Web Intell. **4**(4), 314–335 (2014)
14. Sato, H., Narita, R.: Farthest-point Voronoi region sensitive processing for further precise maximum query results. In: Proceedings of 7th International Conference on Intelligent Interactive Multimedia Systems and Services, pp. 323–334 (2014)

A New Simple Preprocessing Method for MUSIC Suitable for Non-contact Vital Sensing Using Doppler Sensors

Yukihiro Kamiya[✉]

Aichi Prefectural University, Nagakute, Japan
kamiya@ist.aichi-pu.ac.jp

Abstract. Non-contact vital sensing using Doppler sensor is emerging along with interests on its application as a means for monitoring patients in hospitals or the elderly living in home alone. We already proposed a signal processing scheme applicable to Doppler sensor which enables us to monitor multiple persons by a Doppler sensor. As its extended version, we also proposed a scheme measuring the direction of the persons by putting multiple antennas at the receiver. In this paper, we propose to use those schemes as a preprocessing for the multiple signal classification (MUSIC) aiming at the precise identification of the direction of the persons. The proposed scheme is to not only improve the resolution of the identification of directions but also enables MUSIC to identify the directions exceeding the degree-of-freedom. Some numerical examples are provided for performance evaluation.

Keywords: Doppler sensor · MUSIC · Angle of arrival · Array antenna

1 Introduction

Coming aged society urges us to develop new technologies for contributing to the promotion of welfare for the elderly. Non-contact vital sensing using Doppler sensors is one of such technologies aiming at monitoring patients in hospitals, or elderlies at nursing home. It enables us to measure heartbeats, breathing and body motion without putting any sensors on human bodies so that it does not cause any stress for the measurement [1]. This technology is expected to improve the efficiency of operation and management of facilities such as hospitals or nursing homes.

Figure 1 shows the principle of the Doppler sensor. The transmitter embedded in the Doppler sensor radiates an electro-magnetic wave and the reflected waves yielded by human bodies are received by the receiver. The receiver outputs the frequency deviation of the received signal comparing with the transmitted wave. Then the heartbeat can be extracted by digital signal processing of the frequency deviation. If there are multiple human bodies, we need to separate the two heartbeats.

© Springer International Publishing AG 2018
G. De Pietro et al. (eds.), *Intelligent Interactive Multimedia Systems and Services 2017*,
Smart Innovation, Systems and Technologies 76, DOI 10.1007/978-3-319-59480-4_51

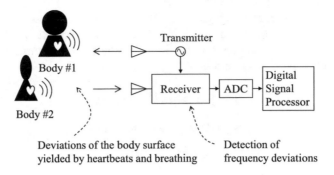

Fig. 1. Configuration of the Doppler sensor.

However, actually, we can hardly find conventional works with respect to the non-contact monitoring of the heartbeat which is capable to cope with multiple bodies. So we proposed a new scheme to cope with multiple bodies in [2]. It should be also noted that it is computationally-light and memory-efficient. This scheme is extended by using multiple receive antennas so that it is capable of estimating direction-of-arrival (DOA) of the signal, i.e., the direction of the target human bodies [3]. Although this scheme is computationally very light, the DOA estimation is not very accurate. So, if we can afford more computational cost, multiple signal classification (MUSIC), a super resolution technique, emerges as a candidate for the DOA estimation method. However, the performance of MUSIC is constrained by the number of antennas. It means that MUSIC cannot estimate DOA if there are multiple signals exceeding the degree of freedom of the array antenna which is defined as $B - 1$ where B is the number of antennas [4].

In this paper, we propose to combine the method in [2] with MUSIC so that we can obtain the following two advantages: (1) the proposed method allows MUSIC to estimate DOAs for multiple signals exceeding the degree of freedom, and (2) the combination with MUSIC brings improvement of the accuracy of the DOA estimation by the method in [2]. In the proposed method, the method in [2] is utilized as a preprocessing for MUSIC.

To the best of our knowledge, there is no conventional works except [5] trying to cope with multiple targets using a Doppler sensor with an array antenna at the receiver. Although [5] does not estimate DOA, i.e., just separating the multiple signals in the space domain, the separation is constrained by the degree of freedom.

In the following, in-depth explanation is provided after the formulations of signals. Then, some numerical examples are given to demonstrate the behavior and the performance of the proposed method.

2   Formulations of Signals

Figure 2 illustrates the vital sensing using the Doppler sensor proposed in this paper. The transmitter sends $s(t)$ which will be reflected by human bodies, yielding the multiple reflected signals $s_n(t)$, $n = 1, \cdots, N$ where N is the number of signals. The signals are received by B antennas combined with noise $\eta_b(t)$, $b = 1, \cdots, B$. The received signals obtained by the multiple antennas are expressed by a vector of size $(B \times 1)$ as follows:

$$\mathbf{x}(t) = \begin{bmatrix} x_1(t) \ x_2(t) \ \cdots \ x_B(t) \end{bmatrix}^{\mathrm{T}} \tag{1}$$

The vector of the received signals $\mathbf{x}(t)$ is sampled as:

$$\mathbf{x}[k] = \mathbf{x}(kT_{\mathrm{S}}) \tag{2}$$

where T_{S} denotes the sample interval while k is an integer as a time index. Hereafter, the samples of the signals formulated above is denoted by replasing (t) with $[k]$.

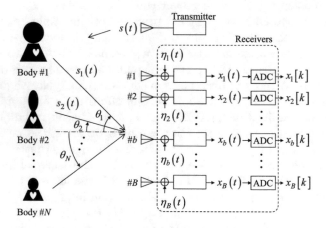

Fig. 2. The vital sensing using the Doppler sensor

So $\mathbf{x}[k]$ is expressed as follows:

$$\mathbf{x}[k] = \mathbf{A}\mathbf{s}[k] + \sqrt{\frac{P_\eta}{2}} \eta[k] \tag{3}$$

where $\mathbf{s}[k]$ is a vectors of size $(N \times 1)$ containing the reflected signals as follows:

$$\mathbf{s}[k] = \begin{bmatrix} s_1[k] \ s_2[k] \ \cdots \ s_N[k] \end{bmatrix}^{\mathrm{T}} \tag{4}$$

The signal $s_n[k]$, $n = 1, \cdots, N$ is defined as

$$s_n[k] = e^{j\varphi_n} \sum_{m=0}^{M-1} \sqrt{P_{\mathrm{S}}^{(n)}} \phi_n[k - mT_n] \tag{5}$$

where $P_{\mathrm{S}}^{(n)}$ denotes the signal power. The unit-power fundamental waveform is given by $\phi_n[k]$ of which the period is defined by T_n. Finally, M is the number of the repetition of the fundamental waveform $\phi_n[k]$ contained in $s_n[k]$ while φ_n denotes the initial phase of $s_n[k]$.

In addition, P_η is the noise power while $\eta[k]$ of size $(B \times 1)$ contains the noise samples, defined as follows:

$$\eta[k] = \begin{bmatrix} \eta_1[k] \ \eta_2[k] \ \cdots \ \eta_B[k] \end{bmatrix}^{\mathrm{T}} \tag{6}$$

where $\eta_b[k]$, $b = 1, \cdots, B$ is the unit-power complex noise defined as follows:

$$\eta_b[k] = \eta_{\mathrm{R},b}[k] + j\eta_{\mathrm{I},b}[k] \tag{7}$$

where $\eta_{\mathrm{R},b}[k]$ and $\eta_{\mathrm{I},b}[k]$ are the independent- and identically-distributed random variables following $N(0,1)$.

Finally, $\mathbf{A}$ is a matrix of size $(N \times B)$ expressing the phase variation caused by the direction-of-arrival (DOA) of the N reflected signals, defined as follows:

$$\mathbf{A} = \begin{bmatrix} \mathbf{a}(\theta_1) \ \mathbf{a}(\theta_2) \ \cdots \ \mathbf{a}(\theta_N) \end{bmatrix} \tag{8}$$

If the antennas are aligned linearly and are spaced equally with d-spacing, $\mathbf{a}(\theta_n)$ is given as the following.

$$\mathbf{a}(\theta_n) = \begin{bmatrix} a^0 \ a^2 \ \cdots \ a^{B-1} \end{bmatrix}^{\mathrm{T}} \tag{9}$$

$$a = e^{-j\frac{2\pi}{d}\sin\theta_n} \tag{10}$$

In this paper, we propose a simple method to estimate the fundamental waveforms $\phi_n(t)$, the number of signals N and the periods of each fundamental waveforms T_n. It is suitable for the vital sensing using the Doppler sensor due to its simplicity and sufficient performance.

3   Proposed Method

Figure 3 shows the configuration of the proposed method. The proposed method is composed by the signal-to-parallel converter (SPC) banks, the spatial combiners, the estimator and the DOA estimator. This configuration is also interpreted that the proposed method is to insert the pre-processor, indicated by the dotted line in the figure, between the receive antennas and the DOA estimator. The following sections explain each of the components.

3.1   Basic Idea

Prior to the details, this section provides an intuitive explanation of the basic ideas toward the proposed method [2]. Figure 4 illustrates the situation that a sample sequence is fed into an SPC whose number of the output port is equal to the period of the input, 8. In this case, we can imagine that the SPC repeatedly

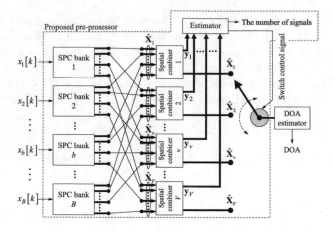

Fig. 3. An overview of the proposed scheme

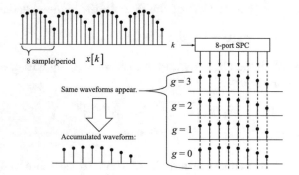

Fig. 4. SPC output: obtained if the SPC output length is equal to the period of the input signal.

outputs the same waveform. Therefore the accumulation of the SPC outputs grows high.

In contrast, Fig. 5 illustrates the case where the number of the SPC output ports is 7 even though the period of the signal is 8. In this case, the accumulated values of the SPC outputs do not grow so high since the waveform of the SPC outputs is never the same on the time axis g.

The proposed method exploits this property for the estimation of $\phi_n[k]$, N and T_n using multiple SPCs.

The following section provides detailed description of the configuration and the procedure of the proposed method.

3.2   SPC Banks

Figure 6 depicts the b-th ($b = 1, \cdots, B$) SPC bank. The sample $x_b[k]$ is copied and is fed into the SPCs. The multiple SPCs are identified by the number #1 to

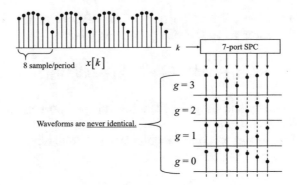

Fig. 5. SPC output: obtained if the SPC output length is NOT equal to the period of the input signal.

$\#V$. Let us call this number as SPC ID. The number of the SPC output ports is the minimum at SPC#1, equipped with with $U_{\min}$ ports. The adjacent SPC #2 is equipped with $(U_{\min} + 1)$ ports. Likewise, the number of the output ports is incremented by 1 till the SPC $\#V$ equipped with $U_{\max}$ ports as the maximum. Therefore, the relationship between the SPC ID $v\,(= 1, \cdots, V)$ and the number of the output ports $u\,(= U_{\min}, \cdots, U_{\max})$ is formulated as follows:

$$u = \Lambda(v) \tag{11}$$
$$\Lambda(v) = v + U_{\min} - 1 \tag{12}$$

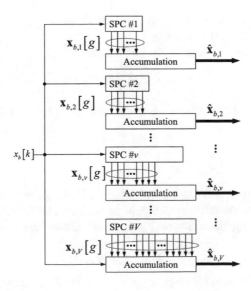

Fig. 6. An overview of the proposed scheme

The g-th $(g = 0, \cdots , G - 1)$ output of the SPC#v at the b-th antenna is expressed by the vector $\mathbf{x}_{b,v}[g]$ of size $(v \times 1)$ as follows:

$$\mathbf{x}_{b,v}[g] = \left[x_1^{(b,v)}[g] \; x_2^{(b,v)}[g] \; \cdots \; x_v^{(b,v)}[g] \right]^{\mathrm{T}} \tag{13}$$

where $x_1^{(b,v)}[g]$ denotes the output of the first output port of the SPC #v connected with the antenna #b.

Then the SPC output $\mathbf{x}_{b,v}[g]$ is accumulated as follows:

$$\hat{\mathbf{x}}_{b,v} = \sum_{g=0}^{G-1} \mathbf{x}_{b,v}[g] \tag{14}$$

3.3 Spatial Combiners

As shown in Fig. 3, the SPC banks are followed by the spatial combiner. Let us focus on the v-th $(v = 1, \cdots , V)$ spatial combiner. The inputs of this spatial combiner are expressed by the following matrix of size $(B \times v)$.

$$\hat{\mathbf{X}}_v = \left[\hat{\mathbf{x}}_{1,v} \; \hat{\mathbf{x}}_{2,v} \; \cdots \; \hat{\mathbf{x}}_{B,v} \right] \tag{15}$$

It is observed that the matrix $\hat{\mathbf{X}}_v$ collects the accumulated SPC outputs whose size is identical to one another, from all antennas. Then the collected vectors are spatially combined as follows:

$$\mathbf{y}_v = \hat{\mathbf{X}}_v \mathbf{w}_v^* \tag{16}$$

where $\mathbf{w}_v$ is the weight vector of size $(B \times 1)$ while $\mathbf{y}_v$ of size $(v \times 1)$ is the output of the v-th spatial combiner.

The weight vector $\mathbf{w}_v$ is obtained as follows:

$$\mathbf{w}_v = (\hat{\mathbf{x}}_{1,v})^{\mathrm{H}} \hat{\mathbf{X}}_v. \tag{17}$$

This weight vector enables us to compensate the phase difference among the antenna.

3.4 Estimator

The estimator in Fig. 3 collects $\mathbf{y}_v$, $(v = 1, \cdots , V)$ in order to estimate NOS, the period and the fundamental waveforms as well as the fundamental waveform, through the following procedures. First, it calculates the power of $\mathbf{y}_v$, namely P_v as follows:

$$P_v = (\mathbf{y}_v)^{\mathrm{H}} \mathbf{y}_v. \tag{18}$$

Second, NOS is estimated by counting the number of P_v exceeding the threshold δ. This number is NOS. Suppose that $P_{v'}$ exceeds δ.

3.5   DOA Estimator Based on MUSIC

As shown in Fig. 3, $\hat{\mathbf{X}}_{v'}$ is selected by the switch controlled by the estimator. Then, calculate the correlation matrix of size $(B \times B)$ as follows:

$$\mathbf{R}_{v'} = \left(\hat{\mathbf{X}}_{v'}\right)^{\mathrm{T}} \left(\hat{\mathbf{X}}_{v'}\right)^{*} \tag{19}$$

where superscripts ^{T} and * denote the transpose and the complex conjugate, respectively.

Performing the eigenvalue decomposition (EVD), we can obtain the eigenvalues and eigenvectors. Suppose that $\hat{\theta}_{v'}$ denotes the estimated DOA of the signal corresponding to $\mathbf{y}_{v'}$. This is obtained as follows:

$$\hat{\theta}_{v'} = \max_{\theta} Q_{v'}(\theta) \tag{20}$$

$$Q_{v'}(\theta) = \frac{\mathbf{a}(\theta)^{\mathrm{H}} \mathbf{a}(\theta)}{\left| \left(\mathbf{e}_{\mathrm{N}}^{(v')}\right)^{\mathrm{H}} \mathbf{a}(\theta) \right|^{2}} \tag{21}$$

where $\mathbf{e}_{\mathrm{N}}^{(v')}$ denotes the eigenvector of $\mathbf{R}_{v'}$ of size $(B \times 1)$ corresponding to the noise subspace.

4   Numerical Examples

This section provides a numerical example of the proposed method under the conditions listed in Table 1. We generate four fundamental waveforms with different period settings. It should be noted that the number of the signals exceeds the degree of freedom. Therefore, in this situation, MUSIC does not work.

Table 1. Conditions

Antennas	(B)	4
Antenna spacing	(d)	$\lambda/2$
SNR [dB]	$P_{\mathrm{S}}^{(1)} = P_{\mathrm{S}}^{(2)} = P_{\mathrm{S}}^{(3)} = P_{\mathrm{S}}^{(4)}$	-20
		1
The number of signals	(N)	4
The number of period	$M_1 = M_2 = M_3 = M_4$	10000
Fundamental waveforms	$(\phi_1[k], \phi_2[k], \phi_3[k], \phi_4[k])$	Impulsive response roll-off filter
Phase	$(\varphi_1, \varphi_2, \varphi_3, \varphi_4)$	Uniformly random $[0, 2\pi]$
Period [sample]	(T_1, T_2, T_3, T_4)	$T_1 = 31, T_2 = 32, T_3 = 33, T_4 = 34$
Direction of arrival [deg]	$(\theta_1, \theta_2, \theta_3, \theta_4)$	$\theta_1 = -10, \theta_2 = 0, \theta_3 = 10, \theta_4 = 20$
Time offset	$\tau_1 = \tau_2 = \tau_3 = \tau_4$	0

The four fundamental waveforms are generated as follows: Firstly, $\phi_1[k]$ is generated and the period is set at 31 samples, as depicted in Fig. 7. This is the impulse response of the roll-off filter of which the roll-off factor is set at 0.5. Secondly, $\phi_2[k]$ is produced by appending one sample of 0 at the tail, so the period is 32 samples. Likewise, $\phi_3[k]$ and $\phi_4[k]$ are generated by appending 0 s at the tail of $\phi_1[k]$ so that the periods of the two fundamental waveforms are 33 and 34 samples, respectively.

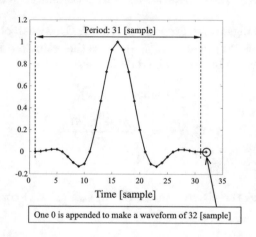

Fig. 7. An example of the fundamental waveform.

By setting the random phase shift φ_n $(n = 1, \cdots, 4)$ following the uniform distribution $[0, 2\pi]$, the fundamental waveforms are successively connected to form chains of the fundamental waveforms, as formulated in (5). Figure 8 shows the real- and imaginary- part of $s_n[k]$ $(n = 1, \cdots, 4)$. It is seen that they are quite similar even though the period is just slightly different. It should be emphasized

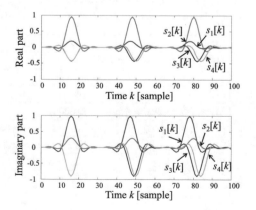

Fig. 8. Examples of the two periodic signals.

that these two signals are overlapping in the frequency domain. Therefore it is impossible to divide those two signals by filtering.

Now, we can estimate the number of signals using Fig. 9, showing the SPC IDs (v) versus the P_v obtained in (18) normalized by the maximum value. It is clearly seen that there are four large values at $v = 31$, 32, 33 and 34, enabling us to estimate the periods of the signals. At the same time, the preprocessing tells us to apply MUSIC for $\hat{\mathbf{X}}_{31}, \hat{\mathbf{X}}_{32}, \hat{\mathbf{X}}_{33}, \hat{\mathbf{X}}_{34}$.

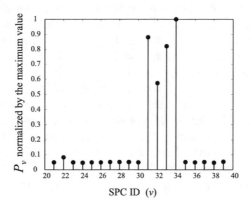

Fig. 9. The number of SPC outputs ports vs. the maximum absolute value of the accumulated SPC outputs.

Finally, Fig. 10 shows the direction of arrival θ versus $Q_{v'}(\theta)$, $v' = 31, 32, 33, 34$ defined in (21). It is clearly observed that the keen peaks of $Q_{v'}(\theta)$ indicate the DOA of the four signals.

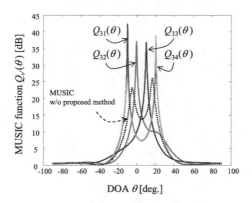

Fig. 10. DOA estimation by the proposed method.

For the sake of comparison, the black dotted line in this figure shows the result obtained by conventional MUSIC. The conventional MUSIC means that the pre-processor part is removed from the configuration in Fig. 3. Since the number of the signal 4 exceeds the degree of freedom 3, MUSIC cannot accomplish the DOA estimation, as the figure shows.

5 Conclusions

In this paper, focusing on the DOA estimation for the vital sensing using Doppler sensors, we proposed to combine our simple parameter estimation method with MUSIC as a pre-processor. The proposed scheme enables MUSIC to estimate DOAs even if the number of signals exceeds the degree of freedom. The numerical example demonstrates its procedure and behavior. In addition, it was confirmed that the proposed method completes the DOA estimation for the environment where conventional MUSIC cannot estimate the DOAs.

Although MUSIC achieves highly accurate DOA estimation, it is computational heavy because it consists of the estimation of the correlation matrix as well as its eigenvalue decomposition. So, the method in [3] is more interesting in terms of the computational load reduction even though its estimation is less accurate than the proposed method in this paper.

Feasibility studies with the signal obtained through experiments are now on going.

References

1. Li, C., Lubecke, V.M., Boric-Lubecke, O., Lin, J.: A review on recent advances in doppler radar sensors for noncontact healthcare monitoring. IEEE Trans. Microwave Theory Tech. **61**(5), 2046–2060 (2013)
2. Yano, K., Kamiya, Y.: A simple signal detection and waveform estimation for biometrics using doppler sensors. In: Proceedings of the 13th IEEE International Conference on Signal Processing (ICSP), Chendu, China, 6–10 November, pp. 1329–1332 (2016)
3. Kamiya, Y.: A simple non-contact vital sensing method using doppler sensors applicable to multiple targets. In: 39th Annual International Conference of the IEEE Engineering in Medicine and Biology Society (under review)
4. Hudson, J.E.: Adaptive Array Principles. Inspec/IEE (1981)
5. Petrochilos, N., Rezk, M., Host-Madsen, A., Lubecke, V., Boric-Lubecke, O.: Blind separation of human heartbeats, breathing by the use of a doppler radar remote sensing. In: IEEE International Conference on Acoustics, Speech, Signal Processing (ICASSP), vol. 1, pp. I-333–I-336 (2007). doi:10.1109/ICASSP.2007.366684

A Comparative Study of Communication Methods for Evacuation Guidance Systems in Disaster Situations

Koichi Asakura[1(✉)] and Toyohide Watanabe[2]

[1] Daido University, 10-3 Takiharu-cho, Minami-ku, Nagoya 457-8530, Japan
asakura@daido-it.ac.jp
[2] Nagoya Industrial Science Research Institute, 1-13 Yotsuya-dori, Chikusa-ku,
Nagoya 464-0819, Japan
watanabe@nagoya-u.jp

Abstract. This paper describes communication methods for evacuation guidance systems in disaster areas. Because network availability depends on the damage situation in disaster areas, three types of communication methods are presented to deal with various evacuation situations: the infrastructure, ad-hoc and hybrid methods. We conduct simulation experiments to compare these communication methods from the viewpoint of sharing information on road conditions among evacuees in disaster areas. Experimental results reveal that the ad-hoc method achieves about 90% the performance of the infrastructure method and thus can be an alternative when communication infrastructure has been damaged or malfunctioned. We also show that the hybrid method obtains results that are almost as good as the optimal results.

1 Introduction

In disaster situations such as big earthquakes, evacuees need to acquire correct map information in order to select appropriate evacuation routes to shelters. Although many map information systems are available such as Google maps [1] and Maps With Me [2], these systems cannot be relied on for evacuation guidance because they cannot provide up-to-the-minute information on road conditions in disaster areas. In disaster areas, some roads become impassable due to the collapse of surrounding houses and damage to roads themselves. In such situations, evacuees are not always better off taking the shortest route to the nearest shelter.

For evacuation guidance in disaster areas, we have already proposed a map construction system on the basis of an ant colony system (ACS) [3]. In this system, mobile devices of evacuees record passed roads and passed time during evacuation to the shelters. Thus, the condition of disaster areas, particularly which roads are passable for evacuation, can be recorded in an up-to-date manner. A road that many evacuees have used to evacuate has a high possibility of being passable. Additionally, the possibility is higher when evacuees have used

© Springer International Publishing AG 2018
G. De Pietro et al. (eds.), *Intelligent Interactive Multimedia Systems and Services 2017*,
Smart Innovation, Systems and Technologies 76, DOI 10.1007/978-3-319-59480-4_52

the road more recently. Evacuees can select appropriate routes for evacuation to shelters by sharing the information on road conditions recorded on their mobile devices.

In order to develop the above evacuation guidance system, a communication method is essential for sharing the information on road conditions appropriately. In disaster areas, network infrastructure such as access points and base stations may be destroyed or malfunctioning, so information may not be able to be shared via the internet. Ad-hoc network technologies are one of the most attractive technologies for such severe conditions [4,5]. In ad-hoc networks, all mobile devices communicate with each other and construct network routes without any communication infrastructure. Thus, ad-hoc networks are suitable for communication systems in disaster areas [6,7].

In this paper, we consider communication methods in the evacuation guidance system. We compare several communication methods by using ad-hoc networks and infrastructure networks. By this comparative study, the most appropriate communication method is decided for the evacuation guidance system in disaster areas.

2   Evacuation Guidance System

The evacuation guidance system provides evacuees with a safe-road map for rapid evacuation to shelters. The system is a kind of collaborative mapping system [8], in which evacuees themselves collect information on road condition in disaster areas as mobile sensors. Information collected on mobile sensors is reassembled by a method based on an ant colony system. By this reassembly method, the recentness of information can be taken into account. First, we describe ant colony systems in Sect. 2.1. Then, we describe the application of ant colony systems to the generation of safe-road maps in Sect. 2.2.

2.1   Ant Colony Systems

An ant colony system (ACS) is a kind of swarm intelligence algorithm inspired by the behavior of ants [9,10]. When ants find food, they move back to their colony. While moving back to the colony, ants lay pheromones on the ground. Thus, trails of pheromones are generated between the colony and food. Ants tend to follow a trail of pheromones while moving and thus can find food. In the same way, ants lay down pheromones from the food to the colony. Thus, the paths used by many ants have high concentrations of pheromones. Since a pheromone is a volatile liquid, it evaporates as time advances. By this mechanism, ants can find the shortest path between the food and the colony.

2.2   Generation of Safe-Road Maps

In our evacuation guidance system, an evacuation route from the current position of an evacuee to the shelter is generated by the concept of pheromones in the ACS [3]. When an evacuee selects a road for evacuation and reaches an intersection of the road, pheromones are dropped onto the road virtually. When many evacuees pass the same road for evacuation, the road comes to have a high density of pheromones. Similarly, when an evacuee has passed a road recently, the road also has a high density of pheromones. Thus, the density of pheromones represents the appropriateness of using the road for evacuation. In our evacuation guidance system, first, an evacuation route to the shelter is generated from only the roads with pheromones. If the route cannot be generated, an alternative route is generated from all roads. To calculate the alternative route, passing costs of roads are weighted with the density of pheromones. Evacuation routes generated by using pheromones may enable evacuees to reach the shelters fast and safely.

3   Communication Methods

In this section, we describe communication methods for sharing safe-road map information. In disaster areas, communication infrastructure may be damaged and malfunctioning. Thus, we present three types of communication methods:

1. Infrastructure method: the communication infrastructure is undamaged and information can be transferred via a radio access network such as LTE and WiMAX.
2. Ad-hoc method: no communication infrastructure can be used, and mobile devices share safe-road information by sending information periodically to neighboring devices in an ad-hoc network manner.
3. Hybrid method: mobile devices can use both an infrastructure network and an ad-hoc network.

The following sections explain these communication methods.

3.1   Infrastructure Method

In this situation, infrastructure networks can be used for communication between mobile devices and a map server. However, network load must be decreased in disaster areas since networks in disaster areas have extra heavy traffic. In order to decrease network traffic, in this method, we use infrastructure networks not for gathering information on road conditions, but for sending safe-road map information to evacuees. Specifically, information on a safe-road map is transferred periodically to mobile devices via an infrastructure network, whereas information on road conditions (pheromones) on mobile devices is gathered when the evacuee reaches a shelter by a wired connection such as a USB or by short range wireless network such as Bluetooth.

Figure 1 shows an overview for collecting information on road conditions in mobile devices. In this figure, lines stand for roads, black bullet points stand for intersections, and white bullet points stand for evacuees. The evacuee moves

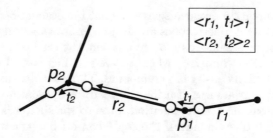

Fig. 1. Recording sequence for passed road and time

along the road r_1 and reaches the intersection p_1 at time t_1. In the mobile application, positional information of an evacuee is measured by a global positioning system (GPS) and the passing road is determined by map matching algorithms [11]. When the evacuee reaches an intersection, the mobile application records a pair of the passed road and arrival time at the intersection with a unique identifier: $<r_1, t_1>_{id}$. This pair is an element of pheromones in the ACS.

Figure 2 shows an algorithm for gathering pheromone information in the map server when an evacuee reaches the shelter. In the map server, pheromone information is stored as a set of pairs of a passed road and a passed time. The identifiers of the pairs ensure uniqueness of elements in the set.

ALGORITHM UpdatePheromoneInformation
Input
    PSet: pheromone set in the map server.
    ESet: pheromone set in the mobile device.
Output
    PSet: updated pheromone set in the map server.
begin
    for each element *e* in *ESet*
        id ← identifier of *e*.
        if element with *id* is not included in *PSet* **then**
            add *e* into *PSet*.
        endif
    endfor
end

Fig. 2. Algorithm for gathering pheromones information

Figure 3 shows an algorithm for calculating pheromone information. In ACS, pheromone information is calculated by using two parameters: α and β [3]. α is the initial value of pheromones, and the β is the decreased value of evaporating pheromones. For each pair $<r_i, t_i>_{id}$, the pheromones information is calculated as follows:

$$\max(0, \alpha - \beta \cdot (t_{current} - t_i)), \tag{1}$$

where $t_{current}$ stands for the current time. Mobile devices receive the set of elements of pheromones periodically and construct safe-road maps.

```
ALGORITHM CalculatePheromones
Input
    PSet: pheromone set.
Output
    Pheromones: set of pairs of road and pheromones.
begin
    Pheromones ← ∅.
    t_current ← the current time.
    for each element e in PSet
        r ← road in e.
        t ← time in e.
        p ← max(0, α - β · (t_current - t)).

        if p ≠ 0 then
            if <r, p_now> is included in Pheromones then
                delete element <r, p_now> in Pheromones.
                add element <r, p_now + p> in Pheromones.
            else
                add element <r, p> in Pheromones.
            endif
        endif
    endfor
end
```

Fig. 3. Algorithm for calculating pheromones of roads

3.2   Ad-Hoc Method

In this situation, infrastructure networks may malfunction and evacuees share information between themselves. No map servers are provided. Thus, mobile devices communicate with neighboring devices periodically and exchange the sets of pheromone elements.

Figure 4 shows a processing flow of a communication protocol for exchanging elements of pheromones. Figure 4(a) shows the positional relationship of mobile devices, which are represented by bullet points. The dashed circle expresses the communication range of the device S. Figure 4(b) shows a communication sequence in the ad-hoc method. Each mobile device sends a PREQ (Pheromone REQuest) packet periodically for requesting pheromones information. The PREQ packet has no destinations and is delivered to all surrounding devices in the communication range. When mobile devices receive a PREQ packet, they return PREP (Pheromone REPly) packets for sending pheromones as replies. A PREP packet consists of the packet destination and series of pheromone elements. When a mobile device receives a PREP packet, the PEXG (Pheromone EXchanGe) packets are sent to the sender of the PREP packet for exchanging pheromones information. The PEXG packet has the same contents as the PREP packet. By this three-phase packet exchange, mobile devices share pheromone information.

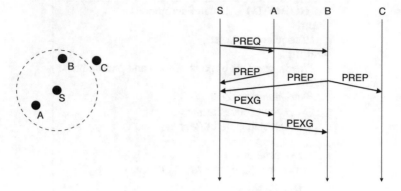

(a) Positional relationship of devices   (b) Communication sequence among devices

Fig. 4. Communication sequence in the ad-hoc method

To collect information on road conditions and construct a safe-road map, the same algorithms as in Sect. 3.1 are used.

3.3   Hybrid Method

In this situation, both the infrastructure and ad-hoc methods are activated in the system. In other words, to gather pheromone information, information acquired from both the map server and neighboring devices is used. All algorithms from Figs. 1 to 4 work cooperatively in the hybrid method.

3.4   Comparative Analysis

This section analyzes the pros and cons of the above three methods comparatively. In the infrastructure method, information on road conditions is not generated until evacuees reach the shelter. That is, it takes a longer time to construct a safe-road map because information is gathered from evacuees who can reach the shelter safely. However, the constructed safe-road map is robust for evacuation guidance. In the ad-hoc method, evacuees can share information on road conditions rapidly. However, quickly constructed safe-road maps consist of only local information and may not be reliable for evacuation. Of course, in the hybrid method, ideal safe-road maps can be constructed by both robust information from the map server and up-to-the-minute local information from neighboring devices. However, the communication frequency becomes high, which causes batteries in mobile devices to quickly deplete.

4   Experiments

In this section, we compare communication methods described in Sect. 3. We analyze performances of the methods in a simulation.

4.1   Simulation Setting

We have developed a simulation system for disaster areas [12]. The simulation system is developed in Java. We can develop and evaluate communication protocols and movement algorithms of evacuee agents in the simulation system.

For the simulation area, we provide a real road map with 2.0 km east to west and 1.5 km north to south. Figure 5 shows the simulation area, which has one shelter to which all evacuees try to move. In order to simulate damages to roads, 1% of roads in the area are selected randomly and become impassable every 5 min during the simulation.

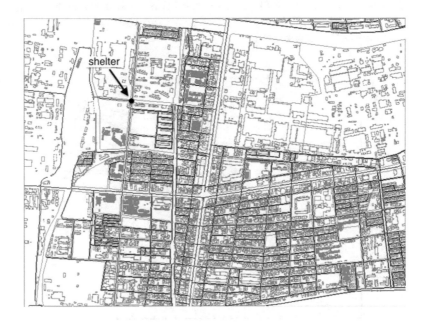

Fig. 5. Simulation area

Five-hundred evacuee agents are deployed on roads randomly. To begin with, 100 agents are randomly selected and start to evacuate. After that, the remaining evacuee agents start to escape to the shelter at a rate of 0.2 persons/second in accordance with a Poisson distribution.

In this experiments, we observe trends in the number of evacuees who can reach the shelter as time goes on. We conduct 10 experiments for each communication method.

4.2   Experimental Results

Figure 6 shows the experimental results. The graph shows the number of evacuees who can reach the shelter by using the three communication methods. Additionally, the graph contains an ideal results in which all evacuees have a god's eye view, i.e., all evacuees understand the disaster situation fully and can select the optimal routes to the shelter. In this graph, the X axis denotes the elapsed time, and the Y axis denotes the number of evacuees who have reached the shelter by that time.

From the comparison between the infrastructure and ad-hoc methods, we can observe that the infrastructure method obtains better results. However, the performance of the ad-hoc method reaches about 90% of that of the infrastructure method. Especially, in the first 40 min of the simulation, the two methods perform almost equally. This is because the infrastructure method does not gather a sufficient amount of pheromone information at an early time since few evacuees have reached the shelter. From this result, we can conclude that the ad-hoc method works sufficiently for evacuation guidance when communication infrastructure has malfunctioned.

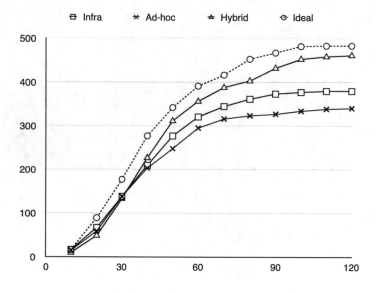

Fig. 6. Experimental result

From the comparison among the three methods, we can observe that the hybrid method achieves the best results. This is because the hybrid method inherits advantages from both the infrastructure and the ad-hoc methods. In the hybrid method, information on pheromones propagates rapidly and widely: a global situation can be provided by the infrastructure method, and local situations can be provided by the ad-hoc method.

From the comparison between the hybrid method results and the ideal results, the hybrid method results are about 90–95% of the ideal results. Thus, we can conclude that the evacuation guidance system with the hybrid communication method works the most appropriately in disaster situations.

5   Conclusion

In this paper, we developed communication methods for an evacuation guidance system. We proposed three types of communication methods: infrastructure, ad-hoc and hybrid. Their performances were analyzed in simulation experiments. From the experimental results, we concluded that the ad-hoc method is good enough to be an alternative to the infrastructure method when communication infrastructure has malfunctioned. We also found that the results of the hybrid method were comparable to the optimal results.

For our future work, we have to consider communication frequency for the evacuation guidance system. Although the hybrid method achieves excellent performance, it has a much larger communication load than the other two methods. Thus, we have to take appropriate intervals of communication into account. Furthermore, in the ad-hoc method, an efficient algorithm for exchanging information on pheromones among neighboring devices has to be developed in order to decrease the frequency of communication and the amount of transferred data.

Acknowledgements. This work was supported by the Hibi Science Foundation.

References

1. Google Inc.: Google Maps. https://maps.google.com/
2. My.com: MAPS.ME. http://maps.me/
3. Asakura, K., Watanabe, T.: Construction of navigational maps for evacuees in disaster areas based on ant colony systems. Int. J. Knowl. Web Intell. **4**(4), 300–313 (2013)
4. Toh, C.-K.: Ad Hoc Mobile Wireless Networks: Protocols and Systems. Prentice Hall, Upper Saddle River (2001)
5. Murthy, C.S.R., Manoj, B.S.: Ad Hoc Wireless Networks: Architectures and Protocols. Prentice Hall, Upper Saddle River (2004)
6. Midkiff, S.F., Bostian, C.W.: Rapidly-deployable broadband wireless networks for disaster and emergency response. In: The 1st IEEE Workshop on Disaster Recovery Networks (2002)
7. Meissner, A., Luckenbach, T., Risse, T., Kirste, T., Kirchner, H.: Design challenges for an integrated disaster management communication and information system. In: The 1st IEEE Workshop on Disaster Recovery Networks (2002)
8. Balaram, S., Dragicevic, S.: Collaborative Geographic Information Systems. Idea Group Pub., Brandon, Hershey (2006)
9. Dorigo, M., Stutzle, T.: Ant Colony Optimization. Bradford Company, Cambridge (2004)
10. Blum, C.: Ant colony optimization: introduction and recent trends. Phys. Life Rev. **2**(4), 353–373 (2005)
11. Asakura, K., Takeuchi, M., Watanabe, T.: A pedestrian-oriented map matching algorithm for map information sharing systems in disaster areas. Int. J. Knowl. Web Intell. **3**(4), 328–342 (2012)
12. Asakura, K., Watanabe, T.: A simulation system of disaster areas for evaluating communication systems. In: The 2nd International Symposium on Intelligent Decision Technologies, pp. 495–506 (2010)

Research View Shift for Supporting Learning Action from Teaching Action

Toyohide Watanabe[(✉)]

Nagoya Industrial Science Research Institute, 1-13 Yotsuya-dori,
Chikusa-ku, Nagoya 464-0819, Japan
watanabe@nagoya-u.jp

Abstract. The teaching action takes an important role to define the educational technology, and various functions/methods to support the teaching actions have been investigated since the research field of educational technology was proposed at first. However, now that we can count 40 or more years since the start-up of the activities in the educational technology, our learning environments were changed. The teaching support technology has enforced the learning process to change drastically from the teaching action, by which a teacher or a tutor instructs at once many learners synchronously in the centralized classroom, to the learning action, which learners themselves must perform their own learning works. In this paper, we address this technological transition from the teaching action to the learning action. Our viewpoint for the technological transition is to discuss the learning process with respect to knowledge management. Also, our concluding remark focuses on the support functionality and the framework for making the learning action clear.

Keywords: Knowledge transfer scheme · Knowledge understanding model · Learning cycle · Educational technology · Learning science · Learning action · Teaching action

1 Introduction

The research fields in the educational engineering have been growing up with the developments of tools, investigations of methods, accumulations of technologies, adjustments of social systems, innovational managements of national government, etc., though the name of research field was replaced by the educational technology. The first important work in establishing the educational technology was to define the research objective and activity goal. For example, the committee in AECT (Association for Educational Communications and Technology), USA defined in 1977 [1]:

"Educational technology is a complex, integrated process involving people, procedures, ideas, devices, and organization, for analyzing problems and devising, implementing, evaluating, and managing solutions to those problems, involved in all aspects of human learning. In educational technology, the solutions to problems take the form of all the Learning Resources that are designed and/or selected and/or utilized to bring about learning; these resources are identified as Messages, People, Materials, Devices,

© Springer International Publishing AG 2018
G. De Pietro et al. (eds.), *Intelligent Interactive Multimedia Systems and Services 2017*,
Smart Innovation, Systems and Technologies 76, DOI 10.1007/978-3-319-59480-4_53

Techniques, and Settings. The processes for analyzing problems, and devising, implementing and evaluating solutions are identified by the Educational Development Functions of Research-Theory, Design, Production, Evaluation-Selection, Logistics, Utilization, and Utilization-Dissemination. The processes of directing or coordinating one or more of these functions are identified by the Educational Management. .... Educational technology is a theory about how problems in human learning are identified and solved. Educational technology is a field involved in applying a complex, integrated process to analyze and solve problems in human learning. Educational technology is a profession made up of an organized effort to implement the theory, intellectual technique, and practical application of educational technology".

Of course, some leading persons in such academic associations for educational technology in Japan defined individually: for example, in the dictionary of educational technology, edited by Japan Society for Educational Technology and published by Jikkyo Shuppan Co., Ltd. in 2000, "educational technology" is described as [2]:

"Educational technology is an academic research field which can focus on the issues about the theory, the method, and the environment design under the fundamentals of educational improvement and development, and contribute to perform the practical implementation, and also has an objective to develop and improve the systematic technology. Many kinds of such technical 'how-to' knowledge can make the effectiveness and efficiency of educations more valuable. At least, the pioneers in the teaching/learning research fields, first of all, focused on the definition of educational technology and have been endeavoring to make the characteristic research domains successful and fruitful together with many researchers and educational experts of the same or similar interests under the definition" (translated by T. Watanabe).

In the initial step to establish the educational technology as an advanced academic research field, the main viewpoint for research activity focused on the teaching action (or instructional action). So, the point is to assist or help teachers' works effactually and effectively since the instructional means for many learners who were gathered at once in the same classroom, are manipulated by only a teacher. In order to reduce the teacher's loads in this situation and instruct each learner successfully, the tools, methods, systems, authoring texts (or contents), procedures, etc. have been investigated and improved with the technology, social system, educational policy, etc. The concept of CAI (Computer Aided Instruction) is the most typical instance which solved the situation along such an objective research.

However, today is neither similar nor the same as the situation in the initial stage of educational technology. The current learning environments have been constructed under the cyber space, and individual learners can manipulate various kinds of IT (Information Technology) -based tools/media ubiquitously. At least, 40 or 50 years ago these IT-based tools/media were not available for every learner, but were usable for only limited teachers from 30 or less years ago. The change of evolutional technology, in particular IT-based change, influenced to various approaches in the learning style, learning method, and learning environment in addition to the teaching style, teaching method, and teaching environment. This trend or evolutional flow changed the target to be recognized from the teaching actions to the learning actions, more or less. In case that we consider our learning method, or our learning system, our learning environment from a viewpoint of perspective in the future, it is important to focus on the learning actions of learners in him/her-self rather than the teaching action of teacher. Of course, the relationship between teaching action and learning action may be front-back or pairwise relationship.

In this paper, we address the issue of supporting the learning action in our current IT-based cyber society from a viewpoint of environmental transition from the teaching action to the learning action. Our characteristic point is to focus on the activeness-oriented learning as the most necessary/desirable learning style, and discuss it under our learning cycle, which we developed [3]. Our idea in this discussion is to explain our learning process with respect to the knowledge management and take into consideration "thinking" in our learning cycle.

2   Framework of Knowledge Understanding

Until today, we have insisted that the learning is a kind of knowledge transfer and it can be defined as our knowledge transfer scheme, shown in Fig. 1 [4]. Namely, the basic and important mechanism is to transfer the knowledge in our out-world into our brain/notes, regarded as the in-world. The systematic process to support this transfer mechanism can be better organized by three stepwise phases such as knowledge composition, knowledge acquisition, and knowledge understanding. The knowledge is moved mechanically from the knowledge resources or learning contents in the out-world as "rote memorization", but the moved knowledge resources cannot always become knowledge if they could not be used or has never been used. This is because the moved information is knowledge if it should be applied successfully to the next action/work of acquired learner. However, if not so, the moved information is not never regarded as "knowledge".

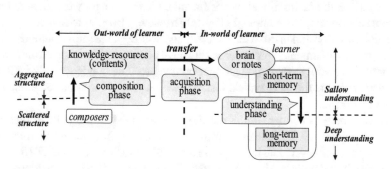

Fig. 1. Knowledge transfer scheme

Hereafter, we interpret analytically the knowledge understanding phase in our knowledge transfer scheme with respect to our knowledge understanding model, shown in Fig. 2 [5]. Figure 2 indicates three different stages such as short-term stage, middle-term stage, and long-term stage, though two-levels memory hierarchy between short-term memory region and long-term memory region is used in the production system.

We introduced third memory region "middle-term stage" newly with a view to representing our understanding process more successively. Of course, whether this model is true or not is not an important factor for us, but it is necessarily satisfied to take into consideration such discriminated manipulation of knowledge. We can use the memory management logically by some practical explanations.

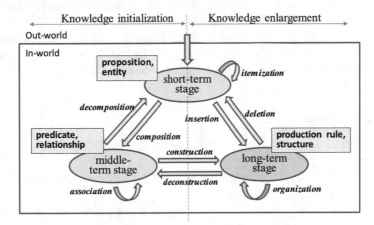

Fig. 2. Knowledge understanding model

These different stages keep inherently different classes of knowledge fragments/ structures individually. Figure 3 arranges such forms or structures of knowledge fragments in each memory stage. The short-term stage keeps individual instances such as propositions: P, Q, R(a), S(s), …; the middle-term stage does predicate types: R(x), S(y), G(u, v), R(x)∧S(y), …; and the long-term stage organizes the structured knowledge of several related-type fragments: R(x) → S(y), R(x)∨S(y) → W(y, x), ….

As for these three-levels hierarchy in Fig. 2 the flow is not only effective from the short-term stage to the long-term stage through the middle-term stage, but also the direct procedure to the long-term stage from the short-term stage is allowable.

3   Thinking in Learning Cycle

The phrase "from memorization learning to thinking learning" is one paradigm shift in the policy of teaching/learning methodology. The practical cases proposed and promoted the activeness-oriented learning such as flip teaching (or flipped classroom), active learning, etc. These point out that learners should learn by themselves firstly before their teacher will instruct in their classroom. The flip teaching may be similarly looked upon as the pre-learning in the traditional learning process, which must be done by him/herself before the scheduled lecture; and the active learning depends on learner own willingness and relates to the collaborative work with other learners. In the activeness-oriented learning paradigm the important factor is "thinking": the learning is not only to memorize knowledge simply, but also to think the meaningful structure among related knowledge deeply, or to derive the analogical and inferential relationships among derived knowledge fragments, newly acquired knowledge fragments and existing knowledge fragments. An important problem in our learning action with the help of teaching action, if necessary, is how we focus on thinking procedure or thinking step in the learning process.

The learning cycle represents learner's activity, derived extensively from our knowledge transfer scheme [3]. The learning cycle consists of three knowledge works such as acquisition, understanding, and utilization. This framework in the learning cycle is consistent to explain the activeness-oriented learning such as active learning, flip teaching, etc. conceptually, as shown in Fig. 4. The important view is that the acquired/understood knowledge should be always used or be meaningfully kept so as to be usable in the next chances.

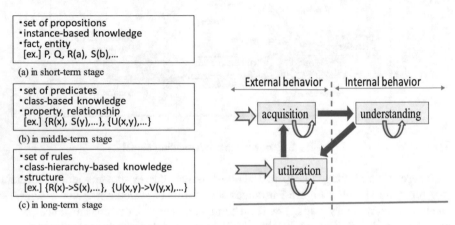

Fig. 3. Knowledge forms

Fig. 4. Learning cycle

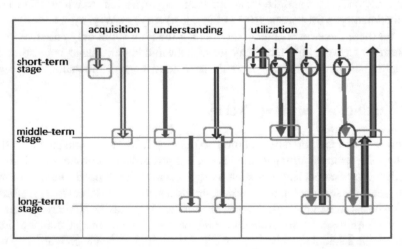

Fig. 5. Thinking procedures in each phases of learning cycle

Figure 5 shows the thinking procedures over individual knowledge works in our learning cycle. The triggers are indicated from the out-world, and also the works are directly observable in the out-world. However, the understanding work is not observed. The reference of learning contents by learners, and teaching start by teacher are the triggers in the acquisition, while the questions, exercises, examinations and so on are so in the utilization. In Fig. 5, the connected/circled parts in individual stages illustrate that thinking procedures are performed in each work. Thus, it is important to take into consideration how to promote thinking procedures effectively and effectually with a view to making the acquired knowledge more fruitful.

4   Supporting Learning Action

The learning action may be approximately defined as the learning procedure that the learner should do by him/her-self and can look upon as the important work for learning activity in the current information society. So, in this cyber world many changes are observed in various situations:

(1)   Distributed self-learning-oriented environment from centralized classroom-oriented environment;
(2)   Asynchronous learning in different open learning spaces from synchronous learning in the same closed learning space;
(3)   Horizontal diversity for evaluating learner's ability from vertical diversity.

Of course, all learning situations are not replaced at once but their ratios are to increase with the time. These changes have been enforcing to improve, enhance, and evolve various learning methodologies and technologies. This trend requests to re-design the learning policy, learning procedure, learning support means/methods, etc. and also to re-consider more powerful usage, more applicable management and more successful evaluation. Namely, the learning paradigm, based on the activeness-oriented learning of learners, enforced to make a role of learning action clear in comparison with the teaching action; the learning action takes a "main (front)" role but the teaching action does a "sub (back)", though the teaching action was focused strongly in the definition of educational technology.

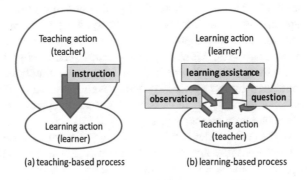

Fig. 6. learning action and teaching action

Figure 6 shows the front-back relationship between the teaching action and the learning action. The currently desirable relationship is that the learning action is "front" and the teaching action is "back": for example, the teaching action can be looked upon as a kind of learning means in the acquisition phase as we referred to our knowledge transfer scheme in Fig. 1.

Next, under our recognition for the learning action we discuss what and how to support the learning action. As for this issue we must pay attentions to two different views: what is the learning activity of learner?; and how to support the learning activity? Of course, these two views are very closed mutually.

4.1   What Is a Learning Activity of Learner?

Because the learning activity was specified as the learning cycle with acquisition, understanding and utilization phases, we must grasp the following situations:

(1)   Learning action (or activity) is performed by learner in him/her-self;
(2)   Learning activity is not transient, but repeatedly continued;
(3)   Learning process/procedure is not similar among individual generations, learners' properties, learners' interests, and others;
(4)   All learners do not always want to learn, but it is rather true to avoid learning if almost many of them were not enforced.

Although researchers must sufficiently recognize these requirements for learning activities, it is very difficult to catch up them directly as it is. These four requirements are very fundamental in the learning, but are difficult to make the goal explicit.

Concern to (1): this issue relates to the goal achievement. The goal achievement consists of performance goal and understanding goal [6, 7]. The performance goal is that learners should attain the predefined learning units completely; and the understanding goal is that learners should pass the understanding criteria, preset for these learning units. The currently developed/investigated learning support tools/systems focused mainly on the performance goal. This is because the support of performance goal is easier than that of understanding goal. The currently IT-supported work is almost successful in the performance goal. The understanding goal in the learning activity is accomplished only by the learner him/her-self, but not absolutely done by other persons or other means/methods [7]. The understanding goal is one of high-level requirements in the learning-specific work. In the learning activity, the performance goal and the understanding goal must be investigated, but the methods/means or functions/mechanisms are not always successful. Currently, many learning support systems are unidirectional control from the system to learners. Even if they are bidirectional the interactive mechanism is controlled by the system: system-driven interaction. The performance goal can be controlled not only by the learner, but also by system, while the understanding goal is dependent on learner's intention.

Even if the systems could understand the learning units, what meaningful is it? The performance goal is commonly appreciated by the progress degree, and the progress degree can be explicitly calculated by the ratio of the currently executed learning units for all learning units to be performed. While, the understanding goal is likely evaluated by the understanding degree, but the degree may not be able to be desirably estimated through checking up partially with examinations, exercises, Q&A, interview, etc. Thus, it is easy to support the performance goal: learning support systems/tools provide with functions/mechanisms, which can promote the learning progress effectively or effectually. Namely, it is at least necessary and important that the course-ware which can be derived from syllabus is arranged, and then individual learning units are successively prepared under their predecessor-successor relationships.

The learning support system can interact stepwise with learners under the well-arranged course-ware and well-composed learning units. Learning support systems/tools have mainly focused on this performance goal and many successful systems/tools have been proposed or developed until today. On the other hand, it is not easy for learning support systems/tools to implement the understanding goal successfully. This is because whether the learner could understand the knowledge about learning units or not is not observable in the explicit form, and in the current strategy we have not any methods beyond the means that checks up the output reactions for the actions, input from the out-world of learner. In this case, whether actions are correct or whether the reactions represent all interactive results is questionable.

Concern to (2): this issue requires that the learning support tools/systems should be designed or developed under the well-arranged course-ware or well-composed learning units. Of course, every approach/investigation for learning style and learning practice should be variedly evaluated under the successive relationship between the pre-assigned situation and current situation. The knowledge to be available in the practical application must be refined and re-generated under the repeated spiral (namely, learning cycle) in knowledge management paradigm because individual knowledge fragments have been re-constructed continuously, and have been growing up one by one. Therefore, we must pay attentions to the flow of course-ware in which learning units are available and also compatible for successive learning units. In such a viewpoint, the currently usable or used learning support systems are typically CMS, LMS, etc.; and we can understand that the management functionality is important.

Concern to (3): this issue says that there are not unified methods for supporting all learners. Even learners in the same generation have differently their own characters, interests, experiences, skills, etc., though almost the learners are different individuals by generations. Thus, in case that we develop some new learning tools/systems or we investigate the characteristics of learners, we must develop or investigate so as to be consistent to the difference, derived from learners as possible as we can. So, when the learner uses learning support tools/systems newly, he/she must generally acquire another knowledge, know-how, skill, and so on to use them smartly or grow up the learning effectiveness by using them. This may be a kind of drawbacks for some learners, but not so for other learners. The drawbacks attended additionally to the inherent work are looked upon as a cognitive load [8, 9]: this type of cognitive load is regarded as the extraneous load. The cognitive load, associated with subject work, is called as the intrinsic load. The cognitive loads are different learner by learner. The learner with well-experience may have not unnecessary extraneous loads; and the learner without rich skills may have comparatively more extraneous loads.

Concern to (4): this issue is the most important and radical view. Namely, this indicates what should be done when learners are not always want to learn. In this case, we must consider this issue from a viewpoint of human activity. This is at all different from the viewpoints of issues (1), (2), and (3). The learning is better to be pleasant and the learning is wonderful to be always supported timely and kindly by others when he/she wants the help [10]. The support function is to be useful and effective so as to be consistent to his/her abilities, interests, progresses, characters, and so on. Such viewpoints are necessary for this issue and the support functionality, systems, and environments must be investigated corresponding to the viewpoint.

4.2   How to Support Learning Activity?

In this research field for educational technology and learning science (or engineering), the researchers with original experiences are classified into three fields [11]: engineering (mainly, information engineering: hereafter, we call this the information field), social science (mainly, education: hereafter, we call this the education field), and humanities (mainly, cognitive science and psychological science: hereafter, we call this the cognitive field). The researchers, based individually on these three fields, are naturally different in their approaches as well as their interesting points in the learning activity. The information field focuses on the development or improvement of processing functions and information systems for supporting learning actions with a view to making learning progress effective and effectual on the basis of information technology; the education field concentrates to data analysis and proposition validation (or hypothesis verification) through the investigation and experiment from a viewpoint of observing learners' behaviors; and the cognitive field contributes to the modelling and conceptualization by making the transitional phenomena of learner's willingness/intention analytically clear through the interview and observation. Namely, the learning actions were analytically researched in the information field by the direct utilization/application of information technology, done in the education field by the indirect observation estimated from learner's activity, and done in the cognitive field by the indirect inferences of learner's willingness/intention.

Table 1 arranges the differences among these three approaches.

Table 1.   Characteristics in cognitive, education and information fields

	Existing field	Methodology	View point	Learner	Research interest
Cognitive field	Humanities (cognitive science, psychological science, etc.)	Observation, interview, etc.	Modelling, conceptualization, etc.	Internal behaviors of learners	Thinking process
Education field	Social science (education, etc.)	Investigation, observation, experiment, etc.	Data analysis, experiment survey, etc.	External results of learners	Working process
Information field	Engineering (information engineering, etc.)	Development, improvement, design, etc.	Information technology, etc.	Support of learners' activities	Activity support progress

5   Conclusion

We can regard that the teaching action and the learning action may have front-back relationship, or we can interpret that they are not so. Historically, the teaching action was a key-phrase in our education technology for 50 or more years until today. For example, if we refer to the category of submission sessions in ED-Media conferences [11], these technical terms in the categories are definitely assigned under the teaching action of the educational technology. However, now that today is called the high-level information society (or knowledge infrastructure society), we must focus on the learners' behaviors or learners' activities directly and consider the learning actions.

References

1. AECT Task Force on Definition and Terminology. The Definition of Educational Technology. In: AECT, p. 16 (1977)
2. Japan Society for Educational Technology (ed.): Educational Technology Dictionary. Jikkyo Shuppan Co. Ltd (2000)
3. Watanabe, T.: Architectural framework in next learning support environment. In: Proceedings of ACM/IMCOM 2017, #S1–6 (2017)
4. Watanabe, T.: Learning support specification, based on viewpoint of knowledge management. In: Proceedings of E-Larn 2012 (AACE), pp. 1596–1605 (2012)
5. Watanabe, T.: Simulation analysis of knowledge understanding. In: Proceedings of ED-Media 2007 (AACE), pp. 2479–2486 (2007)
6. Dweck, C.S.: Motivational processes affecting learning. Am. Psychol. **41**, 1–9 (1986)
7. Watanabe, T.: A framework of information technology supported intelligent learning environment. In: Proceedings of ACM/IMCOM 2016, #11–4 (2016)
8. Sweller, J., van Merrienboer, J.J.G., Paas, F.G.W.C.: Cognitive architecture and instructional design. Educ. Psychol. Rev. **10**, 251–296 (1998)
9. Watanabe, T.: Analysis of learning-support scheme, based on cognitive load. In: Proceedings of E-Learn 2015 (AACE), #45716 (2015)
10. Watanabe, T.: Feature analysis of computer-game to make learning process pleasant. In: Proceedings of IEEE/IGIC 2013, pp. 263–266 (2013)
11. Watanabe, T.: Trends in teaching/learning research through analysis of conference presentation articles. In: Proceedings of KES/IIMSS 2016, pp. 309–322 (2016)

Video Saliency Using Supervoxels

Rahma Kalboussi[(✉)], Mehrez Abdellaoui, and Ali Douik

Networked Objects Control and Communication Systems Laboratory,
National Engineering School of Sousse, Pôle technologique de Sousse,
Route de Ceinture Sahloul, 4054 Sousse, Tunisia
Rahma.kalboussi@gmail.com

Abstract. Physiology and neural systems researchers revealed that the visual system is attracted by some parts of an image more than others. Different computational models were developed to simulate the visual system. In this paper we propose a video saliency model that helps to predict and detect the regions of interest in each video frame. We use a supervoxel segmentation as an indicator of dynamic objects. Based on the observation that dynamic objects attract attention when an observer watches a video sequence, supervoxel segmentation provides a first estimation for what belongs to foreground and background. Then, a saliency score is attributed to each supervoxel according to its motion distinctiveness. Experiments over two benchmark datasets, using several evaluation metrics have shown that our proposed method outperforms five saliency detection methods.

1 Introduction

Naturally, the human visual system is attracted by a specific region in the visual scene that we call region of interest. In that region of interest, one specific object attracts attention more than others called Salient object. Visual Saliency can be defined as the distinct subjective perceptual quality which makes some items in the scene stand out from their neighbors and immediately attracts our attention [9].

The first visual image saliency model was developed by Koch and Ullman [14], based on the feature integration theory, it extracts feature maps by decomposing the visual scene into elementary attributes, then fuses all these maps into one master map called saliency map. Yang et al. [27] introduced a model where the image is represented by a close-loop graph with superpixels as nodes which are ranked according the similarity to background and foreground queries. Klein and Frintrop proposed in [13] a saliency method based on the standard structure of cognitive visual attention models. The model proposed by Garcia et al. [6] is based on two biological mechanisms: the decorrelation and the distinctiveness of local responses, a graph based bottom-up saliency model developed by Harel et al. [7] consists of two steps: the activation maps generation over feature channels, and their normalization. Cheng et al. proposed in [3] a regional contrast based salient object detection model which assumes that human cortical cells preferentially respond to high contrast stimulus in their receptive fields, Hou

© Springer International Publishing AG 2018
G. De Pietro et al. (eds.), *Intelligent Interactive Multimedia Systems and Services 2017*,
Smart Innovation, Systems and Technologies 76, DOI 10.1007/978-3-319-59480-4_54

et al. introduced in [8] a spectral based method by analyzing the spectrum of the input image in order to extract the residual spectrum n the spectral domain, then the corresponding saliency map is constructed in the spatial domain. Scharfenberger et al. proposed an original saliency model in their paper [23] based on the image texture, first the input is divided into coherent regions according to the texture to elaborate the textural model, which will be used for saliency ranking. Lou et al. introduced a regional color contrast model [16] where they used low-level and medium level cues of a visual scene. Low-level cues are used to detect the main features and medium level cues are incorporated to detect more structural information.

Some models address saliency detection in the spatial domain, some others in the spatio-temporal domain.

In this paper, we will present a video saliency detection approach. First, for each video sequence, a supervoxel segmentation is performed to lead segmented frames. Then, for each segmented frame, we define a graph where vertices are supervoxel nodes and edges are the spatial connection between each two supervoxels. Finally, we assign to each vertex a saliency score according to its motion distinctiveness.

In Sect. 2 we will present related works on video saliency and video segmentation. In Sect. 3 we will focus on our proposed video saliency method. In Sect. 4 we will present and discuss experimental results. Finally, a conclusion will be given in Sect. 5.

2 The Related Work

2.1 Saliency in Videos

Video saliency aims to detect object of interest in a video frame. While shape and appearance cues play an important role in defining the region of interest in images, moving targets attract video watchers. In this context, several methods have been developed to predict saliency in videos. Singh et al. presented in [24] a method that extracts salient objects in video by integrating color dissimilarity, motion difference, objectness measure, and boundary score feature, they used temporal superpixels to simulate attention to a set of moving pixels. Mauthner et al. proposed an approach in [19] based on the Gestalt principle of figure-ground segregation for appearance and motion cues. Wangua et al. [25] video saliency object segmentation model based on geodesic distance where both spatial edges and temporal motion boundaries are used as foreground indicators. Zhong et al. [28] proposed a novel video saliency model based on the fusion of a spatial saliency map which is inherited of a classical bottom-up spatial saliency model and temporal saliency map which is the result of a new optical flow model. Based on Bottom-up saliency, Mancas et al. introduced a motion selection in crowd model in [18] where optical flow is used to determine motion region. A Spatio-temporal Visual Saliency Model introduced by [21] where a resultant saliency map is the fusion of static saliency map and dynamic saliency map. The research in the spatio-temporel saliency field is quite new and the few existing models are in their early stages. Kim et al. tried to separate background from foreground to highlight the salient object in their paper [12], this separation comes from the observation that foreground which contains high dimensional features attracts attention more than low and medium level cues. First, an over segmentation is realized then, a high level color transform is applied to compute saliency.

2.2   Video Segmentation

Video segmentation can be used in divers video analysis tasks like object tracking, action recognition, etc. While superpixels were widely used in image segmentation, researchers tried to explore this important technique for video segmentation: the model proposed by Chang et al. [2] and Galasso et al. [5]. Some others tried to develop the dynamic version of superpixels which was introduced as supervoxel segmentation. Supervoxel hierarchies provide a multi-scale decomposition of a giving video sequence into small space–time tubes which can be used in different tasks in video analysis. In Their paper, Xu et al. [26] proposed a method that attens the supervoxels hierarchy into a single segmentation. They group supervoxels from different hierarchies into slides by using their semantic information. Jain and Grauman [11] proposed a semi-supervised foreground segmentation using supervoxel potential. This approach takes as input a labeled video frame where the foreground object is outlined and the output is a segmented frame as output where foreground object is highlighted in the rest of the input video frames. Lu et al. [17] proposed a human action segmentation approach using supervoxel consistency. Their Markov Random Field model takes into account low-level video fragments and high-level human appearance and motion to cover all static and dynamic parts of the human body; then supervoxels consistency through different levels of the hierarchy is used to improve segmentation. Oneta et al. [20] propose a supervoxel segmentation method which uses superpixels, optical flow estimation and edge detection.

3   Video Saliency

Given a video sequence, the main purpose of our paper is to compute the saliency of each frame and determine the object that attracts viewer's gaze. First, we perform a supervoxel segmentation using [20], then elaborate the associated graph, from which a saliency score will be attributed to each supervoxel by computing its motion distinctiveness.

3.1   Graph Construction

In their paper, Oneata et al. [20] proposed a supervoxl segmentation approach which aggregates connected superpixels temporally and spatially by using hierarchical clustering. The main advantage of Oneata's supervoxel method is that hierarchical supervoxels contain probable objects. Unlike the method proposed by Xu et al. [26], Oneata's method provide different segmentation levels which will serve us for a good saliency prediction. Therefore, we tried to get for each video sequence the appropriate segmentation level.

We define G, a complete graph $G = \{U, E\}$ where U is the number of l vertices which is the number of supervoxels in the frame and E is the set of $l(l-1)/2$ edges of the graph which connect a pair of supervoxels. For each video sequence we fix the segmentation level to 200, then construct the graph associated to each frame in the video sequence. If, the graph G does not have a minimum number of vertices equal to two, then the next level is considered. This action is repeated until we get a number of vertex superior or equal to two. The main reason for choosing two as a minimum number of vertices is that a perfect segmentation provides two supervoxels one for the background and another for the dynamic object.

3.2   Motion Distintiveness

In video sequence, the observer does not have enough time to examine the whole scene, that's why his gaze is always directed to the object which is distinct from the others and changes position (dynamic object). In this paper, we consider a salient object, the one that have a unique and distinctive movement compared to the rest of the scene.

Thus, we introduce a new metric inspired from recent works on image saliency proposed by [23] which computes the uniqueness of an object in a video frame using a supervoxel segmentation of that frame. A supervoxel is salient if it has a low motion commonality compared to the rest of supervoxels. Given a segmented frame, we will extract the most revealing and informative supervoxels. Therefore, we introduce a metric which quantify the uniqueness and the distinctiveness of the dynamic characteristics of a supervoxel. The concept of motion distinctiveness is introduced to subtract the region of interest from the rest of the scene. A salient region is characterised by a low motion commonality with regards to the rest of the video frame.

Given a segmented video frame, let be V_i and V_j two supervoxels in the frame $F(x)$. We define the probability of V_i could have the same motion characteristics as V_j as:

$$P(V_i|V_j) = \prod_k P(V_{i,k}|V_{j,k}) \tag{1}$$

where $V_{i,k}$ is the set of supervoxels that have the same motion cues and are segmented using the same color. $P(V_i|V_j)$ can be explained as the statistical commonality between two supervoxels.

Since we would like to determine the distinctiveness between two supervoxels for saliency detection, we define another metric which computes the probability that two supervoxels do not have the same motion features.

$$\rho_{i,j} = 1 - P(V_i|V_j) = 1 - \prod_k P(V_{i,k}|V_{j,k}) \tag{2}$$

Giving this definition, as more as two supervoxels becomes similar as more as $\rho_{i,j}$ decreases and vice versa.

3.3 Saliency Computation

If we consider the graph G = {U, E}, we define the weights over each edge $e_{i,j}$ as to value of motion distinctiveness $\rho_{i,j}$ between a pair of supervoxels V_i and V_j which is equal to $\rho_{i,j}$.

Using the supervoxel segmentation and the computed motion distinctiveness of each supervoxel, saliency map of each video frame can be computed using the assumptions that a supervoxel is salient if its has a distinct movement from the rest of the scene (**motion distinctiveness**). Previous works on saliency detection like Achanta et al. [1,7] have proved that gaze is focus on objects in the center of the scene. Thus, a supervoxel can be associated to a salient object if it is spatially closed to the center of the video frame (**center surround constraint**).

The saliency of a supervoxel can be computed as the product of the expected motion distinctiveness of the supervoxel in the video frame F(x) and the pixel's weighted spatial proximity, which belong to the supervoxel V_i.

For each supervoxel the saliency S_i can be computed fallowing Eq. 3

$$S_i = \sum_{j=1}^{l} \rho_{i,j} P(V_i|F(x)) exp(\frac{-1}{x_{Vi}} \sum_{x \in V_i} \frac{(x - x_c)^2}{\beta^2}) \tag{3}$$

where $P(V_i|F(x))$ is the probability that V_i appears in the video frame and x_{Vi} is the total pixel's number in supervoxel V_i from the video frame $F(x)$, x_c is the center of the frame and β is a controlling parameter.

From the supervoxels' saliency, we can compute the saliency of each pixel in the video frame F(x) according to Eq. 4

$$\alpha(x) = S_i, if x \in V_i \tag{4}$$

The main advantages of using motion distinctiveness for video saliency is that the complexity of these operations depends on the number of supervoxels, which means that the total number of pixels in the video frame does not have any effect on it. As a consequence, the computational complexity of the video frame saliency is linear O(m) where m is the number of supervoxels in tha video frame. Also, the probability that a supervoxels belongs to a video frame $P(V_i|F(x))$ should be computed once per video frame.

4 Experiments

4.1 Results

Our approach detects automatically the object that attracts human attention in a video sequences. In this section, we will compare our results to five state-of-art approaches on two datasets (SegTrack v2 and fukuchi).

SegTrack v2 dataset [15] is used for video segmentation and tracking. It has 14 videos with a total of 976 frames. Some videos contain one dynamic object, some others contain more than one. Each video object has specific chracteristics like Motion blur, Appearance change, Complex deformation, Occlusion, Slow motion and Interacting objects.

Fukuchi dataset [4] has 10 video sequences with a total of 936 frames. Each video contain one object.

We evaluate the performance of our approach by computing the F-score (see Eq. 7) and plotting the precision-recall and the ROC curves.

PR-curve Given a saliency map M, we binarize it to S with a fixed set of thresholds variant from 0 to 255, if the value of the pixel in M is greater than the threshold value then it is marked as 1 else it receives 0. Then the precision is computed (see Eq. 5) and recall see Eq. 6) by comparing the binarized map S to the ground-truth G

$$precision = \frac{\sum_{x,y} S(x,y)G(x,y)}{\sum_{x,y} S(x,y)} \tag{5}$$

$$recall = \frac{\sum_{x,y} S(x,y)G(x,y)}{\sum_{x,y} G(x,y)} \tag{6}$$

Receiver operating characteristics (ROC) curve plots the false positive rate against the truth positive rate by varying a fixed threshold from 0.255.

Even if PR and ROC curves can be informative about the obtained saliency map, only the **F-score** can evaluate the quality of resultant saliency map. It can be defined as a harmonic mean of precision and recall, and is given by

$$F_\beta = \frac{(1+\beta^2) \cdot precision \cdot recall}{\beta^2 \cdot precision + recall} \tag{7}$$

According to different saliency detection works [1,3] we set β^2 to 0.3 in order to increase the precision importance. Precision metric can be more important than recall, because 100% of recall can be obtained by setting all the detected region to foreground.

Table 1 illustrates the different values of F-score over the five state-of-the-art methods on SegTrack v2 and Fukuchi datasets. On both datasets we outperform the other saliency methods.

Figure 2 shows our Precision-Recall curves on the benchmark datasets against state-of-the-art curves. On Segtrack v2 dataset we have the best precision score. On fukuchi dataset we have the second best precision value after the GVS [25] method.

Figure 1 illustrates our ROC curve against state-of-the-art methods over the two evaluation datasets. On SegTrack v2 dataset we have best shape all over the curve. On Fukuchi dataset, we have best shape in the beginning then the GVS method [25] curve exceeded ours.

4.2   Discussion

On **SegTrack v2** and **Fukuchi** datasets we outperform all other approaches with a big gap in term of F-score. The nature of video sequences of the SegTrack v2 dataset differs, we have some videos with one moving object, some others with two, three or more. The advantage of using the supervoxel segmentation is that all the dynamic objects are highlighted and it is up to the proposed motion distinctiveness metric to decide which dynamic object is more salient than the others. As we explained in the Sect. 2.2, GVS [25] is based on computing spatial and temporal edges of each dynamic object in the video frame. This method is very efficient when the video frame has one moving object and the camera is stable which explains the good results on Segtrack dataset. But with the SegTrack v2 dataset which includes video frames with different conditions (as we explained in the last paragraph), results decreased. On Fukuchi dataset results are better because the video frames contain only one object and the camera motion is not intense.

GB [7] does not include motion as a saliency cue which explains the bad PR, ROC curves and F-score values. This method is more suitable for image saliency than video saliency.

RR [18] uses optical flow to select motion in a crowd, this method is very efficient in videos with static camera such as surveillance cameras, but does not produce good results with video frames of SegTrack v2 and Fukuchi datasets where camera motion causes noise.

RT [22] combines CRF (Conditional Random Field) model with a statistical framework and local feature contrast to segment the salient object from its background. Motion features are considered to select saliency in videos. ITTI [10] detects surprising events in videos by fusing different saliency maps to get the final saliency which includes motion, color, intensity, orientation and flicker features. Surprising location can be affected by one of these features. The main cause that the last two methods does not perform very well on the Segtrack v2 and Fukuchi dataset, is that both use spatial features beside motion features, so a static pixel which belongs to background can be considered salient.

Precision-Recall curves are reported from Fig. 2 and our proposed method outperforms other methods. The recall values of **RR** [18] and **GVS** [25] are very small when we varied the threshold to 255 and even can go down to 0 for

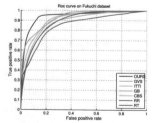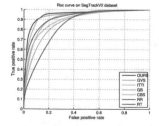

Fig. 1. ROC curves on Fukuchi andn Segtrack v2 datasets

ITTI [10], **RT** [22] and **GB** [7] because their saliency maps do not bring out the salient objects. On Segtrack v2 our method achieves the best precision rate above 0.75 which shows that our results (saliency maps) are more significant and precise. On Fukuchi dataset, we have the second best precision value after **GVS** [25] which also indicate that our saliency maps are informative of the region on interest. In both datasets, the minimum value of recall does not go down to zero which means that in its worst cases, our method detects the region of interest with a good response values. We presented in Fig. 3 a visual comparison of the saliency maps produced by our approach against state-of-the-art methods.

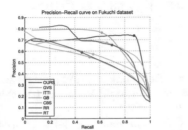 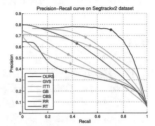

Fig. 2. Precision-Recall curves on Fukuchi and Segtrack v2 datasets

Table 1. F-score values

Dataset	OURS	GVS	RR	RT	ITTI	GB
Sagtrack v2	0.7401	0.6330	0.5701	0.3673	0.4295	0.4807
Fukuchi	0.7638	0.7243	0.5514	0.6634	0.5667	0.5393

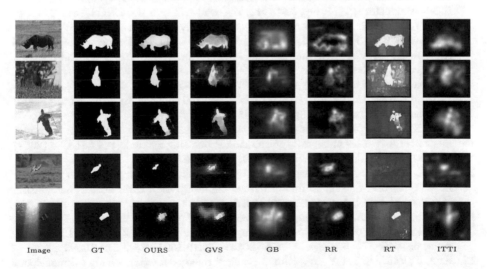

Fig. 3. Visual comparison of saliency maps generated from 6 different methods, including our method, GVS [25], GB [7], RR [18], RT [22] and ITTI [10]

5   Conclusion

We present an unsupervised video saliency method which have been shown to be effective in detecting the salient object in each frame of a video sequence. Our approach uses supervoxel segmentation. A video frame is translated to a graph where each node is a supervoxel and edges are the spatial connection between them. Then, we introduce a a new metric to compute motion distinctiveness of each supervoxel. Saliency score are attributed to each node in the graph by computing its motion distinctiveness. The experimental results have shown that our approach has achieved good results on two benchmark datasets used in video saliency evaluation and outperforms other video saliency methods.

References

1. Achanta, R., Hemami, S., Estrada, F., Susstrunk, S.: Frequency-tuned salient region detection. In: IEEE Conference on Computer Vision and Pattern Recognition (CVPR 2009), pp. 1597–1604. IEEE (2009)
2. Chang, J., Wei, D., Fisher, J.W.: A video representation using temporal superpixels. In: Proceedings of the IEEE Conference on Computer Vision and Pattern Recognition, pp. 2051–2058 (2013)
3. Cheng, M.-M., Mitra, N.J., Huang, X., Torr, P.H., Hu, S.-M.: Global contrast based salient region detection. IEEE Trans. Pattern Anal. Mach. Intell. **37**, 569–582 (2015)
4. Fukuchi, K., Miyazato, K., Kimura, A., Takagi, S., Yamato, J.: Saliency-based video segmentation with graph cuts and sequentially updated priors. In: IEEE International Conference on Multimedia and Expo, pp. 638–641. IEEE (2009)
5. Galasso, F., Cipolla, R., Schiele, B.: Video segmentation with superpixels. In: Lee, K.M., Matsushita, Y., Rehg, J.M., Hu, Z. (eds.) ACCV 2012. LNCS, vol. 7724, pp. 760–774. Springer, Heidelberg (2013). doi:10.1007/978-3-642-37331-2_57
6. Garcia-Diaz, A., Fdez-Vidal, X.R., Pardo, X.M., Dosil, R.: Decorrelation and distinctiveness provide with human-like saliency. In: Blanc-Talon, J., Philips, W., Popescu, D., Scheunders, P. (eds.) ACIVS 2009. LNCS, vol. 5807, pp. 343–354. Springer, Heidelberg (2009). doi:10.1007/978-3-642-04697-1_32
7. Harel, J., Koch, C., Perona, P.: Graph-based visual saliency. In: Advances in neural information processing systems, pp. 545–552 (2006)
8. Hou, X., Zhang, L.: Dynamic visual attention: searching for coding length increments. In: Advances in Neural Information Processing Systems, pp. 681–688 (2009)
9. Itti, L.: Visual salience. Scholarpedia **2**, 3327 (2007)
10. Itti, L., Baldi, P.: A principled approach to detecting surprising events in video. In: IEEE Computer Society Conference on Computer Vision and Pattern Recognition (CVPR 2005), vol. 1, pp. 631–637. IEEE (2005)
11. Jain, S.D., Grauman, K.: Supervoxel-consistent foreground propagation in video. In: Fleet, D., Pajdla, T., Schiele, B., Tuytelaars, T. (eds.) ECCV 2014. LNCS, vol. 8692, pp. 656–671. Springer, Cham (2014). doi:10.1007/978-3-319-10593-2_43
12. Kim, J., Han, D., Tai, Y.-W., Kim, J.: Salient region detection via high-dimensional color transform. In: Proceedings of the IEEE Conference on Computer Vision and Pattern Recognition, pp. 883–890 (2014)

13. Klein, D.A., Frintrop, S.: Center-surround divergence of feature statistics for salient object detection. In: International Conference on Computer Vision, pp. 2214–2219. IEEE (2011)
14. Koch, C., Ullamn, S.: Shifts in selective visual attention: towards the underlying neural circuitry. In: Vaina, L.M. (ed.) Matters of Intelligence: Conceptual Structures in Cognitive Neuroscience. Synthese Library: Studies in Epistemology, Logic, Methodology, and Philosophy of Science, vol. 188, pp. 115–141. Springer, Dordrecht (1987)
15. Li, F., Kim, T., Humayun, A., Tsai, D., Rehg, J.M.: Video segmentation by tracking many figure-ground segments. In: Proceedings of the IEEE International Conference on Computer Vision, pp. 2192–2199 (2013)
16. Lou, J., Ren, M., Wang, H.: Regional principal color based saliency detection. PloS one **9**, e112475 (2014)
17. Lu, J., Xu, R., Corso, J.J.: Human action segmentation with hierarchical supervoxel consistency. In: IEEE Conference on Computer Vision and Pattern Recognition (CVPR), pp. 3762–3771. IEEE (2015)
18. Mancas, M., Riche, N., Leroy, J., Gosselin, B.: Abnormal motion selection in crowds using bottom-up saliency. In: 18th IEEE International Conference on Image Processing, pp. 229–232. IEEE (2011)
19. Mauthner, T., Possegger, H., Waltner, G., Bischof, H.: Encoding based saliency detection for videos and images. In: Proceedings of the IEEE Conference on Computer Vision and Pattern Recognition, pp. 2494–2502 (2015)
20. Oneata, D., Revaud, J., Verbeek, J., Schmid, C.: Spatio-temporal object detection proposals. In: Fleet, D., Pajdla, T., Schiele, B., Tuytelaars, T. (eds.) ECCV 2014. LNCS, vol. 8691, pp. 737–752. Springer, Cham (2014). doi:10.1007/978-3-319-10578-9_48
21. Rahman, A., Houzet, D., Pellerin, D., Marat, S., Guyader, N.: Parallel implementation of a spatio-temporal visual saliency model. J. Real Time Image Process. **6**, 3–14 (2011)
22. Rahtu, E., Kannala, J., Salo, M., Heikkilä, J.: Segmenting salient objects from images and videos. In: Daniilidis, K., Maragos, P., Paragios, N. (eds.) ECCV 2010. LNCS, vol. 6315, pp. 366–379. Springer, Heidelberg (2010). doi:10.1007/978-3-642-15555-0_27
23. Scharfenberger, C., Wong, A., Fergani, K., Zelek, J.S., Clausi, D.A.: Statistical textural distinctiveness for salient region detection in natural images. In: Proceedings of the IEEE Conference on Computer Vision and Pattern Recognition, pp. 979–986 (2013)
24. Singh, A., Chu, C.-H.H., Pratt, M.: Learning to predict video saliency using temporal superpixels. In: 4th International Conference on Pattern Recognition Applications and Methods (2015)
25. Wang, W., Shen, J., Porikli, F.: Saliency-aware geodesic video object segmentation. In: Proceedings of the IEEE Conference on Computer Vision and Pattern Recognition, pp. 3395–3402 (2015)
26. Xu, C., Whitt, S., Corso, J.J.: Flattening supervoxel hierarchies by the uniform entropy slice. In: Proceedings of the IEEE International Conference on Computer Vision, pp. 2240–2247 (2013)
27. Yang, C., Zhang, L., Lu, H., Ruan, X., Yang, M.-H.: Saliency detection via graph-based manifold ranking. In: Proceedings of the IEEE Conference on Computer Vision and Pattern Recognition, pp. 3166–3173 (2013)
28. Zhong, S.-H., Liu, Y., Ren, F., Zhang, J., Ren, T.: Video saliency detection via dynamic consistent spatio-temporal attention modelling. In: AAAI (2013)

A Rehabilitation System for Post-operative Heart Surgery

Giuseppe Caggianese[1(✉)], Mariaconsiglia Calabrese[2,3], Vincenzo De Maio[4],
Giuseppe De Pietro[1], Armando Faggiano[4], Luigi Gallo[1], Giovanna Sannino[1],
and Carmine Vecchione[2,5]

[1] Institute for High Performance Computing and Networking, National Research
Council of Italy (ICAR-CNR), Naples, Italy
{giuseppe.caggianese,giuseppe.depietro,luigi.gallo,
giovanna.sannino}@icar.cnr.it
[2] Department of Medicine and Surgery, University of Salerno, Baronissi, Italy
{macalabrese,cvecchione}@unisa.it
[3] S Giovanni di Dio e Ruggi d'Aragona Hospital, Salerno, Italy
mac.calabrese@virgilio.it
[4] Computer Science Department, University of Salerno, Fisciano, Italy
vinc.demaio@gmail.com, armando.faggiano@gmail.com
[5] I.R.C.C.S. Neurological Mediterranean Institute NEUROMED, Pozzilli, Italy
vecchione@neuromed.it

Abstract. Supervised exercise programs are an important aspect of the
rehabilitation process of patients after heart surgery. A large number of
factors must be taken into account before implementing a rehabilitation
program. These mainly consist in the patient's cognitive and physical
capabilities after the operation and the expectations of recovery. A rehabilitation program should also be designed in relation to the stage of the
healing process, with the therapist selecting the best sequence of exercises while taking into account the most appropriate effort level for the
patient. This paper describes a customizable rehabilitation system for the
early post-operative period, useful for the performance of an assessment
of the patients, through an evaluation of their cognitive and motor abilities, and for a dynamic personalization of the therapy sessions focused
on patient needs.

Keywords: Rehabilitation · Telemonitoring · Microsoft kinect · Unity
3D · Serious game

1 Introduction

The gradual transformation of the demographic profile in the European population shows a continuous average age rise accompanied by new patterns of chronic
disease, resulting in an increasing demand on health services [1,2].

This trend is clearly evident in the hospital sector. In fact, modern hospitals are implementing new public healthcare models directed towards a gradual

© Springer International Publishing AG 2018
G. De Pietro et al. (eds.), *Intelligent Interactive Multimedia Systems and Services 2017*,
Smart Innovation, Systems and Technologies 76, DOI 10.1007/978-3-319-59480-4_55

reduction in both the length of hospital stay and the access to hospital facilities. This adaptation is to a certain extent dependent on the inability to sustain increases in public health care spending. However, it is also determined by technological and operational achievements that have significantly reduced the time periods necessary for diagnosis and post-operative hospital stay. In accordance with this development, hospital activities could benefit from Information and Communication Technology (ICT) applications that could become the right instrument to improve the quality of the healthcare service, also by responding effectively to the emergent needs of managing growing-up complexity and privacy problems [5–8].

In detail, in relation to rehabilitation, the use of low-cost technologies is accelerating the introduction of ICT systems for assisted rehabilitation in both rehabilitation centers and patients' homes. In fact, many examples of rehabilitation platforms directed towards, for instance, the care of post-stroke patients [4,27], cognitive rehabilitation [11,22], motor function recovery [16,23], and the improvement of the quality of life in mature age [25] have been described in literature. These solutions contribute to an intensification and diversification of rehabilitation treatment during hospitalization. In this way, patients can either return home after completing the rehabilitation process or they can start the process in hospital and continue it in their home environment.

Regarding cardiac surgery patients, post-operative rehabilitation is very important in order to prevent or reduce possible complications and to improve outcomes [24]. Currently, therapists prescribe physical therapy treatments for cardiac surgery patients in the first days after their surgery. Thus, patients can perform the therapy until they are discharged from hospital. In order to design a personalized physical treatment, the therapist needs to make an initial assessment of the patient. However, for this kind of patient, it may be inappropriate and ineffective to carry out only one health assessment or to prescribe only a single physical program. In fact, each day several factors can differently affect the overall health condition of the patient. Additionally, any improvements achieved during the rehabilitation program should be considered in the planning of new sessions. In this situation, a personalized rehabilitation program, dynamically focusing on individual patient needs and capabilities, becomes crucial in order to facilitate the recovery of the patient's previous functional level.

For this reason, in this paper we present a customizable rehabilitation system for the early post-operative period, useful for the performance of an assessment of patients, through an evaluation of their cognitive and motor abilities. Moreover, this kind of rehabilitation is also functional for the prescription of the sequence of exercises considered the most appropriate in relation to the patient's recommended effort level. Therefore, this proposed system can support therapists in providing patients with a personalized therapy. In addition, the system shows continuously to the patients their own performance results, so giving useful feedback. Finally, in order to provide a motivational support for the patient and to reduce the dropout rate, all the exercises have been developed as serious games.

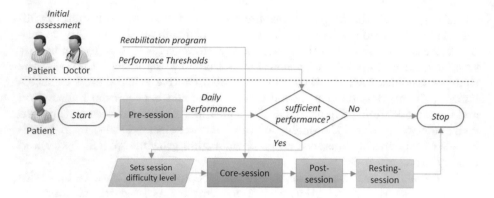

Fig. 1. The post-operative heart surgery rehabilitation process.

The paper is organized as follows. Section 2 describes the main system characteristics in terms of the rehabilitation protocol designed for cardiac surgery patients. The system architecture, technological equipment, human-machine interface and main system features are reported in Sect. 3. Finally, in Sect. 4, our conclusions are reported.

2   Post-operative Heart Surgery Rehabilitation

In the first period after surgery, cardiac surgery patients need to be helped to strengthen their heart muscle, reduce anxiety and improve their quality of sleep. In fact, they get tired easily and, because of the chest wounds, they are not usually able to breath without pain. Such problems also have a negative affect on their recovery.

Essential rehabilitation activities for cardiac surgery patients can start between the 5th and 7th day after the surgery, when the after-effects of the operation are still strong. The therapist designs an individual program focused on patient needs and capabilities. It generally includes daily activities such as walking, simple exercises and self-care practices. Overall, the main goals of post-operative heart surgery rehabilitation can be summarized as: (1) restoring and maintaining muscle strength and joint flexibility; (2) evaluating the heart's response to mild activity; (3) preventing the negative effects of bed rest; and (4) educating the patient and her/his family about the requirements of this new situation.

The proposed rehabilitation system is governed by a protocol specifically designed for cardiac surgery patients with sub-acute symptoms. Thus, this protocol divides the entire daily program into different phases that will be executed sequentially. An overview of the rehabilitation process is shown in Fig. 1.

The rehabilitation program has a duration of between 50 and 60 min, in which the patient performs a combination of physical and cognitive exercises. Before starting the program, the therapist establishes some personalized thresholds for each patient: *sufficient, good* and *optimal*. The program allows the system to classify the performances achieved by each patient during the exercises. Moreover, these thresholds allow the identification of four different levels of performance achieved by the patient and may also be used to initialize the difficulty levels of the subsequent exercises. These levels are, from the lowest to the highest: "insufficient", when the performance is below the sufficient threshold, forcing the patient to discontinue the session; "low", if the performance is below the good threshold; "medium", whenever the patient reaches a score below the optimal threshold; and "high", when the achieved performance is above the optimal threshold.

There are four main phases of the proposed rehabilitation protocol, namely:

- **Pre-session:** before starting the physical rehabilitation program, it is important to evaluate the patient's condition, both cognitively and physically:
  - For the cognitive assessment, the patient needs to execute at least 3 min of cognitive exercises, achieving a result considered acceptable by the therapist.
  - For the physical assessment, the system provides a six minute walking test [9] combined with a mild static exercise. This ensures a greater increase in both heart rate and blood pressure if compared to dynamic exercises [15,17]. In more detail, these two types of exercise have been combined to create a new one, easy to perform in a limited space. Thus, the patient's performance can be monitored by the range of action of sensors embedded in the proposed system. This new exercise combines sitting and standing up from a chair, interspersed with the execution of a biomechanical step forward and then backward to the initial position close to the chair.

This phase is crucial since it allows an assessment of the patient's initial health state. In addition, it provides an evaluation of the performance indicators, used to customize the subsequent phases or to terminate the session, if the performance is considered not sufficient for the rehabilitation program. For instance, a total failure in the proposed cognitive exercises can indicate that the patient is in a state of mental confusion, which can also lead to fainting during the session. This situation makes the continuation of the rehabilitation session ineffective and even dangerous. In fact, in this situation, the system proposes a termination of the session and recommends a period of rest. Moreover, the session will be terminated, if the patient, during the physical assessment, executes a low number of repetitions, meaning she/he has a low tolerance to physical effort. In this case also, the performance of the physical rehabilitation program can be dangerous for the patient; therefore, the system will again advise a termination of the activities and a period of rest.

- **Core session:** the patient is directed toward the performance of a set of exercises which aims at an improvement of her/his physical and motor skills. The combination of the proposed exercises is defined by the therapist responsible for choosing the most appropriate exercises for the patient. Moreover, each of these is further customizable by the therapist, considering the results of the pre-session phase, after modulating the static exercise intensity in accordance with one of the levels defined above. In this phase, the patient will be engaged overall for 20 to 40 min, alternating active periods with resting. The exercises included in this phase are all light static exercises that the patient executes to her/his full capacity, for the required time and always avoiding running out of breath.

- **Post-session:** this phase is used to evaluate the global performance of the patient through the performance of her/his personalized version of the six minute walk test presented in the pre-session phase.
- **Resting session:** after the execution of the entire session, the patient is guided in a respiratory activity in order to normalize the levels of all parameters. The total duration of this last phase should not be less than 10 min.

Finally, throughout the rehabilitation program, the patient's biomedical parameters are continuously monitored using non-invasive sensors, e.g. a wearable electrocardiogram sensor. In fact, while the biomedical monitoring is being performed, the system analyzes in real time the patient's vital signs. In addition, the system alerts the therapist or the medical personnel and interrupts the rehabilitation session, if any possibly dangerous situations are detected

3   System Overview

The Cardiac Surgery Rehabilitation System (CSRS) represents a first prototype of a rehabilitation system that allows cardiac patients to execute rehabilitation activities starting a few days after surgery. The purpose of the system is twofold: to facilitate the rehabilitation process for the patient; and to provide a support tool for therapists and doctors. These objectives become achievable through the realization of a specific infrastructure that also promotes coordination between clinicians [13].

The prototype realized has been designed to be placed in a hospital room without being invasive for the patient. The patient performs the rehabilitation session simply by standing in front of the system monitor and naturally interacting with it. At the start of each session, the system follows the rehabilitation protocol, previously described, carefully customized by the therapist in accordance with the patient needs.

3.1   System Architecture Design

The conceptual design of the CSRS is illustrated in Fig. 2, where the main actors and their principal activities are illustrated. On the horizontal axis, the patient and the set of sensors find a place at the beginning and end, respectively. Both sides represent a continuous input data stream for the system.

The patient, without using any particular input device, executes her/his rehabilitation program performing cognitive and physical exercises. On the other side, a set of sensors continuously updates the system with patient data related to

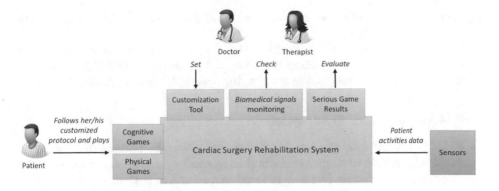

Fig. 2. The conceptual design of the cardiac surgery rehabilitation system.

both her/his movements and vital signs (biomedical signals). In the middle, the CSRS is in charge of: managing the patient's specific rehabilitation protocol; elaborating the input stream of data received by the sensors; and recognizing both gestures and body movements, mapping them to game activities, and providing feedback about the level achieved.

Moreover, the system collects all data elaborated during the rehabilitation session and makes them available to the doctors and therapists responsible for monitoring the patient's activities. In particular, the proposed system stores a file with the monitored vital signs (biomedical signals) of the patient and a final report, in which the level achieved in each serious game is collected at the end of the session.

Finally, the system allows the doctors and therapists to personalize the number, type and duration time of the exercises that the patient should execute in her/his program through the Customization Tool.

3.2 Hardware Components

The proposed system is based on sensor devices integrated in a PC-based station, in order to collect physical and physiological measurements during the performance of the serious games. A Microsoft Kinect v2 sensor is used to trace and measure patient movements and poses. The noise in the acquired data due to the possible different light conditions has been mitigated by using a temporal filtering approach [14]. The reason for this choice can be explained in terms of its limited cost and on the basis of evidence on the use of this device in many similar clinical studies [10,18,20]. Moreover, in order to monitor the biomedical signals we have used a wearable sensor, the BioPatch™ (Zephyr Technology Corporation, Annapolis, Maryland, United States) [26], a small and comfortable device that can be attached to the patient's chest by means of traditional disposable ECG electrodes. This sensor can be easily worn by the patient and it can monitor one lead of the electrocardiogram (ECG) signal, the heart rate (HR), and the accelerometer data in real time, without any impediments to movement.

Additionally, all the monitored data can be transmitted by using a Bluetooth connection in real time.

Finally, the component integration and the realization of the serious games have been accomplished by using the Unity 3D platform [3] and C#. In respect of all the other components, the monitoring system has been realized in Java.

3.3 User Interface Description

In order to simplify the system interaction and allow the patient to be as independent as possible during the rehabilitation session, we decided to exploit for the interaction the same technology used to gather the patient's physical measurements. In this way, the patient, by standing in front of the acquisition sensor, can start to naturally interact with the system only by moving her/his hands.

The main interactions the system requires are those used to start the rehabilitation program and to execute the proposed exercises. In particular, the patient needs to lift her/his right arm over her/his head to start the rehabilitation program. The system will recognize the gesture, replying with a message on the screen and starting the program. Meanwhile, with regard to the exercises, in the physical exercises the patient should repeat the same movements, shown in a short video guide, in the assigned time.

The system, by tracking her/his movements, will update the proposed interface with feedback about the results achieved in terms of the exercise repetition and the time remaining. In the cognitive exercises, the patient needs to interact with some game objects in the scene. With this aim, the interface proposed to the patient allows her/him to control a pointer in the 2D space and to perform a selection action in order to interact with the game objects, a functionality implemented by adopting the wait to click metaphor since this has proved to be the most widely accepted solution [12].

Essentially, the proposed interface means that the patient does not need to make any change in the interaction configuration when moving between the cognitive and physical exercises, so preventing her/him from getting bored and neglecting to perform the prescribed exercises [19, 22].

Finally, we decided to avoid showing any feedback related to the physiological measurements in order not to influence the patient's attitude during the session.

3.4 System Decisional Process

The proposed system collects data from different sources and, by following the protocol organized in the consecutive sessions, leads and monitors the patient activities. This section presents the decisional process applied, reported in Fig. 3.

At the start, the patient needs to log in to the system in order to load her/his customized rehabilitation program. After loading, the system is ready to start the session but waits until the patient executes the starting gesture.

After the starting gesture, the system leads the patient through the rehabilitation program automatically managing the sequence of exercises, the duration

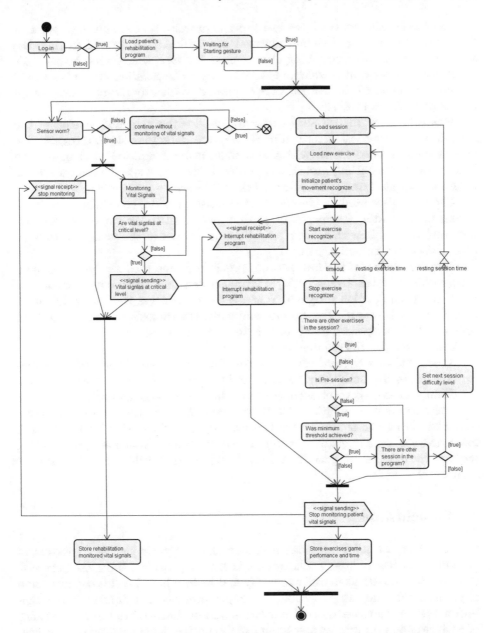

Fig. 3. Activity diagram showing the main decision points of the algorithm.

time of each exercise, and the recovery period between two consecutive exercises or sessions. At the same time, the system starts to monitor the vital signs of the patient by verifying the Bluetooth connection with the sensor [21].

As aforementioned, there are two types of exercise that are given to patients for rehabilitation: cognitive and physical exercises. At the time of writing, the system supports two types of cognitive games: *Memory*, which tests the patient's short-term memory, and *Multiple Features Target Cancellation*, which is a visual search task, in which the subjects are requested to identify a target item within an array of distractors.

Meanwhile, for the physical exercises, besides the extended version of the six minute walking test described in Sect. 2, our system is able to recognize five other exercises: knee lifting (with and without arm lifting), weight change (with and without arm lifting) and thrust. For each exercise a specific recognizer has been implemented and executed to track and interpret the user's movements.

The recogniser used for the cognitive exercises, besides recognizing and measuring the patient's movements, maps that movement to the game dimensionalities in order to allow the control of a floating pointer in the scene and thus an interaction with the other game objects.

After execution of the first session, previously presented as the pre-session, the system evaluates the patient's performance deciding whether to abort or continue the rehabilitation program. When the user achieves the threshold established by the therapist, the system configures the difficulty levels of the next session (the core session) and proceeds to load a new session, after allowing the patient to wait for the required resting time.

At the end of all the scheduled sessions the system stops the patient, monitoring and storing both the vital signs and exercise performance data collected.

Finally, as already mentioned, the rehabilitation session, and in particular the patient movement recognition, can be interrupted in the case of any anomalies detected during the monitoring of the vital signs of the patient. The interrupt message leads to an immediate termination of the rehabilitation program, storing anyway the data relating to the exercises performed and the biomedical signals monitored.

4   Conclusions

In this paper, we have described a system that supports the rehabilitation of patients after heart surgery. The system is able to provide a supervised exercise program to the patient in order to verify and restore both her/his cognitive and physical capabilities. In particular, the paper presents a customizable rehabilitation system for the early post-operative period that, before the performance of each session, performs an assessment of the patient by evaluating her/his cognitive and motor abilities and dynamically personalizes the therapy difficulty level on the basis of these results. Moreover, the system integrates a marker-less motion tracking, used for both the physical exercise recognition and the patient-system interaction, with a set of wearable physiological devices used to monitor the patient's vital signs and a graphical interface based on serious games which should improve the patient's neurological response.

Finally, in the next few months, we plan to verify the efficacy of the proposed system with real patients. This process will also be useful to qualitatively evaluate the usability of the system.

Acknowledgments. This work has been supported by the project "eHealthNet: Ecosistema software per la Sanità Elettronica" (PON03PE_00128_1).

References

1. Ageing report 2015. http://europa.eu/epc/pdf/ageing_report_2015_en.pdf
2. Key figures on Europe - 2015 edition. http://ec.europa.eu/eurostat/web/products-statistical-books/-/KS-EI-15-001
3. Unity 3d. https://unity3d.com/
4. Adinolfi, F., Caggianese, G., Gallo, L., Grosso, J., Infarinato, F., Marchese, N., Sale, P., Spaltro, E.: SmartCARE—an ICT platform in the domain of stroke pathology to manage rehabilitation treatment and telemonitoring at home. In: Pietro, G., Gallo, L., Howlett, R.J., Jain, L.C. (eds.) Intelligent Interactive Multimedia Systems and Services 2016. SIST, vol. 55, pp. 39–49. Springer, Cham (2016). doi:10.1007/978-3-319-39345-2_4
5. Amato, F., Colace, F., Greco, L., Moscato, V., Picariello, A.: Semantic processing of multimedia data for e-government applications. J. Vis. Lang. Comput. **32**, 35–41 (2016)
6. Amato, F., Moscato, F.: A model driven approach to data privacy verification in e-health systems. Trans. Data Priv. **8**(3), 273–296 (2015)
7. Amato, F., Moscato, F.: Pattern-based orchestration and automatic verification of composite cloud services. Comput. Electr. Eng. **56**, 842–853 (2016)
8. Amato, F., Moscato, F.: Exploiting cloud and workflow patterns for the analysis of composite cloud services. Future Gener. Comput. Syst. **67**, 255–265 (2017)
9. ATS Committee on Prociency Standards for Clinical Pulmonary Function Laboratories, et al.: ATS statement: guidelines for the six-minute walk test. Am. J. Respir. Crit. Care Med. **166**(1), 111 (2002)
10. Bao, X., Mao, Y., Lin, Q., Qiu, Y., Chen, S., Li, L., et al.: Mechanism of kinect-based virtual reality training for motor functional recovery of upper limbs after subacute stroke. Neural Regener. Res. **8**(31), 2904 (2013)
11. Bogdanova, Y., Yee, M.K., Ho, V.T., Cicerone, K.D.: Computerized cognitive rehabilitation of attention and executive function in acquired brain injury: a systematic review. J. Head Trauma Rehabil. **31**(6), 419–433 (2016)
12. Brancati, N., Caggianese, G., Frucci, M., Gallo, L., Neroni, P.: Touchless target selection techniques for wearable augmented reality systems. In: Damiani, E., Howlett, R.J., Jain, L.C., Gallo, L., De Pietro, G. (eds.) Intelligent Interactive Multimedia Systems and Services. SIST, vol. 40, pp. 1–9. Springer, Cham (2015). doi:10.1007/978-3-319-19830-9_1
13. Della Vecchia, G., Gallo, L., Esposito, M., Coronato, A.: An infrastructure for smart hospitals. Multimed. Tools Appl. **59**(1), 341–362 (2012)
14. Essmaeel, K., Gallo, L., Damiani, E., De Pietro, G., Dipanda, A.: Comparative evaluation of methods for filtering kinect depth data. Multimed. Tools Appl. **74**(17), 7331–7354 (2015)
15. Hietanen, E.: Cardiovascular responses to static exercise. Scand. J. Work Environ. Health **10**, 397–402 (1984)

16. Karashanov, A., Manolova, A., Neshov, N.: Application for hand rehabilitation using leap motion sensor based on a gamification approach. Int. J. Adv. Res. Sci. Eng. **5**(2), 61–69 (2016)
17. Mitchell, J.H., Wildenthal, K.: Static (isometric) exercise and the heart: physiological and clinical considerations. Ann. Rev. Med. **25**(1), 369–381 (1974)
18. Mousavi Hondori, H., Khademi, M.: A review on technical and clinical impact of microsoft kinect on physical therapy and rehabilitation. J. Med. Eng. **2014**, 846514 (2014)
19. Rego, P.A., Moreira, P.M., Reis, L.P.: Natural user interfaces in serious games for rehabilitation. In: 6th Iberian Conference on Information Systems and Technologies, pp. 1–4 (2011)
20. Ren, W., Pu, F., Fan, X., Li, S., Sun, L., Li, D., et al.: Kinect-based skeleton-matching feedback for motor rehabilitation: transient performance effect of shoulder training. J. Mech. Med. Biol. **16**(03), 1650037 (2016)
21. Sannino, G., De Pietro, G.: An evolved ehealth monitoring system for a nuclear medicine department. In: Developments in E-systems Engineering, pp. 3–6 (2011)
22. Shapii, A., Mat Zin, N.A., Elaklouk, A.M.: A game system for cognitive rehabilitation. BioMed Res. Int. **2015**, 493562 (2015)
23. Standen, P., Threapleton, K., Richardson, A., Connell, L., Brown, D., Battersby, S., Platts, F., Burton, A.: A low cost virtual reality system for home based rehabilitation of the upper limb following stroke: a randomised controlled feasibility trial. Clin. Rehabil. **31**, 340–350 (2016)
24. Stephens, R.S., Shah, A.S., Whitman, G.J.: Lung injury and acute respiratory distress syndrome after cardiac surgery. Ann. Thoracic Surg. **95**(3), 1122–1129 (2013)
25. Sáenz-de Urturi, Z., Zapirain, B.G., Zorrilla, A.M.: Kinect-based virtual game for motor and cognitive rehabilitation: a pilot study for older adults. In: 8th International Conference on Pervasive Computing Technologies for Healthcare, pp. 262–265 (2014)
26. Zephyr Techonology: New Zephyr BioPatch Monitoring Device for Human Performance. http://zephyranywhere.com/
27. Zheng, H., et al.: SMART rehabilitation: implementation of ICT platform to support home-based stroke rehabilitation. In: Stephanidis, C. (ed.) UAHCI 2007. LNCS, vol. 4554, pp. 831–840. Springer, Heidelberg (2007). doi:10.1007/978-3-540-73279-2_93

Evaluation of the Criteria and Indicators that Determine Quality in Higher Education: A Questionnaire Proposal

Fouzia Kahloun[1(✉)] and Sonia Ayachi Ghannouchi[2]

[1] Laboratory RIADI-GDL, ENSI, Mannouba, Tunisia
fouziakahloun@gmail.com
[2] High Institute on Management of Sousse, Sousse, Tunisia
s.ayachi@coseleam.org

Abstract. Higher education is one of the important fields for development in the economic world. Therefore, it must be based on excellence methodologies and managed by a reliable quality approach. That is why; we must encourage the culture of quality within higher education institutions to make management transparent and understandable by all stakeholders (students, teachers). However, despite the changes and the importance in literature, defining the quality concept in higher education remains vague and unclear. Furthermore, leading a quality approach to allow these institutions to adapt to change and the current needs remains difficult. To remedy to these problems, we expect in this paper to define quality in higher education. After that, we identify a set of typical and consistent criteria, with specific indicators for each criterion, considered as substantial, focusing on the needs of stakeholders with the aim of achieving a good quality. These requirements are established by first passing through a preliminary study based on a questionnaire designed to assess the importance attributed to the criteria mentioned. The responses given to this questionnaire will be analyzed in order to identify the quality indicators and to measure from the experts point of view their relative importance.

Keywords: Quality · Criteria · Indicators · Higher education

1 Introduction

The importance of education for the development of excellence and knowledge contributes directly to the social and economic development of a country [1].

For this, universities must be adapted in order to realize technical and strategic change. Furthermore, higher education must be focused on good methodologies, and guided by a reliable quality approach.

Recent works for the implementation of a quality approach in higher education are based on predefined referential (such as ISO, The Malcolm Baldrige National Academy Award, the Deming Prize for Quality, the European quality Improvement System: EQUIS, the model European Foundation for quality management: EFQM...), or a panoply of other models and strategies to a quality service.

© Springer International Publishing AG 2018
G. De Pietro et al. (eds.), *Intelligent Interactive Multimedia Systems and Services 2017*,
Smart Innovation, Systems and Technologies 76, DOI 10.1007/978-3-319-59480-4_56

In fact, many higher education institutions throughout the world have implemented or are in the process of creating their own quality management (QM) systems, their aim is securing and improving teaching and learning [2].

In this context, our contribution in this paper is to collect criteria for the higher education domain and their relationships with indicators as a first step toward their ranking in an order of importance, according to the opinion of the stakeholders. As a result, criteria, measures and analysis present an ideal means to determine the defaults in this field and correct them in order to improve its quality level.

This article is organized as follows: in Sect. 2, we look to define the term quality in higher education institutions. In Sect. 3, we outline the criteria and the indicators as the most important factors of the quality of service. In Sect. 4, we define a study and we illustrate its results. In Sect. 5, we cited some recommendation for improve the quality in higher education domain. We end up with the sixth section as our conclusion and future works.

2   Definition of Quality in Higher Education Domain

The higher education system has been changing positively over the last ten years. These developments are based on the concept of quality through several strategies:

The Development of the national agencies of quality in Africa.

The organization of a number of international conferences and workshops (Conferences on Quality Assurance in Higher Education in Africa-ICQAHEA).

The Development of the African network quality assurance (African Quality Assurance Network-AfriQAN).

The Association of African Universities (AUA) through a number of programs. And other developments that promote quality in higher education....

But despite that, defining the concept of quality in higher education remains blurry. Its concept is still poorly represented, misunderstood, or both, by many universities [3, 4]. In addition, determining a quality approach seems difficult to adapt to change and current needs.

In fact, UNESCO defines quality in higher education as a dynamic, multi-dimensional and multi-level approach. It relates to the contextual settings of an educational model, missions and goals of institutions, as well as specific references in a system, establishment, training or discipline (UNESCO, 2004).

Nicolescu et al. (2010). in accordance with the Commission of the European Communities (2003) identified that improving quality in higher education should concentrate on three directions: (a) ensuring that European universities have sufficient and sustainable resources and use them efficiently; (b) consolidating excellence in teaching and research and (c) opening up universities to a greater extent to the outside and increase their international activities".

In fact a lot of ink has flowed in recent years on the development of quality in higher education. Several definitions have been proposed. However, the definition that is most commonly adopted today is that of "adaptation objectives" [5]. It means that quality in higher education describes the treatment goals set at the beginning by the establishment.

Indeed, defining a reliable quality approach in higher education institutions in order to ensure continuous improvement and further the satisfaction of internal and external university actors cannot do without determining first of all, the criteria and specific measures associated with each of these.

3 Selection of Criteria and Quality Indicators in Higher Education Domain

There are various tools that can be used to assess the quality of the university process. These tools identify the criteria that influence quality. In higher education, these criteria are difficult to identify because of various aspects and stakeholders.

In this part, we try to select some typical criteria of the quality of a higher education system that must be considered in our work, focusing on the needs of stakeholders with the aim of achieving good quality. Also, we mention if possible some specific indicators related to each criterion. In fact, the development of PIs (Performance Indicators) in higher education can be tracked back to the manufacturing industry and relates to the way in which inputs are transformed into outputs [6].

3.1 Efficacy Criterion

A training system needs to be effective to achieve a good quality. This means that it should achieve its objectives [7]. The efficacy concept is described as a comparison of the results we want to achieve (outputs) and those declared (inputs). In other words, efficacy is the ratio of output to input; it is the ratio of the actual outcome to the possible or ideal outcome [8].

The efficacy criterion is presented by two categories: internal and external.

The internal efficacy focuses specifically on educational or pedagogic criteria [9], for instance, the indicators and the questions used mostly for this concept are:

- Repetition rates in each university year.
- Success rates following a training period.
- Access rate for a first diploma in higher education [10].
- The Ratio between the number of graduates and those enrolled.
- Should the Course objectives be understandable?
- Should the Course objectives be attainable?

Whereas the external efficacy has more importance than the needs and expectations of education systems, it lets you know in which measure education responds to the society objectives and responds to the needs of the labour market [11]. Some of the indicators and questions are the following:

- Unemployment rates of higher education graduates in different studies.
- The ratio between the number of job offers and the number of graduates.
- Employment Rates of higher education graduates per disciplines.
- Should the universities respond to the expectations of the labour market?

The difficulties for these indicators are related to the sources of information [10].

3.2 Efficiency Criterion

This criterion determines a link between the expected results and entries taking into account presented resources, which can be financial, human, or physical.

Besides, we can define, as the extent to which the knowledge/skills learned are applicable in the future career of graduates [12].

There are two categories, internal efficiency if we take into account the short-term results and external efficiency if we take into account the long-term results [13]. To determine this criterion, we considered the following indicators and questions:

- The Ratio between the number of creators of new entrepreneurs and expenses related to training modules focusing on entrepreneurship [9].
- The ratio between the number of successful students and supervisors [9].
- Ratio between the number of actual jobs and investment in the education system [9].
- Must the obtained results meet the set objectives?
- Must the available recourses correspond to the obtained set objectives?

3.3 Pertinence Criterion

The pertinence of higher education must be measured as one of the adequacies between what society expects from institutions and what they do [5] (World Declaration on Higher Education, Paris, 1998).

Consequently, this concept describes the compliance report between the declared objectives, expected results and needs.

This criterion presents the degree of dependence between the results obtained and the needs to satisfy or between the pursued objectives and the needs to satisfy [11].

Some authors mentioned for this criterion that the current quality of higher education means nothing without its pertinence. It searches to answer the following questions:

- Should the objectives of the higher education system correspond to high priority needs of the labor market?
- Is it important that higher education institutions are designed and organized to produce needed diplomas? [5].

3.4 Consistency Criterion

Coherence presents the compliance link between resources used and the expected objective, we need to answer questions like these:

- Should the means (resources, strategies…) must be those that we provided?

3.5   Conformity Criterion

This criterion defines the ratio between the internal fundamental components of the action (objectives, resources and constraints, strategies, results) and existing regulations or regulations created for a given action [13]. The question we seek to answer is:

- Is it important to have conformance between activities within the education system and the existing regulations?

3.6   The Balance Criterion

To be of good quality, higher education must be balanced. This is achieved through the involvement of various dimensions of knowledge. This criterion is defined as the ability of the system to develop harmoniously all types of objectives related to knowledge [7].

In fact, there are three types of knowledge:

- Know-reproduce: repetition of what has been retained.
- Know-how: requiring a work of transformation facing a structural situation different from that of learning [7].
- Interpersonal skills: they relate to skills, which are usually awarded. They become spontaneous behavior and so demonstrate what "is" the person [7].

It seeks to answer the question:

- Must the higher education system provide all dimensions of knowledge?
- Must the degree of involving of students in improving the quality of the establishment be important?

4   Study and Analysis of Results

A preliminary study was conducted where the above criteria have been simplified and used to form an investigation. This study through a survey will treat the problem of quality in real processes where contact was established with involved actors and experts. The observation in context will determine a good set of essential and consistent criteria that must be considered in our work and decide on the most appropriate and the most important among those cited in literature. Added to that, it determines, for each criterion, specific measures that determine their values.

For this investigation, we chose the likert scale: 1 = strongly disagree, 2 = disagree, 3 = neutral, 4 = Agree, 5 = Strongly Agree. The participation of actors and experts from higher education is completely anonymous. The total number of collected responses is 83 persons, representing a response rate of 66% (see appendix A, shows more detail for the level of respondents).

We analyzed the results using the classification algorithm of data mining known as J48 more commonly named C4.5. The latter is able to provide a hierarchical decomposition of decision problem in the form of a tree. It contributes to a better understanding and a clear simplified vision of all decision data (Fig. 1).

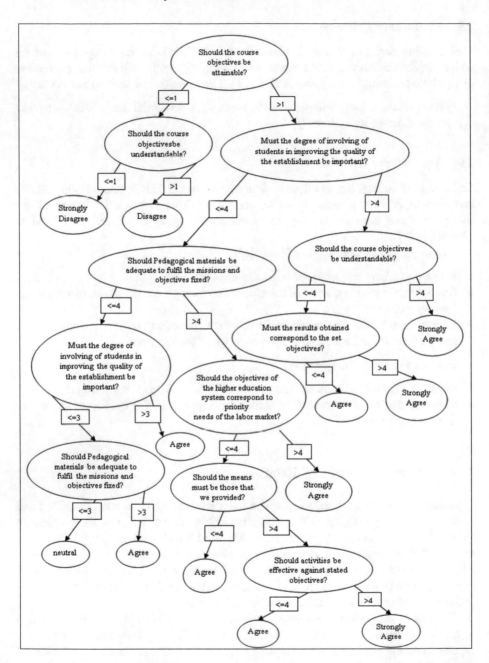

Fig. 1. Classification of quality criteria in higher education following RJ48 data mining algorithm

As presented in Fig. 1, the nodes of this graph represent the indicators, which are presented in the questionnaire, and the edges are the answers of participants following likert scale.

This analysis revealed some interesting facts from the various relevant stakeholders in the analysis of the performance in higher education domain. In fact, the assessment of quality requires to consider criteria of economic nature (efficacy in terms of the obtained products) as the most important, this not only to be effective but also to be balanced presented as the second important criterion, considered as the ability of the system to develop in a harmonious way all the different objectives linked to knowledge. On the third place, we can find the aspect of efficiency that is to say their ability to get products for lower cost next the pertinence criterion, then the criterion of consistency and finally the Conformity aspect.

We can conclude that the quality in the higher education domain is an essential requirement, but relatively complex. It requires a steering taking into account all components of the system to ensure it met the objectives that emerge from a social environment, consistent with available resources and defined strategies.

For instance, relevance is an important quality early in the development of an action plan; effectiveness is a quality to be taken into account when the final assessment or at a regulation meeting. Efficiency concerns more those who has the responsibility to attribute means…

Speaking about measures, we developed another questionnaire that determines them and the information was collected from two institutions of higher education. We aim to make a comparison between them and to identify the factors which directly influence the quality of universities.

5   General Recommendations for Ensuring Quality in Higher Education Domain

Following detailed analysis of the questionnaire, a certain number of recommendations can be created based on results reached:

- An effective education system must be dynamic by the openness of the socio economic environment, adapted to the needs and attempts presented to the outcome and operation level. This is manifested by trainings characterized by flexibility, multiplicity, permeability and interconnection.
- Teachers may receive training to improve their communication skills, these criteria have been observed with great importance among the students.
- To develop a process of change towards a culture of quality with formal guarantees, incentives and sanctions applied to obtain the desired behaviour at organizational, collective, and individual level;
- To introduce means for public acknowledgement of academic staff with good teaching performance (formal reward, prizes etc.) [14].
- To define and implement quality strategies, plans, and assurance goals, in other words to characterize things about quality;

- To organize meeting involving graduates and employers so that they can regularly debate the quality of the studying programs and their relevance;
- To introduce periodical overall internal reviews at institutional level in order to monitor quality [14].
- To create quality assurance collectives at faculty level and quality assurance units at university level that processes and formal structures are conducted.

6 Conclusion and Future Works

To conclude, the quality of an education system in general and especially in higher education is an essential requirement, but relatively complex. It requires steering by taking into account all the components of the System to ensure that it achieves the objectives that emerge from a Multi-faceted social environment. It also has to be consistent with the existing resources and defined strategies. As a result, the notion of quality in higher education is mainly measured through its mode of operation, and quantitative and qualitative results of learners. For this reason, through this paper we presented a number of criteria and indicators that are considered as important to describe the various aspects of quality. Then, we defined a case study that was conducted through a questionnaire distributed to experts, to determine a list of criteria we have to consider.

In future research work, we aim to apply these criteria on two real models describing two different processes: the first describes the case of the tracking of curriculum offer process, more commonly named "habilitation process", and the second defines the Process of tracking master's degree theses in order to evaluate their quality.

Appendix A: Table of Questionnaire Responses

Questions	Frequent distribution	Reponses					Total
		Strongly disagree	Disagree	Neutral	Agree	Strongly agree	
Should the course objectives be understandable?	Frequency	3	1	3	16	60	83
	Percent	3.6	1.2	3.6	19.3	72.3	100
Should the course objectives be attainable?	Frequency	4	0	3	32	44	83
	Percent	4.8	0	3.6	38.6	53	100
Should the universities respond to the expectations of the labour market?	Frequency	2	1	5	16	59	83
	Percent	2.4	1.2	6	19.3	71.1	100
Should activities be effective against stated objectives?	Frequency	2	3	12	27	39	83
	Percent	2.4	2.4	14.6	32.9	47.9	100
Should Pedagogical materials be adequate to fulfil the missions and objectives?	Frequency	2	4	11	18	48	83
	Percent	2.4	4.8	13.3	21.7	57.8	100

(continued)

(continued)

Questions	Frequent distribution	Reponses					Total
		Strongly disagree	Disagree	Neutral	Agree	Strongly agree	
Must the available recourses correspond to the obtained set objectives?	Frequency	19	21	22	18	3	83
	Percent	22.9	25.3	26.5	21.7	3.6	100
Must the obtained results meet the set objectives?	Frequency	1	6	14	33	29	83
	Percent	1.2	7.2	16.9	39.8	34.9	100
Should the objectives of the higher education system correspond to high priority needs of the labor market?	Frequency	6	4	7	27	39	83
	Percent	7.2	4.8	8.4	32.5	47	100
Is it important that higher education institutions are designed and organized to produce needed diplomas?	Frequency	4	2	16	28	33	83
	Percent	4.8	2.4	19.3	33.7	39.8	100
Should the means (resources, strategies...) must be those that we provided?	Frequency	2	10	15	39	17	83
	Percent	2.4	12	18.1	47	20.5	100
Is it important to have conformance between activities within the education system and the existing regulations?	Frequency	0	10	19	35	19	83
	Percent	0	12	22.9	42.2	22.9	100
Must the higher education system provide all dimensions of knowledge?	Frequency	3	5	15	27	33	83
	Percent	3.6	6	18.1	32.5	39.8	100
Must the degree of involving of students in improving the quality of the establishment be important?	Frequency	5	1	14	33	30	83
	Percent	6	1.2	16.9	39.8	36.1	100

References

1. Dragan, M., Ivana, D., Arba, R.: Business process modeling in higher education institutions. Developing a framework for total quality management at institutional level. Procedia Econ. Finance 95–103 (2014)
2. Pratasavitskaya, H., Stensaker, B.: Quality management in higher education: towards a better understanding of an emerging field. Qual. High. Educ. **16**, 37–50 (2010)
3. Safi, H.: Le management socio-économique et la mise en œuvre d'une démarche de qualité intégrale dans un établissement d'enseignement supérieur en Tunisie (2012)
4. Tsinidou, M., Gerogiannis, V., Fitsilis, P.: Evaluation of the factors that determine quality in higher education: an empirical study. Qual. Assur. Educ. **18**, 227–244 (2010)
5. Nicolescu, L., Dima, A.M.: The quality of educational services-institutional case study from the romanian higher education. Transylvanian Rev. Adm. Sci. **6**, 100–108 (2010)
6. Bouzid, N., Berrouche, Z.: Module II–Assurance qualité dans l'enseignement supérieur (2012)

7. Tam, M.: Measuring quality and performance in higher education. Qual. High. Educ. **7**, 47–54 (2001)
8. Gerard, F.: L'évaluation de la qualité des systèmes de formation. Mesure et Évaluation en Education **24**, 53–77 (2001)
9. Cowan, J.: Effectiveness and efficiency in higher education. High. Educ. **14**, 235–239 (1985)
10. Sall, H.N., Ketele, J.M.: Évaluation du rendement des systèmes éducatifs: apports des concepts d'efficacité, d'efficience et d'équité. Measure et Évaluation en Éducation **19**, 119–142 (1997)
11. Martin, M., Sauvageot, C.: Construire un tableau de bord pour l'enseignement supérieur: un guide pratique. Unesco, Institut international dc planification de l'education (2009)
12. Loua, S.: Efficacite interne de l'enseignement supérieur malien (2012)
13. Owlia, M.S., Aspinwall, E.M.: A framework for the dimensions of quality in higher education. Qual. Assur. Educ. **4**, 12–20 (1996)
14. Ketele, D., Marie, J., Gerard, F.M: La qualité et le pilotage du système éducatif (2007)

Toward a Personalized Recommender System for Learning Activities in the Context of MOOCs

Marwa Harrathi[✉], Narjess Touzani, and Rafik Braham

PRINCE Research Unit, ISITCom, Hammam Sousse, Tunisia
marwa.harrathi@yahoo.fr

Abstract. Massive Open Online Courses have brought a revolution in the field of e-learning. However, the lack of support and personalization drives learners to lose their motivation and surrender the learning process. One of issues that MOOC should address in personalization of learning according to learners needs to reinforce motivation. The potential of learning activities to motivate learners in enhancing learning cannot be denied. Therefore, we focus on adapting learning activities to learners through a recommender system in order to suit individual learners' diverse needs. In this paper we outline a set of dimensions that distinguish, describe and categorize learning activities based on existing categorizations. We propose a classification of these recommended learning activities according to Bloom's taxonomy. These learning activities are integrated into and overall a rule based recommender system with modular architecture.

Keywords: Learning activities · E-learning · MOOC · Personalization · Recommender system

1 Introduction

The development of e-learning courses in the context of Massive Open Online Courses (MOOCs) is relatively a recent area of research.

MOOC provides a new way of learning, with is open, distributed, massive and life-long [1]. Therefore, MOOCs have challenges of its own due generally to the large number of simultaneous participants [2]. In fact, providing efficient learning according to the personal needs of each learner in really a big challenge, because learners are not only different in their behaviors and learning approaches, but also in their intelligence and abilities [3].

MOOC design should favor a learner-centered approach [4]. Thus, a large part of the development problem of MOOCs is concentrated on the taking into account of learner. Some learners left the course after starting, some did not intend to follow it entirely, and others dropped out for professional or personal reasons. In addition other learners could not keep their concentration in the optimal function and lose their motivation against difficulties they encountered during the learning process. Based on the consideration mentioned above, the aim of our research work is focused on this last category of learners and we tried to surrounding them during the learning process.

© Springer International Publishing AG 2018
G. De Pietro et al. (eds.), *Intelligent Interactive Multimedia Systems and Services 2017*,
Smart Innovation, Systems and Technologies 76, DOI 10.1007/978-3-319-59480-4_57

In this context, several research on recommendation [5–9] in e-learning courses generally, and MOOCs in particular, have emerged in recent years. The main objective of recommender systems is to assist learners in their learning process, facilitate learners' choices and ensure their motivation [10] during the whole learning process.

For instance, few of e-learning recommender systems are interested on recommending learning activities, especially in the context of massive learning. Therefore we offer our personalized recommender system of learning activities. The main objective is to prevent learners demotivation through adapted learning activities. The choice of recommending learning activities is due to their importance in the follow-up of the learning process. In fact, learning activities perform a key role in knowledge acquisition and maintaining motivation.

Our customized recommender system of learning activities follows a rule-based approach to an automatic and dynamic proposition of learning activities inside a basic scenario. The proposal of learning activities is thus carried out as the learning process on an ad hoc basis timely with learning activities of the basic scenario. So, we distinguish two types of activities throughout learning process: the activities of the basic scenario included during the learning design and the recommended learning activities proposed during the learning process of courses customized for learners at the appropriate time. In this current research, we focus more closely on recommended learning activities. We propose a rich and varied categorization of learning activities with different forms. In addition, we classify these recommended learning activities, with their different level of difficulties, on basis of the consistent taxonomy of Bloom [10].

This paper is divided into a number of sections as follows. In Sect. 2, we look to define the term of learning activities and their categorization with some related works. In Sect. 3, we present our categorization and classification of recommended learning activities for the aim to use it in our personal recommender system. In Sect. 4, we present the architecture of a proposed rule based recommender system of learning activities. We end up with the fifth section as our conclusion and future work.

2   E-learning Activities Categorization

2.1   Learning Activities

A learning activity refers to didactic situations in which the learner is expected to acquire knowledge and/or competences.

Suggesting a personalized activity amounts to filling gaps for learners and respecting his/her needs and interests so that engaging his/her to learning process. It's about identifying requirements of learners and adapting learning using the appropriate learning activities. In fact, learners lose their motivation in front of activities that goes beyond his cognitive abilities or too easy ones compared to his/her level.

E-learning activities are designed according to a number of objectives. These objectives are expressed by the cognitive classification of Bloom. In the context of MOOC, the course consists of units of learning spread over 5 or 6 weeks. For each week a set of learning activities is designed in order to help learners to achieve learning objectives and then acquire knowledges.

Bloom characterizes the representation of knowledge according to six dimensions of the taxonomy, from simple to more complexes [10]:

- Knowledge: manipulate data or information in a basic way.
- Comprehension: ensures learner understanding of learning concepts. In this level, the learner can reformulate what has learned.
- Application: applies a method or a rule, generally applies what was learned into a novel situation.
- Analysis: decompose information into its components parts so that its organizational structure may be understood.
- Synthesis: offer the possibility of creating a new meaning or structure from diverse elements.
- Evaluation: ability to judge the value of materials or ideas for a given purpose.

2.2 Related Works

Learning activities include a set of tasks offered to learners throughout their learning processes. According to [11] several functions are attributed to these learning activities: (1) support to remedy the deficiencies of learners in basic skills. For example remediation of the learner weakness in reading as it's necessary in the access to main sources of knowledge; (2) Information processing that allow learners to understand better and acquire the course concepts; (3) Didactic relationship where activities would be a substitute of the didactic conversation between teacher and learner in the class; (4) action and reflection guide that focus on the orientation of learners in their actions on their environments; (5) Motivation for provoking or keeping attention and curiosity of the learner; (6) learning assessment and control, either by allowing self-assess or by being returned to the tutor to evaluate his progress.

According the different function assigned to them, learning activities differ in several forms. In this work the author gives an overall view about learning activities. It provides practically no details about the forms of learning activities except of open questions, multiple choice questions and exercise statements.

In [12] the author presents a set of learning activities more precisely. He is interested by interactive activities in the context of a specific programming course deployed by Edx [13]. These learning activities are more varied and by different tools. They are multiple choice questions, multiple answer-questions, text entry questions, drop down questions and drag and drop questions. In addition, the use of simple games provided by Blocky tool [14], small programming exercises provided by the CodeBoard development environment [15], and games for understanding object-oriented concepts using Greenfoot [16].

In this article too, the author discusses learning activities in a general way. The set of learning activities used in the context of his programming course remains limited. Moreover, most of these learning activities focus on a low level of Bloom's taxonomy as they focus on testing the learner level comprehension of course concepts.

Based on survey carried out about the proposed learning activities, we note a varied satisfaction of learners about activities difficulty levels. This led us to conduct that learners have different needs and prefer learning activity forms compared with others. In fact, the variety of responses in term of satisfaction regarding a particular activity is due to heterogeneity of learners, which indicate learners' dissatisfaction by this unique repair of activities throughout the course.

This conclusion reinforces our approach of personalized recommended learning activities.

3   Proposed E-learning Activities Categorization and Classification

To promote learning in Massive Open Online courses, learning activities should offer a deep information treatment. To do so, and based on learning activities functions and forms cited above, we propose our own categorization of learning activities according their nature.

We propose to list recommended learning activities in five categories according to learners requests work as above: perception activities, redaction activities, production activities, gamification activities and reflection activities.

- Perception activities: these are activities that present the materiel supports prepared in advance in the form of written or audio-video documents. In this case, the learner is a receiver of information: he hears and/or observes a learning resource and has nothing to deposit in response.
  One of the key of success and understanding in to practice what is exposed by perception activities through learning activities that acquire interaction either individually with the e-learning platform or collaboratively with learners. These activities are redaction, writing, production, gamification and reflection.
- Redaction activities: they request learners about writing a response to a posed question. At this level, learners use basic knowledge and skills to perform an activity. They learn to explain, to narrate, to justify, to write and to describe.
- Production activities: production activities ensure the development of learners' cognitive functions through the evolution of their abilities of abstraction and reasoning. They allow learners to think, to reason, to relate and to produce by means of diagrams and/or coherent words: it's by producing activities that the learner is led to structure and acquire new knowledge and skills.
- Gamification activities: the exploitation of serious games offers a better presentation of knowledge with less fatigue and more fun. They imply a strong commitment of learners in an interactive situation. These activities allow learning by playing. Learners called in this case players, become more qualified in term of information retention and knowledge and skills acquisition.
- Reflection activities: these are activities that awaken and simulate the reflection of learners. A reflection activity may deepen understanding, but also it can be used in measuring level of knowledge understood. In addition, it enables learners to become aware of their strengths and to realize about concepts they must progress on. This kind of activities allows learners to take on new challenges and consequently simulate learners' motivation and encouragement to complete the course.

For each category of activities, several forms have been identifies. Table 1 summarizes this categorization and show different forms of learning activities for every category.

Table 1. Categorization of learning activities and different forms

Learning activities categories	Learning activities forms
Perception	Simulation, reading, video, forums (checking messages), tutorial
Redaction	Question and answer task, journaling, literature review, forums (replay to a message), open short questionnaire, critical of solutions, muddiest point, bookmark notes, essay, advice letter, one-sentence summary, discussion, solution evaluation, objective check, opposites, what/how/why outlines, peer review writing tasks, definitions and applications
Production	Problem solving, simple case study, complex case study, modeling, project, wiki, open discussion, conceptualization, brainstorming, power point presentations, documented problem-solution, concept mapping, panel discussion, build from restricted components, charting
Gamification	Guided discovery, matching game, crossword puzzles, labyrinth, earning points by passing a level in the game, pass the chalk, tournament, scrabble, who am I? Pictionary
Reflection	Drill and practice, one minute paper, study session, on the spot questioning, picture prompt, goal ranking and matching, choral response, problem recognition tasks, ranking alternatives, association, true-false questionnaires, multiple choice questionnaires, single choice questionnaires, debate

All this recommend learning activities with their different forms are divided into collaborative activities and individual activities:

- Collaborative activities: they have been recognized by its ability to motivate learners. It promotes constructive learning based on mutual help and cooperative contributions between learners. In this way, learner is no longer considered as a passive receiver of information, thus the knowledge will be build mutually.
- Individual activities: is assigns the responsibility of learning to learners and guide them in the knowledge acquisition throughout learning process. Individual activities promotes the development of autonomy, written expression skills, synthesis capacity and knowledge acquisition and management.

In the first phase of our research work, we tried to provide a rich and a varied base of learning activities at the aim of recommending them to learners via a personalized recommender system.

The second step consists at identified attributes that we take into account for a consistent description of learning activities. So, we have developed a set of attributes that we consider the most relevant. These attributes are useful for our recommender system in the choice of suitable recommended learning activities based on learners' characteristics and needs.

The most useful attributes used in our description of recommended learning activities are: objective or targeted level in the Bloom's taxonomy, degree of difficulty, type, starting time, duration and priority. They are detailed as follow:

- Objective or targeted level in Bloom's taxonomy: each learning activity is related to a learning objective. We chose to define these objectives based on Bloom's taxonomy. Thereby, recommended learning activities are classified according to knowledge, comprehension, application, analysis, synthesis and evaluation. The targeted level in Bloom's taxonomy of a particular recommended activity is determined by calculating the gap between the prior knowledge of learner and the execution result of basic scenario's activity.
- Difficulty level: it's a qualitative variable taking values as "easy", "average" and "difficult". For each cognitive level of Bloom's taxonomy several activities are classified based on difficulty level. Only knowledge and understanding level have only easy activities.
- Type: all recommended learning activities will be listed based on the targeted level in Bloom's taxonomy according to two types: individual activities and collaborative ones. A learning activity has as value "collaborative" if it requires a group work and it has an "individual" value if it does not. Table 2 provides an overview of a classification of recommended learning activities according to their types and objectives.

Table 2. Classification of learning activities according to their types and Bloom's taxonomy.

Heading level	Individual activities	Collaborative activities
Knowledge	Perception	Perception
Comprehension	Redaction, gamification, reflection	Redaction, reflection
Application	Redaction, production, gamification	Redaction, gamification, production, Reflection
Analysis	Redaction, reflection	Redaction, reflection
Synthesis	Redaction	Redaction
Evaluation	Redaction	Redaction

- Starting time: this is the time at which the learner begins to work on the recommend learning activity.
- Duration: learning activities have different durations depending on their categories and forms.
- Priority: for each learning activity category we distinguish easy activities, average activities and difficult ones. These activities will be proposed considering a priority order and based on learners' information.

4 Architecture of a Proposed Rule Based Recommender System of Learning Activities

The proposed recommender system provides personalized learning activities for learners, exploiting characteristics grouped in a learner profile. The learner profile that we employ consists mainly of three learners' dimensions: level of knowledge, learning style and preferences in term of pedagogical methods. The main objective of our system of recommended learning activities is to maintain, at a satisfactory level, and increase the motivation of learners. So that they follow to massive course to the end. To do this, we use different forms of learning activities proposed above. In fact, we can satisfy learners throughout the variation of recommended and personalized learning activities and increase their motivation and attention.

In accordance with objective we have set, we present the architecture of the proposed system of learning activities recommendations in the context of massive learning, as presented in Fig. 1.

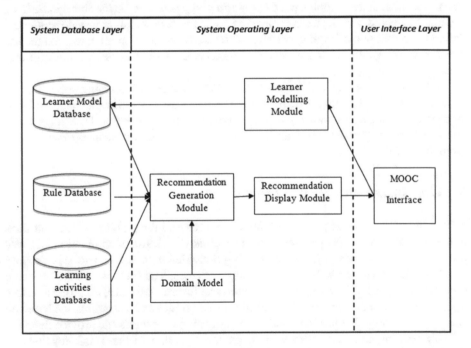

Fig. 1. General recommender system architecture of learning activities in the context of MOOC.

The system is based on a modular architecture organized into separate layers: User interface layer, System function layer, and System data layer. Each layer is composed of one or more component cited as follows:

- The User interface layer ensures interactions and communication between learners and system.

- The system operation layer consists of the learner modeling module, the domain model, the recommendation generation module and the recommendation display module.
- The system database layer includes three components which are learner profile database, rules database and learning activities database.

The operating principle of our system is as follow: after a general presentation of the MOOC, acquisition of learners' information in necessary through the learner Modeling Module. These information recorded for each learner in the learner profile will be used later in the process of activities personalization.

Tutors provide a pedagogical strategy that's the set of rules indexed in the rule database. These rules take as condition the values of learner characteristics collected in the learner model and as a treatment the recommended learning activities.

Then, the recommendation generation module generates the appropriate learning activities for a particular learner. Once the recommended activities are generated, they will be displayed using recommendation display module and shown in a pop up window. The proposition of the recommended learning activity is realized at the appropriate time. It should be noted that the learner have the choice to follow the proposed recommendation or to follow the learning activity of the basic scenario. In all cases, the recommended learning activity is recorded and remains accessible for learners at all times during the learning process.

For the representation and the execution of the pedagogical learning process in the context of massive learning, we propose the use of IMSLD specification standard [17] since this specification standard puts the activities in the center of the learning process. IMSLD have a way to describe learning activities and specially used in representing the Domain Model.

5   Conclusion and Future Works

At the end of the various points presented, we discussed the problem of dropout rates in the context of massive open online courses because of learners' motivation loss. We propose a recommender system of learning activities as a solution to this dropout problem and we choose to focus on learning activities. The synthesis of the literature revealed limits use of learning activities types and forms. Therefore, we propose in this research work, a rich categorization and forms of activities and a classification based on Bloom's taxonomy. It's useful to incite the motivation of learners and provide them the suitable learning activities. Then we talk about the position of these classification and categorization in the proposed system. For that, the architecture of our rule based system and his operating mode are briefly described.

The perspective of this research work is to refine our approach of personalization for proposing recommended learning activities based on learners information and the set of rules.

References

1. Clarke, T.: The Advance of the MOOCs (Massive Open Online Courses): the impending globalisation of business education? Educ. + Train. **55**(4/5), 6 (2013)
2. Guàrdia, L., Maina, M., Sangrà, A.: MOOC design principles: a pedagogical approach from the learner's perspective. eLearn. Pap. (33) (2013)
3. Daneman, M., Carpenter, P.A.: Individual differences in working memory and reading. J. Verbal Learn. Verbal Behav. **19**(4), 450–466 (1980)
4. Daradoumis, T., Bassi, R., Xhafa, F., Caballé, S.: A review on massive e-learning (MOOC) design, delivery and assessment. In: 2013 Eighth International Conference on P2P, Parallel, Grid, Cloud and Internet Computing (3PGCIC), pp. 208–213. IEEE, October 2013
5. Bansal, N.: Adaptive recommendation system for MOOC. Doctoral dissertation, Indian Institute of Technology, Bombay (2013)
6. Alario-Hoyos, C., Estévez-Ayres, I., Pérez-Sanagustín, M., Leony, D., Kloos, C.D.: MyLearningMentor: a mobile app to support learners participating in MOOCs. J. UCS **21**(5), 735–753 (2015)
7. Tarus, J.K., Niu, Z., Mustafa, G.: Knowledge-based recommendation: a review of ontology-based recommender systems for e-learning. Artif. Intell. Rev. 1–28 (2017)
8. Mahajan, R.: Review of data mining techniques and parameters for recommendation of effective adaptive e-learning system. In: Collaborative Filtering Using Data Mining and Analysis, p. 1 (2016)
9. Li, Y., Zheng, Y., Kang, J., Bao, H.: Designing a learning recommender system by incorporating resource association analysis and social interaction computing. In: Li, Y., et al. (eds.) State-of-the-Art and Future Directions of Smart Learning, pp. 137–143. Springer, Singapore (2016)
10. Bloom, B.S.: Taxonomy of Educational Objectives, vol. 2. Longmans, Green, New York (1964)
11. Bilodeau, H., Provencher, M., Bourdages, L., Deschênes, A.J., Dionne, M., Gagné, P., Lebel, C., Rada-Donath, A.: Les objectifs pédagogiques dans les activités d'apprentissage de cours universitaires à distance. DistanceS **3**(2), 33–68 (1999)
12. Alario-Hoyos, C., Kloos, C. D., Estévez-Ayres, I., Fernández-Panadero, C., Blasco, J., Pastrana, S., Villena-Román, J.: Interactive activities: the key to learning programming with MOOCs. In: Proceedings of the European Stakeholder Summit on Experiences and Best Practices in and Around MOOCs (EMOOCS 2016), p. 319 (2016)
13. https://edx.org. Accessed Nov 2016
14. https://blockly-games.appspot.com. Accessed Dec 2016
15. https://codeboard.io. Accessed Dec 2016
16. Kölling, M.: The greenfoot programming environment. ACM Trans. Comput. Educ. (TOCE) **10**(4), 14 (2010)
17. IMS: IMS Learning Design Information Model, IMS Global Learning Consortium (2003). http://imsglobal.org/learningdesign/

Author Index

Printed in the United States
By Bookmasters